Modeling and Simulation of Functional Nanomaterials for Forensic Investigation

Allah Rakha
Department of Forensic Medicine/Medical Jurisprudence, University of Health Sciences, Pakistan

Anam Munawar
Department of Forensic Medicine/Medical Jurisprudence, University of Health Sciences, Pakistan

Virat Khanna
Department of Mechanical Engineering, Maharaja Agrasen University, India

Suneev Anil Bansal
Department of Mechanical Engineering, Bharat Institute of Engineering and Technology, India

IGI Global
PUBLISHER of TIMELY KNOWLEDGE

A volume in the Advances in Digital Crime, Forensics, and Cyber Terrorism (ADCFCT) Book Series

Published in the United States of America by
IGI Global
Information Science Reference (an imprint of IGI Global)
701 E. Chocolate Avenue
Hershey PA, USA 17033
Tel: 717-533-8845
Fax: 717-533-8661
E-mail: cust@igi-global.com
Web site: http://www.igi-global.com

Library of Congress Cataloging-in-Publication Data

Names: Rakha, Allah, 1978- editor.
Title: Modeling and simulation of functional nanomaterials for forensic investigation / edited by Allah Rakha, Anam Munawar, Virat Khanna, and Suneev Anil Bansal.
Description: Hershey, PA : Information Science Reference, [2023] | Includes bibliographical references and index. | Summary: "Modeling and Simulation of Functional Nanomaterials for Forensic Investigation focuses on multiple applications related to forensics and provides information linked with nanoparticles. This book provides nanotechnology results in improving the sensitivity of established forensic techniques. It further focuses on different fabrication and characterization techniques of nanomaterials and relates their characteristics with forensic applications. Covering topics such as explosive detection, nano- forensic testing, and nano-trackers, this premier reference source is a comprehensive resource for material engineers, chemical engineers, nanotechnologists, biotechnologists, forensic scientists, students and educators of higher education, researchers, and academicians"-- Provided by publisher.
Identifiers: LCCN 2022062303 (print) | LCCN 2022062304 (ebook) | ISBN 9781668483251 (hardcover) | ISBN 9781668483268 (paperback) | ISBN 9781668483275 (ebook)
Subjects: LCSH: Forensic sciences. | Nanostructured materials. | Criminal investigation.
Classification: LCC HV8073.5 .M63 2023 (print) | LCC HV8073.5 (ebook) | DDC 363.25--dc23/eng/20230519
LC record available at https://lccn.loc.gov/2022062303
LC ebook record available at https://lccn.loc.gov/2022062304

This book is published in the IGI Global book series Advances in Digital Crime, Forensics, and Cyber Terrorism (ADCF-CT) (ISSN: 2327-0381; eISSN: 2327-0373)

British Cataloguing in Publication Data
A Cataloguing in Publication record for this book is available from the British Library.

For electronic access to this publication, please contact: eresources@igi-global.com.

Advances in Digital Crime, Forensics, and Cyber Terrorism (ADCFCT) Book Series

Bryan Christiansen
Southern New Hampshire University, USA
Agnieszka Piekarz
Independent Researcher, Poland

ISSN:2327-0381
EISSN:2327-0373

Mission

The digital revolution has allowed for greater global connectivity and has improved the way we share and present information. With this new ease of communication and access also come many new challenges and threats as cyber crime and digital perpetrators are constantly developing new ways to attack systems and gain access to private information.

The **Advances in Digital Crime, Forensics, and Cyber Terrorism (ADCFCT) Book Series** seeks to publish the latest research in diverse fields pertaining to crime, warfare, terrorism and forensics in the digital sphere. By advancing research available in these fields, the **ADCFCT** aims to present researchers, academicians, and students with the most current available knowledge and assist security and law enforcement professionals with a better understanding of the current tools, applications, and methodologies being implemented and discussed in the field.

Coverage

- Identity Theft
- Network Forensics
- Vulnerability
- Database Forensics
- Information Warfare
- Malware
- Telecommunications Fraud
- Criminology
- Cryptography
- Digital Crime

IGI Global is currently accepting manuscripts for publication within this series. To submit a proposal for a volume in this series, please contact our Acquisition Editors at Acquisitions@igi-global.com or visit: http://www.igi-global.com/publish/.

The Advances in Digital Crime, Forensics, and Cyber Terrorism (ADCFCT) Book Series (ISSN 2327-0381) is published by IGI Global, 701 E. Chocolate Avenue, Hershey, PA 17033-1240, USA, www.igi-global.com. This series is composed of titles available for purchase individually; each title is edited to be contextually exclusive from any other title within the series. For pricing and ordering information please visit http://www.igi-global.com/book-series/advances-digital-crime-forensics-cyber/73676. Postmaster: Send all address changes to above address.

Titles in this Series

For a list of additional titles in this series, please visit: www.igi-global.com/book-series

Handbook of Research on War Policies, Strategies, and Cyber Wars
Fahri Özsungur (Mersin University, Turkey)
Information Science Reference • ©2023 • 463pp • H/C (ISBN: 9781668467411) • US $315.00

Global Perspectives on the Psychology of Terrorism
Nika Chitadze (International Black Sea University, Georgia)
Information Science Reference • ©2023 • 330pp • H/C (ISBN: 9781668453117) • US $215.00

Aiding Forensic Investigation Through Deep Learning and Machine Learning Frameworks
Alex Noel Joseph Raj (Shantou University, China) Vijayalakshmi G. V. Mahesh (BMS Institute of Technology and Management, India) Ruban Nerssison (Vellore Institute of Technology, India) Ang Yu (Carolina University, USA) and Jennifer Gentry (Carolina University, USA)
Information Science Reference • ©2022 • 273pp • H/C (ISBN: 9781668445587) • US $250.00

Dark Web Pattern Recognition and Crime Analysis Using Machine Intelligence
Romil Rawat (Shri Vaishnav Vidyapeeth Vishwavidyalaya, India) Shrikant Telang (Shri Vaishnav Vidyapeeth Vishwavidyalaya, India) P. William (Sanjivani College of Engineering, Savitribai Phule Pune University, India) Upinder Kaur (Akal University, Talwandi Sabo, India) and Om Kumar C.U. (School of Computer Science and Engineering (SCOPE),Vellore Institute of Technology, India)
Information Science Reference • ©2022 • 281pp • H/C (ISBN: 9781668439425) • US $250.00

Using Computational Intelligence for the Dark Web and Illicit Behavior Detection
Romil Rawat (Shri Vaishnav Vidyapeeth Vishwavidyalaya, India) Upinder Kaur (Akal University, Talwandi Sabo, India) Shadab Pasha Khan (Oriental Institute of Science and Technology, Bhopal, India) Ranjana Sikarwar (Amity University, Gwalior, India) and K. Sakthidasan Sankaran (Hindustan Institute of Technology and Science, India)
Information Science Reference • ©2022 • 336pp • H/C (ISBN: 9781668464441) • US $250.00

Media and Terrorism in the 21st Century
Elnur Ismayil (Istanbul Medeniyet University, Turkey) and Ebru Karadogan Ismayil (Uskudar University, Turkey)
Information Science Reference • ©2022 • 299pp • H/C (ISBN: 9781799897552) • US $215.00

World Politics and the Challenges for International Security
Nika Chitadze (International Black Sea University, Georgia)
Information Science Reference • ©2022 • 427pp • H/C (ISBN: 9781799895862) • US $215.00

IGI Global
PUBLISHER of TIMELY KNOWLEDGE

701 East Chocolate Avenue, Hershey, PA 17033, USA
Tel: 717-533-8845 x100 • Fax: 717-533-8661
E-Mail: cust@igi-global.com • www.igi-global.com

Table of Contents

Detailed Table of Contents

With the increasing demand of nanoparticles in various applications, many challenges are also noticed in terms of synthesizing accurately designed nanomaterials. The characteristics to be customized in order to make these nanomaterials useful need characterization and verification using different tools that have certain limitations and significant strengths. The purpose of this chapter is to provide the description of measurement methods that may be applied to nanodimension in order to study their most important physical and chemical features. This chapter will provide a detailed introduction, mechanism of working, and limitations of various characterization tools. Focus will be on giving detail about widely applicable and commonly approachable methods to measure properties that can be categorized in terms of size, size distribution, aggregation, shape, dimension, surface area, composition, surface chemistry, charge, surface potential, and functionally active interactive groups.

This chapter discusses the designing and modeling of nanomaterials through different updated technologies. It also covers the chemistry behind the nanoparticles and their prominent properties, which are required for different applications, along with an overview of the history of nanoparticle creation and applications. But the basic discussion is about different physical, chemical, and biological techniques for nanomaterial designing and different and novel characterization approaches to find the morphological and chemical traits of the designed nanoparticles. It also discusses the methods that are cost effective and eco-friendly for developing different nanoparticles. Future scope of designing nanoparticles and how the availability of nanomaterial helps in coping with the energy depletion and utilizing minute particles in massive applications are also covered.

Sensors are effective instruments for finding illegal drugs. In the past several years, sensor use and development have garnered a lot of interest. The lack of an effective tool that offers the specialist all the necessary features, including cost-effective manufacturing, on-site analysis, instrumental simplicity, and correct findings, might be blamed for this thirst for the creation of new detectors. Nanotechnology has the potential to significantly advance forensic science and criminal investigation. In forensic nanotechnology, tiny chip materials are employed in place of large apparatus, which simplifies the techniques of analysis and improves the accuracy, precision, timeliness, and suitability of investigations. The reader is provided with up-to-date studies and publications as this chapter describes the use of nano sensors in forensics investigation. Additionally, it details the key distinctions between the three major categories of sensors that provide information on the potential applications of these sensing materials in the future.

Currently, in the field of biomedical engineering and biological applications, the use of soft florescent nanomaterials has increased because of their excellent biocompatibility, easy biofunctionalization, and high brightness properties. This chapter summarizes the current developments of nano-sized fluorescent soft biological imaging agents. Many fluorescent soft nanoparticles like biomaterial-based NPs, vesicles, micelles, nanogels, small-molecule organic NPs, semiconducting polymer NPs, and dye-doped polymer NPs are mentioned briefly starting from the preparation methods, their structures, their optical properties, as well as their functionalization. Depending upon the nano-sized imaging agents' functional as well as optical properties, their uses are briefly described in relation to Vivo imaging, cellular process imaging, and in vitro imaging by using nonspecific and specific targeting.

Analysis of body fluids is an important aspect of forensic investigation; therefore, rapid and sensitive detection is required for their analysis as well as characterization. This chapter presents an overview of the applications of different nanomaterials with respect to the biological material detection. If we talk about forensic biological samples, we especially need to handle them with care, because such samples from crime scenes are always in traces. This chapter briefly discusses the use of functional nanoparticles for the detection of biological materials. Genetic material obtained from crime scenes is one of the most important pieces of evidence in forensic investigations, and the chapter focuses on the application of functional nanomaterials for their detection.

Nano forensics utilizes nano-based sensors in matters of crime or where the legality of the process is challenged. There exist many conventional techniques, but considering the approach of criminals and the modern-day standard of investigation, there is a need for nanotechnology-based sensors for their sensitive, selective, and rapid detection of illicit drugs. Illicit drugs are psychoactive in nature. They act upon the CNS that results in different perceptions and long-lasting responses. This chapter covers the drugs that

Preface

Forensic science is a field that has seen significant advancements in recent years, particularly in the use of nanomaterials. Nanomaterials, with their unique physicochemical properties, have revolutionized the way forensic evidence is collected, analyzed, and interpreted. From crime scene investigation to the analysis of trace evidence, nanomaterials have found applications in various aspects of forensic science.

This book aims to provide a comprehensive overview of the current state-of-the-art in the use of nanomaterials in forensic science. It covers a wide range of topics, including the synthesis and characterization of nanomaterials, their applications in fingerprint analysis, DNA profiling, drug detection, and trace evidence analysis, among others. The book also discusses the challenges and limitations associated with the use of nanomaterials in forensic science and highlights the future directions in this exciting and rapidly evolving field.

The purpose of the introductory chapter is to define the concepts of functional nanomaterials in science and technology. It will also highlight the fundamental characteristics of these materials in research. It also aims at emphasizing role and importance of functional nanomaterials in various useful aspects. As research is the main tool of creating knowledge, different methods used by researchers are also briefly discussed here. This chapter will also develop a frame of reference for studying various chapters that follow. These chapters form modified core of functional nanomaterials and its applications in forensic investigation.

Technology envisaging sustainable environment friendly development with huge potential of using particles with size in nanometer is the nanotechnology focusing nanomaterials that are either source or incidental, natural or artificial. Chapter 2 presents an overview of positive and negative nanotechnology impacts discussed on a broader plane to get rational universe view. Nanotechnology's central tenet is the utilization of sub-micrometer-sized particles for use in a wide variety of disciplines, including medicine, cosmetics, agriculture, and the food industry. Nanotechnology's advantages includes better agricultural production, quality in pharmaceuticals, targeted drug delivery, and sensor applications. In order to move forward with its growth, it is crucial to correctly recognize the potential benefits and unintentional threats of nanomaterials to our surroundings. Knowing the fact that nano-issue is more complex more focused and balanced interpretation of risks and benefits of nanotechnological impacts is very important to be analyzed.

With increasing demand of nanoparticles in various applications, many challenges are also noticed in terms of synthesizing accurately designed nanomaterials. The characteristics to be customized in order to make these nanomaterials useful need characterization and verification using different tools which have certain limitations and significant strengths. The purpose of Chapter 3 is to provide the description of measurement methods that may be applied to nanodimension in order to study their most important physical and chemical features. This chapter will provide a detailed introduction, mechanism of working and limitations of various characterization tools. Focus will be on giving detail about widely applicable and commonly approachable methods to measure properties that can be categorized in terms of size, size distribution, aggregation, shape, dimension, surface area, composition, surface chemistry, charge, surface potential and functionally active interactive groups.

In Chapter 4, we shall discuss the designing and modeling of nanomaterials through different up-dated technologies also cover chemistry behind the nanoparticles and their prominent properties which are required for different application along with a bit overview of history of nanoparticles creation and application. But the basic discussion is about different physical, chemical and biological techniques for nanomaterial designing and different novel characterization approaches to find the morphological and chemical traits of the designed nanoparticles. Modification of nanoparticles surface in order to enhance their properties for different application purpose. Also discuss the methods which are cost effective and eco-friendly for developing different nanoparticles. Future scope of designing nanoparticles and how the availability of nanomaterial helps in coping the energy depletion matter and utilizing minute size particle in massive applications.

The sensors are effective instruments for finding illegal drugs. In the past several years, sensors' use and development have garnered a lot of interest. The lack of an effective tool that offers the specialist all the necessary features, including, cost effective manufacturing, on-site analysis, instrumental simplicity and correct findings, might be blamed for this thirst for the creation of new detectors. Nanotechnology has the potential to significantly advance forensic science and criminal investigation. In forensic nanotechnology, tiny chip materials are employed in place of large apparatus, which simplifies the techniques of analysis and improves the accuracy, precision, timeliness, and suitability of investigations. The reader is provided with up-to-date studies and publications as Chapter 5 describes the use of nano sensors in forensics investigation. Additionally, it details the key distinctions between the three major categories of sensors that provides information on the potential applications of these sensing materials in the future.

Nanomedicine deals with the usage of nanotechnology for medical purposes as studied in Chapter 6. Due to their identical size to the majority of biological molecules, nanomaterials utilized in nanomedicine are advantageous for in vivo applications. The aim of applying nanotechnology in medicine is to effectively diagnose and treat diseases. Although nanomedicine remains in its early stages, many analytical tools, diagnostic devices, biosensors, drug delivery vehicles, and physical therapy applications are being developed under this branch of medicine to treat diseases. It is anticipated that a growing number of medicines will use nanotechnology in the coming years as a result of the progress in the field. However, as nanotechnology develops, it is essential to consider both benefits and limitations of the technology including its potential risks. Current problems in the development of nanomedicine involve the toxicity and environmental effects of nanoscale compounds, but as the field develops, its impact on the economy is expected to be significant in future.

The most vital part of a medico-legal investigation is the estimation of the period since death. The procedures now in use for determining this time period depend on a wide range of variables that are frequently out of the forensic examiner's control. Therefore, using nanosensors to analyse changes in biological components has provided a special benefit to precisely estimate the PMI. The evaluation of post-mortem intervals using novel functional nanomaterials techniques is compared in detail to conventional methods in Chapter 7. It also discusses the difficulties in making an accurate assessment due to changing environmental conditions as well as the benefits of using nanomaterials.

Chapter 8 presents the potential of application of nano-sensors for trace explosive detection. A major challenge faced in detecting trace explosives is the detection of ultra-low quantities of trace compounds. This chapter discusses the selective and sensitive detection of trace explosives by employing nano-sensors, limitations of detecting trace explosives and reevaluate the currently present technologies that are inefficient and inaccurate for trace explosive detection. Application of nano-sensore for the exposure of low concentrations explosive compounds is an effective platform in the advancement of trace explosive detection.

Currently, in the field of biomedical engineering and biological applications, the use of soft florescent nanomaterials has increased because of their excellent biocompatibility, easy biofunctionalization, and high brightness properties. Chapter 9 summarizes the current developments of nano-sized fluorescent soft biological imaging agents. Many fluorescent soft nanoparticles like biomaterial-based NPs, vesicles, micelles, nanogels, small-molecule organic NPs, semiconducting polymer NPs and dye-doped polymer NPs are mentioned briefly starting from the preparation methods, their structure, optical properties and functionalization. Depending upon the nano-sized imaging agent's functional as well as optical properties, their uses are briefly described in relation to Vivo imaging, cellular process imaging, and in vitro imaging by using nonspecific and specific targeting.

Analysis of body fluids are an important aspect of forensic investigation; therefore, rapid and sensitive detection is required for their analysis as well as characterization. Chapter 10 presents an overview of the applications of different nanomaterials with respect to the biological material detection. If we talk about forensic biological samples, we especially need to handle them with care, because such samples from crime scenes are always in traces. This chapter briefly discusses the use of functional nanoparticles for the detection of biological materials. Genetic material obtained from crime scenes is one of the most important pieces of evidence in forensic investigations and the chapter focuses on the application of functional nanomaterials for their detection.

Nano forensics utilizes nano-based sensors in matters of crime or where the legality of the process is challenged. There exist many conventional techniques but considering the approach of criminals and the modern-day standard of investigation, there is a need for nanotechnology-based sensors for their sensitive, selective and rapid detection of illicit drugs. Illicit Drugs are psychoactive in nature they act upon the CNS that results in different perceptions and long-lasting responses. Chapter 11 covers the drugs that can be the potential source of causing abuse and those commonly detected in forensic lab testing. The conventional techniques have a few drawbacks time-consuming, highly expensive and some of these techniques are also destructive. Nanosensor is a nanoparticle-based probing technique that constitutes a Sensing platform for the detection of the biological substance present in the evidence sample collected from the crime scene. These nanosensors can work with limited quantities and the quality of the sample yields valuable information for preparing the scientific report.

Being a multidisciplinary field, forensic science has been greatly evolved over the previous decades. The recent advances in nanotechnology has opened new windows for forensic sciences as it offers cutting-edge methods for simplifying the laborious forensic analysis of the questioned documents. In this regards, various nanoparticles are most widely exploited and coupled with the traditional methods for resolving cases marked with forgery. These methods are now emerging to be the very extensively used strategies for the future analysis of questioned documents to obtain the analytical, quantitative as well as qualitative profile of the ink on real samples. Chapter 12 thus explores the trending (traditional as well as novel nano-based) methods used so far in the questioned documents analysis.

Volatile poisons are the volatile organic carbon based compounds which easily release from indoor surroundings (for example paints, furnishings, building materials) or industry productions that are because of their relatively low boiling points. The detection of volatile toxins can be done in a variety of ways. Various kinds of sensors have been created to analyze the volatile organic toxins. Volatile poison sensors are generally easy to use, inexpensive, programmable and portable. Future developments will include innovations or combinations of advanced technologies for volatile poisons sampling, detection and analysis, as well as validation and standardization of these methods for practical clinical use. Chapter 13 provides an overview and critical analysis of the sampling, collecting, detection, and analytical techniques linked to volatile substances, with a focus on current advancements in nanomaterials such quantum dot-based sensors.

Nanotechnology includes different kinds of nano-materials including the nano-composites, nano-hybrid and nanostructured materials. There are immense and broader views based on pragmatism and cross-fertilization of nanotechnology. Just because of the development in nanotechnology the progression in technical sector is also increasing day after day. Similarly, role of nanotechnology is increasing in the field of Forensic Sciences. There are numerous branches of Forensics covered by nanotechnology including. The trend of nanotechnology enhancing already existing technologies in Forensics. In upcoming era advanced nanotechnology in instruments for illicit drug screening and fingerprint detection is increasing because Nano-based techniques hold an immense future in fingerprinting and security features. Therefore, Chapter 14 is offering an insight application of nanoparticles in numerous areas of Forensics.

Nanotechnology has constant to prove its supremacy with enormous applications to outdated methods in medical, electronic, engineering, and other fields. In forensic science, nanotechnology research has delivered a new viewpoint for real-time crime investigation and established progressive Nano sensors, Nano-manipulators, and Nano-imaging tools for visualization. Often, nanotechnology helps in augmenting and improving the productivity of already existing and applied forensic techniques with high precision, sensitivity, and reducing the time required. So, this chapter reviews International Activities in Nano forensic in different branches of forensic science, e.g., for inspecting questioned documents, age of bloodstains, and time since death, along with its application in improving PCR efficiency, DNA analysis, and explosive detection. Therefore, in Chapter 15, an effort has been made to highlight the new outlook or knowledge of nanotechnology with its applications in several divisions of forensic science at international level.

The book is intended for researchers, academics, and professionals in the fields of forensic science, nanomaterials, and materials science. It is also suitable for graduate and postgraduate students who are interested in the application of nanomaterials in forensic science. The book provides a valuable resource for those looking to gain a deeper understanding of the potential of nanomaterials in forensic science and their impact on the future of the field.

Allah Rakha
Department of Forensic Medicine/Medical Jurisprudence, University of Health Sciences, Pakistan

Anam Munawar
Department of Forensic Medicine/Medical Jurisprudence, University of Health Sciences, Pakistan

Virat Khanna
Department of Mechanical Engineering, Maharaja Agrasen University, India

Suneev Anil Bansal
Department of Mechanical Engineering, Bharat Institute of Engineering and Technology, India

Acknowledgment

This book is a testament to the collaborative efforts and collective wisdom of authors, and we are truly grateful for their contributions.

Prof. Dr. Allah Rakha would like to extend his deepest gratitude and heartfelt appreciation to the Caretaker Health Minister, Punjab, Pakistan Prof. Dr. Javaid Akram and Vice Chancellor, University of Health Science, Lahore, Pakistan Prof. Dr. Ahsan Waheed Rathore who have profoundly influenced and supported throughout the creation of this book. Their wisdom, guidance, and unwavering encouragement have been instrumental in shaping our professional journey.

Dr. Virat Khanna is greatly indebted to his wife Anisha and his daughter Avishka who helped to keep him sane during the whole journey by providing a motivating and supporting environment at home.

Chapter 1
Overview of Functional Nanomaterials

Naseem Akhtar
The Women University Multan, Pakistan

Malika Rani
The Women University Multan, Pakistan

Sumara Ashraf
The Women University Multan, Pakistan

Sara Musaddiq
The Women University Multan, Pakistan

Anam Munawar
University of Health Sciences, Pakistan

Waseem Abbas
Bahauddin Zakriya University, Multan, Pakistan

ABSTRACT

The purpose of this introductory chapter is to define the concepts of functional nanomaterials in science and technology. It will also highlight the fundamental characteristics of these materials in research. It also aims at emphasizing the role and importance of functional nanomaterials in various useful aspects. As research is the main tool of creating knowledge, different methods used by researchers are also briefly discussed here. This chapter will also develop a frame of reference for studying the various chapters that follow. These chapters form a modified core of functional nanomaterials and its applications in forensic investigation.

DOI: 10.4018/978-1-6684-8325-1.ch001

PREVIEW

The purpose of this introductory chapter is to define the concepts of functional nanomaterials in science and technology. It will also highlight the fundamental characteristics of these materials in research. It also aims at emphasizing role and importance of functional nanomaterials in various useful aspects. As research is the main tool of creating knowledge, different methods used by researchers are also briefly discussed here.

This chapter will also develop a frame of reference for studying various chapters that follow. These chapters form modified core of functional nanomaterials and its applications in forensic investigation.

Scientific Objectives

1. Describe overview of functional nanomaterials.
2. Describe aim and scope of functional nanomaterials as a field of scientific study.
3. Explain the structure and analyze the importance of functional nanomaterials for applications.
4. Evaluate common methods of simulation and modeling of functional nanomaterials.

Functional Nanomaterials

Over the last decade, the prefix "nano" has been increasingly prevalent in a variety of study. Nanoscience, nanomaterials, nanotechnology and nano chemistry are only a few of the new nano-related terminology that have become widely known, even among non-experts, in scientific studies, popular literature, and newspapers. The prefix nano is derived from the ancient Greek, which is copied from the Latin nanus, which way dwarf, by extension, and little. It is used to represent a 109-fold reduction factor according to the International System of Units (SI) agreement. As a result, the nanoscale universe is measured in nanometers (1nm = 10-9 m) and includes systems that are larger than molecular dimensions but smaller than macroscopic ones. Nanotechnology is the study of very small things. It change the use and influence of materials on a very small scale. Molecules and atoms acquit differently at this size, allowing for several unhoped and provocative applications. Studies in nanotechnology and nanoscience have exploded in popularity in recent years across a wide range of product categories. It opens up possibilities for the production of new substantial, including those for medical uses, when traditional methods may be limited. Nanotechnology should not be regarded as a single approach with limited applications. Nanotechnology does not only relate to exceedingly small buildings and products, despite its common moniker of "little science." In bulk materials and huge surfaces, nanoscale characteristics are frequently integrated. Nanotechnology entails the planning, development, and manufacture of nanoscale objects.

Nanoparticle Classification

Nanomaterials can be classified using a variety of methods. Nanoparticles are divided into different dimensions: one, two, and three

Figure 1. Unique characteristics of functional nanomaterials (FNM)

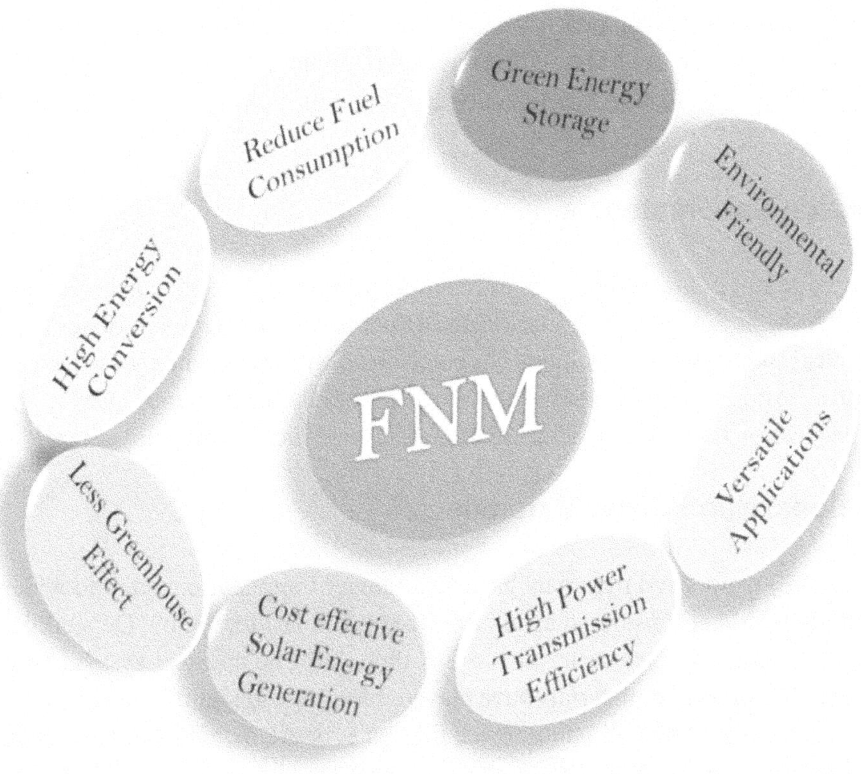

Nanoparticles

A nanoparticle is a tiny atom accompanying partly one proportion lowers than 100nm. Nanoparticle research is presented a field of wide scientific research due to extensive diversity of utilization in photoelectric field, ocular or optical, pharmaceutical and energy applications.

Nanoscience and nanotechnology are at the cutting edge of current exploration. The quickly developing economy here requires specialists who have an exceptional information on nanoscience in mix with the aptitudes to use this information in new items. A different logical training is pivotal to furnish industry and exploration establishments with top choice specialists who have a conventional foundation in the diverse subdisciplines, for example, gadgets, material science, science, material science, biotechnology, and simultaneously be specialists in a single specific field. This is the thing that is offered in this expert programmed.

These infant logical controls are arranged at the interface between physical science, science, materials science, microelectronics, organic chemistry, and biotechnology. Control of these orders hence requires a scholarly and multidisciplinary logical schooling.

Two dimensional (2D) nanomaterials are broadly examined for electronic, biomedical, drug conveyance and biosensor applications.

Categorisation of Nanostructured Materials

Nanomaterials are nano sized structures with Low-level dimensional and encountered as basic thread of nanotechnology. Pokropivry and Skorokhod characterized the nanomaterials as 0D, 1D, 2D and 3D D= Dimensional. They investigated them deeply according to types of nanomaterials (Pokropivny & Skorokhod, 2007).

Zero-Dimension Nanomaterials

Term nano is used to imply the number . Significant improvement has been achieved in the study and execution of 0D NSM in this decade. To fabricate 0D nano sized materials in better dimensions, huge strategy of chemical and physical mechanism has been introduced. At recent, by various research batches 0D nanostructures are immensely studied in quantum dots, core shell dots, nano lenses and hollow ssphere. In various single electron transistor devices, 0D nanomaterials such as quantum dots are widely studied.

One Dimensional Nano-Sized Materials

1D (one dimensional) nano-sized materials have vast range of applications, due to which synthesis of large number of 1D NSMs is considered as interesting issue in last few years. Depending upon the size and dimensionality, 1D nano-sized material study is thought to be an ideal system. In manufacturing of energy devices 1D nano-sized materials have played interlink and significant role. Quality impact of 1D nano-sized materials is shown in nano-device fabricated systems, nanocomposites for fabrication, unconventional energy security system and nano electronics.

Two-Dimensional Nano-Sized Materials (NSMs)

2D nano-sized materials are designed with 2D parameters which are responsible for determining its properties such as shape and size. Since 2004, 2D Nano-sized materials are studied and synthesized to show low dimensional properties, which are distinct from other bulk properties. To understand the growth mechanism of nanostructures and extensive applications in nano-reactors, photocatalysts and sensors type 2D nano-sized are explicitly designed.

Three-Dimensional Nano-Sized Materials

Depending upon the large surface area and other unique preferable properties 3D nanomaterials are considered bulky materials in comparison to other quantum sized materials. Many 3D materials have been synthesized which have wide range of applications due to their peculiar performance depending upon size, shape, and morphologies. In different fields of science, 3D- nanostructures are occupied in catalysis, electrode material in batteries and magnetic behavior of materials.

Dendrimers are a new class of nanometric-scale pressurized structured polymers. Dendrimers utilized in drug delivery and imaging typically have a diameter of 10 to 100 nm and numerous functional groups on their surface, making them suitable drug carriers (Wiener et al., 1994). Dendrimers have been extensively researched in terms of their structure and function. Dendrimers can now be extremely tailored, encasing certain functional molecules.

Quantum dots (QDs) are microscopic gadgets containing a single droplet of free electrons. QDs are mixture of semiconducting material nanocrystals with diameters ranging from 2 to 10 nm. Colloidal synthesis or electrochemistry can be used to make QDs from a variety of semiconductor materials. Cadmium selenide (CdSe), cadmium telluride (CdTe), indium phosphide (InP), and indium arsenide are the most often used QDs (InAs).

Quantum dots containerful have as few as a single electron or as many as tens of thousands. Electrons' size, shape, and number may all be precisely regulated. Semiconductors, insulators, metals, magnetic materials, and metallic oxides have all been produced. It has applications in optical and optoelectronic devices, as well as quantum computing and data storage.

Current Trends of Functional Nanomaterials

Nanoparticles on their structure, natural and synthetic features, magnitude, characteristics can be categorized into various classes. Few of these are polymeric tiny particles, semiconductor small sized particles, and Carbon occupying tiny particles. Many issues can be cured by using nanotechnology like disorder and to stop health problems in developing Countries. For water purification nanotechnology is also used for example filtration of water, substantial and distillation cleanup, soil water analysis, dissipation of water analysis. It may also use in industry for packing of food, its processing and production in better way and perseveration of food. Because of their specific structural dimensions and potential material possession, carbon-based nanomaterials are becoming more appealing materials. Carbon nanotubes, graphene, graphene oxide, and carbon and graphene quantum dots have a wide range of uses in biosensing, drug/gene delivery, tissue engineering, imaging, regenerative medicine, diagnostics, and cancer therapy, to name a few. Cancer continues to be one of the world's major health issues, and several therapeutic approaches are focusing on designing targeted anticancer drug delivery nanosystems using benign and less hazardous resources with high biocompatibility, ease of functionalization, remarkable targeted therapy issues, and low adverse effects. This chapter examines the potential and prospective diagnostic and therapeutic uses of carbon-based nanomaterials in the realm of cancer treatment.

With a focus on biomass based renewable products the quest for novel, cost-effective, and practical biosorbents have been generated. Lignin and nanomaterial decorated lignin derived entities are capable of removing organic/inorganic pollutants from aqueous environments efficiently and cost effectively. Future research should focus on refining and improving the economic feasibility of low-cost, green functional nanomaterials, as low preparation costs are critical for their widespread applicability in water treatment. Manufacturing and utilization of nanomaterials of nanomaterials is one of the most effective ways in nanotechnology for the synthesis of hydrogels and nanocomposites.

This study focuses on functional nanomaterials and their applications in water treatment/purification, with an emphasis on surface assimilation as catalytic degradation of a range of contaminants.

Scope of Functional Nanomaterials

The scientific community has recently been focused on nanoscience and technology, with researchers proposing a wide range of utilize for these materials, giving them the moniker "functional nanomaterials." This class of materials is unique because, when compared to their bulk counterparts, they exhibit exceptional transitions in their characteristics in terms of size management. n heterojunction solar cells, pin the improved power conversion. The latter involves the hydrolysis of TiO2 nanoparticles with a significant surface area. ZnO is also included as another wide bandgap semiconductor, the richest family of nanostructures among all materials. zinc oxide polycrystalline nanostructures in the form of fibres, rhombic flakes, and spheres using different polyol solvents is synthesized by Tseng et al. To compare photocatalytic efficiency, the newly generated ZnO structures were used to degrade methylene blue. Two studies on carbon nanotubes (CNTs) are included because of their prospective uses. Zhao and Li described a better layer-by-layer vacuum filtering process for fabricating single-walled carbon nanotube (SWCNT) films with adjustable thickness and density. The multi-layered SWCNT films' electrical transport capabilities have been explored. Mazov et al. describe a straightforward method for brominating multiwall carbon nanotubes (MWNTs) in the vapour phase at moderate temperatures. To make Br-functionalized products, MWNTs with an average diameter of 9 to 3 nm were treated with bromine vapour (Macutkevic et al., 2012).

Materials, are considered pillar industries of the twenty-first century global economy along with energy and information. In the progress of nanotechnology with focusing on the rational synthesis and research of nanoscale materials a new discipline of materials chemistry and physics has developed. Functional nanomaterials are particularly attractive with new or improved properties by combining multiple constituents and leveraging synergistic effects such as electronic, optical, magnetic, catalytic, or bioactivity, selective permeation, adsorption, and so on because they allow the synthesis of materials. So, Functional nanomaterials are a form of highly important materials with possible uses in specialized industries. They have a unique attribute or several exceptional functions.

Nanoparticle Preparation

The best approach for making nanoparticles is observed by the physicochemical possession of the material. The medicine to be loaded, as well as the polymer. The main production processes. Nanoparticles made from premade polymer can be made in a variety of ways, including:

1. Evaporation of Emulsion-Solvent:

This is one of the most used methods for the nanoparticle's synthesis Evaporation of the solvent after emulsification consists of two phases. The first stage entails emulsifying the oil into an aqueous phase with a polymer solution. The second stage entails as the polymer solvent evaporates; polymer precipitation occurs. nanospheres. The nanoparticles are gathered by a device called a nanoparticle collector. To remove ultracentrifugation and rinsed with distilled water lyophilized for storage of stabilizer residue or any free drug (Song et al., 2014). High-pressure emulsification and solvent evaporation are two variations of this process. This approach entails making an emulsion, homogenizing it under high pressure, and then repeating the process. By shaking the mixture thoroughly to eliminate the organic solvent (Pal et al., 2011), The size can be modified by varying the stirring rate, kind, and temperature. and the amount of dispersion agent, organic and aqueous viscosity temperature, and phases (Bodmeier & McGinity, 1987). This technique, on the other hand, Limitations are imposed by liposoluble medications and can be applied to them. Method of double emulsion and evaporation: The process of emulsion and evaporation methods are also affected by this. As a result, encapsulate a medication that is hydrophilic. The process of using two emulsions is known as double emulsion. The addition of aqueous medication solutions is used in this method. This is introduced to the second aqueous phase without the emulsion. The emulsion is a mixture of oil and water. This oil/water emulsion is diluted with enough water or aqueous solution to increase solvent penetration into the aqueous phase, resulting in the creation of nanospheres.

2. Emulsion-Diffusion Approach:

Another extensively used method for preparing nanoparticles is emulsions approach. The encapsulating polymer is dissolved in a partly water-miscible solvent (such as propylene carbonate or benzyl alcohol) to ensure the initial thermodynamic equilibrium. Solvent diffusion into the exterior phase and the creation of nanospheres or Nano capsules can be achieved by the polymer-water saturated solvent phase which is then emulsified in an aqueous solution containing a stabilizer. This is also depending on the oil-to-polymer ratio. Finally, it is removed by filtration and evaporation by controlling the boiling point of the solvent.

The benefits of this approach are high encapsulation efficiency (usually 70%), no requirement for homogenization, ease of scaleup, high batch-to-batch reproducibility, simplicity, and a limited size distribution. The enormous volumes of water that must be removed from the suspension and the leaking of water-soluble drugs into the saturated-aqueous exterior phase during emulsification, which reduces encapsulation efficiency, are disadvantages (Tiruwa, 2016). The approach was used to make drug-loaded sodium glycolate nanoparticles, including mesotetra (hydroxyphenyl) porphyrin-loaded PLGA (p-THPP) nanoparticles (Vargas et al., 2004), doxorubicin-loaded PLGA nanoparticles (Malinovskaya et al., 2017), and cyclosporine (cy-A-) (El-Shabouri, 2002).

3. Method of Solvent Displacement/Precipitation:

From an organic solution solvent displacement entails the precipitation of a preformed polymer in the presence or absence of surfactant into the aqueous medium. Acetone or ethanol, polymers, drugs, and lipophilic surfactants which are semipolar water miscible solvent are dissolved. The solution is poured or injected into an aqueous solution containing a stabilizer under magnetic stirring. The fast solvent causes nano particles to prepare in a millisecond.

The solvent is then extracted from the suspension under reduced pressure. The size of the particles is affected by the pace at which the organic phase is added to the aqueous phase. It was discovered that both particle size and drug entrapment decrease (Quintanar-Guerrero et al., 1996) as the rate of mixing of the two phases rises. Most poorly soluble medicines are well suited to the nano precipitation technique.

Nanosphere size, drug release, and yield is affected by adjusting preparation conditions was demonstrated. Adjusting polymer concentration in the organic phase was found to be effective in producing smaller nanospheres by restricting the polymer to drug ratio to a narrow range (Aleksandrova, 2002).

Study of Structure of Nanomaterials

Different characterization techniques like X-rays diffraction, scanning electron microscopy, transmission electron microscopy, PL spectroscopy and Raman, etc. are used for the analysis of structural and optical properties of nanoparticles.

The average particle sizes, charges and their diameter are influenced the stability of nanoparticles. The structure and overall morphology of nanoparticles are determined using electron microscopy technique. The performance and stability is also effected by the surface charges.

Particle size distribution and shape are the most essential criteria in nanoparticle description.

Size and morphology of nanoparticles is determined by Electron microscopy. Nanoparticles are most commonly used in medicine delivery. On medicine delivery the impact of nanomaterials is great due to its surface area. Majority of nanoparticles have greater surface area for conduction. So, majority of medicine come in contact with the surface which results maximum delivery of medicines. On the other hand, drugs diffuse slowly among bigger particles, which is a disadvantage. As a result, there is great relation between nanomaterial's size and optimum stability.

There is also great importance of nature and intensity of the charge on the surface of an nanomaterials. Nano particles are crucial because they determine how they interact with their electrical surroundings as well as the biological environment. Bioactive chemicals interact with each other. Colloidal stability is high. The zeta potential of nanoparticles was investigated. This potential is enormous to a measure of the surface charge that isn't directly measured. It's the same as difference in potential between the outer Helmholtz plane and the inner Helmholtz plane the shear surface. The degree of hydrophobicity on the surface can then be determined. The zeta potential can be used to also Surface hydrophobicity can be determined by a number of factors. Hydrophobic interaction chromatography, for example, contact angle, biphasic partitioning, probe adsorption measurements, and so on Several powerful analysis tools have recently been released.

FUTURE PROBLEMS AND OPPORTUNITIES

For drug delivery devices nanoparticles have been proven to be effective. Massive amounts of nanoparticles are available. advantages in terms of medicine targeting, delivery, and use one of the key advantages of combining diagnostic and therapy. There are numerous technological challenges in this field, developing the approaches listed below:- Virus-like computer systems intracellular systems, biomimetic polymer architecture, and control functionalities (active drug targeting, sensitive medicines) Systems that are triggered by biochemical signals and interact with my body nanochips for nanoparticle release, transportation for nanoparticle release improved polymers for therapeutic peptide delivery. Techniques for delivering or administering drugs have been developed. Internally, most of the important and well-established formulations and research programmers on medication delivery down to a dispersion containing component are developed.

Applications in Biomedicine

Polysaccharide nanocrystals have unique features, including bioactivity, biocompatibility, biodegradability, and low toxicity, which make them suitable for pharmaceutical and biological applications. Polysaccharide nanocrystals, on the other hand, are derived from natural biomass resources, allowing them to circulate in the bloodstream for longer periods of time and delaying clearance by the mononuclear phagocytic system, making them useful for the environment.

REFERENCES

Aleksandrova, E. (2002). Structural properties of polyimides and the mechanism of carrier photogeneration in polyimide-based molecular complexes. *Optics and Spectroscopy*, *93*(1), 118–125. doi:10.1134/1.1496734

Bodmeier, R., & McGinity, J. W. (1987). Polylactic acid microspheres containing quinidine base and quinidine sulphate prepared by the solvent evaporation technique. II. Some process parameters influencing the preparation and properties of microspheres. *Journal of Microencapsulation*, *4*(4), 289–297. doi:10.3109/02652048709021821 PMID:3504510

El-Shabouri, M. H. (2002). Positively charged nanoparticles for improving the oral bioavailability of cyclosporin-A. *International Journal of Pharmaceutics*, *249*(1-2), 101–108. doi:10.1016/S0378-5173(02)00461-1 PMID:12433438

Macutkevic, J., Seliuta, D., Valusis, G., Adomavicius, R., Krotkus, A., Kuzhir, P., Paddubskaya, A., Maksimenko, S., Kuznetsov, V., Mazov, I., & Simonova, I. (2012). Multi-walled carbon nanotubes/PMMA composites for THz applications. *Diamond and Related Materials*, *25*, 13–18. doi:10.1016/j.diamond.2012.02.002

Malinovskaya, Y., Melnikov, P., Baklaushev, V., Gabashvili, A., Osipova, N., Mantrov, S., Ermolenko, Y., Maksimenko, O., Gorshkova, M., Balabanyan, V., Kreuter, J., & Gelperina, S. (2017). Delivery of doxorubicin-loaded PLGA nanoparticles into U87 human glioblastoma cells. *International Journal of Pharmaceutics*, *524*(1-2), 77–90. doi:10.1016/j.ijpharm.2017.03.049 PMID:28359811

Pal, S. L., Jana, U., Manna, P. K., Mohanta, G. P., & Manavalan, R. (2011). Nanoparticle: An overview of preparation and characterization. *Journal of Applied Pharmaceutical Science*, 228–234.

Pokropivny, V., & Skorokhod, V. (2007). Classification of nanostructures by dimensionality and concept of surface forms engineering in nanomaterial science. *Materials Science and Engineering C*, *27*(5-8), 990–993. doi:10.1016/j.msec.2006.09.023

Quintanar-Guerrero, D., Fessi, H., Allémann, E., & Doelker, E. (1996). Influence of stabilizing agents and preparative variables on the formation of poly (D, L-lactic acid) nanoparticles by an emulsification-diffusion technique. *International Journal of Pharmaceutics*, *143*(2), 133–141. doi:10.1016/S0378-5173(96)04697-2

Song, W.-K., Cui, Y.-J., Tang, A. M., Ding, W.-Q., & Tran, T. D. (2014). Experimental study on water evaporation from sand using environmental chamber. *Canadian Geotechnical Journal*, *51*(2), 115–128. doi:10.1139/cgj-2013-0155

Tiruwa, R. (2016). A review on nanoparticles-preparation and evaluation parameters. *Indian Journal of Pharmaceutical and Biological Research, 4*(2), 27.

Vargas, A., Pegaz, B., Debefve, E., Konan-Kouakou, Y., Lange, N., Ballini, J.-P., van den Bergh, H., Gurny, R., & Delie, F. (2004). Improved photodynamic activity of porphyrin loaded into nanoparticles: An in vivo evaluation using chick embryos. *International Journal of Pharmaceutics*, *286*(1-2), 131–145. doi:10.1016/j.ijpharm.2004.07.029 PMID:15501010

Wiener, E., Brechbiel, M., Brothers, H., Magin, R., Gansow, O., Tomalia, D., & Lauterbur, P. (1994). Dendrimer-based metal chelates: A new class of magnetic resonance imaging contrast agents. *Magnetic Resonance in Medicine*, *31*(1), 1–8. doi:10.1002/mrm.1910310102 PMID:8121264

Chapter 2
Impacts of Nanotechnology

Rubia Shafique
The Women University Multan, Pakistan

Malika Rani
The Women University Multan, Pakistan

Anam Munawar
University of Health Sciences, Pakistan

Maryam Arshad
COMSATS University, Islamabad, Pakistan

ABSTRACT

This chapter presents an overview of positive and negative nanotechnology impacts discussed on a broader plane. Nanotechnology's central tenet is the utilization of sub-micrometer-sized particles for use in a wide variety of disciplines, including medicine, cosmetics, agriculture, and the food industry. Nanotechnology's advantages include better agricultural production, quality in pharmaceuticals, targeted drug delivery, and sensor applications. In order to move forward with its growth, it is crucial to correctly recognize the potential benefits and unintentional threats of nanomaterials to our surroundings. Nanotechnology is more complex, more focused, and a balanced interpretation of risks and benefits is very important to analyze.

It's quite easier to enlist nanotechnology technical uses but how its applications covers energy and health related issues along with information processing is quite interesting. It is noteworthy that windows for new possibilities and opportunities in this field must be meet up by individuals, major factor is environment shaping both as producer and a consumer. Nanotechnology covers fields such as bio-chemistry, computer, materials-science along with ethical to mental, legal to environmental aspects. That's why for long-term survival needs meetup, vision and technological development importance for its promotion in terms of realization is need of era. Because of the small size of nanotechnology, it may be possible to automate tasks that were previously impossible due to physical limits, lowering the amount of work, land,

DOI: 10.4018/978-1-6684-8325-1.ch002

and maintenance that humans are necessary to perform. Environmental, health, and safety concerns are among the potential risks, as are transitional impacts such as the replacement of established sectors as nanotechnology products become more prevalent, which privacy advocates are concerned about. These could be especially relevant if nanoparticles' possible harmful effects are disregarded. Just with prior knowledge of its physics and chemistry we can't achieve without having insight that how its physical behavior affects its chemical behavior for having nanotechnological revolution. That's why it has been referred as invisible, slippery and ubiquitous. The question of whether nanotechnology warrants specific government regulation is a contentious one. The Environmental Protection Agency (EPA) in the United States and the Health and Consumer Protection Directorate (HCPD) of the European Commission have both opened investigations into the risks posed by nanoparticles. No man-made nanoparticles can be included in organic nanomaterials that have been given the seal of approval.

BACKGROUND OF NANOTECHNOLOGY

Structural development and application with basic, new and valuable functional properties on basis of structural size within range 1-100 nanometers constitute nanotechnology by involving creation (manipulation or discovery) of structural materials at atomic/molecular/supermolecular atomic scale level (Mansoori & Soelaiman, 2005). Nanotechnology development phase still continues as it starts from 1960's with biotechnological evolution in 1980's by allowing unlimited characteristic applications discovery making it 21st century technology including manufacturing industry, human health in terms of nanomedicine development, energy conversion by protecting environment to act as economic driving force (Sheeparamatti, Sheeparamatti, & Kadadevaramath, 2007). US national science foundation (NSF) worldwide estimated that by end of year 2015, about products of worth 1 trillion US dollars that are incorporating nanotechnology as a key component using manpower of about 2 million workers (Hulla, Sahu, & Hayes, 2015; Roco, Mirkin, & Hersam, 2010; Williams & Alivisatos, 2013). Research and development industry investment exceeds about 6 billion US dollars by 2006 keeping in view global risk management factor by holding potential applications to solve technical, economical, ecological and social problems (Kay & Shapira, 2009). National research and development programs during last few years have a tendency to revolutionized digitally involving cognitive sciences owing to funding by agencies by regulating impacts on human health and environment as well (Roco, 2011).

Defining the Term "Impact"

As nanotechnology is continuously investigated by considering public-private as well as commercial research institutions but technological innovation is another question. For technological adoption a scenario needs to be developed but for impact factor calculation, what will be the probability of adoption of that scenario is important. From this we can formulate into a relation as:

Actual impact=Theoretical + Probability of adoption

For making a decision being rational regarding innovation adoption decision has to be made on basis of J-value termed as Judgement value introduced to world by Thomas while considering adoption of safety procedures in nuclear industry (Bhushan, 2017; Malik, Muhammad, & Waheed, 2023; Rickerby

& Morrison, 2007). Safety measures designing purpose is to enhance quality while considering that measurement cost can reduces life quality but remember key point that decrease must not be approachable to the increase. In case of nanotechnology, cost means applications development with consideration of benefits enhancement in terms of life quality changing measures.

Defining the Term "Nanotechnology"

Controlled and deliberate evaluation, matter production, replacement, manipulation on nanoscale for materials/devices/systems production with aim of introducing new functionalities and applications is another name of nanotechnology (Charles & Frank, n.d.). Three major aspects includes universal process of fabrication, materials designing, fabrication and modelling on nanoscale, novelty in terms of creativity. Fabrication involves top-down, bottom-to-bottom and bottom-up techniques covering machining, semiconductors processing, additive manufacturing, complex-sophisticated structures designing and materials reproducibility features. Another thing that kept in mind is novelty factor in terms of nanoscale physio-chemical behavior, nano-objects fabrication such as carbon nanotubes, macro-objects with ultra-smallness in terms of both size and cost, nano-enabled technology development (Schulte, 2005). On practical mode, nanomaterials containing nanoparticles are not inherently dangerous. Only a few qualities, such as their increased sensitivity and movement, could make them harmful. We would be dealing with a true threat if nanomaterials specific properties considered harmful for both humanity and the environment. Nanotechnology being combination of three prominent plus emerging technologies such as bio-technology including microorganisms controlled activity alongwith DNA sequencing including bioreactors-technology main production unit, IT (information technology) relying on nano-hardware components designing and CS (cognitive sciences) such as brain nervous activity control via ultra-small electrodes, quantum phenomena understanding (Chan, 2009; Chaudhary, Ghatmale, & Chavan, 2016; Jadhav & Jadhav, 2018).

Science Impact

On atomic scale, nanotechnology structural and procedural investigations are very important incorporating many nanotools (Manoharan, 2008). One such example includes "GENE CHIP" another name of DNA array used for bacteria identification and genetic changings identification but cost effectiveness and ease of availability restricts investigations. For a better comparison, need of measurement by repeated experiments is necessary and it will not be possible without gene chip technology. Another example includes wave technology innovation towards OWLS (optical waveguide light-mode spectroscopy) for attainment of binding coefficients with help of grating couplers. Nanotechnology will have far-reaching consequences for society in areas such as defense, intellectual property, employment, and the power dynamic between individuals and governments as shown in Figure 1.

Engineering Impact

As molecular-biology, bio-technology and medicine are interrelated, same-way nanotechnology and engineering are interconnected. Nanotechnology basic aim is attainment of quality in attainment of nano blocks in terms of specificity and individuality. In case of electronic devices, major contribution is of processing power with maximum circuit components within small chip. This is the revolution brought by

Figure 1. Nanotechnology applications overview

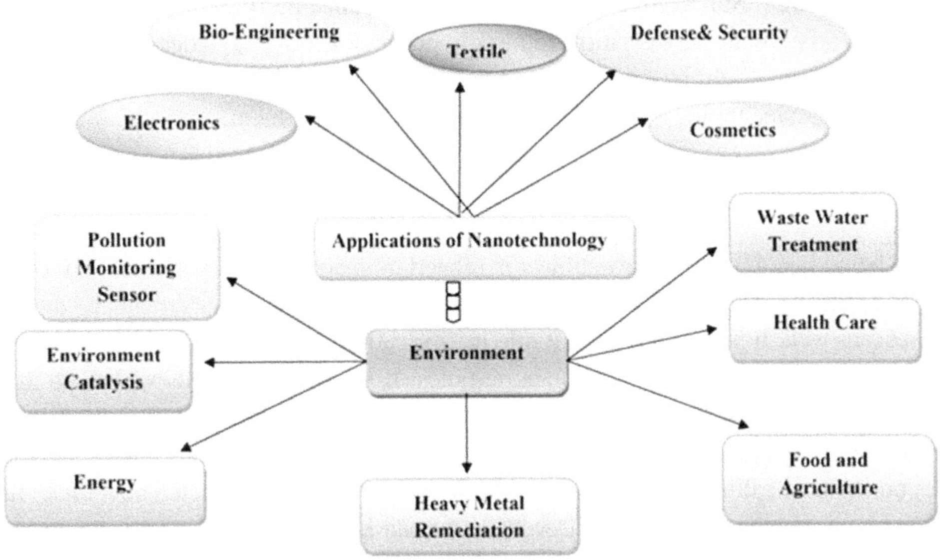

nanotechnology in terms of nanomaterials fitness and applicability until attainment of designing goals (Singh, Dubey, & Dubey, 2019). Nanotechnology-based goods are expected to drive breakthroughs in sectors such as sustainability, health, robotics in the future, according to academics. For example, nanotechnology engineering may be used to develop medications that target particular cells in the body or materials that can be used to construct artificial organs (Seeman, 1999).

Information Technology Impact

ITRS stands for international technology roadmap for semiconductors involves betterment in processing of semiconductors and logic gates designing in making circuits for transistors. Considering this, electron chip used as carrier of information is reward of nanotechnology (Salinas, Barros, Eisenberg, Smith, & Viswanathan, 2001). Another approach involves quantum cellular automaton involves electronic configuration via quantum dots. Quantum computers and all-optical computers with usage of photonics, a three-dimensional nanostructure and quantum states superpositions. Data storage, data processing, presence or absence of single atom for information to be readable all covers nanotechnology (Markovic, Zivkovic, Cvetkovic, & Popovic, 2012). It should also be important to consider that on nanoscale designing of logic gates it is necessary to overcome manufacturing difficulties in terms of reliability (Theis, 2001).

Energy Impact

Energy harvesting in nanotechnology another opportunity for problem solving to overcome the energy gap by providing convenience to humanity involving solar energy, fuel cells, energy storage and energy efficiency (Christian, Adityawarman, & Indarto, 2013; Raina, Sharma, Slathia, Bhagat, & Pathak, 2020; Serrano, Rus, & Garcia-Martinez, 2009). As basic technology of converting solar energy into electrical energy in photovoltaic cells is simple but involves high cost that restricts its usage on large scale.

Another factor is storage of that produced energy chemically by decreasing thickness for enhancing efficiency is under research. Nanotechnology basic aim is to reduce inefficiency, cost, material thickness and better catalyst attainment for oxidation-reduction processes. Fuel cells designing is in sequence of these aspects' attainment. Until attainment of atom-by-atom assembly, catalyst fabrication will remain the special case of nanotechnology for existing technology improvement.

Health Impact

Nanomedicine being nanobiotechnology subject is related to nanotechnology application towards human health. Nanomaterials used for medicine regeneration in terms of tissues replacement and damaged tissues regrowth aspects as well. It plays a significant role in fabrication of such surfaces that acts as bactericidal such as hospital paints consisting of silver nanoparticles. It includes diagnosis with help of nanoscales sensors, labs-on-chips and powerful computers so that large scale data integration for better diagnosis takes place. Health consequences are the potential consequences of using nanotechnological materials and technologies on human health (Dreher, 2004). Because nanotechnology is such a new topic, there is a lot of speculation about whether it will benefit or harm human health. The potential for nanotechnological advancements to have medical applications to cure disease, as well as the potential health concerns provided by nanomaterial exposure, are two aspects of nanotechnology's health impacts. Currently world pandemic situation involving medical researchers, bio-engineers are experimentally trying to resolve and cure the disease to protect communities by vaccination development as mentioned in Figure 2.

Figure 2. Nanotechnology health impacts

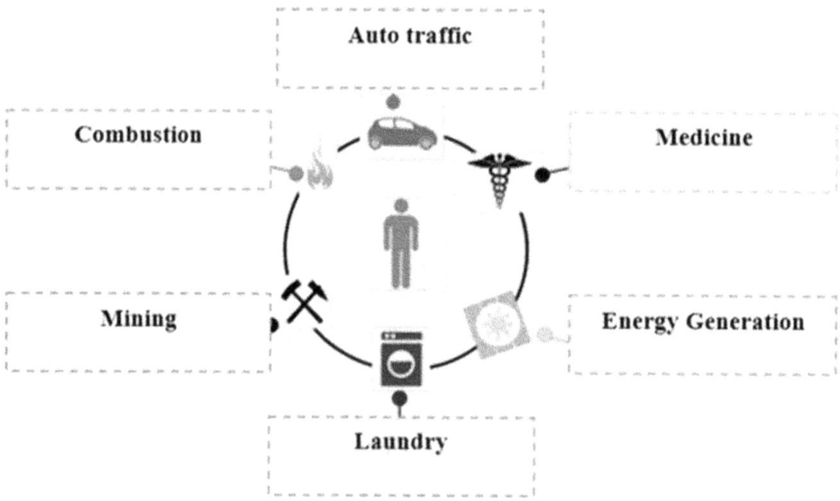

Nanomedicine Refers to the Application of Nanotechnology in Medicine

Such a science relating cure, health restoration and disease prevention using remedial organic and inorganic substances by controlling diet, eating habits is termed as medicine. Nanomedicine can take numerous forms, ranging from medical nanomaterial applications towards nano-electronic biosensors development and molecular nanotechnology usability. In the near future, nanomedicine is the collection of valuable research tools with beneficial technologies at clinical level. Pharmaceutical industry will see new commercial uses such as improved delivery systems of drugs, innovative treatments, and in-vivo imaging according to the national nanotechnology initiative criteria as shown in Figure 3.

Figure 3. Nanotechnology medical applications

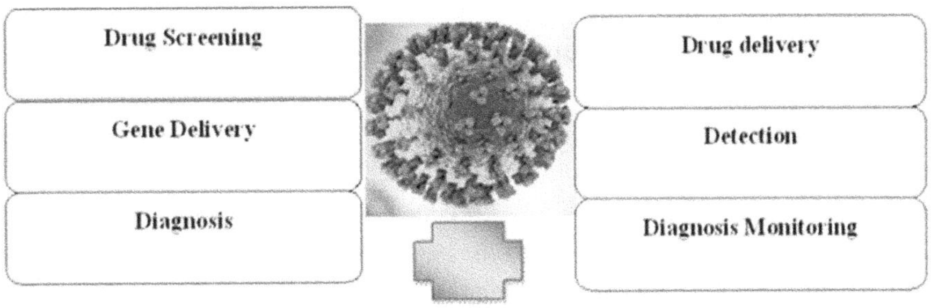

Another vital area of study is the development of neuro-electronic interfaces and other nanoelectronics-based sensors. There is possibility of alteration in medicines using cell repairing machines according to the aims of speculative science of molecular nanotechnology (Patra et al., 2018). The National Institutes of Health in the United States established a five-year plan in 2005 to construct four nanomedicine facilities, and nanomedicine research is funded directly. 130 nanotechnology-based drugs, delivery systems developed around the world as reported by journal Nature Materials till April 2006. Globally, nanotechnology research and development receive annual funding of $3.8 billion from at least 200 companies representing 38 different products. It means expansion in nanomedicine industry have a significant impact on the economy (Farokhzad & Langer, 2009). Drug synthesis is another important area contributed by nanotechnology at microscale level involving drug designing features such as collection of targeted molecules, structurally related but functionally unrelated molecules selection and identification. Surgery is another factor related to nanomedicine by usage of convenient surgical tools and more economical in terms of delivery making it attractive for international heath delivery system. Last but not least, control over diet termed as nutrition via nanomedicines involving fertilizers, pesticides related to agriculture advantage. In terms of humanity, it is related to fresh food provides and to packaged food available commercially. These nanodevices enables temperature monitoring and color change to prevent food being nutritionally damaged. Nanofertilizers advantage is the delivery of required nutrients to plants by releasing active substances needed by humanity for healthier survival (Butt & Naseer, 2020). Detailed analysis of soil, usage of proper water content and water purification by nanomedicine is important (Abeer, 2012; Kargozar & Mozafari, 2018; Pautler & Brenner, 2010).

STUDY OF THE POTENTIAL HEALTH EFFECTS INCLUDING NANOTOXICITY

Because of their exceedingly small size, human body significantly more easily absorbed nanomaterials comparable to larger particles. However, nanoparticles response inside the organism is among primary challenge needed to be overcome nanoparticles interactions with their environments are determined by factors such as their size, shape, and the reactivity of their surfaces (Patel, Nanda, & Sahoo, 2015). For instance, they may produce stress reactions that lead to inflammation and a lowered resistance to various diseases by overwhelming phagocytes, the cells responsible for absorbing and discarding foreign material. Nanoparticles, because of their large surface area, will immediately adsorb onto their surface some of the macromolecules they come into contact with when exposed to tissue and fluids; this poses a problem because non-degradable or slowly degradable nanoparticles can accumulate in organs and interfere with biological processes within the body (Sufian, Khattak, Yousaf, & Rana, 2017). This could have an impact on enzyme and other protein regulatory systems. Since there are so many factors at play it's tough to make blanket statements about the dangers of being exposed to nanomaterials. Instead, researchers must carefully evaluate each new nanomaterial on its own merits (Muhammad, Haroon, Shah, Ullah, & Haleem, 2018). In the workplaces of enterprises manufacturing or employing nanomaterials, as well as in nanoscience and nanotechnology facilities, health and environmental concerns collide.

Early research on nanoparticles interaction within system of body and how workers being exposed to nanoparticles during both fabrication as well as usage was conducted by the National Institute for Occupational Safety and Health nanoparticles. National Institute of Occupational Safety and Health (NIOSH) presented introductory guidelines for working with nanomaterials based on current scientific understanding (Tamers et al., 2020). NIOSH's National Personal Protective Technology laboratory has conducted research on nanoparticle filter penetration along with noncertified dust masks. Penetrating particle size ranging between 30 to 100 nanometers whereas number of nanoparticles found in the test dummies' respirators was determined to be mostly determined by leak size. Nanotoxicity determining factors includes difference in chemical composition, structure, shape, surface, solubility, aggregation and presence/absence of important functional groups as compared to other nanomaterials (Casciano & Sahu, 2009; Chen & Chen, 2012; Fischer & Chan, 2007). It's difficult to make broad comments about the health risks associated with nanomaterial exposure due to variables large numbers that is determinant of toxicity where it is necessary for each new nanomaterial to be examined individually by taking into consideration nanomaterial all properties. According to studies, the release of new nanoparticles being engineered and subsequent personal exposure occurrence throughout a number of work functions. The situation alerts regulatory agencies to the necessity for Nanoscaled workplace safety measures and rules.

ENVIRONMENTAL IMPACT

Nanotechnology major impact on environment includes two factors one is direct and short-term whereas the other is indirect, long-term and sustainable. First one involves nanoparticles usage for remediation of environment such as usage of iron nanoparticles for contaminated soil remediation with chlorinated hydrocarbons. Second one involves usage of technology including water purification such as harmful elements elimination whereas sea and air traffic elements play their role by contributing environmental benefits to nanotechnology. The potential environmental consequences of nanotechnological materials and technologies are referred to as nanotechnology's environmental impact (Dingman, 2008). Because nanotechnology is a relatively young field of study, there is debate on how the use of nanoparticles in industry and commerce may affect animals and ecosystems. Two aspects of nanotechnology's environmental impact also include nanotechnological developmental potential in improving environment as well due to the potential environmental damage that nanotechnological materials pose.

Utilization of nanotechnology in improving the environmental sustainability of processes that cause negative externalities is known as green nanotechnology (Nasrollahzadeh, Sajjadi, Sajadi, & Issaabadi, 2019). It is in other words application of nanotechnology for enhancing nanoparticles sustainability. Green nano product creation and nano product use to increase sustainability are two examples. In another sense "green nanotechnology" refers to the research and implementation of methods to mitigate the negative impacts on the environment and human health throughout the full product lifecycle from raw materials to final disposal as visible in Figure 4.

Figure 4. Green nanotechnology

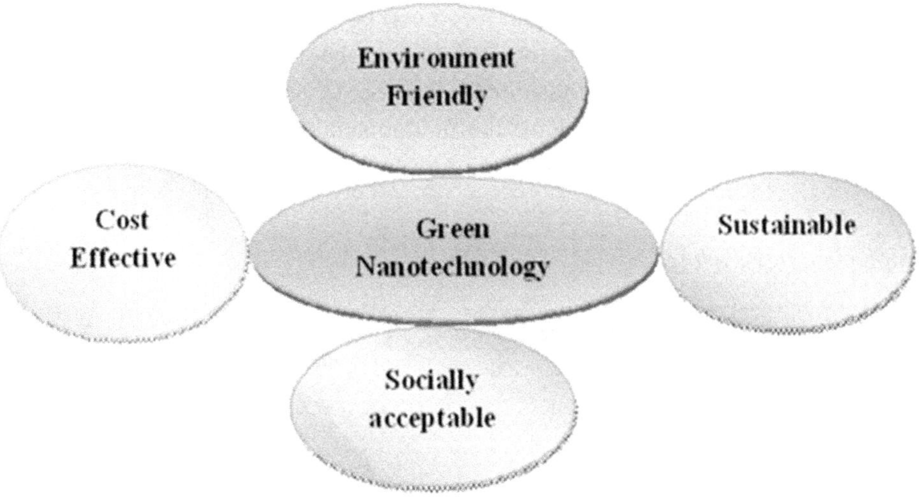

Nanoscale solutions to environmental problems are the focus of "green nanotechnology" which also aims to create nanomaterials and products that are safe for human health and the environment (Smith, 2011). It employs preexisting green chemistry and engineering principles to develop nanomaterials and nano-products that don't contain any harmful substances are produced at low temperatures using only renewable energy sources, and are designed with lifecycle considerations baked into the initial concept and all subsequent iterations of the design modelling, and engineering (S. H. Khan, 2020). A catch-all word for nanodevices waste generated by or during nanomaterials manufacturing is termed as nano-pollution (Jan, Majeed, Ahmad, Lone, & John, 2022). Nano wastage is a phrase that describes particles collection that are either released into the environment or disposed with their goods still attached. Machinery and lighter cars that need less fuel, energy sources and alternative fuel; and its all possible to detect and clean up environmental contaminant's materials. Centre for bio-nano interaction studies the nanoparticles dispersed impact in environmental milieu, where animal matter and decaying plant becomes natural organic matter which are mostly composed of the polysaccharides (Faria et al., 2018). These polysaccharides basically interact with nanoparticles and also examine how the interaction affects organic stability, environmental fate, dispersibility and behavior.

SOCIAL IMPACT

Toxicity is a danger to people's health and the environment. Nanotechnology has a greater societal impact than first-generation nanoparticles and it presents more social concerns (Roco & Bainbridge, 2007a). Nanotechnology societal concerns according to social scientists, should be recognized and valued more than "downstream" hazards or repercussions. Instead, the difficulties should be included in during "upstream" research and decision-making to ensure that technical progress actually meets social goals (Roco & Bainbridge, 2007b). Several researchers in the social sciences and nongovernmental organization shave weighed in on the issue believe that public interaction should be a part of technology appraisal and governance. In order to analyze the significant amount of risk linked with nanotechnology and nano-related products, stakeholder perspectives must be explored. In 2003, roughly 800 patents linked to nanotechnology were granted, and by 2012, the number had climbed to around 19,000 worldwide. Corporations have already filed a considerable number of patents on Nanoscale technology and discoveries. NEC and IBM, for example, have the fundamental patents on carbon nanotubes, one of nanotechnology's present pillars as mentioned in Figure 5.

Figure 5. Nanotechnology cycling

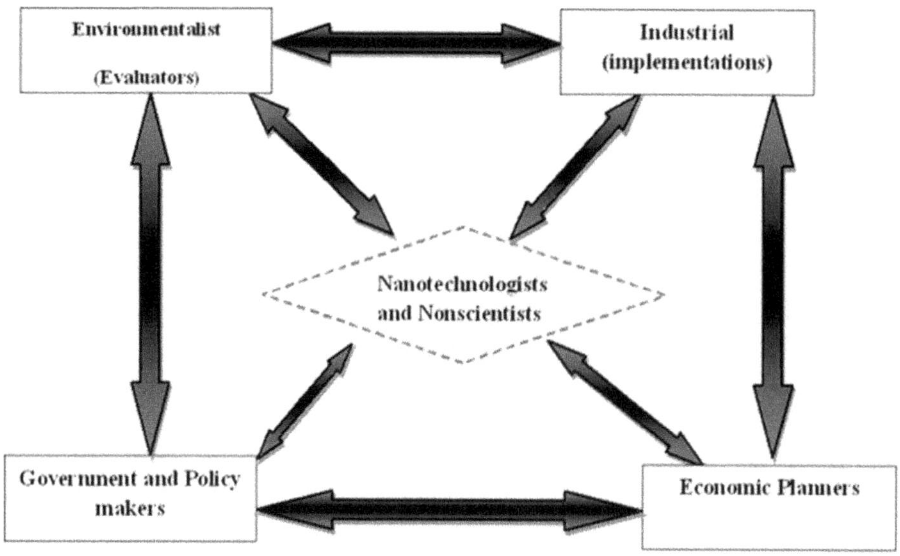

Carbon nanotubes are used in a variety of industries, spanning computers and software, reinforced materials, medication delivery, and medical devices, and are expected to become increasingly popular over time. In underdeveloped nations, millions of people lack access to essential services such as safe drinking water, dependable energy, health care, and education, as well as nanotechnologies might provide new answers (Stern & McNeil, 2008). According to the United Nations Task Force on Science, Technology, and Innovation, nanotechnology provides a variety of advantages, including production with minimal labor, high productivity, land, or upkeep low cost, and low material and energy needs. However, there is widespread concern that the promised advantages nanotechnology will not be made available to the general public evenly, and that any gains (technological and/or economic) is likely to disproportionately favor wealthy countries. Longer-term concerns revolve on the influence of new technologies on society as a whole, and if they will lead to a post-scarcity or post-industrial economy.

INDUSTRIAL IMPACT

The nanotechnology industry will lay the groundwork for incredibly powerful and affordable computers, as well as water treatment and remediation technologies that will handle the world's water issue, as well as cost-effective alternative energy technologies and new medical technology that might save millions of lives. Nanomaterials provide lightweight and durable materials for automotive and construction applications. Nanotechnology in coatings enhances the durability, water resistance, and stain resistance of surface materials, resulting in textile solutions for high-end fashion and high-end athletic goods (Hobson, 2009). Food manufacturing, processing, safety, and packaging may all benefit from nanotechnology. Nanotechnology can tackle a complicated set of technical and scientific difficulties in the food and bio processing business for producing high-quality, safe food in an efficient and sustainable manner (Elcock, 2007).

Applications

- Food processing in agriculture biomedicine
- DNA molecular detection using nanotechnology.
- Nutraceutical bioavailability is improved by nanocapsules.
- Nanobiosensors for cancer agent detection and diagnosis
- Nano chips for tracking and preserving identities.

Nanotechnology, like electricity and computers, may be viewed as a general-purpose technology that is employed in a variety of sectors, including basic chemicals, cosmetics, agriculture, food manufacture, electronics, instrumentation, bioengineering, pharmaceuticals, and textiles as shown in Figure 6.

Figure 6. Industrial factors affecting nanotechnology

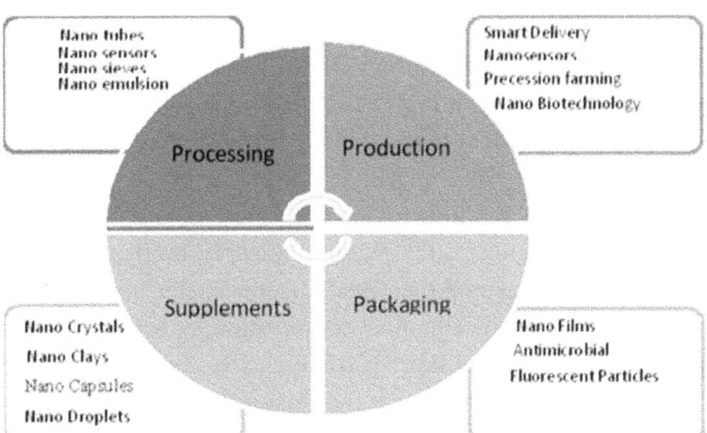

ECONOMIC IMPACT

Commercialization and adoption of new nanoscaled R&D-derived technologies are expected to yield significant economic benefits (Yadav, Khan, & Mishra, 2013). In addition to enhancing our basic quality of life, advances in nanotechnology-related medical, national security, environmental protection, energy generation, and education have all improved our fundamental quality of life. Nanotechnology has the potential to be applied to a wide variety of products, technologies, and applications that protect the environment and climate, from conserving raw materials, energy, and water to reducing greenhouse gas emissions and hazardous waste. Despite the many benefits of nanotechnology, there are potential risks and ethical problems linked with its use. The economic implications of nanotechnology span a wide variety of issues, including ethics, privacy, the environment, and security (Hullmann, 2006).

POLITICAL IMPACT

Nanotechnology is anticipated to expand the gap between the military's and the population's political violence capabilities. Nanotechnology's growth could contribute to terrorism by exacerbating existing asymmetric warfare trends. Inevitably, the gap between the military and the populace in terms of their capacity for political violence will widen thanks to the advent of nanotechnology. As a result, the progress of military nanotechnology applications may indirectly increase the risk of future terrorist attacks (Clunan, Rodine-Hardy, Hsueh, Kosal, & McManus, 2014).

POTENTIAL AND BENEFITS OF NANOTECHNOLOGY'S NEGATIVE EFFECTS

Although nanotechnology has obvious therapeutic and environmental benefits, it, like other technologies, has the potential to have unforeseen environmental consequences, both in the human body and the ecological environment. The health implications of adopting this innovative technology has advantages in terms of health, the environment, and sustainability must be investigated by science. Nanotechnology has recently sparked a lot of attention in the media. Assessing toxicity with the same chemical composition can be difficult since not all nanoparticles formed are more harmful than fine-sized particles. UV exposure, particle surface coatings, and dispersion properties, according to Eva Oberdörster of Southern Methodist University, can all affect particle behavior (Warheit, 2018). The method of creating nanoparticles, such as fumed vs. precipitated silica, and whether particles agglomerate and subsequently disaggregate once they reach lung fluids, appear to be critical in pulmonary research. According to David Warheit of the DuPont Company, the particles' ability to impair macrophage function, induce cell and lung damage, enhance inflammation, and persist in the lungs will be utilized to establish a working hypothesis for detecting particle toxicity (Warheit et al., 2007). Warheit wrapped up by underlining the challenges of research. Researchers have found that some man-made nanoparticles are more damaging (inflammatory) than fine-sized particles of the same chemical composition, as reported by Warheit. The basis of this method is a comprehensive analysis of three particle classes: titanium dioxide, carbon black, and diesel particles. On the other hand, he did note that the existing beliefs are not based on facts. All risks involved within nanotechnology development are mentioned in Figure 7.

Figure 7. Risks in nanotechnology development

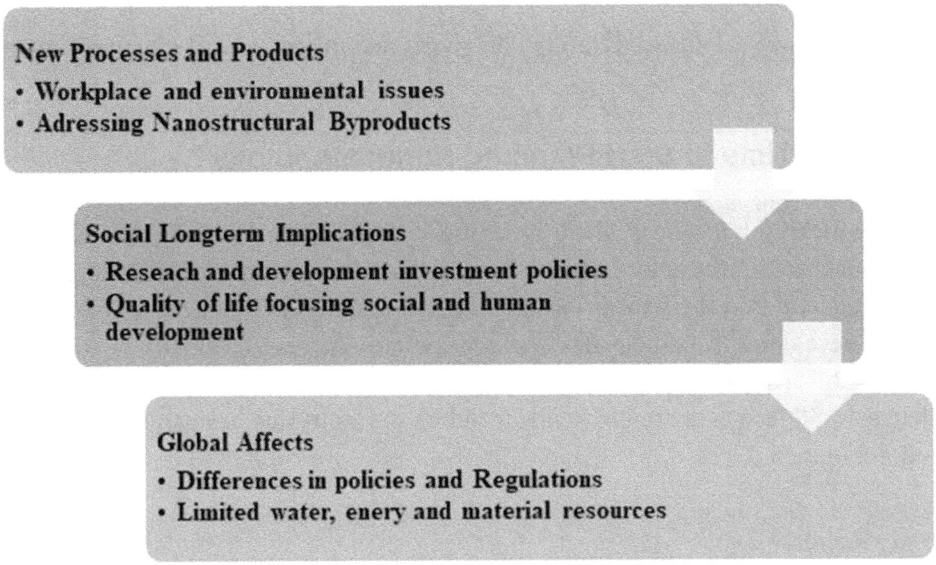

Clean and Cheap Energy

Standard design for converting sunlight into electricity is less affiance than prototype solar panel incorporating nanotechnology. Nanotechnology is now being used in batteries and capacitors. Nanomaterials may enhance hydrogen storage materials and catalysts for fuel cells. By creating lighter storage units and more surface area, energy generation, conversion and energy storage for fuel cells, thermos to electric, solar cell, biomass energy, super capacitors, secondary batteries and thermal storage fluids can enhance by nanotechnology (Pallas, Peijnenburg, Guinée, Heijungs, & Vijver, 2018).

The Protection of Human Right to Have Clean Water

Nanotechnology provides low-cost contaminant detection and low-cost water purification as a result of its quick capabilities. Arsenic may be removed from drinking water by using ultra-small rust magnetic interactions. Nanotechnology may potentially enhance water and air quality monitoring by inventing more sensitive detection systems that can concurrently test a wide variety of contaminants and harmful chemicals (Nagar & Pradeep, 2020). Some scientists claim that ground water contamination can be remedied by utilizing iron nanoparticles as a chemical reductant. Nanoengineered membranes might be utilized to make water purification operations more energy efficient (N. A. Khan et al., 2019).

Brown-Fields Clean Up

Nano silver antimicrobial properties may clean up hazardous chemicals and oil spills.

Reducing Global Disease Burden

In health care improvements enhanced through diagnosis and treatment will increase personal well being worldwide.

When Is the Best Time to Start Planning Nanotechnology?

Rice University's Vicki Colvin agrees. "Now is the greatest time to start looking into the effects of nanomaterials on human health," the author argues. A new sector is emerging, and it is beginning to control risk before large quantities of these materials are released into the environment or onto consumers, in a way that no other developing technology has ever done before (Davies, 2008). This is a once-in-a-lifetime opportunity to establish a new, flourishing science with a wealth of knowledge about the environmental health challenges we'll face in the future, while avoiding the faults that have already plagued chemistry as mentioned in Figure 8.

Figure 8. Other nanotechnology risks

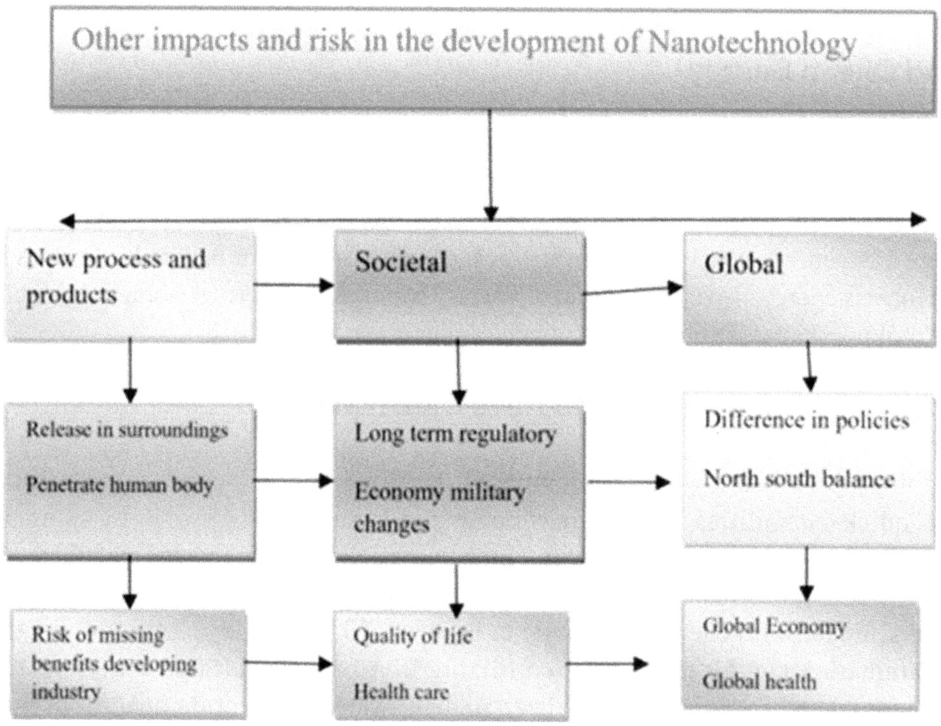

"There is a need for increased levels of collaboration between the sectors involved, government parties to provide financially feasible solutions while simultaneously maintaining the environment and health, and public interest organizations," says Environmental Defense's John Balbus. This is a key goal since it is critical to do nanotechnology development correctly the first time (J. Balbus, Denison, Florini, & Walsh, 2005). Balbus points out that, while contemporary history has seen numerous scientific breakthroughs

Figure 9. Nanotechnology applications in the food industry

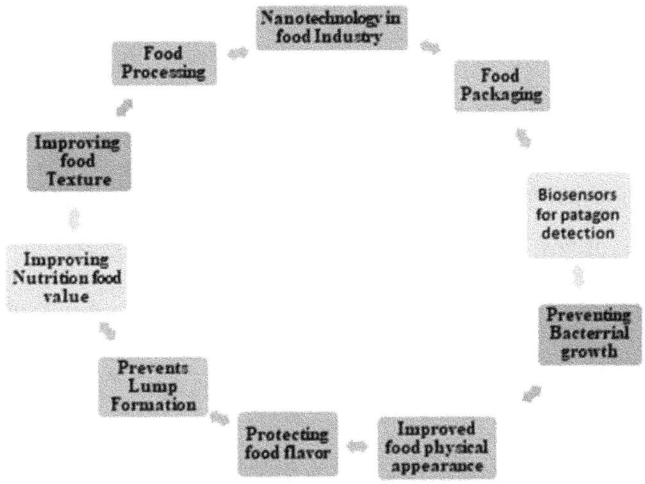

with the potential to change society, these advancements have occasionally come at the expense of safety. The first challenge with nanotechnology is going forward with research in a way that best protects the public while also addressing health and safety issues (J. M. Balbus, Florini, Denison, & Walsh, 2006).

Many of these ideas were reiterated by David Rejeski of the Woodrow Wilson International Center for Scholars, who claimed that, unlike genetically modified organisms, nanotechnology's impacts will be seen across various industries and goods, and risk prevention will be more controllable. In order to stimulate the development of new models that bring together academics, NGOs, and industry to identify principles and best practices, he also advised politicians to consider forging voluntary agreements with industry on responsible nanotechnology use NGOs, and industry to identify principles and best practices (J. M. Balbus, Florini, Denison, & Walsh, 2007). Finally, Balbus emphasized that determining policy goals should be an open process that includes opportunities for public participation as well as research. Food preparation, food packaging, functional food development, food safety, foodborne pathogen detection, food product shelf-life extension are only a few of the nanotechnology applications that have evolved as a result of the increased need for nanoparticles in many sectors of food production and food microbiology (Kumar, Gupta, Dixit, Mishra, & Srivastava, 2019). These applications within food industry are mentioned in Figure 9.

REFERENCES

Abeer, S. (2012). Future medicine: Nanomedicine. *Jimsa*, *25*(3), 187–192.

Balbus, J., Denison, R., Florini, K., & Walsh, S. (2005). Getting nanotechnology right the first time. *Issues in Science and Technology*, *21*(4), 65–71.

Balbus, J. M., Florini, K., Denison, R. A., & Walsh, S. A. (2006). Getting it right the first time: Developing nanotechnology while protecting workers, public health, and the environment. *Annals of the New York Academy of Sciences, 1076*(1), 331–342. doi:10.1196/annals.1371.027 PMID:17119213

Balbus, J. M., Florini, K., Denison, R. A., & Walsh, S. A. (2007). Protecting workers and the environment: An environmental NGO's perspective on nanotechnology. *Journal of Nanoparticle Research, 9*(1), 11–22. doi:10.100711051-006-9173-7

Bhushan, B. (2017). Introduction to nanotechnology. Springer handbook of nanotechnology, 1-19.

Butt, B. Z., & Naseer, I. (2020). Nanofertilizers. *Nanoagronomy*, 125-152.

Casciano, D. A., & Sahu, S. C. (2009). *Nanotoxicity: from in vivo and in vitro models to health risks.* John Wiley & Sons.

Chan, W. C. (2009). *Bio-applications of nanoparticles* (Vol. 620). Springer Science & Business Media.

Charles, P. P., & Frank, J. O. (2003). *Introduction to nanotechnology.* Academic Press.

Chaudhary, L. S., Ghatmale, P. R., & Chavan, S. (2016). Review on: Application of nanotechnology in computer science. *International Journal of Scientific Research, 5*, 1542–1545.

Chen, R., & Chen, C. (2012). *Nanotoxicity. In The nanobiotechnology handbook.* CRC Press.

Christian, F., Adityawarman, D., & Indarto, A. (2013). Application of nanotechnologies in the energy sector: A brief and short review. *Frontiers in Energy, 7*(1), 6–18. doi:10.100711708-012-0219-5

Clunan, A., Rodine-Hardy, K., Hsueh, R., Kosal, M. E., & McManus, I. (2014). Nanotechnology in a globalized world: strategic assessments of an emerging technology. Academic Press.

Davies, J. C. (2008). Nanotechnology oversight. Washington, DC: Project on Emerging Nanotechnologies.

Dingman, J. (2008). Guest commentary: nanotechnology: its impact on food safety. *Journal of Environmental Health, 70*(6), 47–50. PMID:18236938

Dreher, K. L. (2004). Health and environmental impact of nanotechnology: Toxicological assessment of manufactured nanoparticles. *Toxicological Sciences, 77*(1), 3–5. doi:10.1093/toxsci/kfh041 PMID:14756123

Elcock, D. (2007). *Potential impacts of nanotechnology on energy transmission applications and needs.* Academic Press.

Faria, M., Björnmalm, M., Thurecht, K. J., Kent, S. J., Parton, R. G., Kavallaris, M., . . . Boyd, B. J. (2018). Minimum information reporting in bio–nano experimental literature. *Nature Nanotechnology, 13*(9), 777–785.

Farokhzad, O. C., & Langer, R. (2009). Impact of nanotechnology on drug delivery. *ACS Nano, 3*(1), 16–20. doi:10.1021/nn900002m PMID:19206243

Fischer, H. C., & Chan, W. C. (2007). Nanotoxicity: The growing need for in vivo study. *Current Opinion in Biotechnology, 18*(6), 565–571. doi:10.1016/j.copbio.2007.11.008 PMID:18160274

Hobson, D. W. (2009). Commercialization of nanotechnology. *Wiley Interdisciplinary Reviews. Nanomedicine and Nanobiotechnology*, *1*(2), 189–202. doi:10.1002/wnan.28 PMID:20049790

Hulla, J., Sahu, S., & Hayes, A. (2015). Nanotechnology: History and future. *Human and Experimental Toxicology*, *34*(12), 1318–1321. doi:10.1177/0960327115603588 PMID:26614822

Hullmann, A. (2006). *The economic development of nanotechnology-An indicators based analysis.* EU Report.

Jadhav, S. S., & Jadhav, S. V. (2018). *Application of nanotechnology in modern computers.* Paper presented at the 2018 International Conference on Circuits and Systems in Digital Enterprise Technology (ICCSDET). 10.1109/ICCSDET.2018.8821086

Jan, N., Majeed, N., Ahmad, M., Lone, W. A., & John, R. (2022). Nano-pollution: Why it should worry us. *Chemosphere*, *302*, 134746. doi:10.1016/j.chemosphere.2022.134746 PMID:35489464

Kargozar, S., & Mozafari, M. (2018). Nanotechnology and Nanomedicine: Start small, think big. *Materials Today: Proceedings*, *5*(7), 15492–15500. doi:10.1016/j.matpr.2018.04.155

Kay, L., & Shapira, P. (2009). Developing nanotechnology in Latin America. *Journal of Nanoparticle Research*, *11*(2), 259–278. doi:10.100711051-008-9503-z PMID:21170134

Khan, N. A., Khan, S. U., Ahmed, S., Farooqi, I. H., Dhingra, A., Hussain, A., & Changani, F. (2019). Applications of nanotechnology in water and wastewater treatment: A review. *Asian Journal of Water, Environment and Pollution*, *16*(4), 81–86. doi:10.3233/AJW190051

Khan, S. H. (2020). Green nanotechnology for the environment and sustainable development. *Green materials for wastewater treatment*, 13-46.

Kumar, A., Gupta, K., Dixit, S., Mishra, K., & Srivastava, S. (2019). A review on positive and negative impacts of nanotechnology in agriculture. *International Journal of Environmental Science and Technology*, *16*(4), 2175–2184. doi:10.100713762-018-2119-7

Malik, S., Muhammad, K., & Waheed, Y. (2023). Nanotechnology: A Revolution in Modern Industry. *Molecules (Basel, Switzerland)*, *28*(2), 661. doi:10.3390/molecules28020661 PMID:36677717

Manoharan, M. (2008). Research on the frontiers of materials science: The impact of nanotechnology on new material development. *Technology in Society*, *30*(3-4), 401–404. doi:10.1016/j.techsoc.2008.04.016

Mansoori, G. A., & Soelaiman, T. F. (2005). *Nanotechnology—An introduction for the standards community.* ASTM International.

Markovic, D. S., Zivkovic, D., Cvetkovic, D., & Popovic, R. (2012). Impact of nanotechnology advances in ICT on sustainability and energy efficiency. *Renewable & Sustainable Energy Reviews*, *16*(5), 2966–2972. doi:10.1016/j.rser.2012.02.018

Muhammad, W., Haroon, M., Shah, M., Ullah, M. A., & Haleem, I. (2018). Potential application of nanotechnology in health care: An insight. *Nanoscale Reports*, *1*(2), 1–8. doi:10.26524/nr1821

Nagar, A., & Pradeep, T. (2020). Clean water through nanotechnology: Needs, gaps, and fulfillment. *ACS Nano*, *14*(6), 6420–6435. doi:10.1021/acsnano.9b01730 PMID:32433866

Nasrollahzadeh, M., Sajjadi, M., Sajadi, S. M., & Issaabadi, Z. (2019). *Green nanotechnology Interface science and technology* (Vol. 28). Elsevier.

Pallas, G., Peijnenburg, W. J., Guinée, J. B., Heijungs, R., & Vijver, M. G. (2018). Green and clean: Reviewing the justification of claims for nanomaterials from a sustainability point of view. *Sustainability (Basel)*, *10*(3), 689. doi:10.3390u10030689

Patel, S., Nanda, R., & Sahoo, S. (2015). Nanotechnology in healthcare: applications and challenges. *Med. Chem, 5*(12).

Patra, J. K., Das, G., Fraceto, L. F., Campos, E. V. R., Rodriguez-Torres, M. P., Acosta-Torres, L. S., Diaz-Torres, L. A., Grillo, R., Swamy, M. K., Sharma, S., Habtemariam, S., & Shin, H.-S. (2018). Nano based drug delivery systems: Recent developments and future prospects. *Journal of Nanobiotechnology*, *16*(1), 71. doi:10.118612951-018-0392-8 PMID:30231877

Pautler, M., & Brenner, S. (2010). Nanomedicine: Promises and challenges for the future of public health. *International Journal of Nanomedicine*, 803–809. PMID:21042425

Raina, N., Sharma, P., Slathia, P. S., Bhagat, D., & Pathak, A. K. (2020). Efficiency enhancement of renewable energy systems using nanotechnology. *Nanomaterials and Environmental Biotechnology*, 271-297.

Rickerby, D., & Morrison, M. (2007). Nanotechnology and the environment: A European perspective. *Science and Technology of Advanced Materials*, *8*(1-2), 19–24. doi:10.1016/j.stam.2006.10.002

Roco, M. C. (2011). *The long view of nanotechnology development: the National Nanotechnology Initiative at 10 years* (Vol. 13). Springer.

Roco, M. C., & Bainbridge, W. S. (2007a). *Nanotechnology: Societal Implications: Individual Perspectives*. Citeseer. doi:10.1007/978-1-4020-5432-7

Roco, M. C., & Bainbridge, W. S. (2007b). *Nanotechnology: societal implications: maximizing benefits for humanity*. Springer Dordrecht. doi:10.1007/978-1-4020-5432-7

Roco, M. C., Mirkin, C. A., & Hersam, M. C. (2010). *Nanotechnology research directions for societal needs in 2020. Retrospective and Outlook. National Science Foundation*. Sponsor.

Salinas, F. M., Barros, E. F., Eisenberg, W., Smith, D. M., & Viswanathan, S. (2001). Impact of nanotechnology on telecommunications. *Sohail Anwar M. Yasin Akhtar Raja Salahuddin Qazi*, 23.

Schulte, J. (2005). *Nanotechnology: Global strategies, industry trends and applications*. John Wiley & Sons. doi:10.1002/0470021071

Seeman, N. C. (1999). DNA engineering and its application to nanotechnology. *Trends in Biotechnology*, *17*(11), 437–443. doi:10.1016/S0167-7799(99)01360-8 PMID:10511701

Serrano, E., Rus, G., & Garcia-Martinez, J. (2009). Nanotechnology for sustainable energy. *Renewable & Sustainable Energy Reviews*, *13*(9), 2373–2384. doi:10.1016/j.rser.2009.06.003

Sheeparamatti, B., Sheeparamatti, R., & Kadadevaramath, J. (2007). Nanotechnology: Inspiration from nature. *IETE Technical Review*, *24*(1), 5–8.

Singh, A., Dubey, S., & Dubey, H. K. (2019). Nanotechnology: The future engineering. *Nanotechnology, 6*(2), 230–233.

Smith, G. B. (2011). *Green nanotechnology.* Paper presented at the Nanostructured thin films IV. 10.1117/12.893114

Stern, S. T., & McNeil, S. E. (2008). Nanotechnology safety concerns revisited. *Toxicological Sciences, 101*(1), 4–21. doi:10.1093/toxsci/kfm169 PMID:17602205

Sufian, M. M., Khattak, J. Z. K., Yousaf, S., & Rana, M. S. (2017). Safety issues associated with the use of nanoparticles in human body. *Photodiagnosis and Photodynamic Therapy, 19*, 67–72. doi:10.1016/j.pdpdt.2017.05.012 PMID:28552731

Tamers, S. L., Streit, J., Pana-Cryan, R., Ray, T., Syron, L., Flynn, M. A., ... Guerin, R. (2020). Envisioning the future of work to safeguard the safety, health, and well-being of the workforce: A perspective from the CDC's National Institute for Occupational Safety and Health. *American Journal of Industrial Medicine, 63*(12), 1065–1084. doi:10.1002/ajim.23183 PMID:32926431

Theis, T. N. (2001). Information Technology Based on a Mature Nanotechnology: Some Societal Implications. *Societal Implications of Nanoscience and Nanotechnology*, 60.

Warheit, D. B. (2018). Hazard and risk assessment strategies for nanoparticle exposures: How far have we come in the past 10 years? *F1000 Research, 7*, 7. doi:10.12688/f1000research.12691.1 PMID:29636906

Warheit, D. B., Hoke, R. A., Finlay, C., Donner, E. M., Reed, K. L., & Sayes, C. M. (2007). Development of a base set of toxicity tests using ultrafine TiO2 particles as a component of nanoparticle risk management. *Toxicology Letters, 171*(3), 99–110. doi:10.1016/j.toxlet.2007.04.008 PMID:17566673

Williams, R., & Alivisatos, P. (2013). *Nanotechnology research directions: iwgn Workshop Report: vision for nanotechnology in the next decade*. Springer Science & Business Media.

Yadav, S. K., Khan, Z. A., & Mishra, B. (2013). Impact of nanotechnology on socio-economic aspects: An overview. *Reviews in Nanoscience and Nanotechnology, 2*(2), 127–142. doi:10.1166/rnn.2013.1026

Chapter 3
Analytical Techniques for Characterization of Nanomaterials

Khizra Bano
National Institute for Biotechnology and Genetics Engineering, Faisalabad, Pakistan

Anam Munawar
University of Health Sciences, Pakistan

ABSTRACT

With the increasing demand of nanoparticles in various applications, many challenges are also noticed in terms of synthesizing accurately designed nanomaterials. The characteristics to be customized in order to make these nanomaterials useful need characterization and verification using different tools that have certain limitations and significant strengths. The purpose of this chapter is to provide the description of measurement methods that may be applied to nanodimension in order to study their most important physical and chemical features. This chapter will provide a detailed introduction, mechanism of working, and limitations of various characterization tools. Focus will be on giving detail about widely applicable and commonly approachable methods to measure properties that can be categorized in terms of size, size distribution, aggregation, shape, dimension, surface area, composition, surface chemistry, charge, surface potential, and functionally active interactive groups.

INTRODUCTION

There is a significant difference between the physical or chemical properties exhibited by nanoscale materials and bulk materials of the same composition. This is because of obvious variation in surface to volume ratio, surface plasmon resonance, reactivity and electronic interactions. This is why nanotechnology and nanomaterials have attracted researchers around the globe. Nanomaterials are synthesized using different conventional and unique methods which can be mechanical, ionic, biologically assisted, laser induced or chemical in their nature. Miniaturization to nanoscale is crucial in order to successfully

DOI: 10.4018/978-1-6684-8325-1.ch003

conduct experiment without wasting up the resources and using the results for improving the quality of life (Koczkur, Mourdikoudis, Polavarapu, & Skrabalak, 2015).

"Nano" means "a billionth," i.e., in nanotechnology is the study of materials having dimensions of the order of a billionth of a meter. It is actually not clear when humans first began to take advantage of nanosized materials. Back in fourth century, Roman glass makers reportedly fabricated color glass containing nanosized materials as Lycurgus glass which contained silver and gold nanoparticles. Today, nanotechnology is a set of advance scientific methodologies to manipulate the material at nanoscale. This fastest growing area of science and technology is of emerging importance so much that investment in nanotechnology research is reported to increase many folds in fact reaching millions of dollars since it started. On the broader scale, usage/ fabrication/ exploration of objects in nanometer dimensions in order to reveal their applications in chemical, physical or biological worlds comes under the sky of nanotechnology.

Nanoparticles play significant role as medication when their role in biomedical applications in considered. Moreover, the high surface area to volume ratio enhanced colloidal stability make them most attractive for biomedical and sensing applications., In addition, size restriction of materials, imparts absolutely different properties to nanoparticles as compared to their bulk counterparts. For instance, Surface plasmon resonance (SPR) band as exhibited by some metal nanoparticles is not observed in bulk metal. This phenomenon is highly affected by the size, shape, surface chemistry, core electrostatic properties and surrounding temperature. Currently, great innovative products are being synthesized using nanotechnology which is further used in medical, industrial, agricultural and even crime investigation fields. This large number of applications and synthesis call for precise, authentic and up-to-date characterization.

Nanometrology is the science of measuring at nanoscale which is not only limited to measurement of size or length of nanoparticles/ nanomaterials. It also deals with analysis of chemical composition, optical properties, mass, electrical and other physical or chemical properties. There are many nanometrological techniques used for instance SEM, TEM, AFM, XRD, and EDX which are used to obtain morphological data about nanoparticles and various other studies are performed to analyze electrical behavior of nanoparticles. However, the study of nanomaterials using these techniques comes with its own challenges where inherent properties of nanomaterials may hinder accurate study of their features, insufficient production at laboratory scale and the multidisciplinary nature of this research also affect access to characterization tools required for specific purposes. It is, therefore, necessary to study the usability of data obtained from characterization tools, limitations of these technique and strengths of the tool keeping the application in mind (Wang et al., 2013).

Different techniques of characterization can be categorically grouped according to their usage. For example, microscopy-based methods for surface, morphology, size data (TEM, FESEM, AFM, etc.), magnetic techniques to study certain type of groups (SQUID, VSM etc.) and techniques for structural or elemental properties (X-ray spectroscopy, XRD, UV-Vis spectroscopic techniques). Table 1 will summarize some of the important parameters and characterizations tools to measure them.

BACKGROUND

Nanotechnology is focused on using the advantage of unique characteristics of materials at nanoscale. With its vast usability in health, energy, forensic, agriculture, consumables and other industries, it has directed research towards itself from around the globe. Nanomaterials are not defined in terms of a specific

Table 1. Significant properties and related characterization techniques

Properties	Technique for Measurement
Crystal structure	XRD, HRTEM, Electron diffraction
Shape and Size	TEM, XRD, DLS, SEM, AFM, MALDI, NMR, XPS
Size distribution analysis	DLS, Mass spectrometry, FMR.
Crystal growth kinetics	NMR, TEM, Cryo-TEM, liquid TEM
Surface charge	EPM, Zeta potential
Elemental composition analysis	XRD, SEM-EDX, NMR
Trace detection of nanomaterials	SEM, AFM

shape or size or composition, in fact any material with at least one dimension in nanoscale is considered as nanomaterial (Klaessig, Marrapese, & Abe, 2011). With difference in properties from bulk materials, difference in requirement for handling and characterization also comes forth. So, the nanomaterials require difference characteristic tools than the conventional bulk counterparts. Also several physical and chemical terminologies are required to completely describe the nature of a nanomaterial such as size, shape, composition, functional groups, surface properties, crystal lattices, charge and agglomeration etc with challenging handling (Hassellöv, Readman, Ranville, & Tiede, 2008). Every change in laboratory condition, sampling technique or the preservation method can result in altered the features at nanoscale (Pessoa, Sipoli, & de la Torre, 2017). Another challenge is acquiring the right image to gather accurate visual information about the synthesis of nanoparticles (X. He et al., 2013).

An image or optical illustration should be very similar to the real object. According to the description provided by J.C. Maxwell in his article published in 1858, a perfect image is the one which has equivalent point for each similar point of the real object and with similar geometric alignment. Most of the images are not usually ideal because of the image defects found generally. This is guided by the lens properties, focusing strength, lens aberrations and corresponding disk of confusions, orientation of particles, magnification or demagnifications and distortions. A distorted image is characterized by variation in the position of real points in the actual object and these distortions can be of pincushion, barrel and spiral types. These can be identified in relation with magnification factor with the object or image. An increased M away from the axis is pincushion distortion, decreasing M results in barrel distortion and a increase in rotation of the image in termed as spiral distortion. Another problem associated with image formation is its distance from the actual image which means that a lens will capture various points of an object from various angles or distances and this can lead to a curved field of view resulting in curvature of the capturing field. Hence various problems associated with imaging an object are blurriness, distortion and curvature resulted by violating Maxwell's rule 1, 2 and 3 respectively (Egerton, 2005). These problems are solved using various angles, types of lenses and focusing properties applied in various imaging techniques like microscopic examinations.

Nanoparticles toxicological concerns are also directed by the parameters which are necessarily measured while synthesizing these nanomaterials. Physico-chemical properties like size, surface, shape, chemistry influence their usage in biological set-up and the physical or chemical interaction of these nanomaterials with environment also modify their role. Right kind of characterization tool like DLS in such situations can help in studying the interference from complex surrounding and designing the right surface characteristics (Hachenberger et al., 2023).

Microscopy methods like scanning electron microscopy, transmission electron microscopy can be of importance here because nanoparticles do not fall within the diffraction limit of ordinary light. Most of the microscopy methods are destruction however, can be effectively coupled to spectroscopic analysis to obtain elemental analysis information. Dark field microscopy is a revolutionary technique in this regard for enabling high contrast imaging of nanoparticles in biological matrices and complex combinations and this can be exploited in forensic sciences effectively (Fakhrullin, Nigamatzyanova, & Fakhrullina, 2021).

Other methods like spectroscopy are effective in studying the size, concentration and shape of the nanoparticles. Fluorescent quantum dots and SPR based metal nanomaterials can be studied using spectroscopy to determine particle size and interaction capacities. Different characterization tools which can be of immediate help in developing a useful understanding of the nanomaterials are being discussed herein.

Scanning Electron Microscopy

The scanning electron microscopy is one of the most advanced technologies to study and examine micro/nano structures including the morphology, surface topology and chemical structures. Stinzing and Knoll designed this microscopic technique in early 1965. It uses high energy electron beam assisted with electromagnetically operated lenses under vacuum for the study. SEM can be utilized for detailed analysis of surface characteristics in bulk samples. Thus it can be effectively applied to various applications including forensic anthropology and forensic applications like examining hair, fibers or materials collected from crime site. Further SEM can also provide useful information about crystalline structures, elemental compositions and conductive behaviors of material under study. As compared to optical microscope which works best with 1000X magnification, a magnification of up to 100,00,00X is obtained with an SEM.

It is necessary to understand terms like resolution limit and signal generation to get full understanding of SEM functions. Resolution limit is the minimum distance between two objects to make them appear as distinct objects. Ernest Abbe was the first scientist to relate limit of resolution with the wavelength of the light source. He also gave a mathematical relationship to calculate resolution as $d= 0.612\lambda /n\sin\alpha$ which is now known as Abbe's equation and can be used to describe resolution in an ideal optical system theoretically (Bogner, Jouneau, Thollet, Basset, & Gauthier, 2007).

One of the major advantages of SEM is that it can focus on a large surface simultaneously because of its large depth of field, thus enabling study of rough surfaces without destroying the specimen. Although flawed sample preparation can result in undermined image quality however, the sample preparation time is very short as compared to other microscopic techniques. In fact there is no special protocol observed for semi-conductor specimen's preparation prior to study, and it is simply mounted on a SEM stub generally made up of aluminium along with some electrically conducting pad. In exceptional cases, electrical connections are also established (Murtey & Ramasamy, 2016).

The functioning operation starts with a beam of electron in a vacuum chamber. With energy of 2-40KeV, the incident electron is shot from any of the three different types of electron guns: field emission guns, Lanthanum hexaboride filaments and tungsten hairpin filament. Among these, most commonly utilized source is tungsten hairpin filament (also known as cathodes or electron emitters) which is also widely utilized in TEM and other microprobe systems. Optimum electronic beam is emitted from the filament according to the requirement of the device and specified according to the needs. These tungsten filaments are made using high grade tungsten which is known to have highest melting point of 3422°C among all metals and lowest vapor pressure at more than 1650°C of temperature. This filament is heated to 2500°C temperature to enable electron beam emission.

Although lanthanum hexaboride filaments are costly but they last longer and produce brighter beams and work by same thermionic principle as of the tungsten filament (Tunaboylu, Toprak, Binark, Öztürk, & Zaim, 2021). In comparison to both these, field emission guns known as cold cathodes produce the brightest and most coherent beam which provides better spatial resolution utilizing electric field. This type of guns uses quantum mechanical tunneling effect of electrons for aligning and producing a high energy electron beam. However field emission guns require greater than 10-6 Torr of vacuum which is normally required by thermionic cathodes to preserve the tip. This high vacuum increases the operational cost thus making cold imaging expensive procedure of SEM function (Fleck & Humbel, 2019).

The beam which is produced by either ways is converted into a fine probe using electromagnetic condenser lenses which is passed to a raster scan coil setup to the sample surface. Controlled by the elemental composition of sample, the energy and angle of the electron beam, the emitted electrons interact with the specimen and produce different signals i.e. secondary electrons, backscattered electrons, cathodoluminescence, electron beam induced current, auger electrons, x-rays and sometimes light (Mohammed & Abdullah, 2018). All of these signals are detected by appropriate detectors in the chamber. These detectors generate readable signals on the monitor and the magnification is controlled by a ratio of monitor display length with the length of specimen exposed to the raster.

When basic interactions between beam and specimen occurs, elastic or inelastic scatterings happen. Secondary electrons have less than 50eV of energy and are removed from their orbits due to electronic beam interaction. However this escape happens only from a shallow layer thus providing topographic information. Backscattered electrons are another kind of incident electrons on the specimen which nearly collide with the nucleus of the atom and subsequently bounce back with a larger angle. With their higher energies as compared to secondary electrons, these produce a faint image because of the larger area of the signal as compared to probe and the depth from which these are emerged. Backscattered electrons give ample information about strain in semi-conductors etc (Hettler et al., 2015).

Another kind of electronic emission which gives topographical knowledge is emitted auger electrons and x-rays. These are emerged from the surface proximity and are measured and imaged through dedicated equipments which have advanced detecting monitors. Simultaneous measurement of x-ray energies can provide data about elemental composition or chemistry of the sample under study. Electron beam induced current is another imaging technique (Sutter & Sutter, 2014). Every electronic interaction creates multiple electron-hole pairs. A strong electric field applied is used to separate charge and minor carriers and minority carriers diffuse to the depletion layer at p-n junction. These collected carriers generate EBIC which helps in imaging the junction at depth, the length of the diffusion and a quantitative measurement of all these aspects (Conlan, Moldovan, Bruas, Monroy, & Cooper, 2021).

Cathodoluminescence is the emission of UV-Vis or Infrared light which happens because of electronic beam (Yacobi & Holt, 2013). Some organic molecules, a large number of semiconductors and some other elements exhibit this phenomenon which is further utilized by SEM for imaging, gathering analytical data or significant material differences. Both CL and EBIC are combined frequently to image the lattice defects (Seiler, 1983). Thus by using all these imaging techniques accordingly, SEM can be used to gather a good resolution image and information in a range of 1 μm to 1 nm using various signals. Recent advances in the form of high-resolution SEM have given a better control over imaging and resolution although sample conductance is still considered its limitation where thin conductive layer is necessary for it (Cohen Hyams, Mam, & Killingsworth, 2020).

Figure 1. Schematic representation of electron microscopy: (a) scanning electron microscopy, (b) transmission electron microscopy
Source: Malenica et al. (2021)

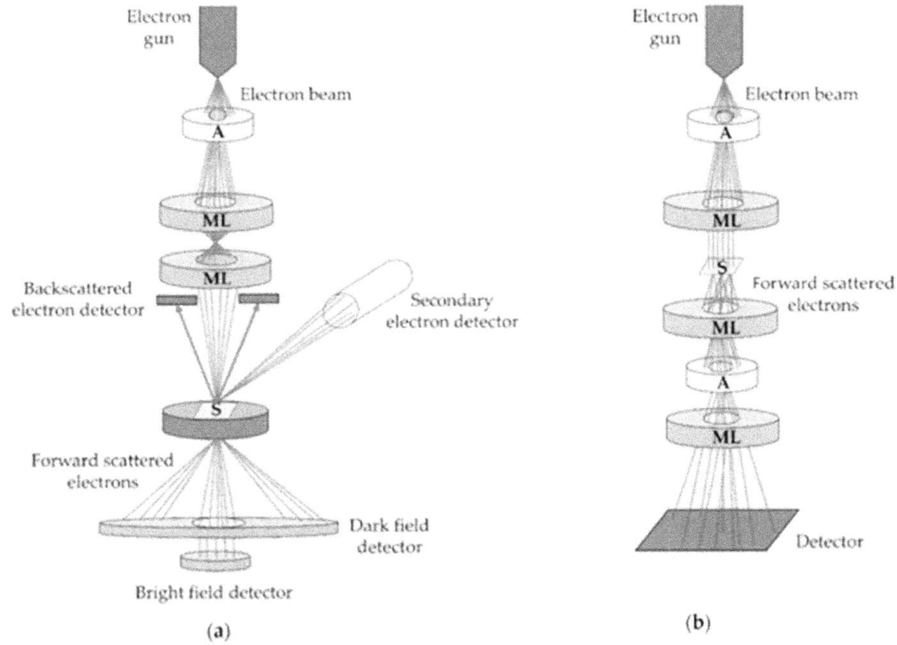

Transmission Electron Microscopy

Transmission electron microscopy is known to be the most commonly used microscopy for studying morphology, elemental membrane information and crystal lattice structure. TEM lets electrons transmit from a specimen to form an image. Ultrathin sample usually in the form of a suspension is placed on a grid. Image is then acquired on imaging device, like fluorescent screen or a sensor. Because of de Broglie wavelength of electrons, it gives higher resolution as compared to light microscope, thus giving fine details of sample. Its spatial resolution is 1- 10 nm, although a sub-nanometer level of resolution can also be obtained due to the shorter wavelength of the electronic beam as compared to photonic wavelength of visible spectrum. TEM generates images using electron beam which is able to transmit through a thinly prepared or spread specimen however biological samples preparation can limit its resolution to up to 2 nm because of various steps like cryofixing and cryosectioning or even resin embedding involved which ultimately increase image noise decreasing the quality of the image.

The main challenge in TEM imaging is the sample preparation according to nature of the sample being observed in order to preserve the nature and morphology of sample and maintain a better resolution. Sample preparations are usually time consuming and require highly skilled person in order to operate the microscopy (Tizro, Choi, & Khanlou, 2019). However the advantage of using this technique in order to accurately visualize the nanostructures and cells or tissues without the need of markers makes this microscopy technique stand prominent among imaging techniques.

Various cellular uptake mechanisms can be visualized through TEM accurately and even the presence of free nanomaterials dispersed in cytosol after escaping the endocytic process can also be visualized through TEM (Mu et al., 2012). This can be attributed to phenomenon like proton sponge effect. However anionic and neutral nanoparticles can also break the norms of staying confined inside endocytic membranes. This phenomenon is specifically observed from gold nanoparticles and nanodiamonds (Petit et al., 2013). These nanodiamonds can be effectively utilized for latent finger printing detection and this procedure can be visualized through TEM (Jung et al., 2020). Thus nanoparticles tracking through TEM can be utilized efficiently in comparison to light microscope or even fluorescent microscope which is unable to provide complete differentiation of non-degradable nanoparticles parts from their entirety. This makes TEM also comprehensively useful in forensic applications although a commercial TEM is expensive equipment to operate with an estimated cost of up to 5 dollars per eV utilized in sample study. Therefore, it is necessary to pre-plan and carefully investigate where the microscopy is needed and also considering the limitations of the technique to avoid excessive burden on the economy while studying the materials accurately. Additionally traditional transmission electron microscopy cannot study nanoparticles crystal growth in solution which is another limitation of this technique.

Liquid TEM is another innovation which helps in counteracting the vacuum system requirements of TEM. Vacuum system is essential in TEM to preserve filament and avoid scattering (Pu, Gong, & Robertson, 2020). Liquid samples can deteriorate the functioning of conventional TEM thus disabling its application in solid-liquid phases. Recent development in formation of sealed liquid cells was revolutionary in TEM using epoxy sealed silicon nitride membranes. Liquid TEM allows tracking of crystal growth while observing it directly. It can also be used to observe growth mechanisms using different pathways with overall same size and morphology yield (Zheng et al., 2009).

High-resolution TEM uses phase contrast image formation which utilize transmitted and scattered electrons. It requires larger objective aperture to effectively use scattered electrons. Phase contrast enables highest resolution imaging and detects arrays in crystal lattices. Thus it can be utilized to study internal structure of nanoparticles. This makes its successful use in studying effects of ligand in nanoparticles growth. HRTEM can also differentiate between anisotropic crystal structures (Eshed, Abd El Majid, Bamberger, & Osovski, 2020).

Atomic Force Microscopy

AFM is a superior technique to study properties of materials at atomic scale. Invented in 1986 by G. Binning, it is a modified version of scanning tunneling microscopy. It is used to study topography and properties at nanoscale. Basic working principle lies in the activation of attractive or repulsive forces when sample interacts with cantilever tip. The cantilever system acts as a way to sense the force. Laser is used for detecting deflections and a high resolution detector which is based on position is used to record these changes. Piezoelectric scanner positions the sample under the tip hence making 3-D imaging possible(Rana, Pota, & Petersen, 2018).

Laser causes deflection in AFM cantilever which is measured by associated electric systems to acquire data using Fourier techniques. Among all working modes of AFM most utilized modes include tapping and contact modes only where the later uses the deflected cantilever reading from sample-probe distance measurements. This reading helps in observing topology and constructing a topographical image by measuring resonance-oscillation relation with distance z. These two measurement techniques are comparable on two major grounds, first is the probe interaction with sample surface and the other feature is

Figure 2. Schematic diagram summarizing tip-sample interactions in AFM
Source: Zhou, Cai, Tong, and Wang (2017)

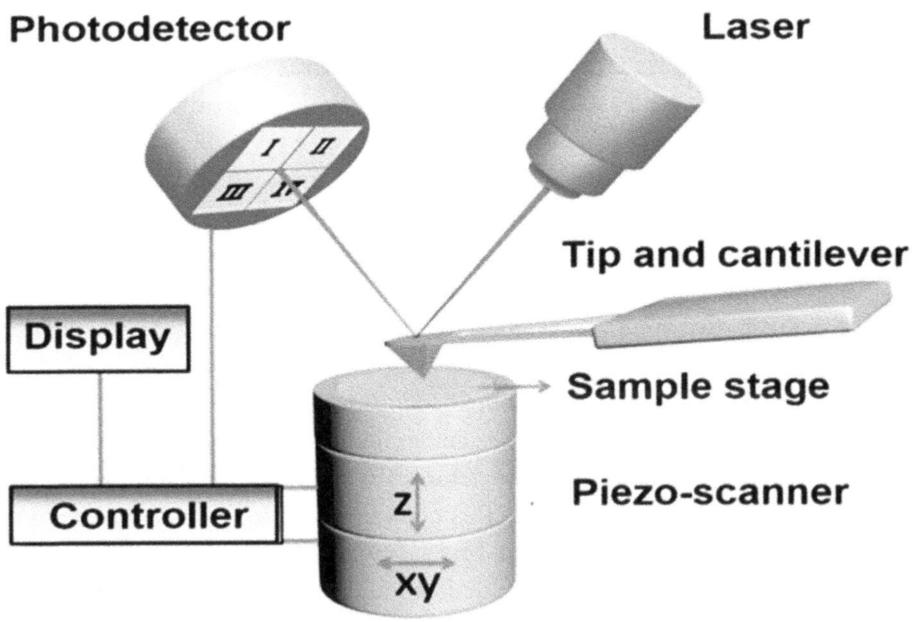

pressure exerted by tip on sample surface while scanning for topographical information (Ratcliff, Erie, & Superfine, 1998). This implies a higher tip pressure and lower sensitivity in contact mode as compared to the tapping mode. This result in preference given to tapping mode for softer samples and preference towards contact mode while studying randomly expected surface changes.

Major advantage of AFM is its ability of magnifying in X, Y and Z planes and also that it does not involve electron beam radiation thus leaving the sample unaffected (Eaton & West, 2010). AFM is considered a crucial and useful method to study the surface characteristics of solids both at nano and micro scale effectively. The scanning tip which is also known as AFM probe has certain interactions with sample surface and this interaction capacity is measured as P i.e. dependent upon distance z between probe and sample. The feedback system is utilized to monitor this distance which in response to changes in z create a differential signal and this signal strength generate morphological characteristics of the sample surface. Interestingly, AFM measurements are not only limited to morphological features study but also help in understanding various electrical or mechanical features such as charge distribution, conductivity and dielectric constant (Osaka et al., 2017).

AFM is also used to gain ample information about mechanical properties like adhesion or surface tension and even for studying the effects of functional groups. Various techniques have been affiliated with classical AFM method such as coupling with infra-red or Raman spectroscopy to gain a further understanding of the nature of the materials. AFM is also sometimes coupled with TEM or SEM to gather quantitative knowledge about the sample like surface roughness, film deposition or size distribution analysis (Efimov, Tonevitsky, Dittrich, & Matsko, 2007).

AFM coupling with IR enhances its uses to chemical analyses of the materials too especially focusing on long chain molecules like polymers. Phase imaging is performed to allow mapping variations for surface study of such materials. AFM-IR principle is based upon combining AFM with tunable IR source of radiation for studying photo-thermal effect utilizing pulse laser and utilizing thermal expansion. For obtaining resolutions greater than 50 nm repetition rates can be tuned and amplitude modulated by mechanical chopping. One of the modes used in AFM-IR is termed as scanning thermal microscopy mode. This utilizes contact mode for AFM with a resistive apex on its tip. This resistive structure changes its resistance with every single change in the thermal properties of the sample being investigated (Dazzi & Prater, 2017).

Sample preparations for AFM study is described as a simple procedure without the involvement of any complex handling. This enables study of various structures and materials in an easy manner. However, to ensure accurate measurements, the sample height should be kept within a specific range of max 10 um which is general limit for distance in the scanner. This information leads to the development of an accurate image in diversified applications however main utilization of this technique is in building a detailed morphological high-resolution image of the sample.

The obtained high resolution of the images obtained is considered one of the major benefits of using AFM simultaneously studying the crystal growths and grain formation. This quality also enhances uses of AFM in characterization of polymeric nanostructures (El Kirat, Burton, Dupres, & Dufrene, 2005).

One of the major uses of AFM in context to this book is its use in forensic studies to assess explosives. Detecting biological or inorganic residual traces from a crime scene for DNA profiling is done in crucial situations. For such type of forensic applications and investigations, AFM etc are utilized. Residues from gunpowder obtained for shape or size analysis help in measuring shooting distance is done using AFM. Interaction of explosive particles with surfaces from which these are obtained can give sufficient information about forensic explosive detection. Even those surfaces or fabrics which had similar indentation, the pull off force varied significantly and adhesion was found best for smooth surfaces like glass. Environmental conditions are also responsible for variation in these interactions and can be predicted using AFM technique. Various other objects like human hair, document forgery, fabrics studies and fingerprints have been significantly studies using AFM which can detect and analyze these materials even in trace amounts (Smijs, Galli, & van Asten, 2016).

Detection of trace evidences is necessary to investigate crime scenes and this is done effectively using AFM. AFM can examine blood stains to study variation in elasticity of blood components over time to estimate time-frame accurately. Similarly finger printing from a crime scene is crucial in investigations due to the uniquely constructed ridges patterns on finger tips which are identified as a strong biometric identification for each person. Chemical identification can also be done from these finger prints which can reveal health conditions or individual habits of that particular person. AFM can also be explored well for de-convolution studies on samples obtained from crime scenes where fingerprints are overlapped by bloodstains (Pandey, Tharmavaram, Rawtani, Kumar, & Agrawal, 2017).

Fluorescence Microscopy

The basic principle of vision lies in contrast. In optical microscopy techniques, contrast has been improved using various techniques resulting in different versions of microscopy. Tyndall effect is most employed phenomenon in dark field contrast, which utilizes scattered light through small particles. Extension in the contrast was done with the invention of staining techniques in histochemical staining where a tissue or biological sample is stained with a chemical dye absorbing a specific wavelength resulting in emerging of a color. Phase contrast, polarization microscopy and differential interference contrast resulting in three dimensional imaging are important innovations in optical microscopy for biological sample studies. However fluorescence technique is the most popular among all these contrasting techniques of microscopy.

Figure 3. Schematic of basic fluorescence principle
Source: Ishikawa-Ankerhold, Ankerhold, & Drummen (2012)

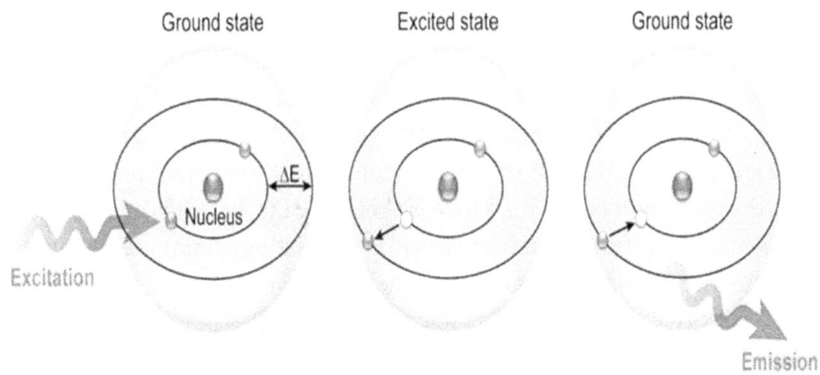

The phenomenon of fluorescence was first discovered by an English mathematician Herschel in 1825 from a quinine solution describing it as epipolic dispersion of light. However, the term fluorescence was used by John Gabriel in 1852. The principle behind fluorescence is that certain kind of fluorphores is used which absorbs a light of certain known wavelength and then emit a beam of low energy photons resulting in shift towards longer wavelength. In short, in this technique, light energy is absorbed and subsequently emitted a few nanoseconds later. This fraction of time creates energy loss, thus it is obvious that a light with longer wavelength is emitted with obvious less energy as compared to the incident energy. This energy change is usually depicted in the form of Jablonski energy diagram.

Figure 4. Jablonski diagram showing the basic processes following irradiation destroying cancer tumor
Source: Kyrkou et al. (2022)

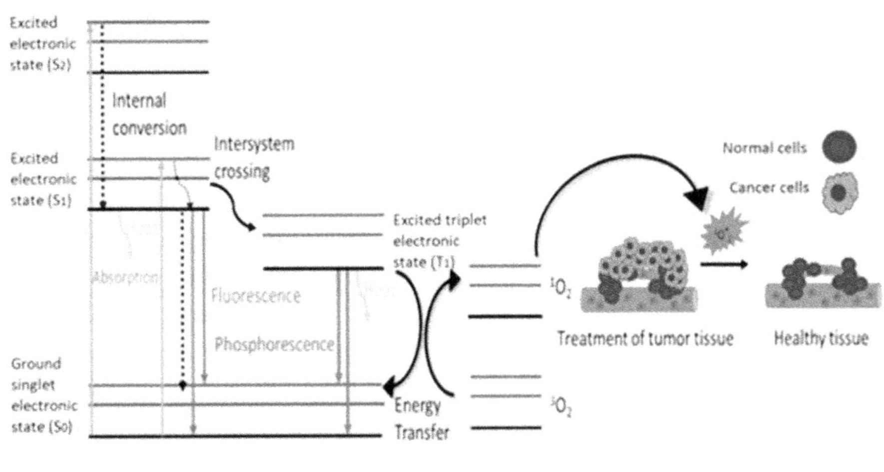

But the exception here lies that although theoretically one photon is absorbed and released as low energy photon, but in reality numerous photons can be absorbed and still release one low energy photon. This can be possible if the absorbed photons individually had lower energy thus red light being converted to green light signal. This phenomenon is known as multiphoton fluorescence and is only possible if there is a dense incident photonic beam obtained using high energy, pulsed and specialized laser beam. Using red light source is better for excitation purpose in terms of less scattering ability, least possible damage to biological tissues and this multi-photon excitation is best indicated in the red region due to photonic fluorescence and emission.

A large number of dyes are being introduced for studying depending upon the absorption range which can range from ultraviolet to near infrared region. Started from the use of fabric stains, introduction of fluorescently labeled molecules, fluorescent heterocyclic probes and other chemical inventions have revolutionized biological applications of this microscopy. Most important innovation in this regard was super resolved fluorescence microscopy for which Nobel Prize in Chemistry was awarded to the scientists.

As, the major advantage of using this microscopy is the distinguishing capability between excitation and emission wavelengths, so this technique is very helpful in studying the dynamic behavior of the live cells and it is highly sensitive, even detects as few as 50 molecules per cubic micrometer. It gives high contrast, with enhanced sensitivity, specified for the purpose and it is selective in nature if any chemical competitors are present. Forensic sciences have utilized this microscopy for various analyses like examining blood stains from crime spots. This can reveal crucial information about samples collected the nature of blood and ultimately leading towards identification of criminal. Comparing the information obtained from sample with available DNA profile can lead to individual identification. This microscopy can help in investigation of the fluorophores within the bloodstain in order to determine time since deposition (TSD) even at the nanoscale. Also it can successfully help in building the phenotypic characterization of sample which makes it a very useful technique in analysis at nanoscale especially in forensic applications ((Saini & Shukla, 2022).

Dark Field Microscopy

This is one of the most powerful imaging techniques for biological samples. Dark field microscopy works on the principle of rejecting excitation light and reading only scattered photons from the sample under study. For its unique reading capacities, it has been widely explored for spectral studies majorly in nanodimension. One of the limitations of this microscopic technique is its sensitivity which is affected by residual light while detecting the sample. It is a background free study technique which focuses on scattering detection just like fluorescence microscopy works on reading fluorescence signals. Difference between these two techniques is illuminating source which is cheaper white light source in dark field microscopy. In dark field microscopy, excitation and scattering photonic spectra are identical and the background is completed rejected by blocking the excitation beam. This results in only scattered photons being detected. Image quality depends upon the brightness of the illuminating source and how efficiently background is rejected which result in generating a good contrast. Various nanodimension structures like spheres, rods, shells etc have been studied and their plasmonic resonance properties have been explored using this microscopy. At single molecular level, this microscopy can be utilized for fast tracking like in vitro or in vivo motor motion at molecular level (Wonner, Evers, & Tschulik, 2018).

One innovation in this microscopy is coupling it with hyperspectral imagery (HIS) which is considered to be a novel method to identify or quantitatively measure specified biological components.

Hyperspectral-enhanced dark field microscopy (HEDFM) has excellent resolution to visualize nanomaterials *in-situ*. Dark field assisted illumination captures images which are good in contrasts with ten times higher signal-to-noise ratio as compared to commonly used dark field microscopy. The main advantage of this technique is real-time analyzing capabilities and implementation of this information to diverse fields like nanotechnology.

Figure 5. Schematic diagram of HEDFM (hyperspectral enhanced dark field microscopy)
Source: Zamora-Perez, Tsoutsi, Xu, & Rivera Gil (2018)

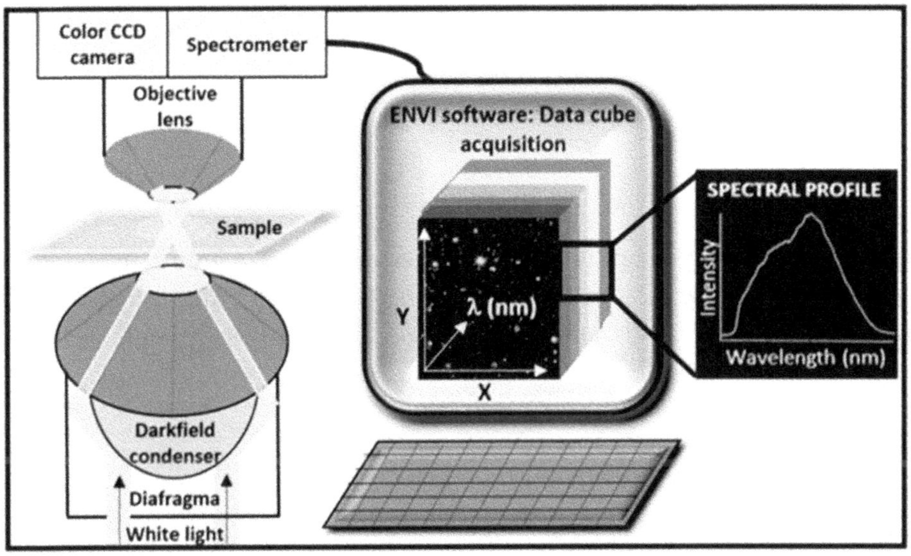

X-Ray Diffraction (XRD)

This important technique is used to elucidate the crystallographic information of a certain material. Main principle of this is measurement of intensities and angles of diffraction of different emitted X-rays that are incident on the material under study. Thus it can identify different materials by studying their diffraction patterns and can also help in revealing the composition of the materials. Deviation study of actual structure from ideal theoretical built and phase identification is also one of the important uses of this technique (Epp, 2016).

X-rays are electromagnetic in their nature and when they interact with array of atoms inside a crystal lattice, elastic scattering happens. The crystal atoms scatter these incident x-rays giving special waves which ultimately cancel each other resulting in destructive interference. In some exceptional arrangements, constructive interference also occurs which can be explained on the basis of Bragg's law. $2d\sin\theta = n\lambda$ where, diffraction pattern is generated by x-rays when wavelength λ is equal to the space d present between the crystal planes of the crystal lattice (K. He, Chen, Wang, Wei, & Chen, 2018).

XRD is non-destructive in its nature and sample is not denatured while investigating. The usual sample is in crystalline powder form in larger quantity as compared to microscopic examination techniques;

Figure 6. (a) X-ray diffraction in x-direction, (b) vector representation of incident or diffracted beams and path difference
Source: Ali, Chiang, & Santos (2022)

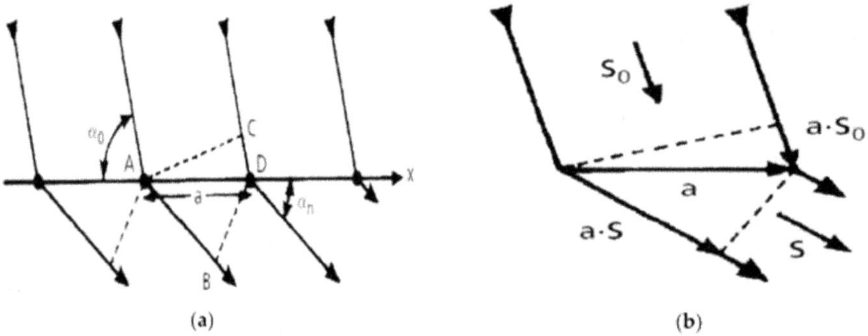

Figure 7. (a) Constructive interference in XRD where red and blue circles are scattering centers, α is x-ray incidence angle. (b) Generalized vector illustration of a (c) vector illustration of conditions leading to constructive interference (d, e) change in θ with change in x-ray wavelength (f) result of changes in incident angle.
Source: Widjonarko (2016)

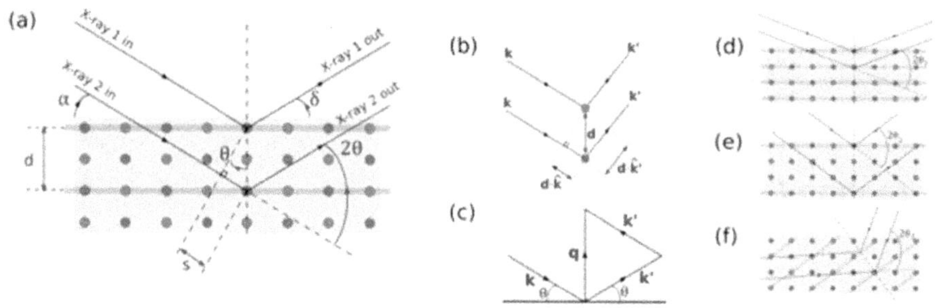

however some of the samples need purification like carbonates rich materials, large granular samples, preparation of smear slides when smaller sample size is to be studied (Holder & Schaak, 2019).

Major applications of XRD technique involve but not limited to identification of crystalline alignment for qualitative or quantitative data retrieval, measuring grazing angle incidence, residual stress measurement, study of structural properties, and measurement of thickness and analysis of atomic arrangement both at micro or nano levels. These applications make it a crucial characterization tool for drug formulation and pharmaceutics (Chauhan & Chauhan, 2014). XRD can be applied to study nanomaterials for conducting forensic analysis for assessment of composition or purity of various substances like drugs. It can be used to study crystal structure polymorphs and any contamination with an analysis of amount (Bunaciu, UdriŞTioiu, & Aboul-Enein, 2015).

Energy Dispersive X-Ray Spectroscopy (EDX)

Energy dispersive X-ray spectroscopy or spectrometry uses a sharp electronic beam bombarded on the sample to be analyzed for specifically chemical analysis where the sample emits a broad X-ray spectrum. It has a wide range of element detection and recognition which ranges from Beryllium to Uranium although for lighter atomic numbers (less than 10) special equipment is required. The data obtained is easy to interpret because of the quick identification and qualitative analysis of x-ray spectral lines. The intensity of each line shows quantitative information about the sample element and composition.

The main principle of this technique is that x-ray beam intensity is displayed and scanned in a raster which is basically a monitor to produce maps of each element showing its distribution in a certain sample. The main principle significantly overlaps with the functions of scanning electron microscopy which can also produce maps of different elements, and if aided with an x-ray spectrometer produce point analysis (Raval et al., 2019). EDX technique is superior in its specific functions for elemental composition investigation. With different modes of operation, it can also yield useful information about topography and difference in atomic numbers. EDX spectrometers use pulse-height analysis with the help of pulse-height analyzer where a detector gives pulse readings in response to variations in X-ray photon energy peak (Shaban, Ibrahiem, El-mongy, & Elshereafy, 2013).

Figure 8. Schematic diagram showing energy dispersive x-ray spectroscopy in combination with scanning electron microscopy
Source: Piccinotti (2018)

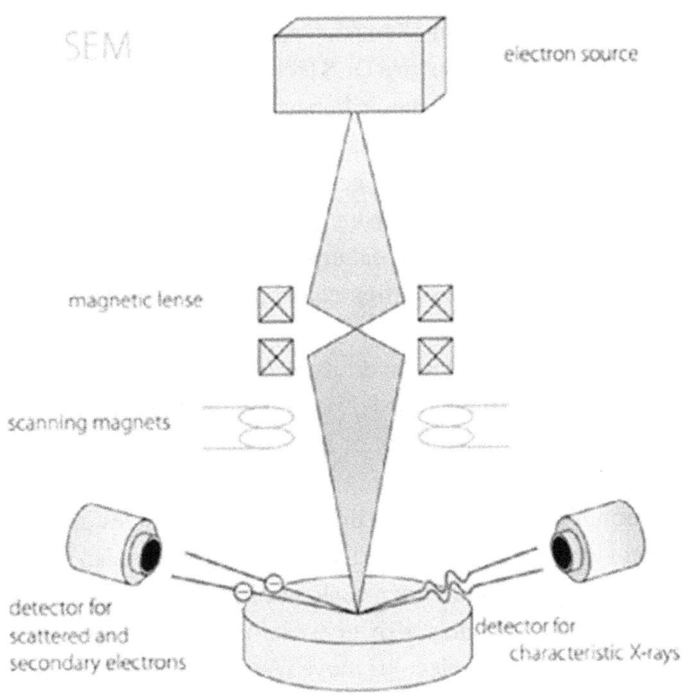

A spectrum obtained this way is a digital representation with x-ray energy or intensity on x-axis and y-axis having the count number. Mono-energetic photons generate X-ray line which can be broad or thin as explained through Gaussian profile. The accuracy of the result is evidently dependent upon the x-ray intensities which is controlled by number of emitted photons. The chance of error in either measurement is limited statistically and the precision is authentic in major cases. Also, signal to noise ratio or background noise is reduced by detaching smaller intensity peaks by the effective use of bombarding electrons and continuous x-ray spectrum. The routine procedures of operating the system, usual detection limits by weight remain about 1000 ppm which can be customized and reduced too (Kaastra & Bleeker, 2016).

One significant feature of spectroscopic techniques is the spatial resolution which is controlled by penetration and field spreading of the incident beam. Ultra thin specimens aid in obtaining a better spatial resolution. To obtain a refined image, sample preparation needs to be carefully accomplished taking in account the shallow depth ability of the electronic probe. Usage of vacuum compatible samples or coating, well polished smooth surface and embedding opaque materials in epoxy resins is essential as in reflected light microscopy. Whereas in transmitted light study, polished thin samples can be studied after placing on glass slide. One essential preparation for non-conductors is to coat the sample with some conducting material like carbon which does not have any noticeable interaction with incident x-rays.

Energy Dispersive X-ray spectroscopy has broad applications either used independently or coupled with SEM or Raman spectrometers. Various case studies from geosciences, minerals, forensic (Reyes et al., 2018), pharmaceutical compound studies, hybrid material studies, and cultural or historical objects have utilized this technique in identifying complex materials(Michalska, Zadora, & Martyna, 2016).

Dynamic Light Scattering (DLS)

Various researchers working on solution chemistry and determining diffusion coefficients of various molecules designed the predecessor of modern DLS technique and apparatus. This analytical technique is popularly employed to determine hydrodynamic behaviors in solution and suggestions about uniformity and agglomeration. Theoretically, all macromolecules are in consistent motion inside solution which is known as Brownian motion. When any light in general, or specifically a monochromatic light due to its intensities, passes such solution, it has a collision with macromolecules inside which in result scatter this light in every direction. If it is a static scattering, a time-averaged intensity is obtained to reveal gyration radius and molecular weight of solution constituents (Carvalho, Felício, Santos, Gonçalves, & Domingues, 2018).

However in dynamic scattering of light, hydrodynamic size is estimated by measuring the diffusion coefficient by measuring intensity fluctuations observed in the light scattered. Dynamic light scattering is also called photon correlation spectroscopy or quasi-elastic light scattering. Its results are obtained based upon measurement of Brownian motion after the collision of solution macromolecules with the incident light. This motion is further related to the size of the particles and depends upon not only size but also temperature and viscosity of the solvent. Viscosity, in return also depends upon the temperature making the accurately regulated temperature a basic need of accurate DLS reading.

Time-related movement of molecules also gives information about size as diffusion is inversely proportional to the size, i.e., smaller molecules move faster and do not remain on the same position for longer. Further, the scattered light generates intensity which is detected through a detector. Here Doppler broadening effect is observed for monochromatic incident light due to the continuous motion of solution particles. Scattered lights also have capacity to cancel each other in response to destructive interference

Figure 9. Block diagram of ex-situ DLS
Source: Jia, Li, Gao, Yang, & Kanaev (2023)

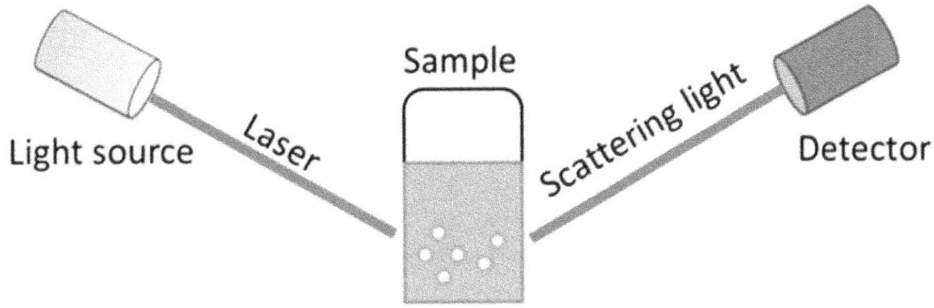

or can generate detectable signal by constructing interference. The autocorrelation function then correlated this intensity of light to the time in order to analyze fluctuations in the intensity.

The detector in most of DLS instruments is employed at 90 degree angle however in backscattering detection system this detector is either at 173 degree or 158 degree making it closer to the incident light. Diffusion behavior of the particles is best obtained by high scattering angle due to ability of neglecting rotational diffusion (Allen, 1997). This also generates significantly better diffusion behavior measurement for highly concentrated samples because multiple scattering is avoided here. Also backscattered detectors avoid reading of larger particles scattering as they tend to generate forward scattering more than back scattering. In order to get a reliable reading of diffusion co-efficient, obtaining multiple readings at a wide range of concentrations is important because of high dependence of diffusion coefficient on concentration.

Size and homogeneity of the particles in suspension is evaluated by various approaches of data analysis systems. Monomodal distribution methods and non-monomodal distribution methods are two main approaches to fit the correlation function which explains diffusion behavior of macromolecules.

However to obtain a reliable reading, sample needs to be purified using any technique that can be applied. Also higher levels of aggregation should be avoided and sample should be dispersed uniformly for a better reading. The software for analysis used in zetasizer is designed to study aggregation by size distribution analysis, using either volume or intensity. Intensity distribution is used to study trace amounts of aggregates and volume distribution can be used to study homogeneity.

DLS is non-invasive analysis technique which requires less amount of sample and can be conducted with a variety of buffers or temperatures. (Stetefeld, McKenna, & Patel, 2016). It can be used to study complex behaviors of nanoparticles in various solvents dependent upon their surface characteristics. It can be used to study surface charges for predicting potential applications of nanoparticles in various biomedical or analytical applications.

Nanotechnology using metal nanoparticles is applied in various fields due to the difference of applications offered by it with variable changes in size or shape. A precisely estimated size or surface charge of these nanoparticles can help in increasing the applications of nanotechnology in forensic sciences. Dynamic light scattering technique can also help in careful evaluation of the stability of nanoparticles over a range of pH or temperature. Zeta-potential measurement associated with same instrument and

Figure 10. Distribution of size structure aggregates in blood serum analyzed through DLS
Source: Velichko, Makarov, Nepomnyashchaya, and Dong (2020)

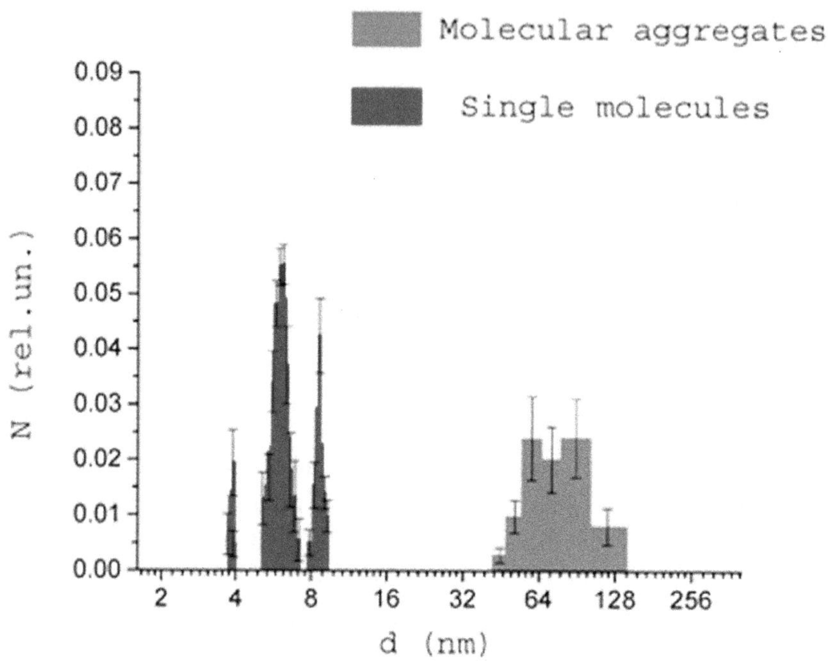

technique is successfully utilized for surface charge studies (Verma & Maheshwari, 2018). These findings make DLS technique a powerful analytical tool for nanomaterial characterization.

In-Situ Characterization

Interaction of nanoparticles with biological or environmental situations requires developing a common interface. This interface can be utilized to study the changes in nanoparticles morphology or chemical structures at nanoscale (Lo Giudice, Herda, Polo, & Dawson, 2016). In-situ characterization is performed with conventional techniques utilizing the features and adapting these to the desired function. Various kinds of techniques are utilized for in-situ characterization which includes wide field SPR microscopy; in-situ localized SPR spectroscopy, in Situ macroscale UV-Vis NIR extinction spectroscopy, liquid cell electron microscopy and in Situ X-ray studies of crystallization kinetics and growth dynamics.

Surface plasmon resonance microscopy (SPRM) is a real-time characterization technique. Recently, this technique has been much improved and innovated for commercialization purposes. This technique helps in developing a better understanding of binding dynamics among various biomolecules at nanoscale. Conventional SPR is not adopted well due to various shortcomings however installing a digital camera to the system has enabled to acquire a proper image thus equalizing each image pixel to independent SPR. This technique is further innovated to develop single molecular sensitivity thus obtaining the snapshot of even a single nanostructure. Similarly, high spatial resolution SPR images are obtained at nanoscale level due to the differences in their brightness (Huang et al., 2022). Wide field SPR microscopy works similarly by creating a optical image from the disturbed surface plasmons of adsorbed nanoparticles.

The image is acquired through specially designed microscopic instruments but the single nanoparticles are unable to give a strong signal which can overcome the background noise which can be subsequently decreased by frame averaging. One of the other causes of noise in SPR imaging is residual light or layer thickness of plasmonic surface. The plasmonic surface here is able to absorb millions of nanoparticles without producing a strong signal. These factors like image noise can be reduced by measuring image at the minimal end of the SPR. The drawback of wide field SPR microscopy is low resolution image and its applications are on analytical detection of nanoparticles preferably in suspension form. A variety of software can be applied to process the images in situ to characterize nanomaterials (Nizamov & Mirsky, 2018). In solution nanoparticles surfaces have been decorated and modified with the preferred material using in Situ localized surface plasmon resonance spectroscopy technique. A deep understanding of this process is necessary to get detailed information of mechanistic involvements in chemical depositions while in the suspension form. A random deposition can lead to red shift in localized plasmonic resonance. Taz et al., investigated the kinetics of CdS thin films on silver nanoparticles and studied the real time shift (Taz et al., 2015). One unique exploitation of plasmonic properties of nanoparticles is nucleic acid detection which can be effectively applied to forensic applications (Fong & Yung, 2013). Various metallic nanoparticles extend or restrict the LSPR band range on different sides. This property is exploited in a number of applications including biosensors which rely on LSPR shift in the spectrum caused by the dielectric changes in the surrounding environment. This can be applied to sensing of biomarkers associated with certain diseases, derivation of certain information in forensic sciences etc (Zhao, Zhang, Yonzon, Haes, & Van Duyne, 2006). Functional nanomaterials are explored well using various techniques including in-situ X-Ray absorption spectroscopic (XAS) studies. Researchers investigated the in situ ionization process using this technique and this spectroscopic technique helped in studying the variations in nanocrystals formation and metallic environment(Pelliccione, Timofeeva, & Segre, 2016).

One intervention related to microscopic techniques was the modification in transmission electron microscopic technology in order to enhance spatial resolution of electron beams. This was done using liquid cell electron microscopic advancements. Conventional TEM required dried samples for analysis and it has been an increased interest for enabling hydrated imaging in situ using TEM.

CONCLUSION

There are numerous characterization techniques used in science to verify the structure, morphology, chemistry, nature of bonding or interactions and functional groups present in any sample. Analytical techniques discussed in this chapter are just the tip of an iceberg currently used in nanomaterial study. Electron microscopic techniques like TEM, SEM, AFM, dark field microscopy are essential tools to study the morphological features of nanoscale objects under examination. The right characterization to ensure required physical or chemical features is necessary for designing appropriate nanomaterials for various applications and especially in forensic sciences where the meticulous nature of experimentation is crucial.

REFERENCES

Ali, A., Chiang, Y. W., & Santos, R. M. (2022). X-ray diffraction techniques for mineral characterization: A review for engineers of the fundamentals, applications, and research directions. *Minerals (Basel), 12*(2), 205. doi:10.3390/min12020205

Allen, G. (1997). *Protein: A Comprehensive Treatise*. Academic Press.

Bogner, A., Jouneau, P. H., Thollet, G., Basset, D., & Gauthier, C. (2007). A history of scanning electron microscopy developments: Towards "wet-STEM" imaging. *Micron (Oxford, England), 38*(4), 390–401. doi:10.1016/j.micron.2006.06.008 PMID:16990007

Bunaciu, A. A., Udriştioiu, E., & Aboul-Enein, H. Y. (2015). X-ray diffraction: Instrumentation and applications. *Critical Reviews in Analytical Chemistry, 45*(4), 289–299. doi:10.1080/10408347.2014.949616 PMID:25831472

Carvalho, P. M., Felício, M. R., Santos, N. C., Gonçalves, S., & Domingues, M. M. (2018). Application of light scattering techniques to nanoparticle characterization and development. *Frontiers in Chemistry, 6*, 237. doi:10.3389/fchem.2018.00237 PMID:29988578

Chauhan, A., & Chauhan, P. (2014). Powder XRD technique and its applications in science and technology. *Journal of Analytical & Bioanalytical Techniques, 5*(5), 1–5. doi:10.4172/2155-9872.1000212

Cohen Hyams, T., Mam, K., & Killingsworth, M. C. (2020). Scanning electron microscopy as a new tool for diagnostic pathology and cell biology. *Micron (Oxford, England), 130*, 102797. doi:10.1016/j.micron.2019.102797 PMID:31862481

Conlan, A. P., Moldovan, G., Bruas, L., Monroy, E., & Cooper, D. (2021). Electron beam induced current microscopy of silicon p–n junctions in a scanning transmission electron microscope. *Journal of Applied Physics, 129*(13), 135701. doi:10.1063/5.0040243

Dazzi, A., & Prater, C. B. (2017). AFM-IR: Technology and applications in nanoscale infrared spectroscopy and chemical imaging. *Chemical Reviews, 117*(7), 5146–5173. doi:10.1021/acs.chemrev.6b00448 PMID:27958707

Eaton, P., & West, P. (2010). *Atomic force microscopy*. Oxford university press. doi:10.1093/acprof:oso/9780199570454.001.0001

Efimov, A. E., Tonevitsky, A. G., Dittrich, M., & Matsko, N. B. (2007). Atomic force microscope (AFM) combined with the ultramicrotome: A novel device for the serial section tomography and AFM/TEM complementary structural analysis of biological and polymer samples. *Journal of Microscopy, 226*(3), 207–216. doi:10.1111/j.1365-2818.2007.01773.x PMID:17535260

Egerton, R. F. (2005). *Physical principles of electron microscopy* (Vol. 56). Springer. doi:10.1007/b136495

El Kirat, K., Burton, I., Dupres, V., & Dufrene, Y. (2005). Sample preparation procedures for biological atomic force microscopy. *Journal of Microscopy, 218*(3), 199–207. doi:10.1111/j.1365-2818.2005.01480.x PMID:15958012

Epp, J. (2016). 4 - X-ray diffraction (XRD) techniques for materials characterization. In G. Hübschen, I. Altpeter, R. Tschuncky, & H.-G. Herrmann (Eds.), *Materials Characterization Using Nondestructive Evaluation (NDE) Methods* (pp. 81–124). Woodhead Publishing. doi:10.1016/B978-0-08-100040-3.00004-3

Eshed, E., Abd El Majid, S., Bamberger, M., & Osovski, S. (2020). TEM and high resolution TEM investigation of phase formation in high entropy alloy AlCrFe2Ni2. *Frontiers in Materials*, *7*, 284. doi:10.3389/fmats.2020.00284

Fakhrullin, R., Nigamatzyanova, L., & Fakhrullina, G. (2021). Dark-field/hyperspectral microscopy for detecting nanoscale particles in environmental nanotoxicology research. *The Science of the Total Environment*, *772*, 145478. doi:10.1016/j.scitotenv.2021.145478 PMID:33571774

Fleck, R. A., & Humbel, B. M. (2019). Biological Field Emission. *Scanning Electron Microscopy*, 2.

Fong, K. E., & Yung, L.-Y. L. (2013). Localized surface plasmon resonance: A unique property of plasmonic nanoparticles for nucleic acid detection. *Nanoscale*, *5*(24), 12043–12071. doi:10.1039/c3nr02257a PMID:24166199

Hachenberger, Y. U., Rosenkranz, D., Kromer, C., Krause, B. C., Dreiack, N., Kriegel, F. L., ... Bierkandt, F. S. (2023). Nanomaterial Characterization in Complex Media—Guidance and Application. *Nanomaterials (Basel, Switzerland)*, *13*(5), 922. doi:10.3390/nano13050922 PMID:36903800

Hassellöv, M., Readman, J. W., Ranville, J. F., & Tiede, K. (2008). Nanoparticle analysis and characterization methodologies in environmental risk assessment of engineered nanoparticles. *Ecotoxicology (London, England)*, *17*(5), 344–361. doi:10.100710646-008-0225-x PMID:18483764

He, K., Chen, N., Wang, C., Wei, L., & Chen, J. (2018). Method for determining crystal grain size by x-ray diffraction. *Crystal Research and Technology*, *53*(2), 1700157. doi:10.1002/crat.201700157

He, X., Ma, Y., Li, M., Zhang, P., Li, Y., & Zhang, Z. (2013). Quantifying and imaging engineered nanomaterials in vivo: Challenges and techniques. *Small*, *9*(9-10), 1482–1491. doi:10.1002mll.201201502 PMID:23027545

Hettler, S., Wagner, J., Dries, M., Oster, M., Wacker, C., Schröder, R. R., & Gerthsen, D. (2015). On the role of inelastic scattering in phase-plate transmission electron microscopy. *Ultramicroscopy*, *155*, 27–41. doi:10.1016/j.ultramic.2015.04.001 PMID:25879156

Holder, C. F., & Schaak, R. E. (2019). *Tutorial on powder X-ray diffraction for characterizing nanoscale materials* (Vol. 13). ACS Publications.

Huang, S., Chen, J., Zhang, T., Dai, X., Wang, X., Zhou, J., Kong, W., Liu, Q., Qu, J., & Shao, Y. (2022). Recent Advances in Surface Plasmon Resonance Microscopy. *Chemosensors (Basel, Switzerland)*, *10*(12), 509. doi:10.3390/chemosensors10120509

Ishikawa-Ankerhold, H. C., Ankerhold, R., & Drummen, G. P. (2012). Advanced fluorescence microscopy techniques—Frap, Flip, Flap, Fret and flim. *Molecules (Basel, Switzerland)*, *17*(4), 4047–4132. doi:10.3390/molecules17044047 PMID:22469598

Jia, Z., Li, J., Gao, L., Yang, D., & Kanaev, A. (2023). Dynamic Light Scattering: A Powerful Tool for In Situ Nanoparticle Sizing. *Colloids and Interfaces*, *7*(1), 15. doi:10.3390/colloids7010015

Jung, H.-S., Cho, K.-J., Ryu, S.-J., Takagi, Y., Roche, P. A., & Neuman, K. C. (2020). Biocompatible fluorescent nanodiamonds as multifunctional optical probes for latent fingerprint detection. *ACS Applied Materials & Interfaces*, *12*(5), 6641–6650. doi:10.1021/acsami.9b19245 PMID:31939655

Kaastra, J., & Bleeker, J. (2016). Optimal binning of X-ray spectra and response matrix design. *Astronomy & Astrophysics*, *587*, A151. doi:10.1051/0004-6361/201527395

Klaessig, F., Marrapese, M., & Abe, S. (2011). Current perspectives in nanotechnology terminology and nomenclature. *Nanotechnology Standards*, 21-52.

Koczkur, K. M., Mourdikoudis, S., Polavarapu, L., & Skrabalak, S. E. (2015). Polyvinylpyrrolidone (PVP) in nanoparticle synthesis. *Dalton Transactions (Cambridge, England)*, *44*(41), 17883–17905. doi:10.1039/C5DT02964C PMID:26434727

Kyrkou, S. G., Vrettos, E. I., Gorpas, D., Crook, T., Syed, N., & Tzakos, A. G. (2022). Design principles governing the development of theranostic anticancer agents and their nanoformulations with photoacoustic properties. *Pharmaceutics*, *14*(2), 362. doi:10.3390/pharmaceutics14020362 PMID:35214094

Lo Giudice, M. C., Herda, L. M., Polo, E., & Dawson, K. A. (2016). In situ characterization of nanoparticle biomolecular interactions in complex biological media by flow cytometry. *Nature Communications*, *7*(1), 1–10. doi:10.1038/ncomms13475 PMID:27845346

Malenica, M., Vukomanović, M., Kurtjak, M., Masciotti, V., Dal Zilio, S., Greco, S., ... Jelovica Badovinac, I. (2021). Perspectives of microscopy methods for morphology characterisation of extracellular vesicles from human biofluids. *Biomedicines*, *9*(6), 603. doi:10.3390/biomedicines9060603 PMID:34073297

Michalska, A., Zadora, G., & Martyna, A. (2016). Optimized sample preparation for glass fragments by scanning electron microscopy–energy dispersive X-ray spectrometry. *Analytical Letters*, *49*(12), 1884–1895. doi:10.1080/00032719.2015.1126598

Mohammed, A., & Abdullah, A. (2018). Scanning electron microscopy (SEM): A review. *Proceedings of the 2018 International Conference on Hydraulics and Pneumatics—HERVEX.*

Mu, Q., Hondow, N. S., Krzemiński, Ł., Brown, A. P., Jeuken, L. J., & Routledge, M. N. (2012). Mechanism of cellular uptake of genotoxic silica nanoparticles. *Particle and Fibre Toxicology*, *9*(1), 1–11. doi:10.1186/1743-8977-9-29 PMID:22823932

Murtey, M. D., & Ramasamy, P. (2016). Sample preparations for scanning electron microscopy–life sciences. *Modern Electron Microscopy in Physical and Life Sciences, 2.*

Nizamov, S., & Mirsky, V. M. (2018). *Wide-field surface plasmon resonance microscopy for in-situ characterization of nanoparticle suspensions In In-situ characterization techniques for nanomaterials.* Springer.

Osaka, M., Mori, D., Benten, H., Ogawa, H., Ohkita, H., & Ito, S. (2017). Charge transport in intermixed regions of all-polymer solar cells studied by conductive atomic force microscopy. *ACS Applied Materials & Interfaces*, *9*(18), 15615–15622. doi:10.1021/acsami.7b00979 PMID:28437063

Pandey, G., Tharmavaram, M., Rawtani, D., Kumar, S., & Agrawal, Y. (2017). Multifarious applications of atomic force microscopy in forensic science investigations. *Forensic Science International*, *273*, 53–63. doi:10.1016/j.forsciint.2017.01.030 PMID:28214756

Pelliccione, C. J., Timofeeva, E. V., & Segre, C. U. (2016). Potential-resolved in situ X-ray absorption spectroscopy study of Sn and SnO2 nanomaterial anodes for lithium-ion batteries. *The Journal of Physical Chemistry C*, *120*(10), 5331–5339. doi:10.1021/acs.jpcc.5b12279

Pessoa, A. C., Sipoli, C. C., & de la Torre, L. G. (2017). Effects of diffusion and mixing pattern on microfluidic-assisted synthesis of chitosan/ATP nanoparticles. *Lab on a Chip*, *17*(13), 2281–2293. doi:10.1039/C7LC00291B PMID:28608886

Petit, T., Girard, H. A., Trouvé, A., Batonneau-Gener, I., Bergonzo, P., & Arnault, J.-C. (2013). Surface transfer doping can mediate both colloidal stability and self-assembly of nanodiamonds. *Nanoscale*, *5*(19), 8958–8962. doi:10.1039/c3nr02492j PMID:23965659

Piccinotti, D. (2018). *Chalcogenide platforms for photonic metamaterials.* Academic Press.

Pu, S., Gong, C., & Robertson, A. W. (2020). Liquid cell transmission electron microscopy and its applications. *Royal Society Open Science*, *7*(1), 191204. doi:10.1098/rsos.191204 PMID:32218950

Rana, M. S., Pota, H. R., & Petersen, I. R. (2018). A survey of methods used to control piezoelectric tube scanners in high-speed AFM imaging. *Asian Journal of Control*, *20*(4), 1379–1399. doi:10.1002/asjc.1728

Ratcliff, G. C., Erie, D. A., & Superfine, R. (1998). Photothermal modulation for oscillating mode atomic force microscopy in solution. *Applied Physics Letters*, *72*(15), 1911–1913. doi:10.1063/1.121224

Raval, N., Maheshwari, R., Kalyane, D., Youngren-Ortiz, S. R., Chougule, M. B., & Tekade, R. K. (2019). *Importance of physicochemical characterization of nanoparticles in pharmaceutical product development. In Basic fundamentals of drug delivery*. Elsevier.

Reyes, L. C., López, C. E., Barrios, A. B., Soto, C. G., Ibáñez, C., & Díaz, F. J. (2018). Development and application of a new nose hairs sample collection device for GSR Particles by scanning electron microscopy with energy dispersive X-ray spectroscopy (SEM-EDS). *Forensic Science International*, *290*, 42–48. doi:10.1016/j.forsciint.2018.06.029 PMID:30015278

Saini, V., & Shukla, R. K. (2022). *Forensic Applications of Fluorescence Microscope. In Forensic Microscopy*. CRC Press.

Seiler, H. (1983). Secondary electron emission in the scanning electron microscope. *Journal of Applied Physics*, *54*(11), R1–R18. doi:10.1063/1.332840

Shaban, S. E., Ibrahiem, N. M., El-mongy, S. A., & Elshereafy, E. E. (2013). Validation of scanning electron microscope (SEM), energy dispersive X-ray (EDX) and gamma spectrometry to verify source nuclear material for safeguards purposes. *Journal of Radioanalytical and Nuclear Chemistry*, *296*(3), 1219–1224. doi:10.100710967-012-2374-x

Smijs, T., Galli, F., & van Asten, A. (2016). Forensic potential of atomic force microscopy. *Forensic Chemistry (Amsterdam, Netherlands)*, *2*, 93–104. doi:10.1016/j.forc.2016.10.005

Stetefeld, J., McKenna, S. A., & Patel, T. R. (2016). Dynamic light scattering: A practical guide and applications in biomedical sciences. *Biophysical Reviews*, *8*(4), 409–427. doi:10.100712551-016-0218-6 PMID:28510011

Sutter, P., & Sutter, E. (2014). Thickness determination of few-layer hexagonal boron nitride films by scanning electron microscopy and Auger electron spectroscopy. *APL Materials*, *2*(9), 092502. doi:10.1063/1.4889815

Taz, H., Ruther, R., Malasi, A., Yadavali, S., Carr, C., Nanda, J., & Kalyanaraman, R. (2015). In situ localized surface plasmon resonance (LSPR) spectroscopy to investigate kinetics of chemical bath deposition of CdS thin films. *The Journal of Physical Chemistry C*, *119*(9), 5033–5039. doi:10.1021/jp512738b

Tizro, P., Choi, C., & Khanlou, N. (2019). Sample preparation for transmission electron microscopy. *Biobanking: Methods and Protocols*, 417-424.

Tunaboylu, B., Toprak, B., Binark, A. K., Öztürk, O., & Zaim, S. (2021). *Manufacturing process development for thin film filaments as a new product.* Paper presented at the Digital Conversion on the Way to Industry 4.0: Selected Papers from ISPR2020. 10.1007/978-3-030-62784-3_44

Velichko, E., Makarov, S., Nepomnyashchaya, E., & Dong, G. (2020). Molecular Aggregation in immune system activation studied by dynamic light scattering. *Biology (Basel)*, *9*(6), 123. doi:10.3390/biology9060123 PMID:32545635

Verma, P., & Maheshwari, S. K. (2018). Preparation of Sliver and Selenium Nanoparticles and Its Characterization by Dynamic Light Scattering and Scanning Electron Microscopy. *Journal of Microscopy and Ultrastructure*, *6*(4), 182–187. doi:10.4103/jmau.jmau_3_18 PMID:30464890

Wang, Y., Wan, D., Xie, S., Xia, X., Huang, C. Z., & Xia, Y. (2013). Synthesis of silver octahedra with controlled sizes and optical properties via seed-mediated growth. *ACS Nano*, *7*(5), 4586–4594. doi:10.1021/nn401363e PMID:23631674

Widjonarko, N. E. (2016). Introduction to advanced X-ray diffraction techniques for polymeric thin films. *Coatings*, *6*(4), 54. doi:10.3390/coatings6040054

Wonner, K., Evers, M. V., & Tschulik, K. (2018). Simultaneous opto-and spectro-electrochemistry: Reactions of individual nanoparticles uncovered by dark-field microscopy. *Journal of the American Chemical Society*, *140*(40), 12658–12661. doi:10.1021/jacs.8b02367 PMID:29995398

Yacobi, B. G., & Holt, D. B. (2013). *Cathodoluminescence microscopy of inorganic solids*. Springer Science & Business Media.

Zamora-Perez, P., Tsoutsi, D., & Xu, R., & Rivera Gil, P. (2018). Hyperspectral-enhanced dark field microscopy for single and collective nanoparticle characterization in biological environments. *Materials (Basel)*, *11*(2), 243. doi:10.3390/ma11020243 PMID:29415420

Zhao, J., Zhang, X., Yonzon, C. R., Haes, A. J., & Van Duyne, R. P. (2006). *Localized surface plasmon resonance biosensors*. Academic Press.

Zheng, H., Smith, R. K., Jun, Y.-w., Kisielowski, C., Dahmen, U., & Alivisatos, A. P. (2009). Observation of single colloidal platinum nanocrystal growth trajectories. *Science*, *324*(5932), 1309-1312.

Zhou, L., Cai, M., Tong, T., & Wang, H. (2017). Progress in the correlative atomic force microscopy and optical microscopy. *Sensors (Basel)*, *17*(4), 938. doi:10.339017040938 PMID:28441775

KEY TERMS AND DEFINITIONS

Auger Electrons: Low energy electrons emitted by incident electron beam.
Cathodoluminescence: Radiations generated in reaction of incident cathode rays.
Fluorescence: Absorption of shorter wavelength and release of longer wavelength light (excitation-emission).
Illumination: Transmittance of light through translucent object.
Magnification: Enlarging the apparent size in an image in relation to the object size.
Resolution: Ability of any instrument to clearly distinguish between two points.
SPR: Electronic charge oscillations excited by incident light.

Chapter 4
Nano-Level Modeling

Marriam Zargham
University of Health Sciences, Pakistan

Muhammad Farhan Khan
University of Health Sciences, Pakistan

ABSTRACT

This chapter discusses the designing and modeling of nanomaterials through different updated technologies. It also covers the chemistry behind the nanoparticles and their prominent properties, which are required for different applications, along with an overview of the history of nanoparticle creation and applications. But the basic discussion is about different physical, chemical, and biological techniques for nanomaterial designing and different and novel characterization approaches to find the morphological and chemical traits of the designed nanoparticles. It also discusses the methods that are cost effective and eco-friendly for developing different nanoparticles. Future scope of designing nanoparticles and how the availability of nanomaterial helps in coping with the energy depletion and utilizing minute particles in massive applications are also covered.

INTRODUCTION AND BACKGROUND

With the advancement of nanomaterial, the scientific industry enters in the new horizon of research, development and technology and brought remarkable revolutions in modern world, from the last few decades number of synthesizing nanoparticles increase by 25 folds because of their increasing demand in every field. Nanoparticles have application almost in every field of science from electronics, mechanics, agriculture, medicine and forensics etc. As the scientific industry revolutionized the demand of nanomaterial is increasing day by day. Incorporation of nano matters in forensic science field brought more potential revolution in field of forensic science. By engineering this nanomaterial according to its requirement, it can be employed from trace detection to macro level detection but somehow its modeling is precise way is necessary for its meaningful usage Because of its minute sizing it provide more tactful particularities of the evidence make more reliable investigation.

DOI: 10.4018/978-1-6684-8325-1.ch004

The term 'nano' coined from Greek word 'nanos' which means 'dwarf' in the conference of International Union of Pure and Applied Chemistry (IUPAC) in the year of 1947 this prefix was embraced in order to explain one billionth (10^{-9}) part of a unit. (Schaming & Remita, 2015b; Joudeh & Linke, 2022). Nanoparticle is defined as the matter particle having atomic size range of < 100 nm. Nano dimension material have more definite properties and discrete ionic make up than bulk material which provide more reactive surface, mechanical, electrical and magnetic properties. Because of these properties nano matter can be mold into various structures like nano films, nanotubes and bulk nano material used for different purposes. Nanomaterials are synthesized using a variety of processes, but energy-intensive approaches and intricate methods are major barriers to their widespread usage. Various methods have been discovered for designing nanoparticles form which some of them we will discuss in this chapter.

Categorization of Nanomaterials

On the basis of shape, origin, size and optical properties nanoparticles are categorized into variety of groups and sub groups. With size reduction, atoms perform better near the surface and stimulate different interactions with their surroundings than they do in the bulk material form. Additionally, the material's electrical structure is changed by decreasing the size, producing unexpected quantum phenomena. Additionally, size affects mobility, which is important for biological and environmental processes and is predominantly governed by Brownian motion for nanoparticles. (Shrestha, Wang, & Dutta, 2020).

Generally, the nanoparticle-based material was divided into two categories on the basis of origin. First type is nature based, like bacteria, proteins and viruses and other sources like nanoparticles created by forest fires, volcanic eruptions, minerals, nanostructured crystals and many more, the other type is human based whether they synthesized unintendingly like nanoparticles produced by the power plants, combustion of diesel or gasoline, and incinerators, as well as those created intentionally using a specified production process (Schaming & Remita, 2015b).

Nanomaterials are also classified on the basis of dimensions like zero-dimension, one dimension, two dimension and three dimensions. This dimensional classification is based on definition that one or more than one extensions of nanomaterial lie within the nanoscale i.e., less than 100 nm (Wani et al., 2022).

Another type of stratification is compositional base. Different composition forms are found in different nan matters like carbon based, organic based, inorganic nanoparticles, bio based nanoforms and metallic and metallic oxide nanoparticles fall in this stratum (Ijaz, Gilani, Nazir, & Bukhari, 2020).

On morphological basis nanoparticles are also classed as isometric and inhomogeneous or scattered and agglomerate depending on homogeneity. This aggregation relies on the electromagnetic, magnetic, and surface charge characteristics of NPs. Different morphological forms of nano matters includes nanohooks, nanoplates, nanostars, nanorods, etc. (Yun, Nam, & Kim, 2020).

Nanoparticles: Solution for Today

Today the world is facing energy crises. Renewable energy resources are depleting every day to cope up with this matter nanoparticles provide a solution. Nanoparticles provide solutions for saving, generating and utilizing energy in appropriate and efficient ways like solar cells, thin film nanocomposite, thin film batteries, semiconductors, transistors, integration of conductive nanomaterial with polymers which enhance the energy generation capabilities.

Figure 1. Nanoparticle modeling

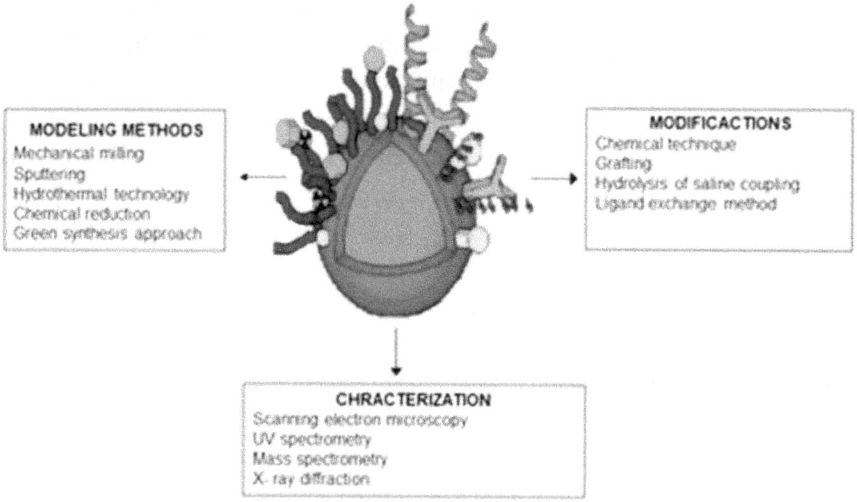

History of Nanoparticles

Our ancestors use carbon of fire and waxes for paintings in caves, some well-known examples of these paintings are found in Indonesia and France. Moreover, the hand stencils in Sulawesi caves are also fall in the category of ancient use of carbon nanomaterials use. These evidences proving that ancient people employed nanomaterial without knowing it (Barhoum et al., 2022).

Figure 2. Prehistoric cave paintings
Source: Tan (2014)

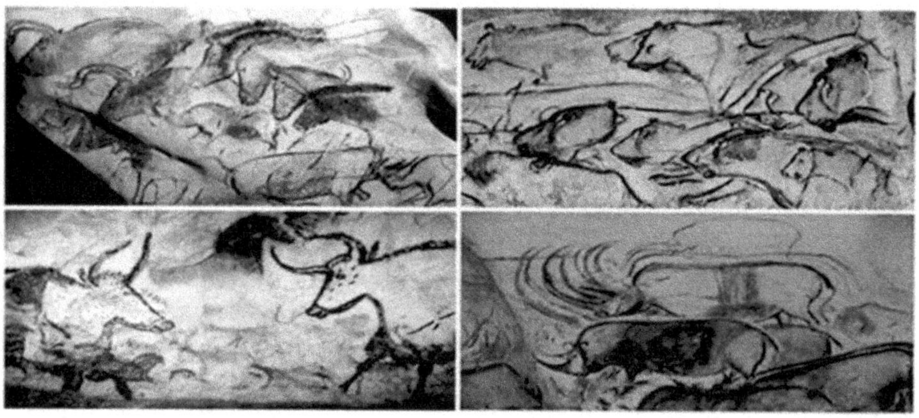

About 5000 BC clay was used as cloth bleaching agent in Cyprus. Carbon nanotubes and cementite nanowires found in wootz steel were manufactured in India about 900 AD (Schaming & Remita, 2015a). About 4,500 years ago, the humans began taking edge of fortifying the capabilities of ceramic molds by incorporating the naturally available nanofiber asbestos. Ancient Egyptians also made use of nanomaterials from almost 4000 years ago. A man-made substance to manufacture PbS based Nanopowders having diameter of 5 nm for hair color. In a similar vein, "Egyptian blue" was the first artificial color which the Egyptians made and used with a flinted glass and quartz mixture with a nanoscale scale during the third century (Montanarella & Kovalenko, 2022). The most well-known example of ancient metallic NPs use is a Roman glass workpiece. A Roman glass from the fourth century called the Lycurgus Cup which was composed of dichroic glass, exhibit distinct colors: red green when a light passes and red when a light comes from behind and green from the front side. According to recent investigations, the Lycurgus cup composed of Ag-Au alloy nanometal in 7:3 ratio. approximately 10% Cu (Salem, Hammad, Mohamed, & El-Dougdoug, 2022). Later, red and yellow stains appeared. Glass used in churches from the Middle Ages was made by adding colloidal silver and gold nanofabricates. Practically the chemical synthesis of material nanoparticles began in 13th and 14th century when Mesopotamians and Egyptians used metals for making glass moreover the use of clay having few millimeters thickness is another example of use of natural nano matter in the history (Jeevanandam, Barhoum, Chan, Dufresne, & Danquah, 2018).

Figure 3. Lycurgus cup: An ancient nanometal composed dichroic glass
Source: Slepička, Slepičková Kasálková, Siegel, Kolská, & Švorčík (2019)

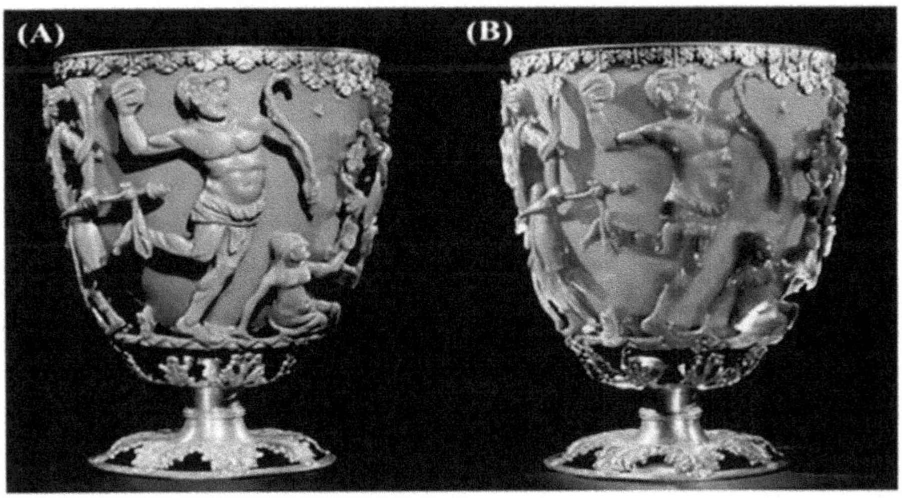

The characteristics of colloids aroused more and more interest in the nineteenth century. The red hue of colloidal gold captivated Michael Faraday. He found that gold colloids have different optical properties than the bulk counterpart metal. In 1857, he published a paper on the first production of gold nanoparticles in solution. The colloids were produced by reducing white phosphorous-containing gold salts. Most likely, this was the first time the reflex of quantum size had been observed. It just marks the beginning of nanoscience in this respect. Later, functional materials provided an explanation for the particular colors of metal colloids in 1908 by Mie (Schaming & Remita, 2015b).

Figure 4. Timeline of nanotechnology beginning
Source: Barhoum et al. (2022)

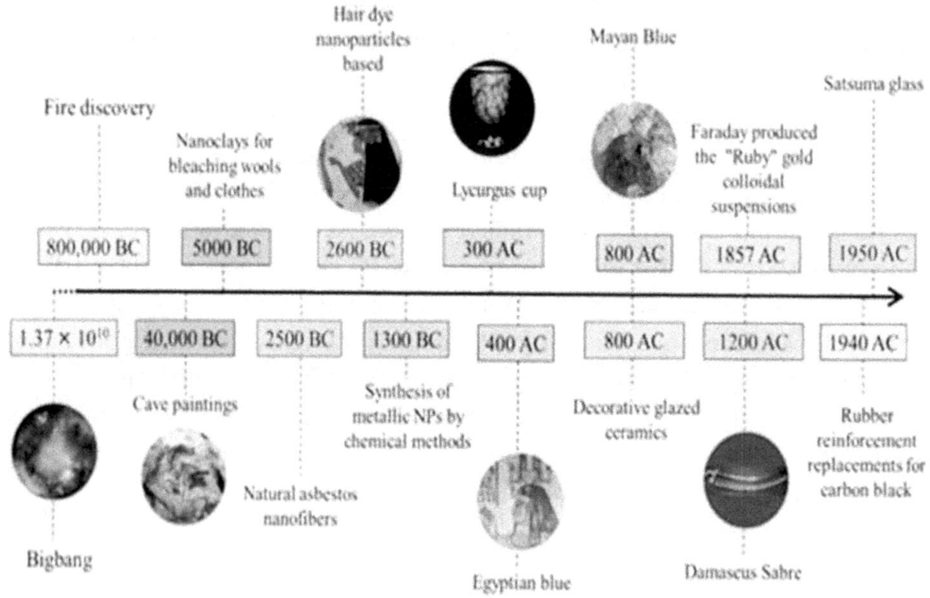

Chemistry of Nanoparticles

When talk about chemistry of nanoparticles it was determined that NPs having excessive surface energies display a duplex nature i.e., great reactivity and poor stability. The elevated reactivity of nanoparticles modeled them unstable, meaning that they can oxidized easily or damaged while using in any application. On the one hand, this makes them easy to turn into desirable utilizable nanomaterials by proper chemical manipulations, such as template synthesis or ion exchange methodologies. This bilateral coordinated complementing behavior is also observed throughout the universe and in nature at many different levels, and therefore recognized as a universal design and fabrication principle for different functional materials innovations (Shrestha et al., 2020).

The nanoparticles have large percentage of atoms on the top. These atoms on the surface offer a great surface tension which produce extreme high pressure leading to the molecular bond length changes in crystal structure which in turn modify the electrical, thermal and mechanical traits. For example, small particles have low melting point as compared to the larger one. The number of nanoparticles is directly correlated to the number of atoms on surface while percentage of atoms present on surface are inversely correlated to the diameter (nm). Consequently the decrease in nanoparticle size increase the proportion of atoms on surface (Ahire et al., 2022), thus making it useful for more reactivity.

MODELING OF NANOPARTICLES

The methodologies of designing nanoparticles are crucial in this modern nanoscience era (Schaming & Remita, 2015b). The choice of synthesis of the nanoparticle process is likewise significant on the ground during particle development processes like kinetics reducing agent attraction with metallic ion and stabilizer adsorption with metal particles, and diversified experimental approaches give a durable impact on its structural morphology, physicochemical properties and stability (Jamkhande, Ghule, Bamer, & Kalaskar, 2019). Nano designing is not so much complex process but their monitoring during synthesis because the application in which they utilize depends upon the dimension and surface properties of synthesized nanomaterials. From last few decades many methods were developed for designing and modeling of required type of nanoparticles but it still requires betterment in this regard. The major two categories in which nanoparticles synthesis methodologies fall are: top down and bottom approaches(Ahire et al., 2022).

Top-Down Methodology

It is a simple method in which bulk material disintegrate into smaller units and then more decomposition leads to the generation of nanoparticles examples include, laser ablation, Mechanical milling, lithography, sputtering, etc.

Bottom-Up Methodology

It's a constructive methodology, converse of top-down method, in this approach nanoparticles developed from simple substance unit, examples are chemical vapor pressure, spinning spry pyrolysis, hydrothermal process etc.

Figure 5. Top down and bottom-up process
Source: Barabadi et al. (2019)

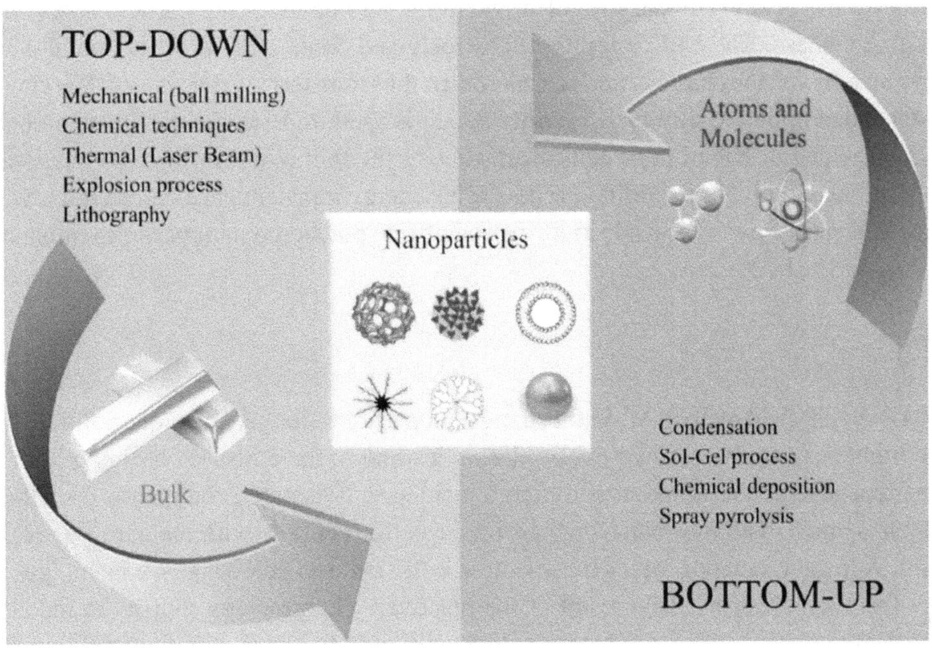

Figure 6. Basic methods of nanoparticles synthesis
Source: Jeyaraj, Gurunathan, Qasim, Kang, and Kim (2019)

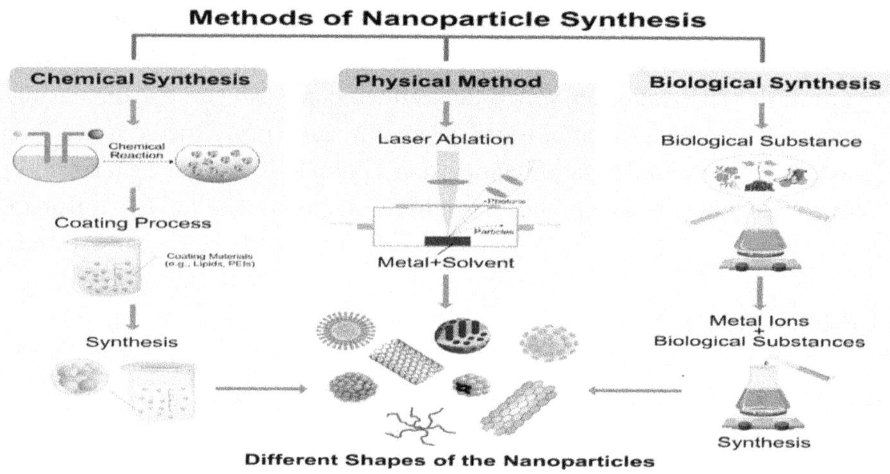

These approaches have further biological, chemical, physical, and based synthesis methods.

Physical Methods

Mechanical Milling

It is type of top- down approach, physical method, cost effective method of designing nanoparticles and other nanocomposite. In this method powder of charge or mixture of powders placed in milling medium having high level of energy. The purpose of this milling is to contract the particle size and mixing of particles in a new phase. The balls can either fall freely and contact the powder and also balls below them or they can retrench the chamber surface parallelly, thus transferring the energy. The energy transfer kinetics is depended upon the mill type, powder which is used to drive the milling chamber, the milling speed, the size and distribution of balls size, whether the mill is wet or dry, the temperature of the milling process, and length of the milling process. This energy transfer cause deformation which in turn change shape and size of particles leads to formation of fine powdered nanoparticles or nanocomposite (Yadav, Yadav, & Singh, 2012).

Sputtering

It is also top-down physical method of synthesizing nanoparticles. In this technique solid surface is bombarded with high energy particles like gas or plasma leading to the emission of tiny clusters of atoms, the emission depends upon the gaseous ion energy incidence. Evacuated chamber is used for sputtering process, which is ample with the sputtering gas. Free electrons collide with the gas as a result of a after applying high voltage to cathode target free electron collision occurs with gas, causing gas ionization. Atoms freed from target surface as a result of the positively charged ions spurred in the electric field when they accessed towards the cathode target. It usually used to create nanomaterial thin films (Baig, Kammakakam, & Falath, 2021).

Figure 7. Mechanical milling representation
Source: Joy et al. (2022)

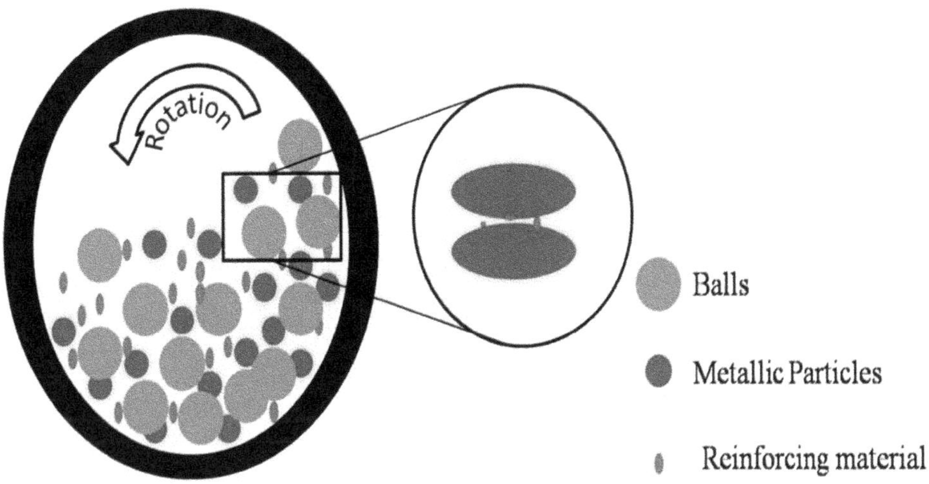

Figure 8. Sputtering method for NP modeling
Source: Chen et al. (2019)

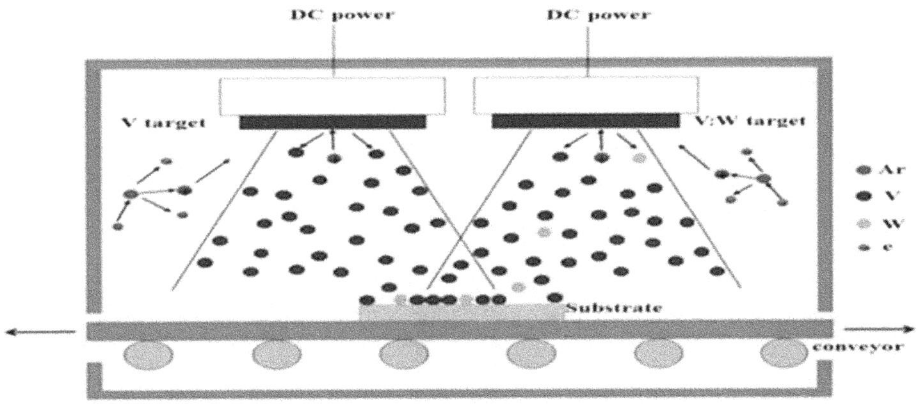

Arc Discharge Method

Usually, this method used to synthesize carbon- based nano matters. In this process a container is filled with constant helium pressure and two graphite rods are adjusted. It is pivotal to fill the chamber completely with helium gas otherwise oxygen or moisture prevent the manufacturing of fullerenes. Arc discharge between the two ends of the graphite rods is what propels the carbon rods vaporization. Under different atmospheric conditions different morphological carbon nanomaterials can be produce with this technique (Baig et al., 2021).

Figure 9. Arc discharge process
Source: Park, Kodama, and Sekiguchi (2021)

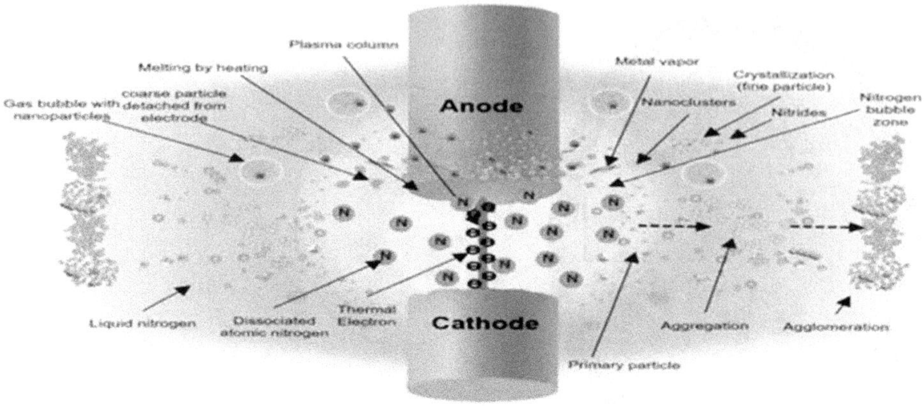

Laser Ablation

This process involves nanoparticles development by hitting the substance with energetic laser beam. Because of high energy of the laser radiations employed in this process, the target material or precursor vaporizes, causing the production of nanoparticles. As no stabilizing agents and other chemicals are essential employed for laser ablation to develop noble metal nanoparticles can be perceived as an environmentally friendly method. Process is feasible to produce semiconductor and magnetic and organic nanoparticles (Amendola & Meneghetti, 2009).

Figure 10. Diagrammatic representation of laser ablation
Source: Jiang, Li, Huang, He, and Ming (2022)

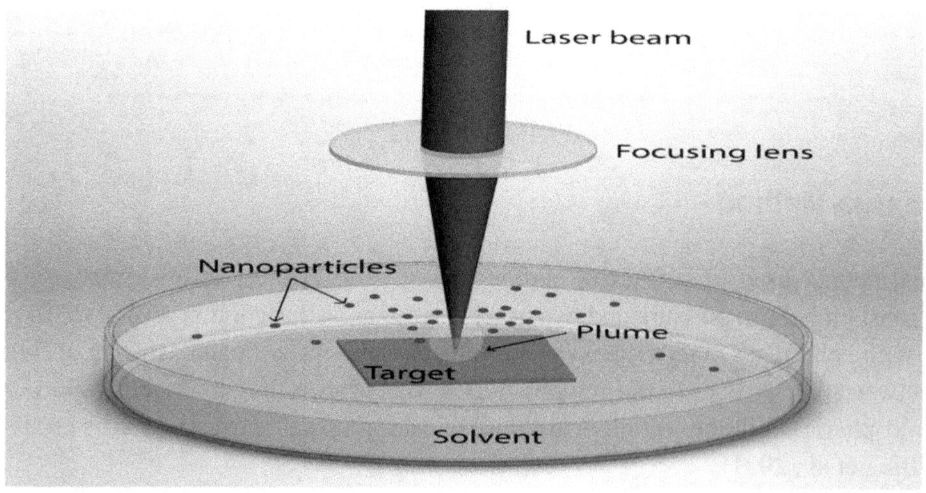

Chemical Methods

Hydrothermal Method

Type of bottom-up approach and a liquid phase method. This method relies on tiny particles deposition caused by aqueous solution vapors reaction with solid material under increased temperature and pressure. This process involves the precipitation of cations in the form of polymeric hydroxides, which are then further dehydrated to expedite the development of crystal structured metal oxide. the second metallic cation that generate help limit particle genesis by hindering the production of complex hydroxide when the base is introduced in a metal salt solution. Well crystallized nano powder is produced through this process (Jamkhande et al., 2019).

Figure 11. Hydrothermal treatment for NP development
Source: Simonenko et al. (2021)

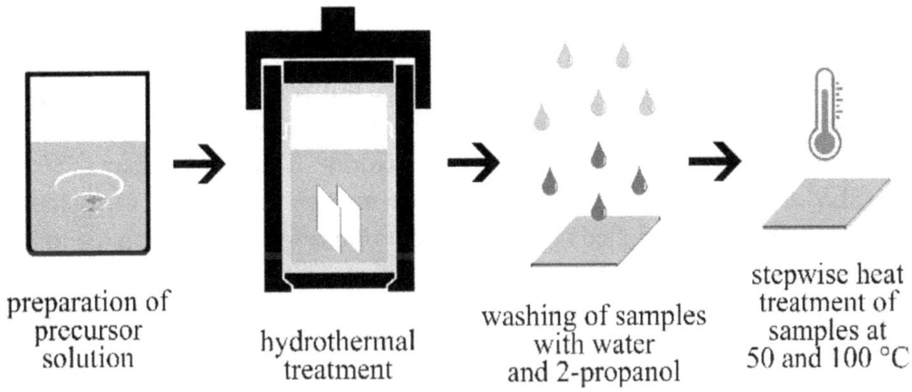

Chemical Vapor Deposition

It is a sound bottom-up phase technique. Here, the chemical reaction between atom-based gaseous molecules is advantageous for the production of films. The target material, which serves as a precursor and is released as a volatile molecule, interacts chemically with the substrate surface, the precursor fragment, and each other to form a thin layer. By using this technique, the surface chemical reaction leads to the creation of thin films by atomic layer deposition (ALD).

In plasma enhanced chemical vapor accumulation process, plasma is created in the barren chamber and is then chemically accumulated as a thin coating on the surface of substrate. This technology also makes use of microwave, radio frequency, inductive coupling, and electromagnetically induced electric current. It's advantageous for broad range of industrial applications, as well as for the creation of nanotubes and graphene nanostructures because it can function relatively at low temperatures (Jamkhande et al., 2019).

Figure 12. CVD chamber
Source: Kim, Bak, and Kim (2020)

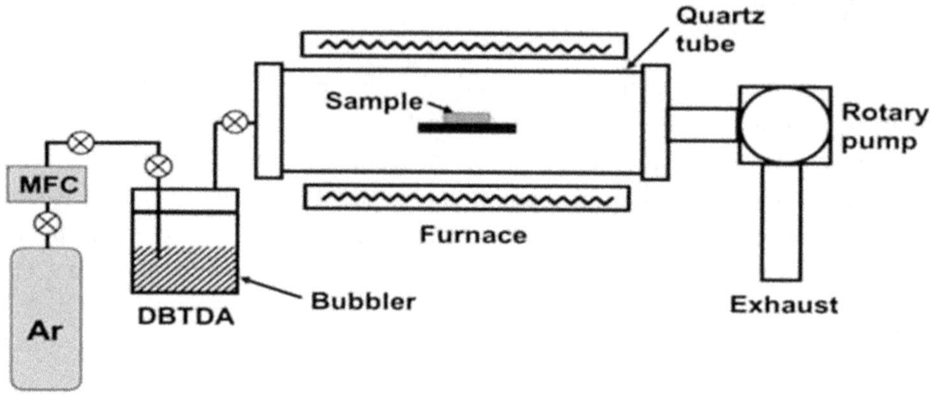

Chemical Reduction

In this liquid state bottom-up approach ionic salt is reduced utilizing various reducing agents in a suitable media with surfactant in the chemical reduction process. Metal nanoparticles are made by reducing a substance in an aqueous solution, such as sodium borohydride. Trisodium citrate (TSC) and Sodium lauryl sulphate (SLS) are employed for cap embodied metal nanoparticles (SLS). Occasionally, reducing agent and stabilizing agent are combined. The assessment of absorbance was applied to keep track the stability of the metallic nanoparticles during dispersion. Most used approach for synthesizing nanoparticles (Jamkhande et al., 2019).

Figure 13. Schematic representation of chemical reduction

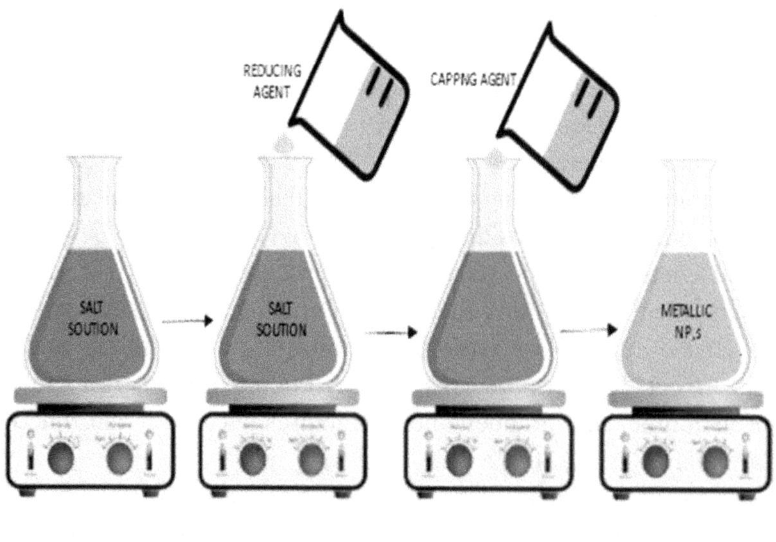

Spray Pyrolysis

It's a gas phase bottom-up approach. The basic principle behind this approach is that metallic salt solution as a precursor subjected to atomization giving the metallic oxide or metal particles and then placed these droplets in a high temperature furnace. Depending upon the final required composition or structure of particles water or different organic salt or metallic salt solution like hydroxide, halides, bromides etc. can be used.

Figure 14. Flame spray pyrolysis representation
Source: Heidinger, Spranger, Dostál, Zhang, and Klaus (2022)

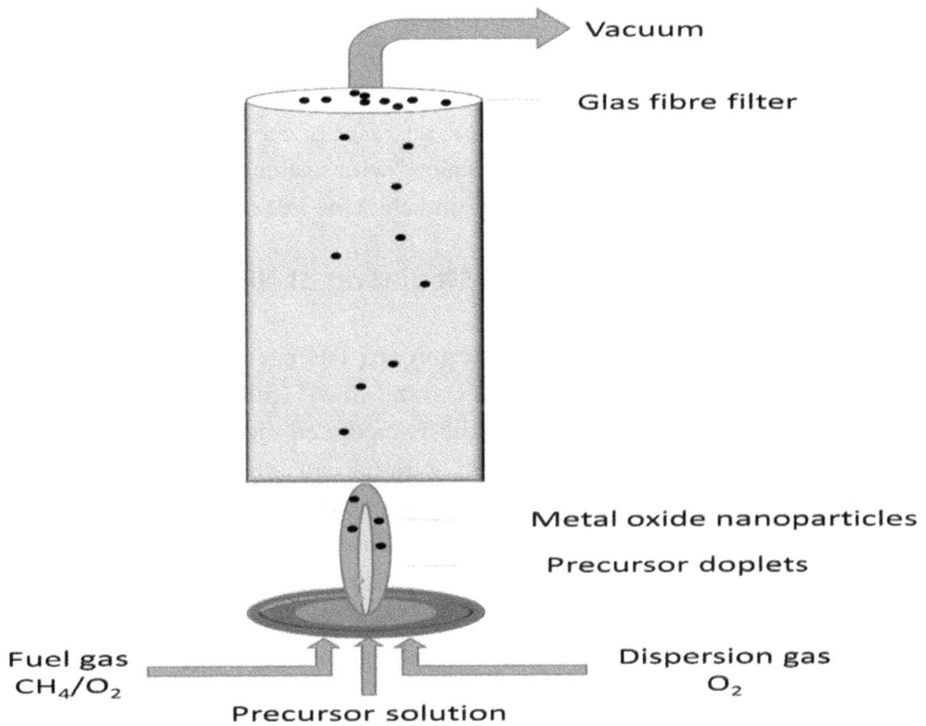

Ultrasonic Spray Pyrolysis

It couples the employment of ultrasonic to disperse the progenitor solution into the droplets form with high-temperature chemical breakdown of the dissolved substance inside the droplets to produce fine powder. The solvent inside the dry droplets in the USP reactor first evaporates, followed by pyrolysis, which chemically breaks down any solvent left inside, and sintering, which produces fine particles. The USP technique has the benefits of easy configuration changes, continuous synthesis of nanomaterials, and the ability to create nanoparticles from a variety of basic materials (Majeri & Rudolf, 2020).

Drawbacks of Chemical and Physical Synthesis

Nanoparticles that are prepared through chemical methods are not appropriate for their use in the medical field because its dangerous chemicals present on the surface moreover, other derivatives produced in chemical synthesis are toxic also to the environment. Physical techniques for producing of Nanoparticles have a few downsides, as well. These techniques require large energy and space and furthermore costly (Pirtarighat, Ghannadnia, & Baghshahi, 2019).

Green Synthesis Approach

A large number of horrendous impacts of chemical and physical methodologies can be controlled by the green synthesis approach of nanopowders utilizing different biological substances. Essentially, the green mediated synthesis of nanopowders, produced by control adjustment and clean-up remediation process help in creating safe toxic free environment. A few fundamental standards of "green synthesis approach" also provide benefits like waste reduction, decrease of subsidiaries/contamination, and employment of more secure (or non-harmful) solvents along with sustainable feedstock. The utilization of essential natural assets and solvent systems is vital for achieving this objective.

Basic Necessities for Green Synthesis Mediation of Nanoparticles

Nanoparticles development by green synthesis approach has been embraced to oblige different organic entities e.g, plant extracts, fungi, bacteria and algae (Singh et al., 2018). Usually, the bottom-up methodology is applied in green synthesis mediated methods where nanoparticles can be produce using reducing and stabilizing agents without causing much toxicity. Manufacturing required few things for process proceeding these are, suitable solvent medium, toxic free reducing, capping and stabilizing agents. Green synthesis approach is also cost effective because the use of natural ingredients for processing nanoparticles development are not much expensive and readily available, adding another benefit in utilizing biogenic source of nanoparticles production. Many type of biomolecules DNA, microbes, plants, fungi and algae have been employed for nanoparticles designing and modeling.

Bacterial-Mediated Approach of Nanoparticle Synthesis

Since bacteria have an ability to reduce and precipitate various metals and metal oxide so it is utilizable in the creation of nanoparticles. They provide certain enzymes and proteins for metal ion reduction process. Basic bacterial which provide reduction power are nitrate dependent reductase and NADH-dependent reductase. It employs both intra and extracellular mediation of nanoparticle synthesis. Various nanoparticles like $TiO_{2,}$ ZnO, CuO nanoparticles can be synthesized using bacterial source, but the bacterial synthesis required culturing technique which is somehow cause difficulty in regulating crystallinity, size and shape of the nano matters that's why this approach has limited use in fabrication of nanoparticles (Gebre & Sendeku, 2019).

Fungi-Based Nanoparticle Synthesis

Fungi have immense intracellular metal absorption capabilities and produce abundant enzymes per unit of biomass makes fungal production of nanoparticles advantageous. Reductases, an extracellular enzyme released by fungi, facilitate the fabrication of metallic nanoparticles with various chemical compositions. On the other hand, the spread of the metal as a catalyst was also successfully achieved by growing fungi on the surface of inorganic substrates. Different fungal species produce nanoparticles that vary in size and shape, which provide opportunity towards broad range of applications. Despite of these benefits fungal based mediation is time taking process and its downstream processing is also difficult. Fungi support the production of monodispersed nanopowders with defined size and shape. Especially gold, titanium oxide particles etc. are produce by this approach (Jeevanandam, Chan, & Danquah, 2016).

Plant-Based Nanoparticle Designing

Plants are considered more reliable source of biogenic designing of nanoparticles than other biomolecules because of their environment friendliness, safety and presence of many phytochemicals which include, terpenoids, flavonoids, phenolic acids, alkaloids, polyphenols, polysaccharides and tannins etc., found in plant extracts of the stem, leaf, or flower, function as bio-reductants also as capping agents in the biogenic synthesis of metal oxide nanoparticles. It is widely known fact that the type of plant extract, the agglomeration of the metallic salt, plant extract content, temperature, the pH and length of contact time all significantly affect how quickly nanoparticles are produced, how many there are, and other properties (Gebre & Sendeku, 2019). Wide variety of nanoparticles are produces using plant based especially for medicinal, therapeutics and pollution remediation purpose. Despite of many advantages the green synthesis fabrications process is still lacking in research and investigation and their proper designing steps are under investigation.

Characterization of Nanoparticles

Now a days reproducibility is a major problem which prevent the use of nanoparticles in industry. However, this issue is largely inherent because the end result of development is consistently predisposed to produce a polydispersion of nanomaterials, often with a wide range of shapes, sizes and flaws. Therefore, in order to understand the genesis of nanoparticle performance and translating its execution advantages from laboratory into specific real-world utilization the nanoparticle characterization is required.

Why Is Characterization Necessary?

Characterization determines the physiochemical properties of the designed nanoparticles and link of these properties to the functionality of the nanomaterials. Different physical techniques have been applied to discover the partial picture of the designed nanoparticles. Few of them we will discuss in this chapter. Certain parameter greatly affects the dimension and hence the functional behavior of the modeled nanomaterial. Here throws a light on the parameters which are required to measure.

Nanoparticles are distinguished from the broader category of nanomaterials, which include nanofabricated equipment's that might have dimensions in milli and micrometer regime, such as nanofilms

Figure 15. Basic representation of phytochemicals in nanoparticles modeling

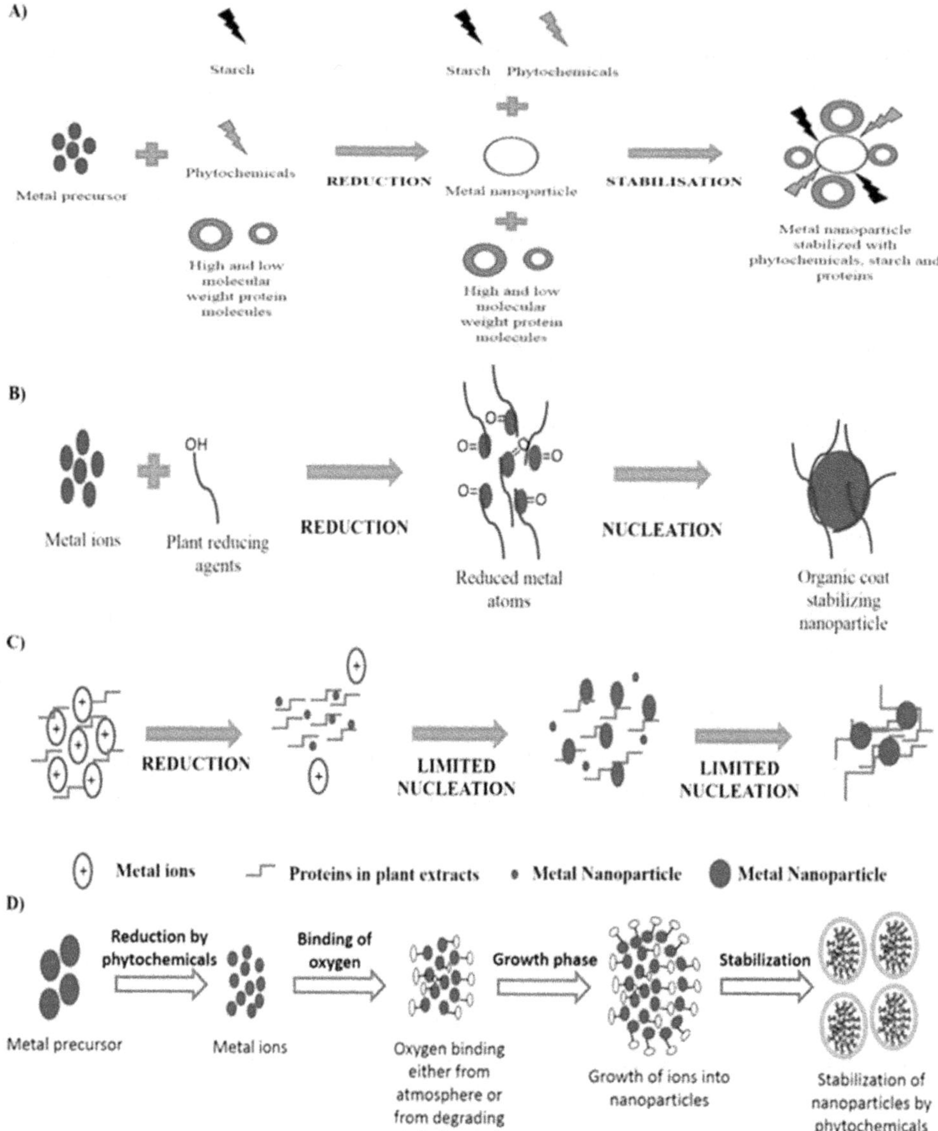

and nanotubes, from their ability to be disseminate into distinct units, which are designated by charge, surface size, porosity and shape. The tendency of a nanoparticle to utilize, diffuse. fluid drag, their optical traits and uptake inside the cell are all encapsulated by architecture and size. Moreover, the administering colloidal +suspension's stability and tendency to cumulate and surface charge is essential in determining the nanoparticles interrelate with surroundings. Lastly, nanoparticles have vast exterior surface area that can optimized for diverse applications due to their augmented surface-to-volume ratio. Additionally, hollow or porous nanoparticles have a large inner surface area that can be more supplemented to add new features, like creation of smart nano matter-based drug delivery systems (Modena, Rühle, Burg, & Wuttke, 2019).

Figure 16. Phytochemical properties revealed by characterization
Source: Todaro and Santi (2022)

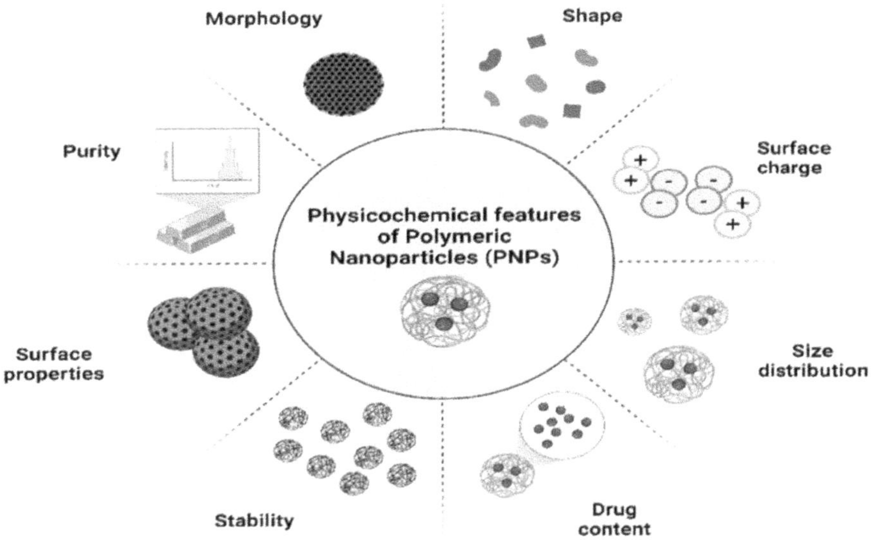

Scanning Electron Microscopy

SEM is used for morphological analysis of the sample material.

Principle and Working

SEM imaged the sample surface using secondary electron detection which emits from the sample after interacting with beam of impinging electrons. High-resolution SEM imaging normally requires conductive substrates, however nonconductive morsels is capped with a thin (5–10 nm) metal coating prior to analysis. When analyzing SEM micrographs, it is important to take into account the nonconductive nanoparticles' altered size and surface structure as a result of sample processing. The constancy of SEM analysis is often limited to >2-3 nm due to the prudent electron energies, however this also significantly reduces the risk of beam-induced sample damage. Moreover, SEM measure at faster pace and also user friendly (Modena et al., 2019).

Transmission Electron Microscopy

Direct imaging of nanostructures is possible with the use of transmission electron microscopy. Additionally, it is frequently utilized to disclose size and morphology, orientation, and crystal impurities as well as surface structures and superstructures.

Principle and Working

A directed electron beam is utilized by TEM to create micrographs of nanoscale matter having appreciable lateral spatial resolution on a thin (usually less than 200 nm) sample. By spatially intensifying and

Figure 17. SEM general representation
Source: Sharma, Jaiswal, Duffy, and Jaiswal (2019)

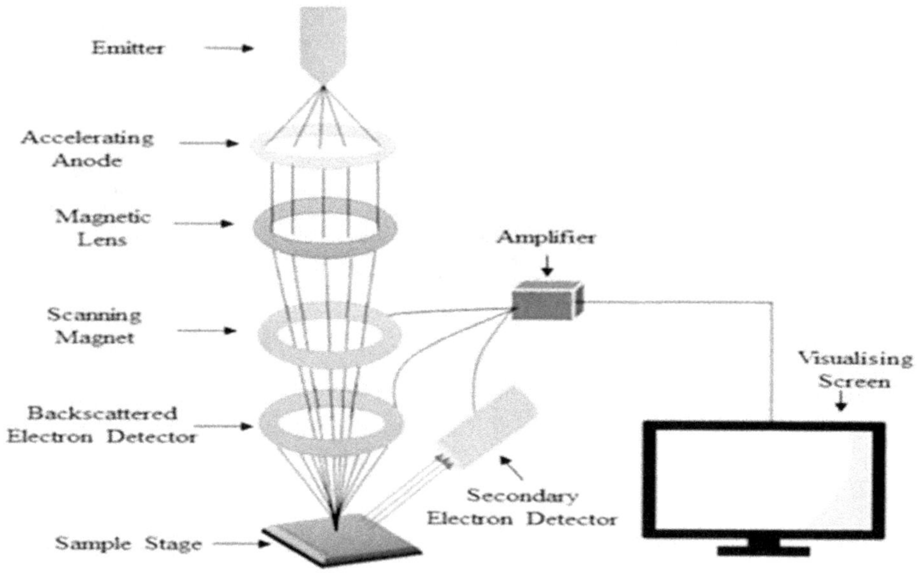

confining the impinging beam, the analysis of the arising electron diffraction pattern, and investigating the crystallinity of specific microscopic sections of crystalline materials, become possible by TEM (Wasson et al., 2020). It is a costly technique as compared to TEM.

X-Ray Diffraction Analysis

XRD peaks help to determine the organization of crystal of nanoforms. It also determines crystallite strains and crystal size.

Principle

The procedure is mainly employed for the characterization and distinguishing of substances on the basis of their diffraction pattern. The primary outcome of an incident monochromatic X-ray beam's interaction with target substance is the scattering of those X-rays from the target substance's atoms. Constructive and destructive interference occurs with the scattered X-rays in crystalline materials with regular structure.

The diffraction of X-rays by crystals is determined by Bragg's. It establishes how a single crystal or grain will be oriented. Finding an unknown substance's crystal structure and determining the size, shape or inner tension of small crystal zones. (Al Thaher et al., 2021).

Atomic Force Microscopy

It's a scanning probe type microscope utilized for examining the surface of nanometer or also atomic-sized matters as well as a number of other force-related parameters.

Figure 18. X-ray diffraction working principle
Source: Widjonarko (2016)

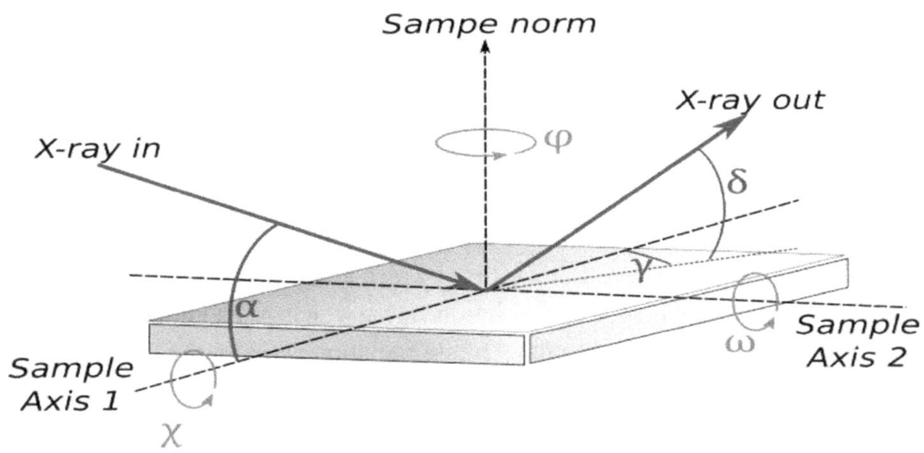

Mechanism

A cantilever with a sharp edge at the end is moved across a sample's surface while being measured, and the forces which cantilever encounters as the consequences of its synergism with the substance is tracked using a laser beam that is reflected off the cantilever's tip and onto photodiode array. Electric conductivity is not required for an AFM standard topography scan, unlike scanning tunnelling microscopy. The size and configuration of the nanoobjects are clearly visible in the scanned image. The necessity thing to remember that, while the lateral size and shape is a combination of nano objects size and configuration of the probe tip, the size in the z-direction is typically fairly exact. Image obtainment can take few minutes, depending on the measurement method and the volume of scan area, and the scanned region is often only a few tens or hundreds of micrometers (Seo & Jhe, 2007).

Figure 19. AFM working representation
Source: Joshi, Homburg, and Ehrmann (2022)

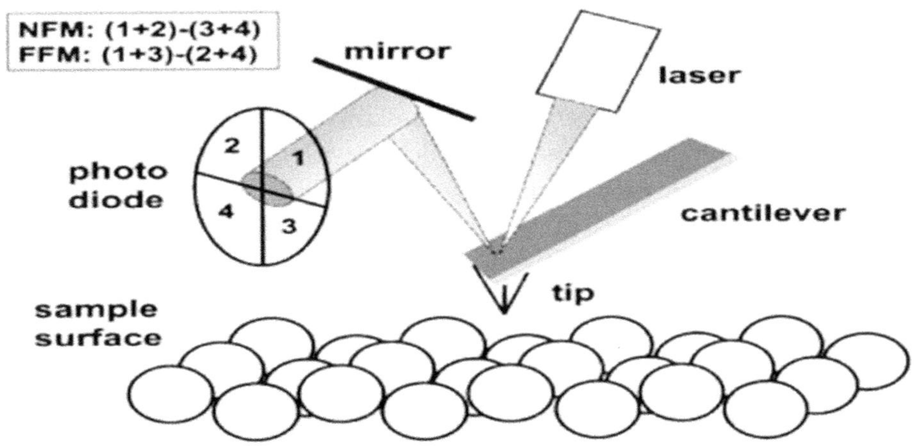

UV Spectroscopy

UV spectroscopy is an absorption type spectroscopy where molecules absorb light in UV region 200–400 nm. When UV rays are absorbed, the electrons are provoked from their lower state into an upper energy state. The amount of UV light absorbed has an energy equal to difference in energy between the low energy and high levels.

Beere-Lambert Law

UV spectroscopy complies to the Beere-Lambert law, which stated that when a pile of mono chromatic light is passed through the absorbing substance solution, rate at which the magnitude of radiation decreases thickness of the solution is proportional to the striking radiation and also the solution's concentration. Law equation representation:

$A = \log (I_0\text{-}I) = ECL$

A is represent absorbance, I_0 is intensity of light incident upon the sample cell, I showed the light intensity, leaving the sample cell, C is the molar concentration of the solute, L is the length of the sample cell (cm), and E is the molar absorptivity.

Figure 20. Beere-Lambert law
Source: Joshi et al. (2022)

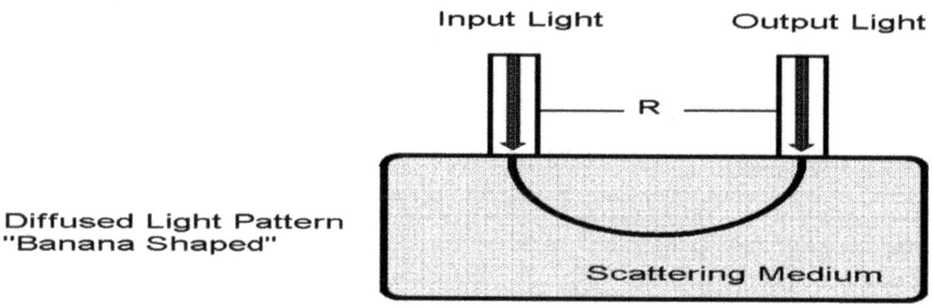

UV visible is used to quantify the quantity and size of Nanopowder. It is employed to assess the level of NP aggregation. The stability of Nanoparticles solutions can be checked using a quick and dependable method called UV Visible spectroscopy. When the particles destabilize, the original extinction peak will often expand or develop a subsequent peak at longer wavelengths, with the peak's amplitude declining as stable nano matters are consumed (settling of aggregates) (Chirayil, Abraham, Mishra, George, & Thomas, 2017).

Mass Spectrometry

By exposing the stoichiometry of their constituent parts after assimilation and disintegration, mass spectrometry was first utilized to characterize the composition of nanoparticles. The utility of mass spectrometry (MS) has expanded and also include the analysis of samples with a megadalton range due to the invention of soft ionization procedures like electrospray ionization (ESI) and other is matrix-assisted laser desorption ionization (MALDI), as well as methods of segregation and identification like ion-mobility spectrometry (IMS), time-of-flight (TOF) analysis, and single particle inductively coupled plasma. Another distinctive feature of MS is the insights it offers about the functionalization layer grafted onto nanoparticle surfaces.

Figure 21. Mass spectrometry
Source: Torres-Sangiao, Leal Rodriguez, and García-Riestra (2021)

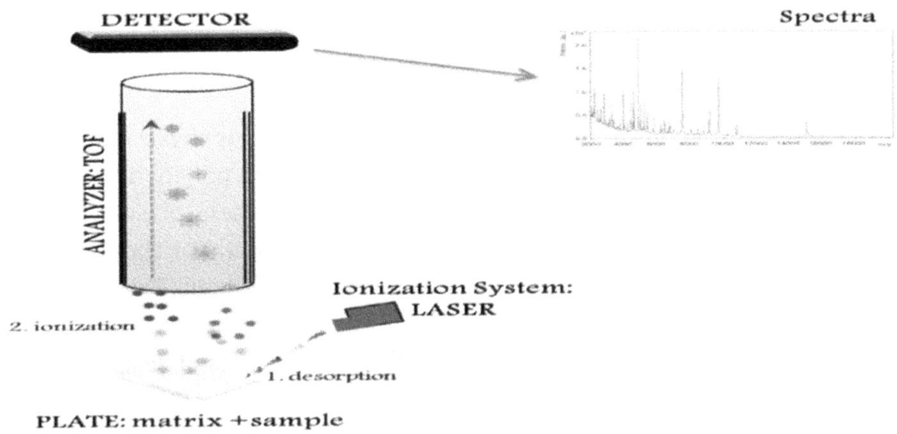

Additionally, MS is frequently used to recognize and measure the architecture of multiple protein coronas on the nanomaterial top following contact to biological fluids in order to study interactions between nanoparticles and biomolecules. Moreover, MS has also been proven to be extremely useful in analyzing the circulation and ingestion of nanoparticles throughout various organs upon administration of nanoparticle based solutions for toxicological analysis due to its capacity to ionize and examine soft materials, like tissues (Costa-Fernández, Menéndez-Miranda, Bouzas-Ramos, Encinar, & Sanz-Medel, 2016; Jeon, Oberreit, Van Schooneveld, & Hogan, 2016).

MODIFICATION OF NANOPARTICLE'S SURFACE

In order to enhance the surface quality and functionality of the designed nanoparticles their surface is modified with different coatings producing nanocomposites which have improved mechanical, chemical, optical, thermal, physical, electric and magnetic properties. Large variety of different organic and inorganic polymers are made to attach with designed nanoparticles according their desired applications. But there is also problem related to the formation of polymeric nanocomposite in which optical and mechanical attributes of nanocomposites are harmed by the nanoparticles' significant propensity for agglomeration and inadequate disintegration in the polymer matrix. In order to increase the disintegration stability of nanoparticles in fluids or polymer matrix, it is imperative that particle surface manipulation by incorporating polymer surfactant molecule. The presence of a second issue in nanocomposites owing to the inorganic material's rigidity, it has a lesser influencing power than in solitary organic progenitor alone, resulting in utilization of elastomeric compounds to boost the tensile strength of composites (Kango et al., 2013).

But despite these few drawbacks, nanocomposites type of modification is valuable and resourceful option for increasing the functionality of the designed nanoparticles. Some methods are discussed here for surface modification of nanoparticles but before method discussion we add some requirement for developing nanocomposites. In-depth modeling, material drafting is required for the creation of these high-performance nanocomposites. It entails choosing polymers and fillers from among the countless polymers and fillers now on the market in order to achieve the desired attribute. Choosing the best strategy is another aspect of it (Akpan, Shen, Wetzel, & Friedrich, 2019).

Physical Techniques

The substratal methods to enhance the surface stability of Nanosubstance is physical surface modification, which can be accomplished by coating the nanosubstance surface with polymeric and ionic surfactant. Both hydrophobic groups and hydrophilic groups are present in the chemical structure of surfactants. The foundation of the surface modification procedure is the electrostatic or chemical bond-based adsorption of a hydrophilic group onto the Nanosubstance surface. Surfactants have the ability to reduce particle interactions and reduce the impact of interfacial forces. Surface tension and the rate of particle agglomeration are reduced by the surfactant's adsorption on the surfaces of suspended particles. As a result, the particles' stability enhances. In this approach the enhancers have hydrogen and weak Vand er Waal forces so this technique is not stable thermally. Surface modification of particles like TiO_2, Al_2O_3 has been reported (Ahangaran & Navarchian, 2020).

Chemical Treatment

The ability of nanoparticles to disperse more steadily in different liquid media can be enhanced by chemically altering their surfaces (for example, by absorbing silane coupling agents). The surface can also be done by reacting with metal alkoxides, epoxides like propylene oxide, and alkyl or aryl isocyanates. The capacity of silane modifiers to form a strong chemical connection linking organic and inorganic matters is their crucial quality. The silane modifiers ($X(CH_2)_nSiR_3$), on the other hand, have two different functional groups. Because of its reactivity and congeniality with organic materials, the organic functional group (X) was chosen. Methoxy, ethoxy, and other hydrolyzable groups (R) are mediators in the production of silanol groups for coupling to inorganic nanoparticles surface. Some functional group in saline structure are: chloropropyl, phenyl, epoxy, methyl, vinyl, mercapto, etc.

Hydrolysis Technique of Saline Coupling

The common technique for saline coupling modification is hydrolysis technique, in this technique polar solvent like ethanol, acetone etc. along with water or simple water solvent is employed. Generally, the water in the solution hydrolyzes the alkoxy groups (-OR; $-OCH_3$, $-OC_2H_5$) of the silane blending assistant to silanol groups (Si-OH). Then, on the surface of NPs, the condensation reaction takes place among the Si-OH and -OH groups. In an aqueous media with a range of pH (acidic, alkaline, and neutral), silanes can be hydrolyzed. The alkoxide group undergoes first protonation in an acidic environment. The condensation events then take place, releasing an alcohol molecule (ROH). Under high acidic condition rate of hydrolysis increase producing stable Salinols.

Condensation Mechanism

The salinization of the Nano matters can be accomplished by condensation of non-polar organic solvents such as xylene, cyclohexane and toluene under the completely dry conditions. During the condensation process, the alkoxy groups (-OR; -OCH3, -C2H5) of the silane mixing particles inadvertently combine with hydroxyl groups on the surface of the nanomaterial to release an alcohol molecule. (ROH). Due to the expense and toxicity of the solvent utilized, the surface manipulation through condensation process is not suited for scaling up manufacturing.

Despite all of this, getting hydrolytically stable silane-modified Nano powders remains a significant difficulty even though the silane layer surrounding the remodeled NPs boosts their sustainability in solution and inhibit or reduces the agglomerates formation. For example, it was studied that excessive moisture and heat exposure can cause the silane layer to break down, leads towards reduction the silane modifier's coupling potency and the stability of the Nanofabricates. The majority of silane groups, on the other hand, perform well at improving the surface qualities of Nanofabricates in dry environments. By raising the degree of crosslinking in the silane layer and enhancing the permutation of hydroxyl bonds to O-SiO bonds on the Nanomaterial surface, the hydrolytic stability of the silane-modified NPs can be made more stable (Ahangaran & Navarchian, 2020).

Grafting of Synthetic Polymer

This is an approach of grafting synthetic polymers to substrate of organic and inorganic particles. This process improves the chemical activity and modifies the surface chorography of the indigenous inorganic and organic matters. That's why they are also known as organic-inorganic nanocomposites. Monomers typically have low molecular weights by nature, allowing them to pass through cumulative nanoparticles and interact with the active locus there. The interstitial volume within the aggregate of nanoparticles partially filled. By starting the graft polymerization from beginner group positioned on the surfaces of the polymer-grafted inorganic particles, a larger proportion of effective grafts can be attained. The grafted polymers spread from the particle's surface during the polymerization operations, which utilize radical, anionic, or cationic polymerization methodologies.

Tsubokawa et al reported that Hyperbranched polymers with pendant azo groups were grafted onto the surfaces of silica nanoparticles, and then they started a radical post graft polymerization of vinyl monomers using the azo groups from those polymer strings. Shirai and his coworkers performed a radical graft polymerization of vinyl monomers on the top of polymethyl siloxane-coated titanium dioxide nanopowder that had been manipulated with alcoholic hydroxyl groups started by the azo groups added on Ti/Si-R-OH.

Ligand Exchange Method

The ligand exchange technique was created to make it easier to use up conversion nanoparticles as energy donors in applications like energy-transfer-based bio detection. This strategy includes swapping

Figure 22. Grafting technique
Source: Díez-Pascual (2021)

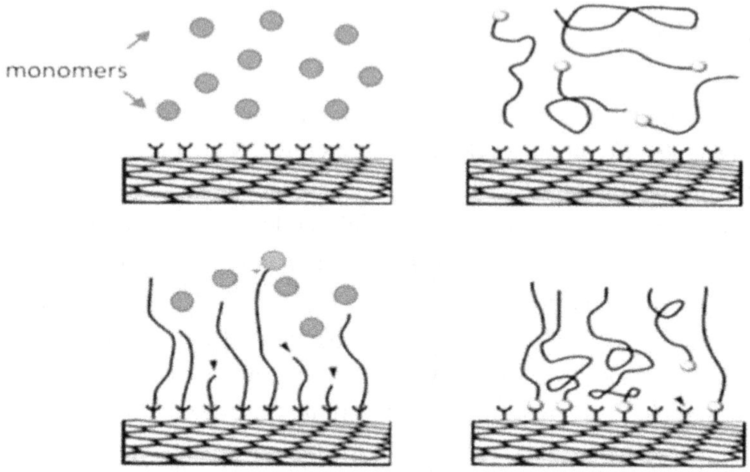

out the original capping ligands for hydrophilic ones. It was revealed that just several nanometers wide ligand sheet is sufficient to screen the nanoparticles and to block the flow of free electrons from NP to NP before fusion can take place; this layer hence considerably inhibits charge relocation.

Additionally, massive organic ligands typically lack functional end groups that could help an NP-polymer blend's charge transport. For inhibiting the amalgam losses and improvised charge transport across the photoactive band, it is necessary to dislodge as many of the synthesis ligands from the NP surface as feasible. This can be done by reducing the gap between NPs and between NP-polymer interactions. While preserving the essential sustainability of the nanoparticle's dispersions, the ideal method would reduce the separation brought on by the organic synthesis ligands encapsulating the NPs in order to improve the transfer of electrons. A better ligand can be used in place of the synthetic ligands to create this solution. Thermal annealing and/or vacuum processes would be ideal for removing the substituted ligand from the NP surface in the active layer of a solar cell in order to improve the conductivity among the nanomaterials and enable effective charge segregation between the NPs and the polymer.

Many high-quality, technically significant Nanomaterials, including semiconductor nanocrystals, may now be produced on a regular basis using several altered thermolytic process variations. The precursors will undergo a chemical transformation into simmering atomic or molecular species when the reaction solution is heated to elevated temperature (usually 150–320 C), at which point they can condense to form NPs. The presence of capping ligands has a significant impact on these NPs' ability to develop (Greenham, Peng, & Alivisatos, 1996; Lu & Yin, 2012).

In Situ Surface Modification

In this method modification of surface occur during nanoparticles synthesis process. Example of this approach is thermal destruction of organometallic compound, micelle method or polyol technique. Here for preventing the agglomeration of nanoparticles, capping agents or other surfactants like trioctylphosphine oxides (TOPO),

Figure 23. Ligand exchange method
Source: Cook, Labiento, and Chauhan (2021)

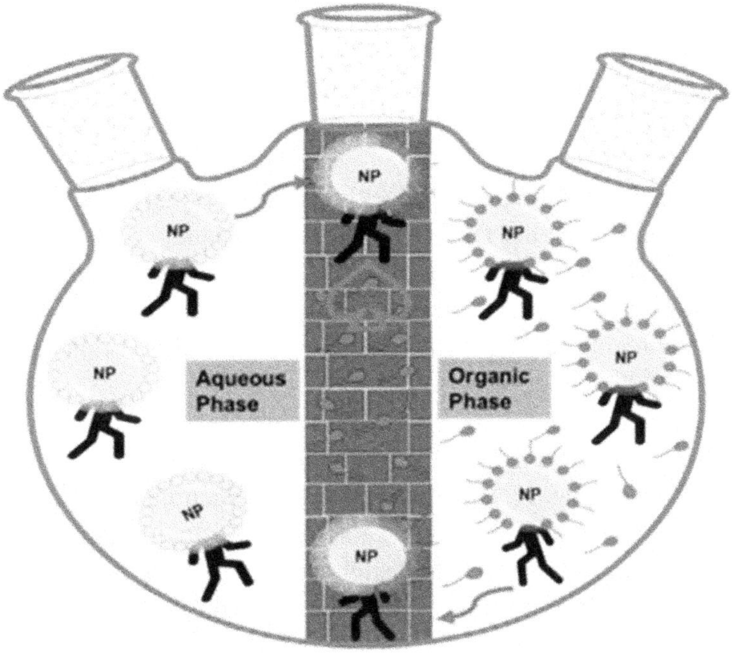

oleic acid, and amines are fused in amalgam. The surface properties of surfactant-capped nanoparticles created by the in-situ surface manipulation approach can be more supplemented (Kango et al., 2013).

FUTURE PERSPECTIVE

Different Nanofabricates are hallmark for future because of their eliciting properties from which the prominent one is their small size providing huge surface activity enable us to replace bulk use of mate-

Figure 24. In situ hybridization synthesis
Source: Darwish, Mostafa, and Al-Harbi (2022)

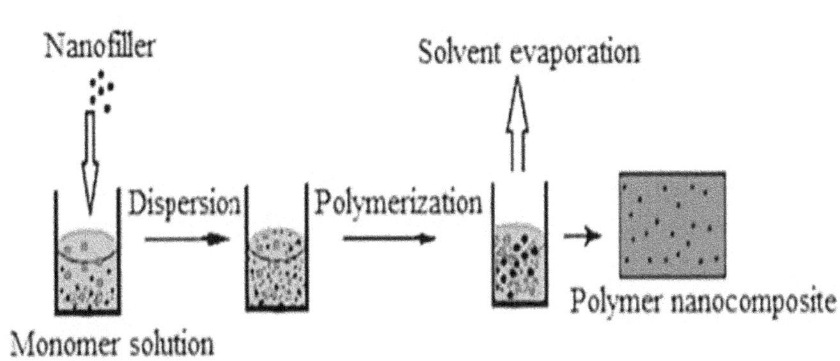

rials. This trait is helpful in reducing the depletion of resources and also the combat the environmental threat which is more prominent with bulk usage because bulk create more waste than these nano matters.

Besides energy conservation their chemical and optical properties are better helping hands in medical, biological and forensic studies and research. Their antimicrobial activity providing their way in using drugs and medicine area provide more effective treatments as compared to bulk chemical-based methods. Their mechanical, electrical and magnetic characteristics also help in lifting other industries like computer, physics, chemicals etc.

Despite of many advancements we are still lacking in this field because of some drawbacks still prevailing in the development of this marvel, for example: nanoparticle cause recombination of holes and electrons effecting its properties but this can be overcome by nanocomposites but on the other hand, nanocomposite also have a drawback which was already discuss above, so this field still require more research and development to bring up their use so that matter of depletion of non-renewable resources.

Moreover, the toxicological hindrance is also lies in nanomaterial fabrication process which is somehow minimized by designing nanoparticles with biogenic methods. But these methods required more standardization for checking the toxicological level of each fabrication method. For this purpose, different regulatory authorities should be established for maintaining the check and balance for chemical dosage and composition ratio of designing and manufacturing of different nanoforms and try to devised protocols which are more environment friendly, cheap and effective motive for different nano fabrications.

The advancements in utilization of these nanofabricates in different industries and applications is enhanced by manufacturing different forms of these novel nanostructure and forms like carbon nanotubes, nanofillers, nanosheets, nanobatteries, nanoflakes, nano cones etc. which help in modifying their morphology and hence their problem of recombination can be avoided. With the establishment of proper standards and protocols for avoiding different manufacturing and manipulating challenges we will be able to use all these nanoforms in our different modernizing industries.

CONCLUSION

In this chapter author discuss the chemistry behind the nanoparticles traits and structure which allow their use in almost all fields of life, how these particles synthesized from bulk materials using other chemical and their salt sources and different physical chemical biological techniques. It is also discussed that how the threat cause by chemical production shall be avoided by their designing using biogenic approaches.

To find the inherit properties of designed nanomaterials different techniques are discussed according to the requirement of observing properties. Lastly to enhance the functionality of the nanoparticles and to further manipulate their property for further increase in their functional capabilities by blending them different organic and inorganic matters and produce nanocomposites enhancing their use in more fields. Light is also shed regarding the future perspective of these nanoforms to make their utilization more reliable and easily approachable.

REFERENCES

Ahangaran, F., & Navarchian, A. H. (2020). Recent advances in chemical surface modification of metal oxide nanoparticles with silane coupling agents: A review. *Advances in Colloid and Interface Science, 286*, 102298. doi:10.1016/j.cis.2020.102298 PMID:33171357

Ahire, S. A., Bachhav, A. A., Pawar, T. B., Jagdale, B. S., Patil, A. V., & Koli, P. B. (2022). The augmentation of nanotechnology era: A concise review on fundamental concepts of nanotechnology and applications in material science and technology. *Results in Chemistry, 4*, 100633. doi:10.1016/j.rechem.2022.100633

Akpan, E., Shen, X., Wetzel, B., & Friedrich, K. (2019). *Design and synthesis of polymer nanocomposites. In Polymer composites with functionalized nanoparticles.* Elsevier.

Al Thaher, Y., Satoof, A., Kamal, A., Almani, D., Shaban, D., Kassab, G., & Ajaleh, S. A. (2021). *Instrumental analytical techniques for physicochemical characterization of bio-nanomaterials. In Handbook on Nanobiomaterials for Therapeutics and Diagnostic Applications.* Elsevier.

Amendola, V., & Meneghetti, M. (2009). Laser ablation synthesis in solution and size manipulation of noble metal nanoparticles. *Physical Chemistry Chemical Physics, 11*(20), 3805–3821. doi:10.1039/b900654k PMID:19440607

Baig, N., Kammakakam, I., & Falath, W. (2021). Nanomaterials: A review of synthesis methods, properties, recent progress, and challenges. *Materials Advances, 2*(6), 1821–1871. doi:10.1039/D0MA00807A

Barabadi, H., Najafi, M., Samadian, H., Azarnezhad, A., Vahidi, H., Mahjoub, M. A., Koohiyan, M., & Ahmadi, A. (2019). A systematic review of the genotoxicity and antigenotoxicity of biologically synthesized metallic nanomaterials: Are green nanoparticles safe enough for clinical marketing? *Medicina, 55*(8), 439. doi:10.3390/medicina55080439 PMID:31387257

Barhoum, A., García-Betancourt, M. L., Jeevanandam, J., Hussien, E. A., Mekkawy, S. A., Mostafa, M., ... Bechelany, M. (2022). Review on natural, incidental, bioinspired, and engineered nanomaterials: History, definitions, classifications, synthesis, properties, market, toxicities, risks, and regulations. *Nanomaterials (Basel, Switzerland), 12*(2), 177. doi:10.3390/nano12020177 PMID:35055196

Chen, X., Wu, M., Liu, X., Wang, D., Liu, F., Chen, Y., Yi, F., Huang, W., & Wang, S. (2019). Tuning the Doping Ratio and Phase Transition Temperature of VO2 Thin Film by Dual-Target Co-Sputtering. *Nanomaterials (Basel, Switzerland), 9*(6), 834. doi:10.3390/nano9060834 PMID:31159352

Chirayil, C. J., Abraham, J., Mishra, R. K., George, S. C., & Thomas, S. (2017). *Instrumental techniques for the characterization of nanoparticles. In Thermal and Rheological Measurement Techniques for Nanomaterials Characterization.* Elsevier.

Cook, E., Labiento, G., & Chauhan, B. P. (2021). Fundamental Methods for the Phase Transfer of Nanoparticles. *Molecules (Basel, Switzerland), 26*(20), 6170. doi:10.3390/molecules26206170 PMID:34684750

Costa-Fernández, J. M., Menéndez-Miranda, M., Bouzas-Ramos, D., Encinar, J. R., & Sanz-Medel, A. (2016). Mass spectrometry for the characterization and quantification of engineered inorganic nanoparticles. *Trends in Analytical Chemistry, 84*, 139–148. doi:10.1016/j.trac.2016.06.001

Darwish, M. S., Mostafa, M. H., & Al-Harbi, L. M. (2022). Polymeric nanocomposites for environmental and industrial applications. *International Journal of Molecular Sciences, 23*(3), 1023. doi:10.3390/ijms23031023 PMID:35162946

Díez-Pascual, A. M. (2021). Development of graphene-based polymeric nanocomposites: A brief overview. *Polymers, 13*(17), 2978. doi:10.3390/polym13172978 PMID:34503017

Gebre, S. H., & Sendeku, M. G. (2019). New frontiers in the biosynthesis of metal oxide nanoparticles and their environmental applications: An overview. *SN Applied Sciences, 1*(8), 1–28. doi:10.100742452-019-0931-4

Greenham, N. C., Peng, X., & Alivisatos, A. P. (1996). Charge separation and transport in conjugated-polymer/semiconductor-nanocrystal composites studied by photoluminescence quenching and photoconductivity. *Physical Review. B, 54*(24), 17628–17637. doi:10.1103/PhysRevB.54.17628 PMID:9985889

Heidinger, S., Spranger, F., Dostál, J., Zhang, C., & Klaus, C. (2022). Material Treatment in the Pulsation Reactor—From Flame Spray Pyrolysis to Industrial Scale. *Sustainability (Basel), 14*(6), 3232. doi:10.3390u14063232

Ijaz, I., Gilani, E., Nazir, A., & Bukhari, A. (2020). Detail review on chemical, physical and green synthesis, classification, characterizations and applications of nanoparticles. *Green Chemistry Letters and Reviews, 13*(3), 223–245. doi:10.1080/17518253.2020.1802517

Jamkhande, P. G., Ghule, N. W., Bamer, A. H., & Kalaskar, M. G. (2019). Metal nanoparticles synthesis: An overview on methods of preparation, advantages and disadvantages, and applications. *Journal of Drug Delivery Science and Technology, 53*, 101174. doi:10.1016/j.jddst.2019.101174

Jeevanandam, J., Barhoum, A., Chan, Y. S., Dufresne, A., & Danquah, M. K. (2018). Review on nanoparticles and nanostructured materials: History, sources, toxicity and regulations. *Beilstein Journal of Nanotechnology, 9*(1), 1050–1074. doi:10.3762/bjnano.9.98 PMID:29719757

Jeevanandam, J., Chan, Y. S., & Danquah, M. K. (2016). Biosynthesis of metal and metal oxide nanoparticles. *ChemBioEng Reviews, 3*(2), 55–67. doi:10.1002/cben.201500018

Jeon, S., Oberreit, D. R., Van Schooneveld, G., & Hogan, C. J. (2016). Nanomaterial size distribution analysis via liquid nebulization coupled with ion mobility spectrometry (LN-IMS). *Analyst (London), 141*(4), 1363–1375. doi:10.1039/C5AN02150B PMID:26750519

Jeyaraj, M., Gurunathan, S., Qasim, M., Kang, M.-H., & Kim, J.-H. (2019). A comprehensive review on the synthesis, characterization, and biomedical application of platinum nanoparticles. *Nanomaterials (Basel, Switzerland), 9*(12), 1719. doi:10.3390/nano9121719 PMID:31810256

Jiang, Z., Li, L., Huang, H., He, W., & Ming, W. (2022). Progress in Laser Ablation and Biological Synthesis Processes:"Top-Down" and "Bottom-Up" Approaches for the Green Synthesis of Au/Ag Nanoparticles. *International Journal of Molecular Sciences, 23*(23), 14658. doi:10.3390/ijms232314658 PMID:36498986

Joshi, J., Homburg, S. V., & Ehrmann, A. (2022). Atomic Force Microscopy (AFM) on Biopolymers and Hydrogels for Biotechnological Applications—Possibilities and Limits. *Polymers*, *14*(6), 1267. doi:10.3390/polym14061267 PMID:35335597

Joudeh, N., & Linke, D. (2022). Nanoparticle classification, physicochemical properties, characterization, and applications: A comprehensive review for biologists. *Journal of Nanobiotechnology*, *20*(1), 262. doi:10.118612951-022-01477-8 PMID:35672712

Joy, J., Krishnamoorthy, A., Tanna, A., Kamathe, V., Nagar, R., & Srinivasan, S. (2022). Recent Developments on the Synthesis of Nanocomposite Materials via Ball Milling Approach for Energy Storage Applications. *Applied Sciences (Basel, Switzerland)*, *12*(18), 9312. doi:10.3390/app12189312

Kango, S., Kalia, S., Celli, A., Njuguna, J., Habibi, Y., & Kumar, R. (2013). Surface modification of inorganic nanoparticles for development of organic–inorganic nanocomposites—A review. *Progress in Polymer Science*, *38*(8), 1232–1261. doi:10.1016/j.progpolymsci.2013.02.003

Kim, J.-H., Bak, J. G., & Kim, C.-K. (2020). Electrical Resistivity of Ni–Fe Wires Coated with Sn Using low-Pressure Chemical Vapor Deposition. *Coatings*, *10*(4), 317. doi:10.3390/coatings10040317

Lu, Z., & Yin, Y. (2012). Colloidal nanoparticle clusters: Functional materials by design. *Chemical Society Reviews*, *41*(21), 6874–6887. doi:10.1039/c2cs35197h PMID:22868949

Majerič, P., & Rudolf, R. (2020). Advances in ultrasonic spray pyrolysis processing of noble metal nanoparticles. *Materials (Basel)*, *13*(16), 3485. doi:10.3390/ma13163485 PMID:32784637

Modena, M. M., Rühle, B., Burg, T. P., & Wuttke, S. (2019). Nanoparticle characterization: What to measure? *Advanced Materials*, *31*(32), 1901556. doi:10.1002/adma.201901556 PMID:31148285

Montanarella, F., & Kovalenko, M. V. (2022). Three millennia of nanocrystals. *ACS Nano*, *16*(4), 5085–5102. doi:10.1021/acsnano.1c11159 PMID:35325541

Park, Y. S., Kodama, S., & Sekiguchi, H. (2021). Preparation of Metal Nitride Particles Using Arc Discharge in Liquid Nitrogen. *Nanomaterials (Basel, Switzerland)*, *11*(9), 2214. doi:10.3390/nano11092214 PMID:34578530

Pirtarighat, S., Ghannadnia, M., & Baghshahi, S. (2019). Green synthesis of silver nanoparticles using the plant extract of Salvia spinosa grown in vitro and their antibacterial activity assessment. *Journal of Nanostructure in Chemistry*, *9*(1), 1–9. doi:10.100740097-018-0291-4

Salem, S. S., Hammad, E. N., Mohamed, A. A., & El-Dougdoug, W. (2022). A comprehensive review of nanomaterials: Types, synthesis, characterization, and applications. *Biointerface Research in Applied Chemistry*, *13*(1), 41. doi:10.33263/BRIAC131.041

Schaming, D., & Remita, H. (2015a). Nanotechnology: From the ancient time to nowadays. *Foundations of Chemistry*, *17*(3), 187–205. doi:10.100710698-015-9235-y

Schaming, D., & Remita, H. (2015b). Nanotechnology: From the ancient time to nowadays. *Foundations of Chemistry*, *17*(3), 187–205. doi:10.100710698-015-9235-y

Seo, Y., & Jhe, W. (2007). Atomic force microscopy and spectroscopy. *Reports on Progress in Physics*, *71*(1), 016101. doi:10.1088/0034-4885/71/1/016101

Sharma, S., Jaiswal, S., Duffy, B., & Jaiswal, A. K. (2019). Nanostructured materials for food applications: Spectroscopy, microscopy and physical properties. *Bioengineering (Basel, Switzerland)*, *6*(1), 26. doi:10.3390/bioengineering6010026 PMID:30893761

Shrestha, S., Wang, B., & Dutta, P. (2020). Nanoparticle processing: Understanding and controlling aggregation. *Advances in Colloid and Interface Science*, *279*, 102162. doi:10.1016/j.cis.2020.102162 PMID:32334131

Simonenko, N. P., Musaev, A. G., Simonenko, T. L., Gorobtsov, P. Y., Volkov, I. A., Gulin, A. A., Simonenko, E. P., Sevastyanov, V. G., & Kuznetsov, N. T. (2021). Hydrothermal Synthesis of Ag Thin Films and Their SERS Application. *Nanomaterials (Basel, Switzerland)*, *12*(1), 136. doi:10.3390/nano12010136 PMID:35010086

Singh, J., Dutta, T., Kim, K.-H., Rawat, M., Samddar, P., & Kumar, P. (2018). 'Green'synthesis of metals and their oxide nanoparticles: Applications for environmental remediation. *Journal of Nanobiotechnology*, *16*(1), 1–24. doi:10.118612951-018-0408-4 PMID:30373622

Slepička, P., Slepičková Kasálková, N., Siegel, J., Kolská, Z., & Švorčík, V. (2019). Methods of gold and silver nanoparticles preparation. *Materials (Basel)*, *13*(1), 1. doi:10.3390/ma13010001 PMID:31861259

Tan, N. H. (2014). *Rock art research in Southeast Asia: a synthesis.* Paper Presented at the Arts. 10.3390/arts3010073

Todaro, B., & Santi, M. (2022). *Characterization and Functionalization Approaches for the Study of Polymeric Nanoparticles: The State of the Art in Italian Research.* Paper Presented at the Micro. 10.3390/micro3010002

Torres-Sangiao, E., Leal Rodriguez, C., & García-Riestra, C. (2021). Application and perspectives of MALDI–TOF mass spectrometry in clinical microbiology laboratories. *Microorganisms*, *9*(7), 1539. doi:10.3390/microorganisms9071539 PMID:34361974

Wani, S. U. D., Ali, M., Masoodi, M. H., Khan, N. A., Zargar, M. I., Hassan, R., ... Osmani, R. M. A. (2022). A review on nanoparticles categorization, characterization and applications in drug delivery systems. *Vibrational Spectroscopy*. doi:10.1016/j.vibspec.2022.103407

Wasson, M. C., Otake, K.-i., Gong, X., Strathman, A. R., Islamoglu, T., Gianneschi, N. C., & Farha, O. K. (2020). Modulation of crystal growth and structure within cerium-based metal–organic frameworks. *CrystEngComm*, *22*(47), 8182–8188. doi:10.1039/D0CE01223H

Widjonarko, N. E. (2016). Introduction to advanced X-ray diffraction techniques for polymeric thin films. *Coatings*, *6*(4), 54. doi:10.3390/coatings6040054

Yadav, T. P., Yadav, R. M., & Singh, D. P. (2012). Mechanical milling: A top down approach for the synthesis of nanomaterials and nanocomposites. *Nanoscience and Nanotechnology*, *2*(3), 22–48. doi:10.5923/j.nn.20120203.01

Yun, K., Nam, H.-S., & Kim, S. (2020). Categorization of atomic mixing patterns in bimetallic nanoparticles by the energy competition. *Physical Chemistry Chemical Physics*, *22*(15), 7787–7793. doi:10.1039/C9CP06967D PMID:32239065

Chapter 5
Nano-Trackers (Nano-Sensors) for Forensics Investigation

Muniba Rahmat
Jiangsu University, Zhenjiang, China

Saba Naz
Government College University, Faisalabad, Pakistan

Shumaila Kiran
Government College University, Faisalabad, Pakistan

ABSTRACT

Sensors are effective instruments for finding illegal drugs. In the past several years, sensor use and development have garnered a lot of interest. The lack of an effective tool that offers the specialist all the necessary features, including cost-effective manufacturing, on-site analysis, instrumental simplicity, and correct findings, might be blamed for this thirst for the creation of new detectors. Nanotechnology has the potential to significantly advance forensic science and criminal investigation. In forensic nanotechnology, tiny chip materials are employed in place of large apparatus, which simplifies the techniques of analysis and improves the accuracy, precision, timeliness, and suitability of investigations. The reader is provided with up-to-date studies and publications as this chapter describes the use of nano sensors in forensics investigation. Additionally, it details the key distinctions between the three major categories of sensors that provide information on the potential applications of these sensing materials in the future.

INTRODUCTION

In every sector, nanotechnology is making a significant impact. Its capability to modify and describe the matter at the level of individual and tiny atoms makes it an extensively utilized method. The word "nano" refers to the "billionth" or a factor of 10-9. The word is derived from the ancient Greek "Nanos," which means "dwarf." To put it simply, 1 nm is equivalent to 3–10 atoms. When compared to the typical size encountered every day, it is quite little. As an illustration, 1nm is 1/1000th the breadth of a human

DOI: 10.4018/978-1-6684-8325-1.ch005

hair. Nanotechnology is the term used to define the component of science, engineering, and technology research done at a size between 1 and 100 nanometers (Shukla et al., 2016). Due to their tiny size, wide surface area, and improved reactivity, nanotechnology has the potential to be used in electronics, diagnostics, biosensing, imaging, optical devices, and drug administration (He et al., 2020; Malefane, 2020; Pavan Kumar et al., 2020).

New paradigms in science and technology are being established by the rapid advancements in nanomaterials. Due to the integration of advances in nanotechnology into the field of forensic science, this area of research is particularly significant. With verified cutting-edge nano-imaging technologies for visualization, nano-sensors, nano-manipulators and nano-forensic provides a novel viewpoint for investigations into crimes in the present moment (Dahiya et al., 2022).

It is now considered a technology with a universal purpose due to its widespread applicability in practically all fields. As a result, there are many uses of forensic science for this universal technology. A completely new field of forensic research, nano-forensic is a beautiful fusion of forensic science with nanotechnology. Identification, appraisal, crime scene investigation, and establishing links between specific evidences and offenders are the core concerns of forensic science. By detecting the presence of biological agents, explosive gases and residues, nano-forensics is a highly advanced branch of forensic sciences that is connected to the development of nano-sensors for crime investigations and inspection of terrorist activity (Mathur et al., 2016; Gopi et al., 2016). In the identification of crimes using nanotechnology, nano-analysis is frequently employed. Transmission electron microscopy, scanning electron microscopy, dynamic light scattering, atomic force microscopy and Raman microscopy are a few of these analysis methods (Lloyd-Hughes et al., 2015).

Nanoforensic and other technologies have important uses for DNA analysis, explosive detection, drug testing, and fingerprint analysis (Kaushik et al., 2017). These methods help forensic investigators in two ways: first, by enabling the analysis of nanoscaled samples, and second, by utilising the unique properties of nanomaterial to locate and gather evidence that would not have been possible with earlier methods. The use of modern techniques such as DNA extraction from fingerprints, gun residues, explosives, and heavy metals makes it easier for forensic experts to provide irrefutable evidence (Arif et al., 2015; Singh et al., 2015).

Due to the significance of forensic science to the legal system, errors cannot be allowed in this discipline. The materials used for analysis must exhibit great specificity and precision, just like fingerprint detection, bodily fluids, DNA, drugs, poisons, and explosives (Mahbub et al., 2020). A list of laboratories have financed in preferring analytical methods and nanostructures to provide findings with the necessary precision (Shaw et al., 2017; Rawtani et al., 2019). As in one of case relevant to fingerprint detection (Rawtani et al., 2019; Chávez et al., 2021) or drug detection being carried out by biosensors and the deployment of nanosensors enables their usage at scene of crime (Montagner et al., 2018; Teymourian et al., 2020) and explosives (Moram et al., 2020).

Among the primary fields concerning by assessment as well as analyzing sample taken from a human corpse and the scene of death is forensic microbial nanotechnology. The range of forensic microbiological techniques is expanding as a result of the advancement of microbial nanotechnology, and related approaches are also growing in popularity. Nowadays, forensic microbial nanotechnology techniques are used as aids in a number of fields, likewise the diagnosis of the deceased, the calculation of the postmortem period, and the characterization of the cause, mechanism, and mode of death. Research defining new approaches in forensic sciences have been driven by advancements in microbial nanotechnology, including next-generation sequencing, amplicon metagenomics as well as mass-spectrometry-based microbiome research. A positive development in the field of forensic microbiology is the ease that non-culturable organisms may now be identified (Yıldırım et al., 2021).

Nano-Sensors

The sensors are described as tools for the identification of certain chemicals. According to Wongkaew et al. (2019), these devices are capable of diagnosis, evaluation of various physical characteristics (likewise temperature and mass), and sensing of electrical as well as optical characteristics of materials. A quantifiable signal is produced in order to carry out the sensing, and being linked to the sensor, this signal can be created and interpreted in a variety of ways (Fadel et al., 2016).

Sensors have been an essential component of daily living since the second half of the 20th century. The glucose test and the female pregnancy or fertility test were two of the first products created with this technology. (Bahadır and Sezgintürk, 2015; Meng et al., 2019). Nanosensors' function in the sensing industry is being enhanced by the rapid growth of technology and the shrinking of materials. On the other hand, nanosensors should have at least one dimension less than 100 nm and share certain traits with macro sensors. The detection of many materials on the nanoscale is made feasible by their incredibly tiny size, huge surface area, and great reactivity. The number of studies and subsequent usage of these devices in the sensing sector has expanded significantly in recent years due to the potential applications and multifunctionality of these nanomaterials (Fikiet et al., 2018; To et al., 2020).

Definition of Nanosensors

The gadget must have two primary parts in order to be categorised as a nanosensor: a substance with detecting properties and a transducer with the purpose of receiving a quantifiable signal. The substance that detects changes in pressure and temperature, an electric signal, the development of a chemical bond and biological responses, among other external stimuli, interact with the material that detects them. The various reactions often result in changes to one or more of the material's inherent characteristics, which are subsequently converted into quantifiable signals and transmitted to the transducer (Fikiet et al., 2018). This describes how nanosensors typically operate. However, each type of such nanomaterials has its own unique characteristics, that will be covered in more detail in this articles.

CLASSIFICATION OF NANO-SENSORS

Chemical Sensors

According to the definition given by IUPAC, chemical sensors are "devices or instruments capable of conducting detection by converting a chemical or physical characteristic of a specific analyte into a quantifiable signal, wherein the magnitude is directly connected to the concentration of the analyte." Construction of a chemical sensor is shown in Figure 1 (Paolesse et al., 2016).

When placed near to particular ions or chemicals, nanosensors may deliver data instantly online. The important attributes for the analytical field are affordability and mobility. As can be seen in Figure 2. these materials have the potential to be used in several aspects of daily life. According to Steinegger et al. (2020) and Ge et al. (2020), chemical detection has played an important role in forensics and can

Figure 1. Chemical structure of sensors: In the first stage, analytes interact with the sensing material changing some of its properties, likewise, temperature (ΔT), mass ΔM, conductivity ($\Delta\sigma$), work function ($\Delta\Phi$), refractive index (Δn), permittivity ($\Delta\varepsilon$). In the second stage, the transducer makes your work, converting some of physical properties into the variation of its electric parameter. In the end, the circuit gives rise to the sensing signal.
Source: Meng et al. (2019).

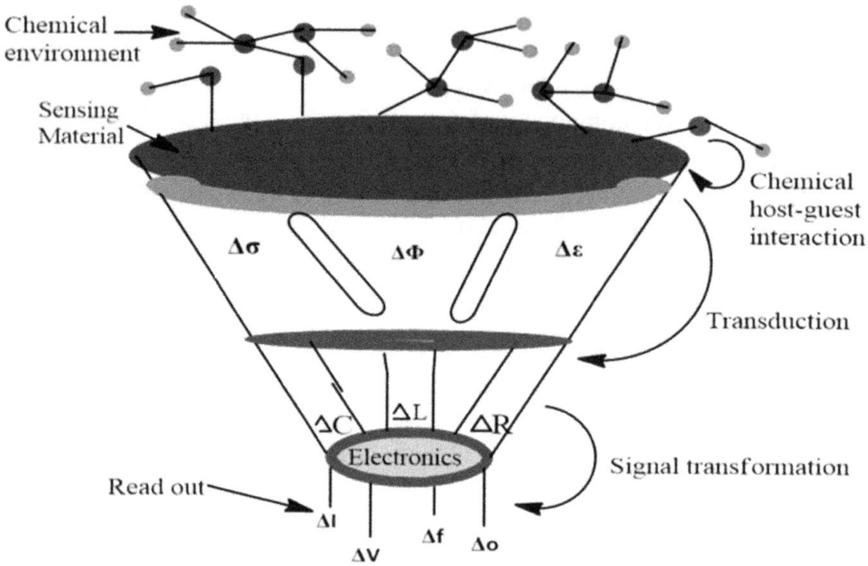

now identify many substances in forensics, comprising narcotics, poisons, gunshot wound residues, explosives, and more (Steinegger et al., 2020; Ge et al., 2020).

Numerous studies and the creation of new tools have focused on the employment of nanomaterials in forensics. The adaptability of these nanomaterials and the potential for detection in quantities that were not even envisaged may be credited for the rising number of research and patents. To, K. C. (2020) provides an overview of explosives used in trace detection. The employment of technology for the detection of explosives in trace levels serve a crucial role in holding on to nationwide security, according to To, K. C. (2020). The author of this paper emphasises the usage of semiconductors with nanostructures as chemical sensors and divides them into five categories depending on their chemical make-up: composites, plasmonic, carbon-based, semiconductors and quantum dots. The author uses p-type and n-type semiconductors including TiO_2, ZnO, SnO, and Cu_2O and highlights other research that link these materials' quick responsiveness to qualities like enhanced charge transfer. The forensic sciences discipline can benefit from the results that have been provided. However, the author cites research by Qu et al. (2016) that created a chemiresistive sensor to detect ammonium nitrate (AN), dinitrotoluene (DNT), trinitrotoluene (TNT), 2,4,6-trinitrophenol (TNP) and1,3,5-trinitroperhydro-1,3,5-triazine (RDX).

There are a lot of publications that mention using plasmonic materials, such gold (Au) (Squire et al., 2020) and silver (Ag) (Ben-Jaber et al., 2017) for explosives detection. Muehlethaler et al. (2016) claim that plasmonic materials have a significant number of active binding sites on their surface that have an

Figure 2. The application of chemical sensors in our everyday lives
Source: Meng et al. (2019)

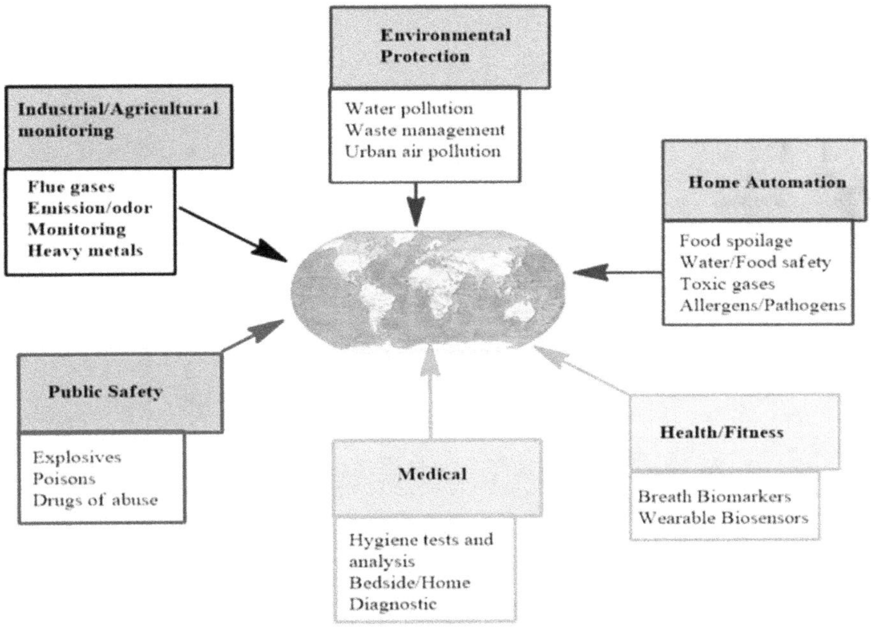

affinity for the chemical groups found in explosives. Explosives having a strong affinity for gold include the Nitroaromatics TNT or DNT. By forming connections with oxygen atoms, these nitrogen molecules attach to the surface of the metal to create a bidentate chelate complex.

It is feasible to comprehend the creation of a sensor that uses gold and silica in a core-shell construction and is used for the detection of vapour DNT in the work by Squire et al 2020. The substance found minute levels of the explosive, at 100 ppm. Additionally, the author claims that this detection happened within 3 minutes at ambient temperature. (Squire et al., 2020). These findings unmistakably demonstrate the possible use of these materials as vapour explosive sensors.

Globally, the use of illegal substances poses a serious danger to public health. Due to the terrible consequences of these chemicals, it is crucial to take steps to stop their sale, transit, and origin from becoming known. In this context, it is accurate to state that one of the primary focuses of investigators and forensic scientists is the investigation and detection of narcotics (Minhas et al., 2020; Zhao et al., 2021; Parrilla et al., 2021).

The scientific community has concentrated on creating sensors that are quick, on-site, easy to use, and have high throughput. Additionally, as observed in studies by Minhas et al. (2020), nanomaterials having a chemical affinity for these medications have been investigated. The Minhas (2020) have shown how to identify athletes who have used illegal performance-enhancing drugs. The samples were sent for examination using a mass spectrometer using biological matrices and the silicon nanostructured device compound (SALDI-MS). When the results were compared to those from other analytical techniques, they demonstrated superior performance. The phases for the creation of a novel device are schematically illustrated in Figure 3 (modified from Minhas et al., 2020).

Figure 3. Scheme of the testing procedure for athlete doping: the biological samples (saliva/urine) are obtained from an athlete and then spotted onto a nano-SI surface, tailored for doping detection
Source: Minhas et al. (2020)

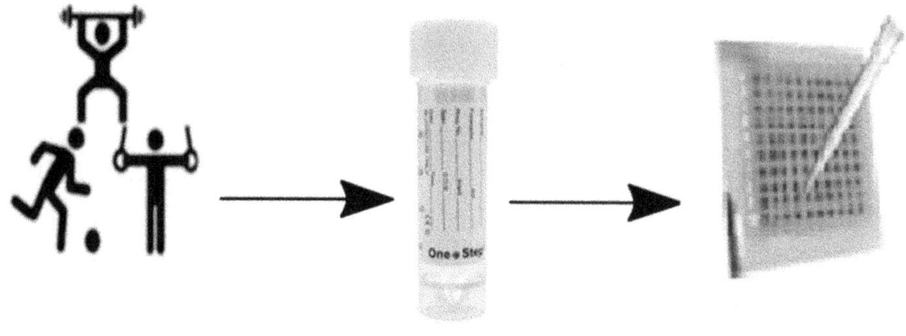

It is important to note that Akhoundian et al. (2019) and Parrilla et al. (2021) have also conducted research that report the identification of illegal narcotics utilising chemical sensors.

A technique for the detection of methamphetamine in trace levels was developed by the authors (Akhoundian et al., 2019) utilising a modified carbon paste electrode and a molecularly imprinted polymer (MIP)/multi-walled carbon nanotube (MWCNTs) electrode. A polymer precipitation method has been used to produce nanopolymers. The sensor has also been tested by human urine and blood. The results showed a linear response limit of $1.0 \times 10–8–1.0 \times 10–4$ mol L^{-1} and a detection limit of $8.3 \times 10–10$ mol L^{-1}. These results were achieved under experimental settings. The limit of detection was smallest ever recorded, according to the author. The creation of this sensor offers a critical tool for the rapid and precise detection of methamphetamine in human urine and blood. On the other hand, Parrilla et al. (2021) created a sensor that uses voltammetric oxidation of a derivatized chemically printed electrode on a graphite screen to identify amphetamines in samples that were confiscated by the police. The outcomes demonstrate the sensor's effectiveness and its capability to quickly and accurately measure amphetamine levels in 20 confiscated samples.

The studies discussed here examine the creation of novel chemical devices with certain qualities, such high-throughput, real-time analysis that are crucial attributes for sensors employed in the forensic field. The creation of chemical sensors to detect explosives and illegal narcotics is the subject of various investigations, as may be seen in this section. Many sensors include multimodal detection capabilities for chemical, biological, optical, and other types of signals. However, the correct classification of sensors is biosensors, which are broadly used when sensors represent combinations of aptamers, biomolecules, enzymes, antibodies, and other biological species.

Biosensors

A transducer, a detector, and biological sensing acceptors (that can be cells, tissues, enzymes, antibodies, proteins, nucleic acids or aptamers) make up biosensors (Figure 4). By attaching to biological receptors,

Figure 4. The schematic structure of biosensor with electrochemical transducer
Source: Selli et al. (2022)

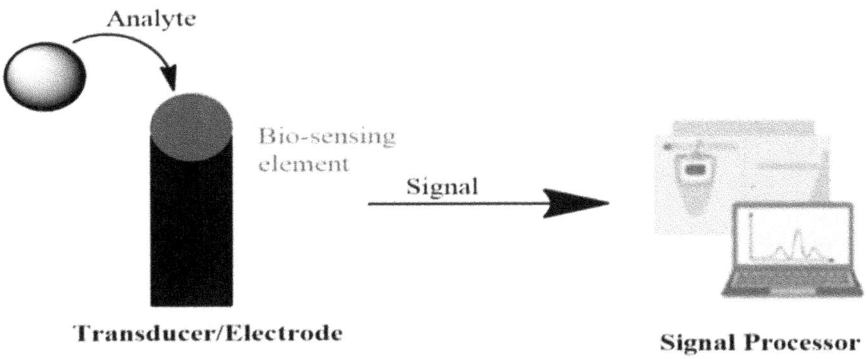

the target analyte generates a detectable signal that is proportional to the target concentration, under inquiry. The advantages of nanostructured biosensors include the ability to detect substances in extremely complex matrices like bodily fluids and samples specified to environment, and reliable analysis with a small quantity of sample (Mao et al., 2020).

It is feasible to quote several sources that discuss the creation and use of nano-biosensors as a potential tool for finding explosives, illegal narcotics, and bodily fluids at crime scenes, among other things (Fan et al., 2020; Xue et al., 2020; Adhikary & Banerjee, 2021). We will go a bit more into biosensors and their uses in this section.

The quantity of these samples at crime scenes and the fact that body fluids are complex matrices make them the subject of several types of research that concentrate on the production of nano-biosensors, such as the study created by Li et al. (2020) that employed human saliva.

It is feasible to quote several sources that discuss the creation and use of nano-biosensors as a potential tool for finding explosives, illegal narcotics, and bodily fluids at crime scenes, among other things (Fan et al., 2020; Xue et al., 2020; Adhikary and Banerjee, 2021). We will go a bit more into biosensors and their uses in this section.

The quantity of samples at crime areas and fact that body liquids are complex matrices make them the subject of several types of research that concentrate on the production of nano-biosensors, such as research done by Li et al. (2020) that employed human saliva. The programme used a smartphone camera to capture and evaluate this shift in hue. The concept behind the app is to gather colour changes and produce results based on the quantity of bacteria found in the saliva. According to Li et al. (2020), the methodology used in (Figure 5) are the processes for the instantaneous detection of *S. salivarius* and *S. sanguinis* in saliva is depend on a sensing system employing a cell phone.

Sweat is a fascinating bodily liquid which demonstrates your non-intrusive character. Sweat was employed in the research produced by Xue et al., (2020), to validate the functionality of the biosensor. The scientists created a sensor using a competitive enzyme-linked immunosorbent assay in conjunction with a microfluidic capillary array (ELISA protocol). With the use of this tool, illegal drugs may be quickly, accurately, and quantitatively detected. The findings showed the sensor's potential for locating

Figure 5. Scheme showing the working of simultaneous detection of S. salivarius and S. sanguini in saliva, applying the smartphone as sensing platform
Source: Li et al. (2020)

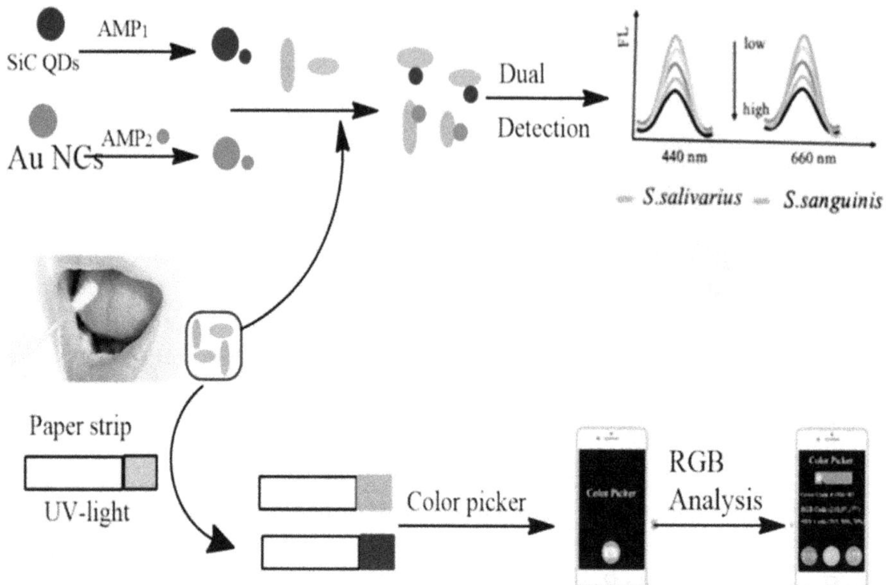

four commonly abused substances—methadone, methamphetamine, amphetamine, and tetrahydrocannabinol (THC)—in fake perspiration. The sensor, according to the author, had remarkable cross-reactivity specificity, with values less than 1%. Furthermore, all the intended outcomes were achieved in around 16 minutes using only 4 L of each medication. The sensor's estimated detection limits for methadone, methamphetamine, amphetamine, and THC were 1.6 pg mL-1, 142 pg mL-1, 35 pg mL-1, and 20 pg mL-1, respectively. The outstanding outcomes proved the method's accuracy because other approaches described in literature show a range of 100–1000 pg mL1 (Xue et al., 2020).

Professionals in border security find it difficult to identify illegal chemicals since these numerous substances have less vapour pressure and very low concentrations of vapour. Additionally, the groups conceal the chemicals in boxes with unusual scents in an effort to fool more precise detection methods. But one of the primary issues with the current detection technologies is the prevalence of numerous false-positive findings and the astronomically expensive maintenance and purchasing costs.

To make a biosensor array, Scorsone et al. (2021) selected several ligand binding proteins (LBPs) and incorporated them into nano-diamond transducers covered with acoustic surface waves. Due to their strong affinity for proteins, explosives and illegal narcotics have been detected using biosensor arrays. The findings show that response in linear way was seen over the ppb–ppm concentration range. The sensor could distinguish between various explosives and drugs by employing Principal Component Analysis. The created biosensor shown a stronger preference for Nitroaromatic chemicals over drugs of abuse. The author believes that bioelectronic noses are a crucial tool for resolving issues with the present detection methods.

As can be seen, biosensors employ aptamers, antibodies and molecularly printed polymers to analyse explosives with extraordinary specificity and sensitivity. A review by Liu et al (2019) discusses current technologies and potential directions for developing biosensors for explosives detection. The review makes references to a number of research, one of which employs aptamer composite sensors to detect TNT. According to Kong et al., work from 2017, a biosensor of chemiluminescent nature made of magnetic beads that contain Co^{2+} and the magnetic compound N-(4-aminobutyl)-N-ethyl isoluminol was produced. The coordination attractions and electrostatic attraction among the Co^{2+} and TNT aptamer held the TNT aptamer to the surface of composite. The compound's chemiluminescence intensity was reduced as a result of this interaction. The author hypothesises that the TNT-aptamer peptide's actions at the catalytic site of Co^{2+} may be responsible for this quenching effect by causing a decrease in intensity of fluorescence. Biosensor's detection limit was 17 pg/mL, and it could detect TNT between 0.05 and 25 ng/mL. A diagrammatic illustration of the biosensor and a comparison of emission curves of the compounds according to the synthesis process are shown in Figure 6 (Kong et al., 2017; Liu et al., 2019).

Figure 6. a) Scheme illustrating of the proposed label-free aptasensor for TNT. B) Comparison of kinetic curves in different moments of compound synthesis.
Source: Kong et al. (2017) and Liu et al. (2019)

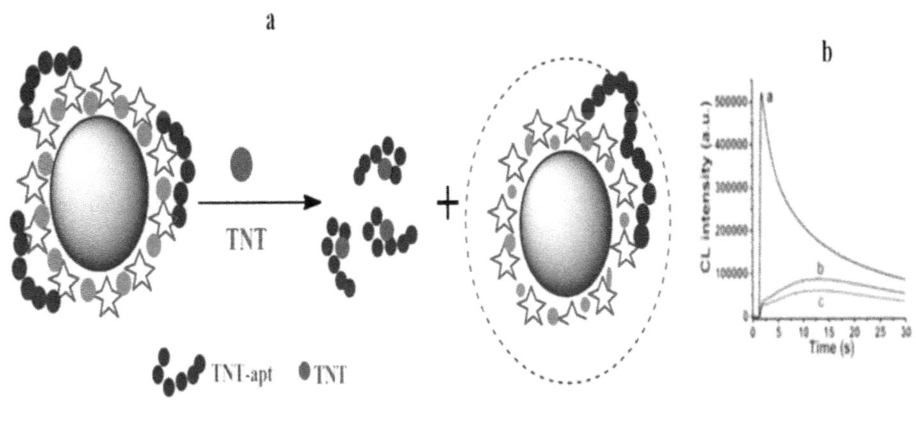

According to the author, there is currently no sensor on the market that combines all the necessary characteristics, such as a low detection limit, stand-off distance, good selectivity, and mobility, into a single unit. However, due to the rapid advancement of both technologies that enhance sensing and developments in the biological field, the future of these materials is quite bright (Liu et al., 2019).

This chapter illustrates the versatility and promise of biosensors. Additionally, we discuss the many traits of biosensors and the potential for improving the devices' performance by adjusting the traits. It is also clear that reaction kinetics, transduction processes, or mass transfer rates will influence how biosensors react.

Optical Sensing

Due to their sensitivity, quick detection, and simplicity of use, optical sensors have been crucial in forensic science (Majdinasab et al., 2019). Additionally, this sensor demonstrates a high degree of simplicity in between substances interactions and components having sensory qualities, delivering crucial information via simple and affordable methods. There are several techniques for detection and analysis employing optical sensors, including immune chromatographic tests, colorimetry, chemiluminescence, surface-enhanced Raman scattering (SERS) and fluorescence (ICAs). Similar to another sensor described in this chapter, it has a similar construction, but it also includes a transducer which takes signal and transforms it into various radiations (Khorablou et al., 2021).

According to Lantam et al. (2020), who created a sensor using the Sol-Gel approach to detect promethazine in lean cocktail and medicinal doses, optical sensors are often produced using basic synthesis techniques. The oxidation of promethazine by potassium persulfate within the sol-gel polymer network is considered as a mechanism for the function of the sensor. In aqueous medium, the reaction changes color from pink to red. Although the samples were examined using a handheld spectrometer, color changes were evident, allowing qualitative analysis to obtain more accurate and quantitative data. According to the author, the gadget was designed successfully because it was inexpensive, simple to use, and portable. The thresholds for quantification and detection were determined to be 16.5 and 48.9 mg L1, respectively. The gadget shown stability for 90 days, and the promethazine detection analysis had an accuracy of between 87 and 105%. The fact that this sensor enables both qualitative and quantitative examination of the target molecule is one of its key features (Lantam et al., 2020). Moreover, how the results were obtained, shown by Figure 7, which was modified from Lantam et al., (2020).

Figure 7. Schematic illustration of device development for promethazine detection using a portable spectrometer
Source: Lantam et al. (2020)

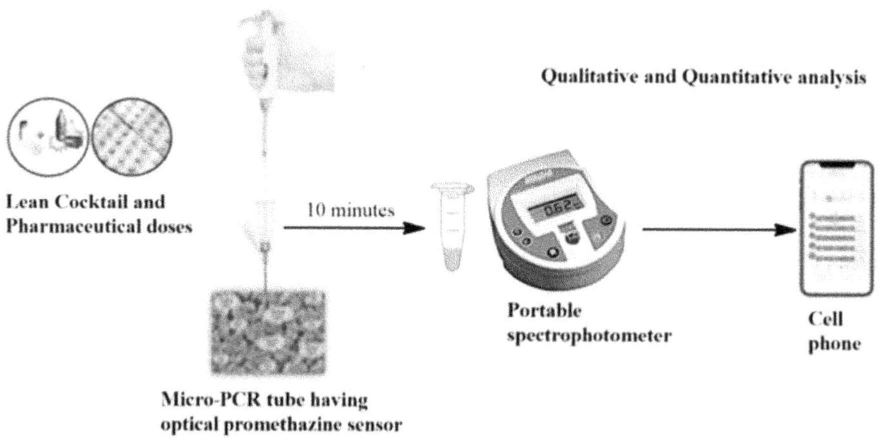

Surface Enhanced Raman Spectroscopy is another frequently used optical sensor-based analytical technique (SERS). As this chapter demonstrates, they are connected to many sensors, including chemical and biological ones. Raman spectroscopy is improved by materials with SERS capabilities. Their capacity to create a confined plasmonic surface, which increases the Raman signal by factors of 103 to 106, is what causes the amplification (Fikiet et al., 2018).

Various studies reporting on the use of SERS for the detection of illegal compounds, namely narcotics like methamphetamine, are reported in the literature (Weng et al., 2018; Hong et al., 2020. Mao et al., 2021) and explosives (Hakonen et al., 2018, Wu et al., 2020). In a review, Khorablou et al. (2021) highlight some developments in creation of the electrochemical and optical sensors for the detection of methamphetamine. The overview introduces us to the four primary techniques for detecting objects using optical sensors: fluorescence, colorimetry, SERS, and ICAs. Because of their non-destructive and easy sampling nature, sensors that exhibit SERS features have demonstrated excellent performance in the detection of illegal drugs, claim Khorablou et al. (2021).

The author Khorablou offers significant research on the characteristics of the materials engineering field, including those done by Hong et al. (2020) and Mao et al. (2021). Both studies offer unique perspectives on the development of sensors. Materials having both photonic and plasmonic characteristics make up optoplasmonic materials. Its primary purpose is to enhance plasmonic characteristics. Based on this, Hong et al. (2020) designed a study aimed at using self-assembled photoplasmonic structures as substrates to detect methamphetamine trace. The sensor was fabricated by self-assembly of gold nanoparticles monolayers and the deposition of dielectric material in form of microspheres. The purpose of the synthetic process is to enhance the electric (E-) field localization strength and relocate the analyte to area close to the electromagnetic hotspot. The author used a variety of calculations and observations with SERS spectroscopy to confirm the accuracy of the information. As can be shown in the study, utilising SiO_2 spheres with a 5 m diameter greatly improved the performance of SERS material. Additionally, Methamphetamine concentrations in the nanomolar range might be detected in bodily fluids (Hong et al., 2020).

However, Mao et al. (2021) constructed a new nanosensor for methamphetamine detection based on aligner-mediated cleavage (AMC). This process is depend on multiple dissociation sequences of aptamers attached to gold nanoparticles leading to aggregate formation and ultimately plasmonic coupling effects. The results obtained indicate that the electric field and SERS effect can be improved. Results showed a linear range of 10 pM to 10 nM and a limit of detection of 7 pM (Mao et al., 2021).

The author concludes by contrasting the benefits and drawbacks of electrochemical and optical sensors as well as highlighting several gaps which should be filled in the sensors development to detect methamphetamine. The optical sensors are often employed in many studies, however they have drawbacks such sensitivity to interference from the environment and complexity depending on the approach. However, it offers benefits such being portable and inert (Khorablou et al., 2021). Applications of nano-sensors in forensic investigation

Forensic Toxicological Analysis

Forensic toxicology uses nanotechnology to examine various hazardous substances from a variety of significant forensic evidences, comprising blood, saliva, hair, vitreous humour, samples of fingerprint and even skeletal remains. According to Redasani VK and colleagues (2015), gold, silver, and TiO_2 nanoparticles are frequently utilized to increase the detection limit (DL). The vitreous humour of the body that contains Lidocaine hydrochloride. The low-cost, active, stable, and time-limited nanosensor produced with these novel methodologies may be employed as an instant spot investigation and a significant replacement for the lab testing processes for the screening of toxicological drug. Forensic nanotechnology has been effectively applied to real samples, shedding light on the applicability of nanosensors for toxicological analysis (Nikolaou et al., 2014; Srividya, 2016; Dahiya et al., 2022).

Forensic Explosive Detection

Globally, terrorist actions have escalated, which has boosted interest in finding concealed bombs. Explosives comprising grenades, bombs and improvised explosive devices (IEDs) are wreaking havoc on a large scale (Rawtani et al., 2019). Different sensing tools have been created during the past year to find explosive traces. Nanomaterials provide the active potential to develop explosives detectors. Nanomaterial is frequently employed to create inexpensive sensors due to its distinctive electrical and optical capabilities. systematic and effective finding of explosives concealed in baggage, cars, and other objects. Explosives tracing is an extremely expensive and challenging process (El-Hussein, 2016). Explosives and other various chemical and biological components are detected using nanostructures as sensors. Tiny gadgets have a lot of sensing potential. Advanced nano-sensor concept devices like nanotubes, nano-mechanical devices and electronic noses are used to detect explosives (Sahu et al. 2018). Biologically based sensors are typically used to detect explosives (Smith et al., 2008). Having a small detection limit as 0.09g/ml, immunosensing approaches offer excellent sensitivity in detecting TNT (Shankaran et al., 2005).

Using an anti-TNT antibody, scientists created a capillary immunosensor for the precise detection of TNT (Charles et al., 2018). The nanomaterials based on carbon like carbon nanoparticles, graphene and carbon nanotubes, have unique electrical characteristics that make them suitable for use as sensors. They are also chemically inert, have low cytotoxicity, and are highly biocompatible. In order to identify nitroaromatic explosives, Chen et al. employed graphene oxide, an oxidised derivative of graphene. Dinitrobenzene (DNB), trinitrobenzene (TNB) and dinitro toluene (DNT) were among the substances they examined using the graphene-based sensor they built (Chen et al., 2011).

Different sensing tools have been created during the past year to find explosive traces. Nanomaterials are frequently employed to create low-cost sensors because to their distinctive electrical and optical capabilities. Biologically based sensors are typically used to detect explosives (Langford & Lurie, 2022). With a less detection limit as 0.09g/ml, immunosensing approaches offer excellent sensitivity in detecting TNT (Rivers & Dahlem, 2022). Due to their chemical inertness, minimal cytotoxicity, high biocompatibility, and distinctive electrical characteristics, carbon-based nanostructures like carbon nanoparticles, carbon and graphene nanotubes are being used as sensors (Zala & Patel, 2022). Canines can detect a variety of smells, including explosives. However, training dogs to find concealed explosives is now too expensive and time-consuming. As a result, dogs may now readily and harmless trace using the electronic nose approach. A network of chemical sensing, like a synthetic neural system, is often present in an electronic nose (Sahu et al., 2016; Ptaszynska-Sarosiek et al., 2016).

FORENSIC DNA ANALYSIS

The civil and criminal judicial systems stand to gain much from DNA analysis (Oosthuizen & Howes, 2022). In forensic investigations, it establishes a person's identify. DNA analysis is now influenced by nanoparticle-based techniques because of its affordability, simplicity of automation, and ease of usage. Because of their improved sensitivity and high DNA yield, magnetic nanoparticles are specifically utilized to extract DNA. Ma et al. extracted nucleic acid from four distinct sources, including bacteria, yeast, human blood, and viruses, using Fe3O4 nanoparticles. The outcomes demonstrated that magnetic nanoparticle-based nucleic acid extraction produced high yields and reasonably pure nucleic acids (Ma et al., 2013).

Applications for the sensor-based DNA detection technology exist in DNA analysis. Due to its optothermal characteristic, gold nanoparticles are mostly employed in sensing systems. Au nanorods were employed by Cheong, et al. (2008) to harvest DNA from cells. Researchers used the optothermal property to convert near-infrared radiation into thermal energy in a microfluidic chip, which caused pathogen lysis and ultimately led to the extraction of DNA (Cheong et al., 2008). A silica nanoparticle-based test also detects DNA with a remarkable sensitivity of 10pM, in addition to gold nanoparticles (Wang and Liu, 2007). Due to its precision and speed, the NanoPCR, a nanoparticle aided PCR, is receiving a lot of interest. The PCR method incorporates several types of nanoparticles, comprising metal nanoparticles, quantum dots and carbon tubes. The effectiveness and specificity of PCR products are being enhanced by these nanoparticles (Ma et al., 2009). The only quantum dot-based nanosensor for the detection of multiple DNA was studied by Zhang and Hu (2010). In this single QD-based nanosensor, the QDs, in addition to the fluorescence-coupled function and fluorescence-resonance-energy-transfer (FRET) donor for coincidence detection, are also localized nanoconcentrators that greatly enhance coincidence-related fluorescence signals and FRET signals. It acts as a localized nanoconcentrator for FRET detection. This single QD-based nanosensor can be used for quick point-of-care screening, gene expression analysis, and other applications using a simple "mix-detect" assays with exceptionally low sample consumption, high sensitivity, and. Short assay times for point-of-use testing, gene expression studies, clinical diagnosis and high-throughput screening (Zhang & Hu, 2010).

Forensic Fingerprint Imagining

Finger markings are the remnants of the finger's ruptured ridges. These fingerprints make it possible to recognise a person. Typically, latent, patent and plastic fingerprints of three categories are seen at crime scenes. Most often, latent fingerprints—those which can't be visualize with human eye or that need additional processing—are discovered at crime scenes. Latent fingermarks are often developed using fingerprint powders (Santos-Oliveira, 2014). Usually, at the crime scene, fingers impart perspiration traces on the objects. Latent prints are made to be colored or bright for simple identification by scattering finger print powder on the region where it would attach to the sweat remains left by the finger. This will provide predictable patterns. The finest materials which have traditionally been utilized in latent fingerprint formation are carbon black and aluminium flake (Menaa, 2014).

Since silver nanoparticles were introduced in 1970, latent fingerprints have been created using nanoparticles. Nanoparticles' small size makes it possible for them to bond with a fingerprint's tiny ridges, improving the ridge features on fingerprints and sweat pores. According to research (Prasad et al., 2020), silver nanoparticles can be employed as physical developers to see latent fingerprints on paper. The organic fingerprint component was converted into silver nanoparticles throughout the reaction, which left a dark grey or black sliver image of the fingerprint on the porous surface (Sodhi & Kaur, 2016). Due to their inertness, excellent selectivity, and sensitivity, gold nanoparticles are useful for latent fingerprint analysis. The benefit is that latent prints made by gold nanoparticles may be preserved for a longer period of time. Due to these characteristics, gold nanoparticles are utilized in multi-metal deposition (MMD) and single metal deposition to increase the visibility of latent fingerprints (Gao et al., 2009; Aggarwal & Chitkara, 2022).

Various nanopowders have been used in forensic investigation procedures to expose latent fingerprints on various surfaces. To improve fingerprint recognition, photoluminescent CdS semiconductor nanocrystals were pre-used with deuctisulfo-succinate. Scientist were capable to take account of amino acids on the non-porous surfaces using gold nanoparticles that would improve latent fingerprint analysis (Vinoda et al., 2015). A team of scientists (Arshad et al., 2015) has created unique ZnO-SiO_2 nano-powder utilizing the time-tested heating technique. Fingerprints were generated using a powder spray and small particle reagent (SPR) technique. This method has been effectively employed to detect latent fingerprints on wet (non-porous) and several dry (non-porous and semi-porous) surfaces. The results displayed that, compared to other commercially available white powders, ZnO-SiO_2 nano-powder has exceptional capability for seeing finger ridge feature at greater levels and exhibits improved discernibility.

Due to their tiny size and exceptional fluorescence intensity, quantum dots and fluorescent materials have attracted a lot of attention. Quantum dots may visualize, according to a study by Dr. Roland Menzel (Menzel, 2001). Bloody fingerprints may also be created using quantum dots. Bloody fingerprints are more likely to spread and be contaminated, which can ruin the ridge details. Quantum dots, which can analyse bleeding prints thanks to their fluorescence feature, were able to defeat it (Zhang & Peng 2022). According to the study's findings (Arshad et al., 2015), adding a little number of ZnO-SiO2 nanoparticles to powder improves the visual enhancement of latent print to third-level ridge features of prints. For the visibility of prints on multicolored or patterned backgrounds, carbon nano powder has been produced.

Nano-Fingerprint Residue Visualization: Microscopy: While perspiration may be identified due to its inorganic components, body creams and sunscreen lotions can be discovered utilizing leftover Titanium dioxide or Zinc oxide nanoparticles (Bhatia, 2022).

Forensic NanoTrackers

Nano trackers are covert patterns incorporated into different products, such as mobile phones, watches, pen drives, and other items, to deter theft. Criminals are having nano trackers injected into their bodies so that, even if they manage to escape from jail, they might still be apprehended using this tracking device. After being released from prison or correctional facilities, criminals may exhibit deviant behavior that has to be watched since they don't always adhere to society norms. This is made feasible by giving them injections of nano trackers, which make the arduous task of monitoring these recently freed criminals much easier. If they commit crimes once again, serial offenders can be located (Pandya et al., 2018). With peptide-based taggants, the items can also be physically marked. These can be applied to an item's surface, and when the item is retrieved, swabs from such taggants exhibit a colour or fluorescence change upon examination. Additionally, in accordance with the Locard's exchange concept, these taggants may also be retrieved from the offender responsible for the offence connected to the tagged item (Gooch et al., 2021).

Along with Nano trackers, barcodes are useful in preventative forensics. It is possible to identify the authenticity of a product using these unique patterns drawn on products, which helps prevent the forging or copying of such goods. The duplicate, forgery, or counterfeiting of any item casts doubt on the validity of a certain product produced by a firm, resulting in socioeconomic losses and harm to those companies' reputations. When determining the authenticity of a product, barcodes and watermarks are dependable indicators that cannot be imitated on counterfeit goods. Nanotechnology has also developed additional novel methods for product identification and verification, including nanofibers and nanodots with various colours. Different forensic light sources, including as UV, IR, and others, can be used to analyse the Nano fibres and Nano dots in order to identify and understand them (Pandya & Shukla, 2018).

Forensics in Illicit Drug Analysis

Drugs that are illegal are becoming in demand. Although new psychoactive compounds have been appearing on the market, the most common illegal drugs remain cannabis, amphetamine-like substances, cocaine, and heroin. The analysis of drug analysis is a significant area of contemporary analytical chemistry in criminal investigations, having numerous implications for law and society. Sensing devices are being utilised to identify drugs due to technological advancements (Lad et al., 2016). Colorimetric, fluorescence, and electrochemical sensors are frequently used in nanoparticle sensing (Chen et al., 2020; Majeed et al., 2020). Gandhi, et al. (2018) described the creation of a dipstick test for the detection of morphine based on the AuNP labelled single-chain fragment variable (scFv) antibody. The created dipstick may be used to detect morphine in a variety of bodily fluids, including blood, urine, and saliva. (Gandhi, et al. 2018). Surface Plasmon Resonance is significantly present in Au nanoparticles. In colorimetric sensors, it is employed. To identify cocaine, Gao et al. established a colorimetric sensor based on aptamer and MoS_2-gold nanoparticles (AuNPs). It has been discovered that this sensor is quick, economical, and extremely sensitive (Gao et al., 2020). Because of their superior quantum yield and fluorescence, several quantum dots approaches are also being exploited for fluorescence-based sensing (Farahani et al., 2018).

Forensic Nano Sprayers and "Lab-on-a-Chip" Device

The preceding literature is also concerning one of most significant aspects of nanotechnology use, namely nano sprayers and "lab-on-a-chip" devices. Nano sprayers are used to discover even the tiniest scraps of evidence at crime scenes, and the lab-on-a-chip gadget can do a number of analyses and tests on a remarkably small sample size (Giannoukos et al. 2016). In that regard, it is projected that trace evidences would shrink, maybe down to the size of a single cell. In forensic research, it would be feasible to identify and analyze a single DNA molecule with the use of nanotechnology (Kong et al., 2017; Cornelis et al., 2019).

The present chemical labs would be replaced by a range of portable instruments and techniques as part of the new analytical tools. It could be able to do complex chemical and biological tests right at the scene of the crime, which would hasten the investigation. Because lab-on-a-chip devices are easy to use and don't require specialized training, they make lab data interpretation even simpler, faster, and more accurate than it already is (2021) (Singh & Samal, 2016). The public will likely be able to access the aforementioned technologies, which can be very beneficial for including a larger no of people in forensic research as well as in general crime prevention, the revealed nanotechnology functions as a forensic tool in and of itself as well as enabling the creation of extra functionality (such as smart devices), establishing a synergy in the convergence of technology. Another notable advantage of nanotechnology in forensics that has been discussed in the literature is the integration of present laboratory techniques into a single device that may resolve laboratory contamination concerns (Lad et al., 2016). All of these nanotools have great significance when taken into account in light of their outstanding advantages of selectivity as well as sensitivity. It derives from careful choice of the molecular receptors in a nanosensor, which function as templates for the object to be detected (Amreen & Goel, 2022).

Cause of Death Investigation

Results of postmortem microbiological tests need to be confirmed using many techniques, and several samples need to be collected. In order to determine bacteremia, it is helpful to collect the samples of blood from a minimum of two separate areas of the body while maintaining sterility, as in cultures of antemortem blood (Sunagawa and Sugitani 2017). Thus, we came to know that a the only culture used following suitable circumstances supposed to be advantageous while accompanied by findings of antemortem that could assist in identifying sepsis (Tajiri et al., 2012; Tang et al., 2013).

During autopsies, histological, toxicological, and microbiological samples are collected for postmortem examinations. When it comes to determining the cause of death from infectious illnesses, forensic microbiology is most helpful. Even common microbial cultures, particularly those from places that though to be sterile in lifecycle and are discovered to maintain sterility for a while within body during the process of postmortem, supposed to be indicative in finding reason for death and occasionally may even be able to identify the way of death (Thorn et al., 2010). Although it is well considered for cultures of microbes are very helpful in determining the reason for death when combined with other postmortem procedures, microbial diagnosis has become increasingly important for analysis of microorganisms (Ventura Spagnolo et al., 2019). Even when there is no infectious cause of death, microbiological examinations are crucial for the exclusion of other diagnoses, the assessment of underlying causes that are not fatal, and the examination of eliminations like sudden infant death syndrome (Saukko and Knight 2016b).

Unfortunately, the profession has not yet witnessed significant advancements in microbial biotechnology over the past several decades or how these advancements have been reflected in contemporary forensic microbiological investigations. Few doctors who do forensic autopsy employ microbiological studies that go beyond standard microbial procedures, according to a survey research (Saegeman et al., 2017). The majority of people with infectious illnesses receive medical care just before they pass away, and most of these deaths take place in hospitals. In postmortem investigations, forensic microbiological testing is required when infectious signs were present throughout the ante-mortem period (Christoffersen, 2015). Cultures, acute phase reactants, enzyme linked immunosorbent assays and specific molecular microbiological tests are performed during the ante-mortem period whether syndromes are suspected prior to death in health care centers, and these tests are crucial in ensuring the accuracy of postmortem diagnoses (Tang et al., 2013).

It's critical to consider the significance of the positive microbiological test results when finding the reason of demise. According to Morris et al., a positive microbial could signify antemortem positivity sparking a factual incursion and/or colonisation (1), infection occurring while following the procedure of postmortem sampling (2), postmortem microbiological translocation, which happens as a result of microorganisms invading sterile sections of the body by interrupted biological defenses after life end (3) finally agonal propagation of microbes likely triggered by mucosal ischemic (Christoffersen, 2015; Morris et al., 2007; Yıldırım et al., 2015).

Additionally classifying reasons of positive microbiological test findings, Palmiere et al. (2016). They essentially categorized all optimistic results into four categories: false positive, true deadly infection, and non-fatal infection (Palmiere et al., 2016). Within this categorization, it was found that contamination, postmortem translocation, or both may have contributed to false positive. It is also interesting that only 35% of the favorable outcomes (Palmiere et al., 2016) were attributable to infectious illnesses that resulted in mortality. In different research (Christoffersen, 2015), 42 instances of varying ages and reasons of death were examined, above of 40% of the cases revealed that microbial biological testing had a substantial positive impact on preventing the cause of death from being discovered.

Results of postmortem microbiological tests need to be confirmed using many techniques, and several samples need to be collected. In order to determine bacteremia, it is helpful to collect blood samples from at least two separate areas of the body while maintaining sterility (Sunagawa and Sugitani, 2017). It should be remembered that molecular approaches may occasionally be the only way to support pinpoint the death reasoning, particularly in cases of diseases brought on by microbs where traditional ways not produce sufficient outcomes (Ziyade et al., 2015).

Sudden infant death syndrome (SIDS) is another form of mortality whose identification depends heavily on microbiological tests. This entity typically manifests as a baby that is discovered dead in his or her cot in the morning in infants younger than two years of age. All other illnesses and injuries that might result in death must be ruled out before a SIDS diagnosis can be made. It is recognized that a list of infectious microbs, including bacterial and viral species, have the potential to result in sudden unexpected infant death (SUDI) (Carter et al. 2017; Saukko and Knight 2016a). So, biological surveys should be preferred to examine SUDI or SIDS.

Postmortem Interval Investigation

A marker is supposed to be simple to perceive, exhibit a clear shape and single way variation over the time period, and this variation must be impacted in the fewest causes, or the parameters encountered may be noticed, in order to be used in postmortem interval estimation. According to postmortem microbiotechnological approaches, microbial activity is hampered by a variety of parameters, including fever, healthiness state during lifetime as well as initial elimination of infectious fences in open injured dead body (Yıldırım et al., 2015). The literature is still suggesting a variety of fresh models, some of which may be helpful in minimizing the postmortem period.

The variations considered to calculate the time interval for postmortem take place in various bodily sections at various periods and in various ways (Chun et al., 2015). With the progress in microbial bioengineering, effective means have recently been recognized that consider more in-depth analysis of biodiversity procedures of sampling from both the internal cavities of the corpses as well as surface (Hauther et al., 2015; Metcalf et al., 2013). Previously, postmortem interval estimation methods relied on microbial alterations and culturable bacterial colonisation detection using traditional techniques. It is now understood that both the presence of microorganisms in human fora and the rate at which they are colonizing the body from the outside vary concerning some body parts and the time of day and these is responsible for the postmortem time duration to be restricted just by examining microbial organisms (Pechal et al., 2018).

According to Saukko and Knight (2016b), the post-mortem toxicological investigation of the corpse's infected maggots is proven to be helpful. They contribute a significant part for the estimation of the postmortem interval through entomological studies. Using the shift of diversity of pathogens from *Fir-*

micutes to *Proteobacteria* discovered by micro-biotechnological investigations using samples of swab acquired from masses of maggot devouring the body to restrict postmortem time is another valuable study with the maggots in remote years (Chun et al., 2015; He et al., 2019).

By doing a microbiological analysis of the soil surrounding the body, the postmortem delay may be estimated. At initial steps of the decaying process, samples of soil collected in surroundings of the dead body failed to show any noteworthy variations, but in the later stages, Firmicutes propagation seen in the mud present in vicinity of the dead body, which could be helpful for estimating the period following death (Adserias-Garriga et al. 2017).

CONCLUSION

The promise of nanotechnology is helping forensic science to solve crimes in a useful way. Forensic science has benefited greatly from nanotechnology. It is particularly beneficial in unearthing concealed evidence that cannot be found without its help. For the identification of distinct forensic materials, several kinds of nanoparticles are utilized. Due of their extreme sensitivity, nanosensors are used in nanoforensics. Furthermore, the research procedure is now quick thanks to nano-forensics. The purpose of this chapter was to showcase several forensics science applications for sensors and trackers. Although there are several materials under development and numerous qualities that need to be enhanced, this field has a promising future. After reading this chapter, it should be clear that combining materials with various qualities is crucial to create a portable detector with the dependability required for forensic investigation (chemical, physical and biological). It is crucial to remember that science is progressing quickly, and every day we learn about new technologies. As a result, it is essential for forensic investigators to stay current with these developments. Future advancements in nanotechnology might make it useful as a preventative tool in several forensic science fields.

REFERENCES

Adhikary, R. R., & Banerjee, R. (2021). Development of smart core-shell nanoparticle-based sensors for the point-of-care detection of alpha amylase in diagnostics and forensics. *Biosensors & Bioelectronics*, *184*, 113244. doi:10.1016/j.bios.2021.113244 PMID:33934052

Adserias-Garriga, J., Hernández, M., Quijada, N. M., Lázaro, D. R., Steadman, D., & Garcia-Gil, J. (2017). Daily thanatomicrobiome changes in soil as an approach of postmortem interval estimation: An ecological perspective. *Forensic Science International*, *278*, 388–395. doi:10.1016/j.forsciint.2017.07.017 PMID:28818754

Aggarwal, P., & Chitkara, M. (2022). Detection of Metabolites in Latent Fingerprints Through Green Nanoparticles for Both Forensic Investigations and Medical Purposes. *ECS Transactions*, *107*(1), 10593–10602. doi:10.1149/10701.10593ecst

Akhoundian, M., Alizadeh, T., Ganjali, M. R., & Norouzi, P. (2019). Ultra-trace detection of methamphetamine in biological samples using FFT-square wave voltammetry and nano-sized imprinted polymer/MWCNTs-modified electrode. *Talanta*, *200*, 115–123. doi:10.1016/j.talanta.2019.02.027 PMID:31036164

Amreen, K., & Goel, S. (2022). Microfluidic Devices with Integrated Conductive Polymeric Electrodes for Biosensing Applications. In Conducting Polymers (pp. 273-286). CRC Press. doi:10.1201/9781003205418-19

Arif, T., Nisa, N., Amin, S. S., Shoib, S., & Mushtaq, R. (2015). Therapeutic and Diagnostic Applications of Nanotechnology in Dermatology and Cosmetics. *Journal of Nanomedicine & Biotherapeutic Discovery*, *5*, 134.

Arshad, A., Farrukh, M. A., Ali, S., Khaleeq-ur-Rahman, M., & Tahir, M. A. (2015). Development of latent fingermarks on various surfaces using ZnO-SiO2 nanopowder. *Journal of Forensic Sciences*, *60*(5), 1182–1187. doi:10.1111/1556-4029.12890 PMID:26348223

Arshad, A., Farrukh, M. A., Ali, S., Khaleeq-ur-Rahman, M., & Tahir, M. A. (2015). Development of latent fingermarks on various surfaces using ZnO-SiO2 nanopowder. *Journal of Forensic Sciences*, *60*(5), 1182–1187. doi:10.1111/1556-4029.12890 PMID:26348223

Bahadır, E. B., & Sezgintürk, M. K. (2015). Electrochemical biosensors for hormone analyses. *Biosensors & Bioelectronics*, *68*, 62–71. doi:10.1016/j.bios.2014.12.054 PMID:25558874

Ben-Jaber, S., Peveler, W. J., Quesada-Cabrera, R., Sol, C. W., Papakonstantinou, I., & Parkin, I. P. (2017). Sensitive and specific detection of explosives in solution and vapour by surface-enhanced Raman spectroscopy on silver nanocubes. *Nanoscale*, *9*(42), 16459–16466. doi:10.1039/C7NR05057G PMID:29063930

Bhatia, T. (2022). Novel nanomaterials in forensic investigations: A review. *Materials Today: Proceedings*, *50*, 1071–1079. doi:10.1016/j.matpr.2021.07.466

Boumba, V. A., & Vougiouklakis, T. (2015). Impact of blood collection tubes on erroneous 1-propanol detection and on forensic ethanol analysis. *Journal of Forensic Toxicology and Pharmacology*, *4*(1). Advance online publication. doi:10.4172/2325-9841.1000134

Carter, D. O., Tomberlin, J. K., Benbow, M. E., & Metcalf, J. L. (2017). Forensic microbiology. *Forensic Microbiol*, *36*(7), 1–391.

Charles, P. T., Wadhwa, V., Kouyate, A., Mesa-Donado, K. J., Adams, A. A., Deschamps, J. R., & Kusterbeck, A. W. (2018). A High Aspect Ratio Bifurcated 128-Microchannel Microfluidic Device for Environmental Monitoring of Explosives. *Sensors (Basel)*, *18*(5), 1568. doi:10.339018051568 PMID:29762499

Chávez, D., Garcia, C. R., Oliva, J., & Diaz-Torres, L. A. (2021). A review of phosphorescent and fluorescent phosphors for fingerprint detection. *Ceramics International*, *47*(1), 10–41. doi:10.1016/j.ceramint.2020.08.259

Chen, C. H., Wang, C. C., Ko, P. Y., & Chen, Y. L. (2020). Nanomaterial-based adsorbents and optical sensors for illicit drug analysis. *Yao Wu Shi Pin Fen Xi*, *28*(4), 654. doi:10.38212/2224-6614.1137 PMID:35696141

Chen, T. W., Sheng, Z. H., Wang, K., Wang, F. B., & Xia, X. H. (2011). Determination of explosives using electrochemically reduced graphene. *Chemistry, an Asian Journal*, *6*(5), 1210–1216. doi:10.1002/asia.201000836 PMID:21387564

Cheong, K. H., Yi, D. K., Lee, J. G., Park, J. M., Kim, M. J., Edel, J. B., & Ko, C. (2008). Gold nanoparticles for one step DNA extraction and real-time PCR of pathogens in a single chamber. *Lab on a Chip*, *8*(5), 810–813. doi:10.1039/b717382b PMID:18432353

Christoffersen, S. (2015). The importance of microbiological testing for establishing cause of death in 42 forensic autopsies. *Forensic Science International*, *250*, 27–32. doi:10.1016/j.forsciint.2015.02.020 PMID:25769131

Chun, L. P., Miguel, M. J., Junkins, E. N., Forbes, S. L., & Carter, D. O. (2015). An initial investigation into the ecology of culturable aerobic postmortem bacteria. *Science & Justice*, *55*(6), 394–401. doi:10.1016/j.scijus.2015.07.003 PMID:26654073

Cornelis, S., Tytgat, O., Fauvart, M., Gansemans, Y., Vander Plaetsen, A. S., Wiederkehr, R. S., & Stakenborg, T. (2019). Silicon µPCR chip for forensic STR profiling with hybeacon probe melting curves. *Scientific Reports*, *9*(1), 1–9. doi:10.103841598-019-43946-5 PMID:31089203

Dahiya, K., Sharma, H., Biswas, L., & Verma, A. K. (2022). *Nanotechnology in Forensic Science: Extensive Applications and New Perspective*. Academic Press.

Dhanyamraju, P., Sharma, A., Dovat, S., Amin, S., & Dhimant, D. (2020). Nanotherapeutics: A mini review. *Modern Approach Drug Design*, *2*(5), 1–3.

El-Hussein, A. (2016). Study DNA damage after photodynamic therapy using silver nanoparticles with A549 cell line. *Journal of Nanomedicine & Nanotechnology*, *7*(1), 1000346.

Fadel, T. R., Farrell, D. F., Friedersdorf, L. E., Griep, M. H., Hoover, M. D., Meador, M. A., & Meyyappan, M. (2016). Toward the responsible development and commercialization of sensor nanotechnologies. *ACS Sensors*, *1*(3), 207–216. doi:10.1021/acssensors.5b00279 PMID:28261665

Fan, M., Andrade, G. F., & Brolo, A. G. (2020). A review on recent advances in the applications of surface-enhanced Raman scattering in analytical chemistry. *Analytica Chimica Acta*, *1097*, 1–29. doi:10.1016/j.aca.2019.11.049 PMID:31910948

Farahani, M., Khademabbasi, K., & Mollatayefeh, N. (2018). A selective morphine nanosensor derived from functionalized CdS quantum dots. *Materials Letters*, *228*, 68–71. doi:10.1016/j.matlet.2018.05.120

Fikiet, M. A., Khandasammy, S. R., Mistek, E., Ahmed, Y., Halámková, L., Bueno, J., & Lednev, I. K. (2018). Surface enhanced Raman spectroscopy: A review of recent applications in forensic science. *Spectrochimica Acta. Part A: Molecular and Biomolecular Spectroscopy*, *197*, 255–260. doi:10.1016/j.saa.2018.02.046 PMID:29496406

Fikiet, M. A., Khandasammy, S. R., Mistek, E., Ahmed, Y., Halámková, L., Bueno, J., & Lednev, I. K. (2018). Surface enhanced Raman spectroscopy: A review of recent applications in forensic science. *Spectrochimica Acta. Part A: Molecular and Biomolecular Spectroscopy*, *197*, 255–260. doi:10.1016/j.saa.2018.02.046 PMID:29496406

Gandhi, S., Banga, I., Maurya, P. K., & Eremin, S. A. (2018). A gold nanoparticle-single-chain fragment variable antibody as an immunoprobe for rapid detection of morphine by dipstick. *RSC Advances*, *8*(3), 1511–1518. doi:10.1039/C7RA12810J PMID:35540925

Gao, D., Li, F., Song, J., Xu, X., Zhang, Q., & Niu, L. (2009). One step to detect the latent fingermarks with gold nanoparticles. *Talanta*, *80*(2), 479–483. doi:10.1016/j.talanta.2009.07.007 PMID:19836507

Gao, L., Xiang, W., Deng, Z., Shi, K., Wang, H., & Shi, H. (2020). Cocaine detection using aptamer and molybdenum disulfide-gold nanoparticle-based sensors. *Nanomedicine (London)*, *15*(04), 325–335. doi:10.2217/nnm-2019-0046 PMID:31976806

Ge, L., Li, S. P., & Lisak, G. (2020). Advanced sensing technologies of phenolic compounds for pharmaceutical and biomedical analysis. *Journal of Pharmaceutical and Biomedical Analysis*, *179*, 112913. doi:10.1016/j.jpba.2019.112913 PMID:31679844

Giannoukos, S., Brkic, B., Taylor, S., Marshall, A., & Verbeck, G. F. (2016). Chemical sniffing instrumentation for security applications. *Chemical Reviews*, *116*(14), 8146–8172. doi:10.1021/acs.chemrev.6b00065 PMID:27388215

Gooch, J., Goh, H., Daniel, B., Abbate, V., & Frascione, N. (2016). Monitoring criminal activity through invisible fluorescent "peptide coding" taggants. *Analytical Chemistry*, *88*(8), 4456–4460. doi:10.1021/acs.analchem.6b00263 PMID:27010696

Gopi, S., Amalraj, A., Haponiuk, J. T., & Thomas, S. (2016). Introduction of nanotechnology in herbal drugs and nutraceutical: A review. *Journal of Nanomedicine & Biotherapeutic Discovery*, *6*(2), 1–8. doi:10.4172/2155-983X.1000143

Hakonen, A., Wu, K., Schmidt, M. S., Andersson, P. O., Boisen, A., & Rindzevicius, T. (2018). Detecting forensic substances using commercially available SERS substrates and handheld Raman spectrometers. *Talanta*, *189*, 649–652. doi:10.1016/j.talanta.2018.07.009 PMID:30086972

Hauther, K. A., Cobaugh, K. L., Jantz, L. M., Sparer, T. E., & DeBruyn, J. M. (2015). Estimating time since death from postmortem human gut microbial communities. *Journal of Forensic Sciences*, *60*(5), 1234–1240. doi:10.1111/1556-4029.12828 PMID:26096156

He, J., Guo, J., Fu, X., & Cai, J. (2019). Potential use of high-throughput sequencing of bacterial communities for postmortem submersion interval estimation. *Brazilian Journal of Microbiology*, *50*(4), 999–1010. doi:10.100742770-019-00119-w PMID:31364013

He, Z., Zhang, Z., & Bi, S. (2020). Nanoparticles for organic electronics applications. *Materials Research Express*, *7*(1), 1–13. doi:10.1088/2053-1591/ab636f

Hong, Y., Zhou, X., Xu, B., Huang, Y., He, W., Wang, S., Wang, C., Zhou, G., Chen, Y., & Gong, T. (2020). Optoplasmonic hybrid materials for trace detection of methamphetamine in biological fluids through SERS. *ACS Applied Materials & Interfaces*, *12*(21), 24192–24200. doi:10.1021/acsami.0c00853 PMID:32351116

Kaushik, M., Mahendru, S., Chaudhary, S., & Kukreti, S. (2017). DNA fingerprints: Advances in their forensic analysis using nanotechnology. *Journal of Forensic Biomechanics*, *8*(1), 1–4. doi:10.4172/2090-2697.1000131

Khorablou, Z., Shahdost-Fard, F., Razmi, H., Yola, M. L., & Karimi-Maleh, H. (2021). Recent advances in developing optical and electrochemical sensors for analysis of methamphetamine: A review. *Chemosphere*, *278*, 130393. doi:10.1016/j.chemosphere.2021.130393 PMID:33823350

Kong, W., Zhao, X., Zhu, Q., Gao, L., & Cui, H. (2017). Highly chemiluminescent magnetic beads for label-free sensing of 2, 4, 6-trinitrotoluene. *Analytical Chemistry, 89*(13), 7145–7151. doi:10.1021/acs.analchem.7b01111 PMID:28551993

Kong, X., Squire, K., Chong, X., & Wang, A. X. (2017). Ultra-sensitive lab-on-a-chip detection of Sudan I in food using plasmonics-enhanced diatomaceous thin film. *Food Control, 79*, 258–265. doi:10.1016/j.foodcont.2017.04.007 PMID:29056826

Lad, A. N., Pandya, A., & Agrawal, Y. K. (2016). Overview of nano-enabled screening of drug-facilitated crime: A promising tool in forensic investigation. *Trends in Analytical Chemistry, 80*, 458–470. doi:10.1016/j.trac.2015.07.016

Langford, J. B., & Lurie, I. S. (2022). Use of micro, capillary, and nano liquid chromatography for forensic analysis. *Journal of Separation Science, 45*(1), 38–50. doi:10.1002/jssc.202100631 PMID:34626162

Lantam, A., Limbut, W., Thiagchanya, A., & Phonchai, A. (2020). A portable optical colorimetric sensor for the determination of promethazine in lean cocktail and pharmaceutical doses. *Microchemical Journal, 159*, 105519. doi:10.1016/j.microc.2020.105519

Li, X., Li, J., Ling, J., Wang, C., Ding, Y., Chang, Y., Li, N., Wang, Y., & Cai, J. (2020). A smartphone-based bacteria sensor for rapid and portable identification of forensic saliva sample. *Sensors and Actuators. B, Chemical, 320*, 128303. doi:10.1016/j.snb.2020.128303

Li, X., Li, J., Ling, J., Wang, C., Ding, Y., Chang, Y., Li, N., Wang, Y., & Cai, J. (2020). Environmental Application of Nanomaterials. *Sensors and Actuators. B, Chemical, 320*, 128303. doi:10.1016/j.snb.2020.128303

Liu, R., Li, Z., Huang, Z., Li, K., & Lv, Y. (2019). Biosensors for explosives: State of art and future trends. *Trends in Analytical Chemistry, 118*, 123–137. doi:10.1016/j.trac.2019.05.034

Lloyd-Hughes, H., Shiatis, A. E., Pabari, A., Mosahebi, A., & Seifalian, A. (2015). Current and future nanotechnology applications in the management of melanoma: A review. *Journal of Nanomedicine & Nanotechnology, 6*(6), 1. doi:10.4172/2157-7439.1000334

Ma, C., Li, C., Wang, F., Ma, N., Li, X., Li, Z., Deng, Y., Wang, Z., Xi, Z., Tang, Y., & He, N. (2013). Magnetic nanoparticles-based extraction and verification of nucleic acids from different sources. *Journal of Biomedical Nanotechnology, 9*(4), 703–709. doi:10.1166/jbn.2013.1566 PMID:23621032

Ma, L., He, S., Huang, J., Cao, L., Yang, F., & Li, L. (2009). Maximizing specificity and yield of PCR by the quantum dot itself rather than property of the quantum dot surface. *Biochimie, 91*(8), 969–973. doi:10.1016/j.biochi.2009.04.020 PMID:19442702

Mahbub, T., & Hoque, M. E. (2020). *Nanofabrication Smart Nanosensor Applications*. Elsevier.

Majdinasab, M., Mitsubayashi, K., & Marty, J. L. (2019). Optical and electrochemical sensors and biosensors for the detection of quinolones. *Trends in Biotechnology, 37*(8), 898–915. doi:10.1016/j.tibtech.2019.01.004 PMID:30777309

Majeed, M. I., Nawaz, H., & Arshad, F. N. (2020). Detecting the Presence of Illicit Drugs Using Biosensors. *Nanobiosensors: From Design to Applications*, 223-253.

Malefane, M. (2020). Applications of nanotechnology towards detection and treatment of HIV/AIDs: A review article. *Research & Development in Material Science, 12*(5), 1315–1321. doi:10.31031/RDMS.2020.12.000796

Mao, J., Kang, Y., Yu, D., & Zhou, J. (2021). Surface-enhanced Raman spectroscopy integrated with aligner mediated cleavage strategy for ultrasensitive and selective detection of methamphetamine. *Analytica Chimica Acta, 1146*, 124–130. doi:10.1016/j.aca.2020.12.028 PMID:33461707

Mao, J., Kang, Y., Yu, D., & Zhou, J. (2021). Surface-enhanced Raman spectroscopy integrated with aligner mediated cleavage strategy for ultrasensitive and selective detection of methamphetamine. *Analytica Chimica Acta, 1146*, 124–130. doi:10.1016/j.aca.2020.12.028 PMID:33461707

Mao, K., Zhang, H., Pan, Y., Zhang, K., Cao, H., Li, X., & Yang, Z. (2020). Nanomaterial-based aptamer sensors for analysis of illicit drugs and evaluation of drugs consumption for wastewater-based epidemiology. *Trends in Analytical Chemistry, 130*, 115975. doi:10.1016/j.trac.2020.115975 PMID:32834242

Mathur, S., Choudhary, B. S. K., & Vyas, C. J. M. (2016). Effect of disguise on fundamental frequency of voice. *Journal of Forensics Research, 7*(327), 2. doi:10.4172/2157-7145.1000327

Menaa, F. (2014). Global financial model for responsible research and development of the fast growing nanotechnology business. *Journal of Business & Financial Affairs, 3*(1), e139. doi:10.4172/2167-0234.1000e139

Meng, Z., Stolz, R. M., Mendecki, L., & Mirica, K. A. (2019). Electrically-transduced chemical sensors based on two-dimensional nanomaterials. *Chemical Reviews, 119*(1), 478–598. doi:10.1021/acs.chemrev.8b00311 PMID:30604969

Menzel, E. R. (2001). Fingerprint detection with photoluminescent. *Advances in fingerprint technology*, 456.

Metcalf, J. L., Wegener Parfrey, L., Gonzalez, A., Lauber, C. L., Knights, D., Ackermann, G., & Knight, R. (2013). A microbial clock provides an accurate estimate of the postmortem interval in a mouse model system. *elife, 2*, e01104.

Minhas, R. S., Rudd, D. A., Al Hmoud, H. Z., Guinan, T. M., Kirkbride, K. P., & Voelcker, N. H. (2020). Rapid detection of anabolic and narcotic doping agents in saliva and urine by means of nanostructured silicon SALDI mass spectrometry. *ACS Applied Materials & Interfaces, 12*(28), 31195–31204. doi:10.1021/acsami.0c07849 PMID:32551485

Montagner, C. C., Sodré, F. F., Acayaba, R. D., Vidal, C., Campestrini, I., Locatelli, M. A., & Jardim, W. F. (2019). Ten years-snapshot of the occurrence of emerging contaminants in drinking, surface and ground waters and wastewaters from São Paulo State, Brazil. *Journal of the Brazilian Chemical Society, 30*, 614–632.

Moram, S. S. B., Shaik, A. K., Byram, C., Hamad, S., & Soma, V. R. (2020). Instantaneous trace detection of nitro-explosives and mixtures with nanotextured silicon decorated with Ag–Au alloy nanoparticles using the SERS technique. *Analytica Chimica Acta, 1101*, 157–168. doi:10.1016/j.aca.2019.12.026 PMID:32029107

Morris, J. A., Harrison, L. M., & Partridge, S. M. (2007). Practical and theoretical aspects of postmortem bacteriology. *Current Diagnostic Pathology*, *13*(1), 65–74. doi:10.1016/j.cdip.2006.07.005

Muehlethaler, C., Leona, M., & Lombardi, J. R. (2016). Review of surface enhanced Raman scattering applications in forensic science. *Analytical Chemistry*, *88*(1), 152–169. doi:10.1021/acs.analchem.5b04131 PMID:26638887

Nikolaou, P., Papoutsis, I., Dona, A., Spiliopoulou, C., & Athanaselis, S. (2014). Beware of 25C-NBOMe: An N-benzyl substituted Phenethylamine. *Journal of Forensic Toxicology and Pharmacology*, *3*(4). Advance online publication. doi:10.4172/2325-9841.1000129

Oosthuizen, T., & Howes, L. M. (2022). The development of forensic DNA analysis: New debates on the issue of fundamental human rights. *Forensic Science International. Genetics*, *56*, 102606. doi:10.1016/j.fsigen.2021.102606 PMID:34710822

Pandya, A., & Shukla, R. (2018). New perspective of nanotechnology: Role in preventive forensic. *Egyptian Journal of Forensic Sciences*, *8*(1), 1–11. doi:10.118641935-018-0088-0

Paolesse, R., Nardis, S., Monti, D., Stefanelli, M., & Di Natale, C. (2017). Porphyrinoids for chemical sensor applications. *Chemical Reviews*, *117*(4), 2517–2583. doi:10.1021/acs.chemrev.6b00361 PMID:28222604

Parrilla, M., Montiel, N. F., Van Durme, F., & De Wael, K. (2021). Derivatization of amphetamine to allow its electrochemical detection in illicit drug seizures. *Sensors and Actuators. B, Chemical*, *337*, 129819. doi:10.1016/j.snb.2021.129819

Pechal, J. L., Schmidt, C. J., Jordan, H. R., & Benbow, M. E. (2018). A large-scale survey of the postmortem human microbiome, and its potential to provide insight into the living health condition. *Scientific Reports*, *8*(1), 1–15. doi:10.103841598-018-23989-w PMID:29636512

Prasad, V., Lukose, S., Agarwal, P., & Prasad, L. (2020). Role of nanomaterials for forensic investigation and latent fingerprinting—A review. *Journal of Forensic Sciences*, *65*(1), 26–36. doi:10.1111/1556-4029.14172 PMID:31454084

Ptaszynska-Sarosiek, I., Charuta, A., Furmanek, M. A., Wardaszka, Z., & Niemcunowicz-Janica, A. (2016). Mercury Poisoning - The Cause of Death Assessed 6 years Posthumously. *Journal of Forensic Toxicology and Pharmacology*, *5*(1), 1. doi:10.4172/2325-9841.1000143

Qu, J., Ge, Y., Zu, B., Li, Y., & Dou, X. (2016). Transition-Metal-Doped p-Type ZnO Nanoparticle-Based Sensory Array for Instant Discrimination of Explosive Vapors. *Small*, *12*(10), 1369–1377. doi:10.1002mll.201503131 PMID:26763156

Rawtani, D., Tharmavaram, M., Pandey, G., & Hussain, C. M. (2019). Functionalized nanomaterial for forensic sample analysis. *Trends in Analytical Chemistry*, *120*, 115661. doi:10.1016/j.trac.2019.115661

Redasani, V. K., Tamboli, P. S., Kalal, D., & Surana, S. J. (2016). Development and validation of spectroscopic methods for the estimation of mupirocin calcium in bulk and in ointment formulation. *Journal of Forensic Toxicology and Pharmacology*, *4*(2). Advance online publication. doi:10.4172/2325-9841.1000139

Rivers, D. B., & Dahlem, G. A. (2022). *The science of forensic entomology*. John Wiley & Sons.

Saegeman, V., Cohen, M. C., Alberola, J., Ziyade, N., Farina, C., Cornaglia, G., & Fernández-Rodríguez, A. (2017). How is post-mortem microbiology appraised by pathologists? Results from a practice survey conducted by ESGFOR. *European Journal of Clinical Microbiology & Infectious Diseases*, *36*(8), 1381–1385. doi:10.100710096-017-2943-6 PMID:28236029

Sahu, G., Nayak, S. R., & Sethi, S. S. (2016). Mutilation of Body after Crime. *Journal of Forensic Toxicology and Pharmacology*, *5*(1), 1. doi:10.4172/2325-9841.1000145

Santos-Oliveira, R. (2014). Pharmaceutical equivalence and bioequivalence of radiopharmaceuticals: Thinking the possibility of generic radiopharmaceuticals and preparing for new technology as nanotechnology drugs. *Journal of Bioequivalence & Bioavailability*, *6*(1). Advance online publication. doi:10.4172/jbb.1000175

Saukko, P., & Knight, B. (2016a). Sudden death in infancy. In P. Saukko & B. Knight (Eds.), *Knights's forensic pathology* (4th ed., pp. 461–473). CRC Press.

Saukko, P., & Knight, B. (2016b). The pathophysiology of death. In P. Saukko & B. Knight (Eds.), *Knights's forensic pathology* (4th ed., pp. 55–93). CRC Press.

Scorsone, E., Manai, R., Cali, K., Ricatti, M. J., Farno, S., Persaud, K., & Mucignat, C. (2021). Biosensor array based on ligand binding proteins for narcotics and explosives detection. *Sensors and Actuators. B, Chemical*, *334*, 129587. doi:10.1016/j.snb.2021.129587

Selli, G. I., Bonatto, A. E. T., Bonatto, F. T., Anzanello, M. J., & Bergmann, C. P. (2022). Nanosensors in Forensic Sciences. In *Environmental Applications of Nanomaterials* (pp. 239–253). Springer. doi:10.1007/978-3-030-86822-2_13

Shankaran, D. R., Gobi, K. V., Sakai, T., Matsumoto, K., Imato, T., Toko, K., & Miura, N. (2005). A novel surface plasmon resonance immunosensor for 2, 4, 6-trinitrotoluene (TNT) based on indirect competitive immunoreaction: A promising approach for on-site landmine detection. *IEEE Sensors Journal*, *5*(4), 616–621. doi:10.1109/JSEN.2005.848150

Shaw, L., & Dennany, L. (2017). Applications of electrochemical sensors: Forensic drug analysis. *Current Opinion in Electrochemistry*, *3*(1), 23–28. doi:10.1016/j.coelec.2017.05.001

Shukla, R., Lodha, A. S., & Pandya, A. (2016). Nanotechnology: An applied and robust approach for forensic investigation. *Forensic Research & Criminology International Journal*, *2*(1), 35–37. doi:10.15406/frcij.2016.02.00044

Singh, R. K., Bansode, F. W., Sharma, S., Singh, P., Bhadauria, S., Singh, S., & Nath, C. (2015). Development of a nanotechnology based biomedicine RISUG-M as a female contraceptive in India. *Journal of Nanomedicine & Nanotechnology*, *6*(4), 1.

Singh, S., & Samal, N. (2021). Nanotechnology: A Powerful Tool in Forensic Science for Solving Criminal Cases. *Arab Journal of Forensic Sciences & Forensic Medicine*, *3*(2), 273–296. doi:10.26735/YDDJ4516

Smith, R. G., D'Souza, N., & Nicklin, S. (2008). A review of biosensors and biologically-inspired systems for explosives detection. *Analyst (London)*, *133*(5), 571–584. doi:10.1039/b717933m PMID:18427676

Sodhi, G. S., & Kaur, J. (2016). Physical developer method for detection of latent fingerprints: A review. *Egyptian Journal of Forensic Sciences*, *6*(2), 44–47. doi:10.1016/j.ejfs.2015.05.001

Spagnolo, E. V., Stassi, C., Mondello, C., Zerbo, S., Milone, L., & Argo, A. (2019). Forensic microbiology applications: A systematic review. *Legal Medicine*, *36*, 73–80. doi:10.1016/j.legalmed.2018.11.002 PMID:30419494

Squire, K. J., Sivashanmugan, K., Zhang, B., Kraai, J., Rorrer, G., & Wang, A. X. (2020). Multiscale photonic crystal enhanced core–shell plasmonic nanomaterial for rapid vapor-phase detection of explosives. *ACS Applied Nano Materials*, *3*(2), 1656–1665. doi:10.1021/acsanm.9b02399

Srividya, B. (2016). Nanotechnology in forensics and its application in forensic investigation. *Res. Rev. J. Pharm. Nanotechnology*, *4*(2), 1–7.

Steinegger, A., Wolfbeis, O. S., & Borisov, S. M. (2020). Optical sensing and imaging of pH values: Spectroscopies, materials, and applications. *Chemical Reviews*, *120*(22), 12357–12489. doi:10.1021/acs.chemrev.0c00451 PMID:33147405

Sunagawa, K., & Sugitani, M. (2017). Post-mortem detection of bacteremia using pairs of blood culture samples. *Legal Medicine*, *24*, 92–97. doi:10.1016/j.legalmed.2016.12.009 PMID:28081798

Tang, M., Ren, G., Zhu, B., Yu, L., Liu, X., Chai, F., Wu, H., & Wang, C. (2019). Correction: Facile synthesis of orange emissive carbon dots and their application for mercury ion detection and fast fingerprint development. *Analytical Methods*, *11*(18), 2502–2502. doi:10.1039/C9AY90065A

Tang, R. K., Liu, Y., Liu, Y. Z., Zhu, S. M., Huang, W., Zhao, P., & Li, J. B. (2013). Evaluation of postmortem heart blood culture in a Chinese population. *Forensic Science International*, *231*(1-3), 229–233. doi:10.1016/j.forsciint.2013.05.020 PMID:23890642

Teymourian, H., Parrilla, M., Sempionatto, J. R., Montiel, N. F., Barfidokht, A., Van Echelpoel, R., & Wang, J. (2020). Wearable electrochemical sensors for the monitoring and screening of drugs. *ACS Sensors*, *5*(9), 2679–2700. doi:10.1021/acssensors.0c01318 PMID:32822166

Thorn, J. L., Gilchrist, K. B., Sobonya, R. E., Gaur, N. K., Lipke, P. N., & Klotz, S. A. (2010). Postmortem candidaemia: Marker of disseminated disease. *Journal of Clinical Pathology*, *63*(4), 337–340. doi:10.1136/jcp.2009.070607 PMID:19939858

To, K. C., Ben-Jaber, S., & Parkin, I. P. (2020). Recent developments in the field of explosive trace detection. *ACS Nano*, *14*(9), 10804–10833. doi:10.1021/acsnano.0c01579 PMID:32790331

Vinoda, B. M., Vinuth, M., Bodke, Y., & Manjanna, J. (2015). Photocatalytic degradation of toxic methyl red dye using silica nanoparticles synthesized from rice husk ash. *Journal of Environmental & Analytical Toxicology*, *5*(336), 2161–0525.

Wang, Y., & Liu, B. (2007). Silica nanoparticle assisted DNA assays for optical signal amplification of conjugated polymer based fluorescent sensors. *Chemical Communications*, (34), 3553–3555. doi:10.1039/b705936a PMID:18080542

Weng, S., Dong, R., Zhu, Z., Zhang, D., Zhao, J., Huang, L., & Liang, D. (2018). Dynamic surface-enhanced Raman spectroscopy and Chemometric methods for fast detection and intelligent identification of methamphetamine and 3, 4-Methylenedioxy methamphetamine in human urine. *Spectrochimica Acta. Part A: Molecular and Biomolecular Spectroscopy, 189*, 1–7. doi:10.1016/j.saa.2017.08.004 PMID:28783586

Wongkaew, N., Simsek, M., Griesche, C., & Baeumner, A. J. (2019). Functional nanomaterials and nanostructures enhancing electrochemical biosensors and lab-on-a-chip performances: Recent progress, applications, and future perspective. *Chemical Reviews, 119*(1), 120–194. doi:10.1021/acs.chemrev.8b00172 PMID:30247026

Wu, J., Zhang, L., Huang, F., Ji, X., Dai, H., & Wu, W. (2020). Surface enhanced Raman scattering substrate for the detection of explosives: Construction strategy and dimensional effect. *Journal of Hazardous Materials, 387*, 121714. doi:10.1016/j.jhazmat.2019.121714 PMID:31818672

Xue, W., Tan, X., Oo, M. K. K., Kulkarni, G., Ilgen, M. A., & Fan, X. (2020). Rapid and sensitive detection of drugs of abuse in sweat by multiplexed capillary based immuno-biosensors. *Analyst (London), 145*(4), 1346–1354. doi:10.1039/C9AN02498K PMID:31967116

Yıldırım, M., & Sinan, S., Akçan, R., Aysun Balseven, O., & Tümer, A. (2015). Use of microbiological methods in postmortem interval estimation. *The Bulletin of Legal Medicine, 20*(1).

Yıldırım, M. Ş., Akçan, R., Kaliszan, M., & Saglam, N. (2021). Microbial Nanotechnology in Reference to Postmortem Diagnosis. *Nanotechnology Applications in Health and Environmental Sciences*, 333-345.

Zala, A., & Patel, H. (2022). Dendrimer Enhanced Fingerprint and Explosive Detection: A Critical Review. *Engineered Science, 20.*

Zhang, C. Y., & Hu, J. (2010). Single quantum dot-based nanosensor for multiple DNA detection. *Analytical Chemistry, 82*(5), 1921–1927. doi:10.1021/ac9026675 PMID:20121246

Zhang, Z., & Peng, D. (2022). Recent Advances in Enhancement Techniques for Blood Fingerprints. *Critical Reviews in Analytical Chemistry*, 1–20. PMID:36045560

Zhao, Y., Xie, Y., Zhou, Q., Wang, P., Chang, Y., & Lin, C. (2020). Automatic Sensing Setup for Methamphetamine Based on the Reactional Wettability Variation Strategy. *ACS Omega, 6*(3), 2045–2051. doi:10.1021/acsomega.0c04995 PMID:33521443

Ziyade, N., Sirin, G., Elgormus, N., & Das, T. (2015). Detection of human bocavirus DNA by multiplex PCR analysis: Postmortem case report. *Balkan Medical Journal, 32*(2), 226–229. doi:10.5152/balkanmedj.2015.15254 PMID:26167351

Chapter 6
Nanomedicine

Midhat Salman
University of Health Sciences, Pakistan

Shanzay Babar
University of Health Sciences, Pakistan

Maida Mariam
University of Health Sciences, Pakistan

ABSTRACT

Nanomedicine deals with the usage of nanotechnology for medical purposes. Due to their identical size to the majority of biological molecules, nanomaterials utilized in nanomedicine are advantageous for in vivo applications. The aim of applying nanotechnology in medicine is to effectively diagnose and treat diseases. Although nanomedicine remains in its early stages, many analytical tools, diagnostic devices, biosensors, drug delivery vehicles, and physical therapy applications are being developed under this branch of medicine to treat diseases. It is anticipated that a growing number of medicines will use nanotechnology in the coming years as a result of the progress in the field. However, as nanotechnology develops, it is essential to consider both benefits and limitations of the technology including its potential risks. Current problems in the development of nanomedicine involve the toxicity and environmental effects of nanoscale compounds, but as the field develops, its impact on the economy is expected to be significant in the future.

INTRODUCTION

The area of medicine known as "nanomedicine" deals with using nanotechnology for medical purposes (Tinkle et al., 2014). Nowadays, nanomedicine is acknowledged as a significant field for developing cutting-edge treatments for unmet medical requirements (Pita, Ehmann, & Papaluca, 2016). Nanomaterials are being used in clinics for diagnosis, control, monitoring and treatment of diseases (Soares, Sousa, Pais, & Vitorino, 2018). For the application of nanomaterials in medicine, the physicochemical properties of the nanoformulation needs to be considered such as distribution, absorption, metabolism and elimination, crossing biological barriers, persistence in body and toxicity (Tinkle et al., 2014).

DOI: 10.4018/978-1-6684-8325-1.ch006

Physicochemical properties of nanomaterials differ from other chemical counterparts mainly because of their size that makes them more suitable to be used in drug development (Soares et al., 2018). Their size provides increased particle surface energy because of increased surface area compared to volume that makes them more reactive (Pino et al., 2014). Nanoparticles thus have the ability to adsorb biomolecules and are designed based on where they enter the body and the kind of bodily fluids they come into contact with (Louro, 2018).

Nanomaterials can be made to engage with certain biological targets by adjusting their shape, size, chemical composition and surface (B. Y. Kim, Rutka, & Chan, 2010). A careful nanoparticle design is thus required for a desired biological outcome. This calls for in-depth understanding of the interaction of nanomaterials with living systems, which requires a better understanding of the pathophysiological nature of disease to design particles that can reach the target site (Albanese, Tang, & Chan, 2012).

To understand the performance of nanoparticle in the body, it is necessary to characterize it. Nanoparticle characterization offers recommendations for process control and safety evaluation. The characterization should ideally be performed at various stages throughout the lifecycle of nanoparticle starting from its design to *in vivo* performance as the different steps of the lifecycle can modify some properties of the nanoparticle and thereby, interfere with measurements. The characterization also helps in informing the potential risks caused by nanoparticles to the human body (Lin, Lin, Wang, & Sridhar, 2014).

Nanomaterials are applied in medicine for nanodiagnosis, nanotherapy and regenerative medicine. They are introducing positive developments in clinical practice because of the following advantages: 1) they allow the use of multiple mechanisms of action. 2) they integrate molecules that could otherwise be highly toxic. 3) they maximize efficacy by reducing dose and therefore toxicity. 4) they have better passage across biological barriers. 5) they provide site specific drug release (Ramalingam & Rana, 2015; Soares et al., 2018). Even though nanomedicine is still in its beginning, there are a wide variety of nanomedicines in market, in clinical trials, or under development (E. M. Kim & Jeong, 2017). Bringing these nanomedicine-based technologies in use will supply strong means for treating diseases. However, there is potential of nanomaterial toxicity and the unique challenges brought by nanomaterials require ethical standards and legal regulations (Zor, Selek, Orlando, & Williams, 2019).

CLINICAL APPLICATION OF NANOTECHNOLOGY

The health sector is currently working to increase output, accessibility, and treatment quality while lowering costs. Nanotechnology has revolutionized the healthcare approaches in recent years that provides hope for a strong influence on offering improved medical services. Nanomedicine involves regulation, fabrication, design and use of treatments and devices with size range in nanoscale (1–100 nm) in this context (Minakshi Prasad et al., 2018). Exclusive countrywide and global organizations and pharmaceutical companies are investing in developing new gene therapies, *in vivo* imaging tools, and drug delivery methods using this technology (M. Prasad et al., 2018). Nanotherapeutics offers fresh opportunities to increase the effectiveness and safety of conventional therapies (Kaspar & Reichert, 2013). Cancers, blood disorders, neurological diseases, orthopaedic issues, infectious diseases and diabetes are among the conditions for which nano-based drug delivery methods hold promise. These methods also hold promise for infectious diseases which present difficulty because the traditional antimicrobial drugs used to treat them have negative impacts and drug resistance. The primary difficulty in achieving the anticipated treatment efficiency from conventional drugs is target specificity. Nanoparticles have demonstrated to be an encouraging technique for controlled and targeted medicine administration. They significantly contribute to overcome the multiple challenges of therapeutic domain due to the fact that formulations based on nanotechnology improve bioavailability, drug targeting and pharmacokinetic characteristics in many ways. In addition to prevention and treatment of diseases, there are additional potential uses for nanomedicine such as biosensors, drug delivery, imaging, tissue engineering and regenerative medicine (M. Prasad et al., 2018) (Figure 1).

Figure 1. Application of nanotechnology in the field of medicine
Source: El-Sayed, Kamel, and Research (2020)

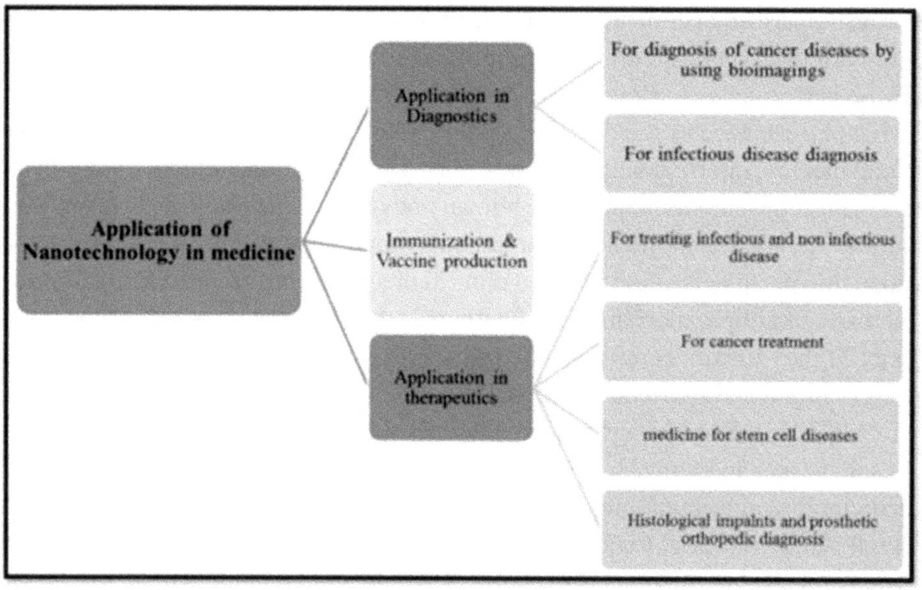

Imaging

Imaging technique is used for diagnosis of diseases inside a body. One of the most essential stages in the medical process is the diagnosis of a condition. A non-invasive technique called *in vivo* imaging allows for the identification of symptoms or indicators within the living tissues of a patient that does not necessitate surgery (Medina, Santos-Martinez, Radomski, Corrigan, & Radomski, 2007). Imaging is likewise essential for evaluating controlled drug release, detecting toxic reactions, monitoring drug circulation and treatment progress. Toxicity of a drug can be lowered by keeping track of how medications are being distributed throughout the body and delivering them when needed (Sakiyama-Elbert & Hubbell, 2001).

Methods of medical imaging have been transformed by the application of nanotechnology. The common imaging techniques include ultrasound imaging, magnetic resonance imaging and tomography (Li, Zhang, Li, & Chang, 2016). However, these methods can just detect alterations on the surface of the tissue somewhat late in the course of the disease. Biological markers are used in imaging technique in order to detect variations in tissues (Figure 2). Some of these markers have now been developed through the use of nanotechnology due to which it has become possible to collect images of biological processes in greater detail. With the use of targeting and contrast agents dependent on nanotechnologies for tracking nanoparticles injected in the body, existing medical imaging techniques are being modified to maximize their potential at the nanoscale (Sim & Wong, 2021). Some of the desired characteristics of these targeted contrast agents include acceptable toxicity profile, extended circulating half-life, low background signal, selective binding to target epitopes, application with readily available commercial imaging methods and production simplicity. The features of nanoparticles, such as blood circulation half-life, tissue penetration, targeting, biodistribution and cellular absorption, are greatly influenced by their size. They have increased permeability and retention effects as well as decreased toxicity in tissues (Sim & Wong, 2021). Liposomes, iron oxide and gold nanoparticles, micelles, nano-bubbles, dendrimers and mesoporous silica are a few types of nanoparticles utilized in imaging technologies. (Li et al., 2016).

Nanoparticles usage in imaging technologies has a number of benefits. By adjusting their compositions, shapes, surface features and sizes, one can readily create nanoparticles with discrete inherent properties (for example, magnetic, optical and electrical capabilities) they can become ideal for particular imaging tasks. The multicomponent structure of nanoparticles also provides a wide foundation for integrating multiple imaging devices (Li et al., 2016).

Figure 2. Advantages and disadvantages of different molecular imaging modalities
Source: Morato, Marciello, Chamizo, Paredes, and Filice (2021)

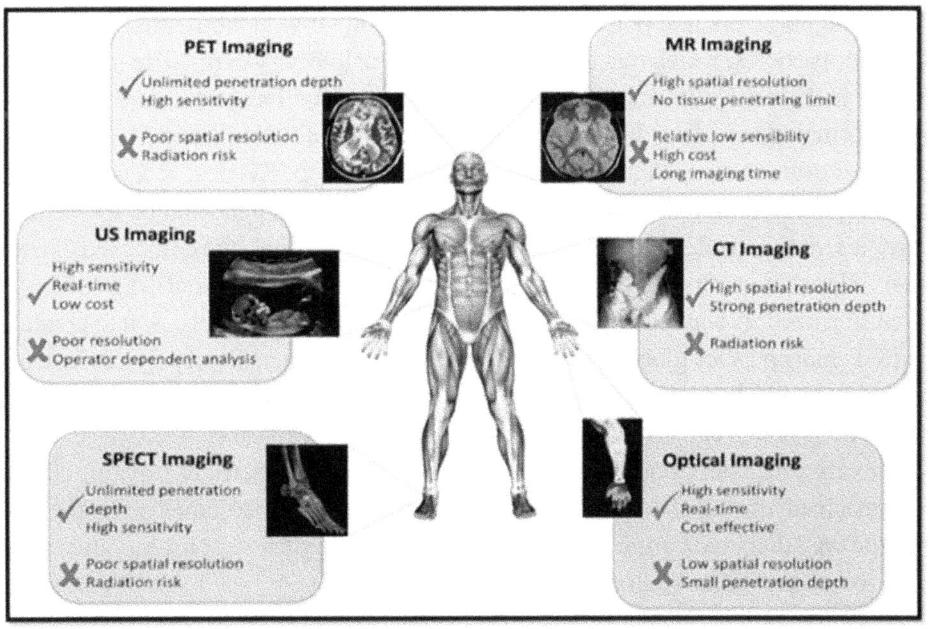

Drug Delivery

With materials in the nanoscale level, drug delivery using nanotechnology is a rapidly growing discipline where therapeutics are delivered to specific targeted regions in a safe method (De Villiers, Aramwit, & Kwon, 2008). Drug delivery using nanotechnology can transform current treatments of diseases such as orthopaedic issues, neurological diseases, infections, cancer, diabetes, and blood disorders (Petros

& DeSimone, 2010). Nanotechnologies are significantly advancing this field by creating cutting-edge medicine delivery methods. The goals of these techniques are to increase therapeutic concentration, drug stability, and drug absorption for more precise drug targeting (M. Prasad et al., 2018).

Control over medication release and the capacity for targeting are essential components of a successful drug delivery system. Specific characteristics of nanotechnology based drug delivery include minimized irritating reactions, longstanding drug release within target site, enhanced penetration within the body and reproducibility (Jain et al., 2010). Nanoparticles aid in the delivery of scarce water-soluble medications to their target areas and prevent drugs from being metabolized in the gastrointestinal tract. Nanodrugs have a higher oral bioavailability due to their common absorptive endocytosis uptake mechanisms. Nanostructures allow combined medications to be released at an appropriate amount because of their prolonged presence inside blood circulatory system. As a result, nanoparticles have less plasma fluctuations and exert fewer negative impact (De Villiers et al., 2008). The enhanced *in vivo* distribution and the advantageous pharmacokinetics are responsible for the effectiveness of nanotechnologies in medication delivery (Ochekpe, Olorunfemi, & Ngwuluka, 2009). Many of the nano-therapeutic items available in market are for oral administration, while the majority are intended for parenteral delivery (Hafner, Lovrić, Lakoš, & Pepić, 2014). For this reason, drug delivery systems are significant in the decision of the delivery path and their associated obstacles.

Nanoparticles should be used for medication delivery depending upon their physicochemical properties. Because they are highly compatible and degradable, natural polymers such as alginate and chitosan as well as synthetic polymers like poly-L-lactic acid and polyvinyl alcohol, are frequently used in the production of nanoparticles, which makes them great drug delivery systems. The same is true for phospholipids, which include micelles and liposomes, and compact lipid nanostructures (Patra et al., 2018). In medication formulations using liposomes, doxorubicin, paclitaxel, vincristine and daunorubicin, represent the most often investigated anticancer drugs (Lombardo, Kiselev, & Caccamo, 2019). Some examples of nanoparticle-based drug delivery systems are nanogels, dendrimers, nanoemulsions, solid lipid nanoparticles, nanocapsules and nanosponges (Figure 3). Nanogels are composed of flexible hydrophilic polymers. Nanoemulsions make up a thermodynamically stable colloidal drug delivery system that may be filtered sterilized. When creating solid lipid nanoparticles, a matrix is used to create oil-in-water nanoemulsions. Dendrimers are globular, hyper-branched, three-dimensional polymeric structures. Drug is put into a liquid or solid core of a nanocapsule, which is then covered by a membrane made of natural or synthetic polymers that is chosen for its specific function (Kothamasu et al., 2012). Nanosponges can load components that are both lipophilic and hydrophilic (Trotta, Zanetti, & Cavalli, 2012). Albeit for the regulatory mechanisms together with safety assessments of nanomedicines being the topic of future development, the way that medications are administered to biological systems has already undergone a revolution due to nanomedicine (Patra et al., 2018).

Biosensors

Biosensors are analytical devices that provide selective and sensitive information for recognition of a biological product. Nanotechnology based biosensors are playing a significantly important role in revolutionizing the health care industry and food industry. They make fast observation and pathogen detection possible in clinical settings; for example, a sensor chip test that uses a drop of blood to identify cancer biomarkers (G. Zheng, Patolsky, Cui, Wang, & Lieber, 2005). In food and drink industry, these are used for multiple purposes including detection of drug residue in food, remote sensing of water quality, de-

Figure 3. Nanoparticle-based drug delivery systems. Various nanotechnology-based drug delivery systems are currently being employed for enhancing drug delivery and promoting better targeting. These systems also reduce the toxicity of conventional drugs.
Source: Sousa, Ferreira, Reis, and Costa (2020)

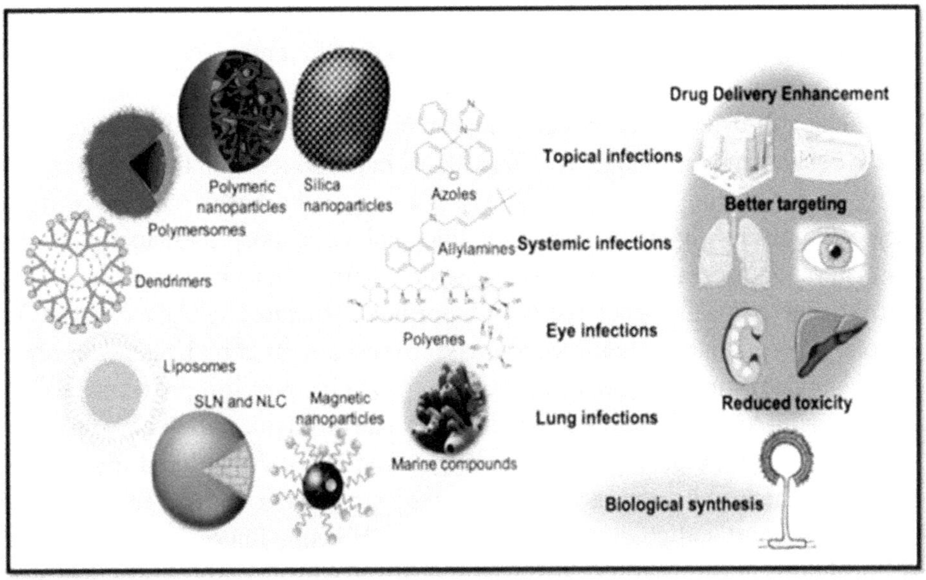

tection of water contaminants, pesticides and heavy metal ions. In environmental applications it allows quick genome analysis of organisms present in the environment (Suravi Pandit, 2016). By effectively turning the nucleotide sequence into an electronic signature to be compared with the signature available database, it enables the examination of nucleic acids (Ganguly, Saha, Bera, & Ghosh, 2017).

Incorporating nanomaterials in the construction of biosensors has improved their performance and sensitivity. The development of processes and tools used for fabricating, measuring and imaging nanoscale objects, has allowed the development of sensitive sensors that interact with small sized molecules that have to be analyzed (Suravi Pandit, 2016).

Regenerative Medicine and Tissue Engineering

Regenerative medicine and tissue engineering have emerged as promising fields for the development of novel therapies for various diseases and injuries. These fields aim to replace or repair damaged tissues and organs by using a combination of cells, biomaterials, and growth factors (Yang, Gong, Li, & Biotechnology, 2021). An enhanced requirement for development exists in the field of regenerative medicine and tissue engineering because of the numerous limitations of tissue and organ transplantation, for example, the requirement for immunosuppression, the scarcity of donors and tissue rejection (Fathi-Achachelouei et al., 2019).

The study of how new tissues and organs can be developed using scaffolds and cells is known as tissue engineering. The scaffolds serve as three-dimensional platforms in which cells develop, multiply, and differentiate to create different tissues and ultimately implantable, working organs. When creating a scaffold to imitate the extracellular matrix composition of a tissue, nanomaterials work best. Thus, in recent years, the integration of nanomaterials into regenerative medicine and tissue engineering has shown significant potential for enhancing the therapeutic efficacy of these approaches. Nanoparticles (molybdenum disulfide, carbon nanotubes, grapheme and tungsten disulfide) are applied as scaffold in tissue engineering (M. Prasad et al., 2018). Nanoparticles provide high control over scaffold properties and can entirely replace current transplant therapies such as artificial implants, for instance by altering their mechanical stability and delivering biological agents in a regulated manner (Cheng, Tietjen, Saucier-Sawyer, & Saltzman, 2015).

Due to their unique size-dependent physicochemical properties, nanomaterials can interact with biological systems at the cellular and molecular level, making them useful tools for applications in regenerative medicine and tissue engineering (Yang et al., 2021). For designed tissues to mature *in vivo*, a carefully timed sequence of actions involving the host immunological, circulatory, and biological components of the implanted material is necessary. Nanoscale structures with various material characteristics can be used to enhance the mechanical characteristics of bioprinted structures, enchance the physicochemical characteristics of tissue - engineered scaffolds for making them antimicrobial, release a variety of growth factors, concentrate contrasting substances for tracking, and give physical properties to induce tissue-specific reactions (Fathi-Achachelouei et al., 2019). They can be engineered to have specific properties, such as mechanical strength, biocompatibility, and bioactivity that are tailored for specific tissue applications. Nanomaterials can be designed to mimic the extracellular matrix (ECM) of native tissues, providing structural support and guiding the growth and differentiation of cells and enable controlled drug delivery. In addition, nanomaterials can be functionalized with bioactive molecules, such as growth factors and cytokines, to enhance cellular responses and promote tissue regeneration (Yang et al., 2021).

One example of nanomaterials used in tissue engineering is graphene-based materials. Graphene is a single-layer carbon nanomaterial with high mechanical strength, electrical conductivity, and biocompatibility. Graphene-based materials have been shown to promote the growth and differentiation of various types of cells, including stem cells, and have been used for the regeneration of different tissues, such as bone, cartilage, and nerve tissue (Yang et al., 2021). Another example is the use of nanofibrous scaffolds for tissue engineering. Nanofibrous scaffolds can be fabricated using electrospinning techniques, which allow the production of fibers with diameters in the nanometer range. These scaffolds can mimic the fibrous structure of the ECM and provide a three-dimensional environment for cell growth and differentiation. Nanofibrous scaffolds have been used for the regeneration of different tissues including bone, cartilage, skin, and nerve tissue (H. Lee, Lee, & Messersmith, 2007) (Figure 4).

Some examples of the applications of nanotechnology in this field are bone and nerve tissue engineering and kidney transplant. Engineering of nerve tissue offers the possibility for the treatment of neurological diseases. Different polymers have been utilized for this purpose and the outcomes of the experiment demonstrate neuronal gap bridging, neurite outgrowth, and human neural stem cells differentiation (Xue et al., 2018). Frequently used substances for creating nerve tissue include hydrogels, polymer scaffolds and nerve conduits. In order to address injuries that need nanomaterial-based implants as adjuvant therapy, tissue engineering for bone marrow is now under development. Present approaches of bone tissue engineering are focused on the assessment of 3D scaffolds that manage bone tissue regeneration in a normal manner (X. Zheng, Zhang, Fu, Meng, & Dai, 2021). The final objective is to develop nano-scale prosthetic kidneys in order to prevent the issues associated with transplant rejection (M. Prasad et al., 2018). However, further research is needed to optimize the design and fabrication of nanomaterials for specific tissue engineering applications.

Figure 4. Skin, cartilage, bone, tendons, nerves, cornea, cardiac tissues, teeth, and ligaments have been especially in focus in regenerative medicine and tissue engineering
Source: Tonndorf, Aibibu, and Cherif (2021)

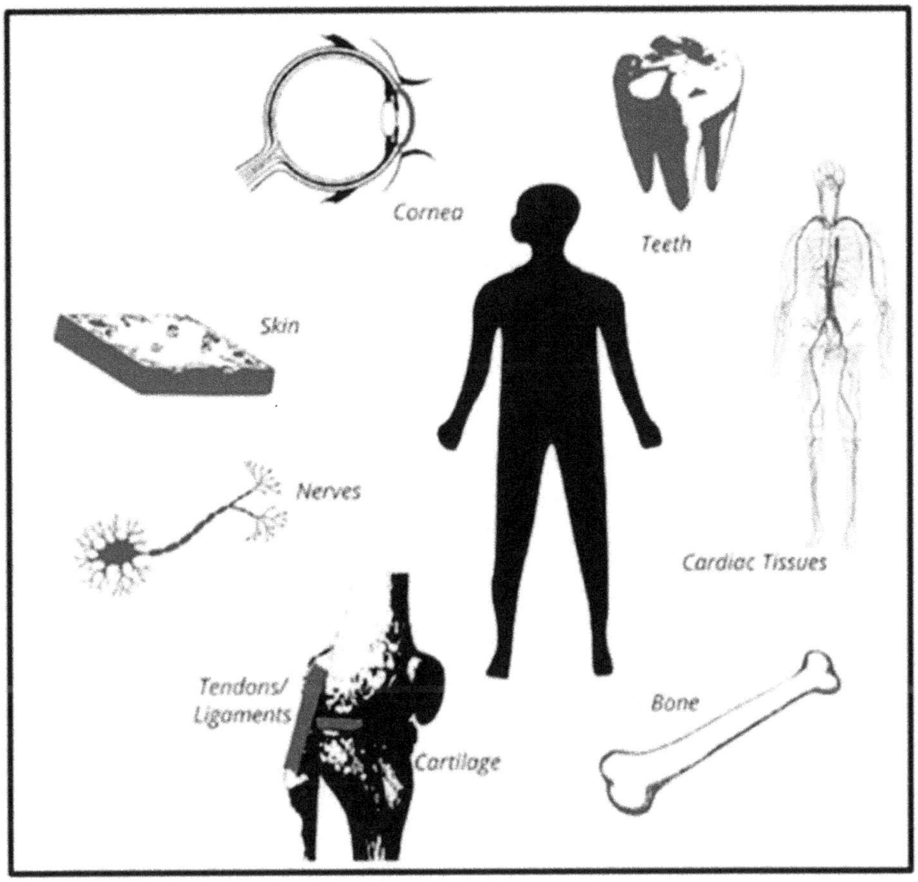

Nanomaterials can be used to deliver therapeutic agents, such as growth factors, to promote tissue regeneration. For example, in cartilage tissue engineering, nanomaterials such as chitosan and hyaluronic acid have been used to deliver growth factors such as transforming growth factor-β (TGF-β) and insulin-like growth factor-1 (IGF-1) to promote the growth and differentiation of chondrocytes (Bhardwaj & Kundu, 2012). Nanomaterials can also be used for drug delivery in regenerative medicine. Nanoparticles can be designed to encapsulate drugs and growth factors, allowing for controlled release and targeted delivery to the site of injury or disease. Nanoparticles can also protect the encapsulated molecules from degradation and enhance their bioavailability. For example, polymeric nanoparticles have been used for the delivery of growth factors for bone regeneration (Nayak, Nanda, & Mishra, 2021).

With regard to current nanotechnology, increasing number of researchers are trying to create novel biomaterials by combining different substances for the repair of tissues. Furthermore, when employing these nanomaterials in regenerative medicine and tissue engineering, issues including implant sensitivity, effect on reproduction, potential toxicity, fetal growth and consequent immune response, must be fully considered (X. Zheng et al., 2021).

BIOCOMPATIBILITY, SAFETY, RISK, AND TOXICITY

Given the potential for greater interactions between biological systems and nanomaterials, their toxicity, risk, and biocompatibility have all been questioned due to the increased usage of nanomaterials in the field of medicine. Nanomaterials may have diverse possible advantages, but their use carries potential hazards as well that must be considered. As a result, a lot of scientists have started especially looking into the branch of nanotoxicology that studies possible toxic consequences of nanomaterials (M. Adabi et al., 2017).

The host-material interactions are influenced by multiple factors including the material characteristics, its point of entrance and time of exposure. Shape, size, surface charge, surface chemistry, surface energy, release of byproducts, concentration, crystallinity, oxidative stress functions, lifetime of particles, and coating are a few variables that have an impact on nanomaterial characteristics (Foldvari & Bagonluri, 2008). For improved understanding of the interactions, pre-clinical assessments of nanoparticles are performed to test possible toxicity through diverse *in vitro* and *in vivo* assays to prove safety (Soares et al., 2018). Pharmaceutical companies, regulatory organizations, academic institutions, and the government are working to develop recommendations for risk assessment of nanomaterials as standardized evaluation criteria for the biocompatibility and toxicity of nanomaterials is lacking (S. M. Hussain et al., 2015).

A nanomaterial's biocompatibility and toxicity may be impacted by multiple factors alone or in combination with others. The surface physicochemical properties, size and shape of nanomaterials determine their pharmacokinetics and dispersion throughout the body. As a result of their size, nanoparticles can reach bloodstream and internal organs to reach cells by penetrating cell membranes. As a result, nanoparticles may be able to access parts of living organisms that larger particles typically are not able to access (Dhawan & Sharma, 2010). For instance, large particles were found in the spleen, liver, and blood, whereas 10 nm nanoparticles were specifically identified in the lung, thymus, testis, heart, brain, and kidney (De Jong et al., 2008). Similar studies have indicated that changing the shape from equiaxed to acicular increases toxicity of nanoparticles, as shown by the fact that lung epithelial cells of human are more sensitive to rod shaped zinc oxide particles than spherical particles (Hsiao & Huang, 2011). The rate of cellular internalization of nanomaterials may also be influenced by their shape particles with a high length-to-radius proportion that are adjacent to cell membrane, spherical particles are easier to internalize into the cell membrane (Decuzzi, Pasqualini, Arap, & Ferrari, 2009). Various surface modification approaches can also be used to reduce or eliminate toxicity of nanoparticle in order to increase their biocompatibility. For instance, pullulan coating significantly reduces the toxicity of superparamagnetic iron oxide nanoparticles (Ai et al., 2011). By altering their surface, the blood circulation time of nanoparticles can be accelerated and their distribution inside the body can be changed (Tiwari & Amiji, 2006). The distribution of nanoparticles in organs and how well they are absorbed by cells is also influenced by their surface. If the surface of a nanoparticle is biocompatible, it does not cause an immunological reaction (Keck, Jansch, & Müller, 2012).

Surface charge impacts the behavior of nanoparticles with organic components, including contacts with cell membranes (Figure 5), stability, protein adsorption and penetration in biological fluid (Maffre, Nienhaus, Amin, Parak, & Nienhaus, 2011). Surface charge and hydrophobicity also have an effect on the rate at which particles are transferred because of their effect on proteins that are adsorbed from plasma (Gessner, Lieske, Paulke, & Müller, 2002). It is consequently essential to evaluate the *in vivo* protein spectrum in order to manage these associations and establish biocompatibility (Keck et al., 2012). Opsonization rates for neutral particles are lower than for charged particles (Owens & Peppas, 2006).

Figure 5. The impact of surface charge on how nanoparticles interact with cells. Cationic nanoparticles are more likely to penetrate cells because of the negative charge of cell membrane.
Source: Mahdi Adabi et al. (2017)

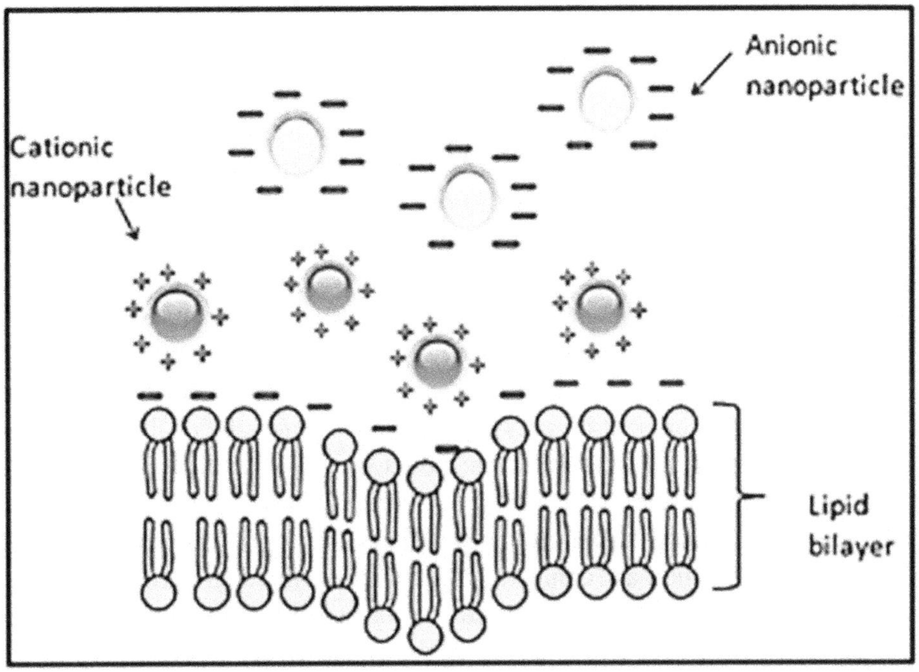

Because of their interaction with the negatively charged cell membranes, positively charged nanoparticles may be more readily absorbed by cells than other nanoparticles (Chen, McCrate, Lee, & Li, 2011). The nature and concentration of biologically-adsorbed substances, typically proteins, are determined by the hydrophobicity of nanoparticles (opsonins). The degree of opsonization and the characteristics of proteins that bind to nanomaterials are both influenced by hydrophobicity (Cedervall et al., 2007). In biological systems, hydrophilic nanoparticles appear to be generally reliable and gentler than hydrophobic particles (Sadaf et al., 2012).

Although, the public has become interested in the quickly evolving field of nanotechnology, but there have also been extensive discussions about its safety and possible dangers. There are significant challenges brought on by the use of nanomaterials, particularly in identifying, interpreting, and controlling any health issues (Sahaym & Norton, 2008). It is vital to remember that only a selected group of nanomaterials particularly those with high reactivity and mobility are regarded as dangerous. The sheer existence of nanomaterials in a laboratory setting won't constitute a concern to people or the environment unless more extensive investigations can demonstrate their harmful impacts (Rickerby & Morrison, 2007).

The unique properties of nanoparticles must be evaluated in medical applications so that improvements can be made to reduce toxicity and enhance biocompatibility. Their unidentified biocompatibility characteristics are the primary obstacle in nanoparticle utilization. Target cells along with the immune system must be considered when considering a nanoparticle's biocompatibility. Moreover, it is necessary to concentrate on a specific application of a particular material and evaluate biocompatibility for that application. To improve our grasp of the physicochemical characteristics of nanoparticles and their unique interplay with the body, more thorough research and comprehension are needed (Zor et al., 2019).

REGULATORY ISSUES, LAW, AND COMMERCIALIZATION

Nanomaterials face the same regulatory issues that are faced by goods utilizing other developing technology. As features of a material define its safety and effectiveness for health and environment, there may be more regulatory hurdles with nanoscale materials. For instance, by scaling down a substance to the nanoscale, its surface area and exposure to the body are increased potentially increasing its toxicity in this form. Due to this, the regulatory bodies need to consider the effects of nanomaterials on health and environment (Coles & Frewer, 2013). Nanomedicines were therefore introduced into the regulatory approval process using the established benefit and risk analysis framework. Although the system yet has to overcome the difficulty of assessing nanomedicines before the patent on the reference medicine expires (T. Singh et al., 2017).

Biological and non-biological medicinal devices are both a part of nanomedicine. Biological nanomedicines are generated from natural sources, whereas non-biological nanomedicines are made up of many synthetic components (Delgado-Ramos, 2014). A number of criteria must be met before a generic drug can be introduced to the market. It is necessary to compare safety, bioequivalence, quality, and efficacy on every step with the reference medication in order to find therapeutic equivalence and ultimately interchangeability (Almeida, Felzenszwalb, Marques, & Cruz, 2020). The European Medicines Agency has developed a regulatory framework for biological nanomedicines in this regard. It contains recommendations for clinical, non-clinical and comparative-quality investigations (Bowman & Hodge, 2006a). There is a lengthy regulatory process to follow before a nanomedicine can be used in clinical settings after receiving marketing clearance. This is due to the fact that each country makes its own judgements about the cost and reimbursement of medications (Bowman & Hodge, 2006b). Health technology assessment using an interdisciplinary approach is also being established to give patients access to safe and effective medications. Information about the cost-efficiency, effectiveness, and safety of medications is produced by health technology assessment to support health and political decision-makers (Arnaldi & Muratorio, 2013).

Pharmacoeconomics research is currently important for nanomedicine to be commercialized. This provides additional value for evaluating the social and economic significance, utilization metrics and life expectancy increased by nanomedicines (Keller, 2007).

BIOETHICS

For understanding the impact of nanotechnology on people, society, and environment, nanoethics examines its social and ethical implications (SEI). As nanotechnology develops, it is essential to consider both benefits and limitations of the technology including its potential risks. Nanoethics considers the effects of nanotechnology on both environment and health at multiple levels such as aspects of personal value systems, responsibilities towards society in informing the public about the issues raised by their work that may affect human health and the surrounding environment (Nordmann, 2007).

One of the major concerns with nanomaterials is their potential to cause harm to human health and the environment. The small size, shape, surface chemistry and the route of exposure of nanomaterials can increase their reactivity, surface area, and ability to penetrate biological barriers, leading to potential adverse effects on human health and the environment. In order to assess the safety of nanomaterials, researchers have developed a variety of methods for testing their toxicity. These include *in vitro* tests using cell cultures, as well as *in vivo* tests using animal models. *In vitro* studies have demonstrated that nanomaterials can induce oxidative stress, inflammation, and DNA damage in cells (Liu, Xu, Zhang, Ren, & Yang, 2010). *In vivo* studies have shown that nanomaterials can cause lung inflammation, liver damage, and neurological effects. Studies have also shown that certain types of nanomaterials can enter the body through inhalation, ingestion, or skin contact and accumulate in various organs, potentially causing damage (Ema, Gamo, Honda, & Pharmacology, 2016). Similarly, studies have been conducted to assess the effects of nanomaterials on the environment. For example, a study was conducted to examine the toxicity of silver nanoparticles on various aquatic organisms. The results indicated that the nanoparticles had toxic effects on algae, crustaceans, and fish, indicating a potential for harm to aquatic ecosystems (Kahru & Dubourguier, 2010).

The protection of human subjects in clinical trials involving nanomaterials is of utmost importance due to potential risks associated with their use. Several guidelines and regulations have been established to ensure the safety of human subjects participating in such trials. To ensure the protection of human subjects in clinical trials involving nanomaterials, informed consent must be obtained from all participants and their privacy and confidentiality must be maintained (Spagnolo & Daloiso, 2009). Additionally, researchers should consider the potential risks and benefits of using nanomaterials in their trials and design appropriate safety measures accordingly.

According to the International Council for Harmonisation of Technical Requirements for Pharmaceuticals for Human Use (ICH), nanomaterials should be thoroughly characterized and evaluated for their potential toxicity before use in clinical trials (ICH, 1997). The US Food and Drug Administration (FDA) has also provided guidelines for the use of nanomaterials in clinical trials, emphasizing the need for safety evaluations and risk assessments (de Vlieger et al., 2019). In addition, the National Institutes of Health (NIH) has established the Nanotechnology Characterization Laboratory (NCL), which provides standardized testing and characterization of nanomaterials for use in clinical trials (Ke et al., 2022).

The ethical use of nanomaterials requires strict adherence to privacy and confidentiality regulations to protect individuals' personal information and ensure data security. Researchers and practitioners working with nanomaterials must respect the privacy of individuals, as well as any proprietary or confidential information related to their work. One of the most important ways to protect privacy and confidentiality is through proper data management practices. This includes using secure storage and transmission methods, limiting access to sensitive information, and obtaining informed consent from participants before collecting data. Researchers and practitioners should be open about their data management practices and ensure that participants understand how their data will be collected, used, and protected (Jones, Caballero, & Davidov-Pardo, 2019).

While Nano medicine offers tremendous promise to revolutionize the way we diagnose, treat, and prevent diseases, there are also concerns about its social and economic impact. One potential social impact of Nano medicine is the potential for unequal access to healthcare. The high cost of developing and producing nanomedical products could lead to limited access to these products for those who cannot afford them (Massaro, Lorenzoni, & Change, 2021). This could exacerbate existing health inequalities, particularly in developing countries. Another potential social impact is the potential for unintended consequences. The use of nanoparticles in medicine could lead to unintended toxic effects on human health and the environment (S. Hussain, Hess, Gearhart, Geiss, & Schlager, 2005). There is a need for careful regulation of Nano medicine to minimize these risks and ensure that the benefits of Nano medicine outweigh the risks. The development of Nano medicine has also led to the creation of new companies and the expansion of existing ones, which has generated new jobs and economic opportunities (Cusato,

Allegra, Nicolò, Calcagno, & D'Avolio, 2018). The development of nanotechnology in medicine has also created new avenues for research and development, which can lead to further innovation and economic growth. The potential for unequal access to healthcare and unintended consequences must be carefully managed, while the economic benefits of Nano medicine must be leveraged to ensure that everyone benefits from this promising field of research.

Intellectual property and patent issues are common concerns in the field of nanotechnology, particularly with the use of nanomaterials. As new and innovative nanomaterials are developed, there is a growing need to protect these materials through patents and other forms of intellectual property rights. One of the primary challenges in securing patents for nanomaterials is defining the scope of the invention. As nanomaterials are often complex and multifunctional, it can be difficult to clearly define the boundaries of the invention. This is particularly true for materials that exhibit unique properties at the nanoscale, such as quantum dots and carbon nanotubes (R. Bawa, 2007). Another issue is the potential for infringement of existing patents. Since nanotechnology is a rapidly evolving field, it can be difficult to determine whether a new nanomaterial is similar enough to an existing patented material to constitute infringement. This challenge is compounded by the fact that many patents related to nanotechnology are broadly written and may not specifically address the unique properties of nanomaterials (H.-J. Lee et al., 2015). By keeping the composition and manufacturing processes of nanomaterials confidential, companies may be able to protect their products without disclosing the full details of their invention (Siri, Fernando, & De Silva, 2020). In addition to patent protection, there are also concerns around the ethical implications of intellectual property in nanotechnology. Some researchers argue that the high costs of securing patents can limit access to nanomaterials and prevent researchers in developing countries from participating in the field (Hulla, Sahu, Hayes, & toxicology, 2015). Despite these concerns, the importance of intellectual property and patent protection in nanotechnology cannot be overstated. As the field continues to evolve, it will be crucial for researchers and companies to navigate these complex issues in order to protect their inventions and promote innovation.

Nanomaterials have shown great potential for addressing environmental problems and achieving sustainability goals in various sectors due to their unique properties (Roco, Mirkin, & Hersam, 2011). Nanomaterials are being used to minimize environmental impact and promote sustainability such as through water purification. Nanomaterials such as carbon nanotubes, graphene, and metal-organic frameworks (MOFs) have been shown to be effective in removing contaminants from water through adsorption, catalysis, and membrane filtration (K. K. Singh, Singh, & Rai, 2022). Nanomaterials have also been used to develop more efficient and cost-effective renewable energy technologies such as solar cells and fuel cells (Ong, Zheng, & Antonietti, 2021). The use of nanoparticles in solar cells has led to higher energy conversion efficiency due to their ability to absorb a broader range of wavelengths of light (Klimov, 2017). Nanomaterials have the potential to revolutionize waste management by making it more efficient and sustainable. For example, nanoparticles have been used in the synthesis of biodegradable plastics that can be easily broken down in the environment (S. Singh et al., 2021). Additionally, nanomaterials have been used in the development of sensors that can detect and monitor pollutants in the environment, thereby facilitating proper waste management practices (Das, Sen, Debnath, & Research, 2015). Despite the potential benefits of nanomaterials in achieving sustainability goals, there are concerns about their potential adverse effects on the environment and human health (Buzea, Pacheco, & Robbie, 2007). Therefore, it is important to conduct thorough research and development of safe and sustainable nanomaterials.

Nanomedical technologies can revolutionize healthcare by enabling highly targeted and personalized treatments for a wide range of diseases. However, the equitable distribution and access to these technologies remains a significant challenge. One major concern is that the development of nanomedical technologies is driven primarily by profit motives, leading to high costs and limited availability, especially in low-income countries. In addition, there is a lack of regulatory frameworks to ensure that these technologies are safe and effective for all populations, further exacerbating existing health disparities (Roco et al., 2011). To address these challenges, it is important to promote policies and initiatives that prioritize equity and access to nanomedical technologies. This can include increasing investment in research and development of affordable and accessible nanomedical technologies, as well as implementing regulatory frameworks that prioritize safety and effectiveness for all populations (Elijah & Odiyo, 2019). Another important consideration is the need to ensure that all communities have equal access to information about nanomedical technologies, as well as the potential benefits and risks. This can be achieved through increased public education and outreach, as well as the development of culturally sensitive and inclusive messaging (McGee, 2009).

Nanomedical researchers and industries have a crucial responsibility in ensuring the safety and efficacy of nanotechnology-based products and therapies. As a rapidly evolving field, it is imperative that nanomedical researchers and industries work together to develop responsible guidelines and regulations that prioritize safety and ethical considerations. One important responsibility of nanomedical researchers and industries is to thoroughly investigate the potential risks and benefits of nanotechnology-based products and therapies before they are introduced to the market. This includes conducting extensive preclinical and clinical trials, as well as assessing the long-term effects of these products on both human health and the environment. In addition, nanomedical researchers and industries must prioritize transparency in their research and development processes. This includes openly sharing information about their products, as well as being transparent about any potential risks associated with their use. Finally, nanomedical researchers and industries must also prioritize ethical considerations when developing and commercializing nanotechnology-based products and therapies. This includes ensuring that their products are accessible and affordable to all, regardless of socioeconomic status, and that they do not perpetuate social and economic inequalities (R. J. N. L. Bawa & Bus, 2008).

Nanomedical products and practices require careful regulation and governance due to their unique characteristics and potential impact on human health and the environment. In this regard, regulatory bodies such as the US Food and Drug Administration (FDA) and the European Medicines Agency (EMA) have established guidelines for the development and commercialization of nanomedicines. The FDA has issued a guidance document on the use of nanotechnology in FDA-regulated products, including medical products, which provides recommendations for the characterization, quality control, and safety evaluation of nanomaterials used in such products (FDA & Drug Administration, 2014). Similarly, the EMA has published a reflection paper on the development of medicinal products containing nanomaterials, which outlines the regulatory requirements for the development, manufacture, and authorization of nanomedicines in the European Union (Committee for Medicinal Products for Human Use %J European Medicines Agency, 2013). In addition to regulatory guidelines, several ethical and societal concerns related to the use of nanomedicines have been identified. Moreover, the equitable access to nanomedicines and the potential for the widening of health disparities have also been discussed. To address these concerns, there have been calls for the development of a comprehensive governance framework for nanomedicines that take into account their unique characteristics and potential risks (Rycroft, Trump, Poinsatte-Jones, & Linkov, 2018) Such a framework should consider the potential benefits and risks of nanomedicines and involve the participation of all stakeholders, including patients, healthcare professionals, regulators, industry, and civil society organizations.

FUTURE IMPLICATIONS

Future developments in nanotechnology are expected to be rapid and substantial. In the coming times, nanomedicine is predicted to make a considerable worldwide contribution to employment and economic expansion. There will be four distinct generations of nanotechnology development, of which we are presently employing the initial or second generation of nanomaterials (Grimshaw, Eccles, & Tetroe, 2004).

The primary generation is exclusively focused on material science and the enhancement of characteristics by means of inert nanostructures, which can be attained by including coatings or carbon nanotubes. The second generation contains active nanostructures for instance those that are bioactive and deliver a drug to a particular target cell. Such nanomaterials may be covered in specific proteins to achieve target delivery. The final two generations are more complex with the fourth generation with an advanced nanosystem, like nanorobotics and molecular nanosystem for the formation of artificial organs (Giunipero, Hooker, Joseph-Matthews, Yoon, & Brudvig, 2008).

Nanomedicine can revolutionize healthcare more strongly than numerous developments of the past by improved health and reduced environmental pollution. Significant research is taking place in academia and industry in nanomedicine since past two decades. The research is both of *in vitro* and *in vivo* categories. Clinical translation of *in vitro* applications is far less complicated compared to *in vivo* applications. Use of *in vitro* technology is for the purpose of diagnosing diseases and studying biological processes using biological samples. In contrast to *in vivo* applications, their clearance by health bodies has less stringent standards. For instance, a diagnostic tool based on nanotechnology must exhibit better clinical accuracy or precision compared to available methods for approval. Consequently, there are less significant scientific obstacles in the development of nanomaterials for *in vitro* applications. As a result, the *in vitro* application of nanotechnology is predicted to accelerate in the next couple of decades. In contrast, the injected nanoparticles will engage with biological fluids, tissues and cells inside the body making their application more challenging due to problems with target specificity and potential toxicity. Clinical translation of *In vivo* applications thus could take much longer than the *in vitro* applications (Cheon, Chan, & Zuhorn, 2019).

CONCLUSION

Numerous types of nanomedicines have been introduced into the global healthcare market over the years. The progress being made in the field of nanoparticles can be partially attributed to their distinctive traits. Researchers can alter the natural biology of diseased cells by producing nanoparticles with medications that can specifically target the diseased cells. Furthermore, it has been demonstrated that adding pharmaceuticals to nanoparticles enhances medication bioavailability, for instance by shielding the medicines from the internal environment and enhancing pharmacological efficacy. This might lessen issues with noncompliance that are related to medicine effectiveness (Lloyd-Parry, Downing, Aleisaei, Jones, & Coward, 2018). The creation of novel drugs as well as the reformulation of current ones has been greatly aided by the expansion of nanomedicine research. Medical devices to nanopharmaceuticals are only a few examples of how nanomedicines have been applied in clinical practice over the past few decades (Soares et al., 2018).

The possible toxicity of the particles themselves is one of the most significant issues. It is well recognized that these synthetic, foreign compounds can have harmful effects on human cells to varying degrees. Clearly, a new conceptual framework to describe pharmacokinetics and biodistribution is necessary due to the intricate interaction between the physicochemical characteristics and *in vivo* performance of nanomaterials (Mast et al., 2021). Now, more research must focus specifically on developing methods to create nanoparticles at a reasonable cost and with the least amount of specialized equipment required. This would increase access to such treatment in the future (Lloyd-Parry et al., 2018). From the adoption of standardized definitions to the creation of characterization protocols and assessment, there remains a long way to go before nanomedicines are fully regulated. Additionally, as a result of ongoing research, more complex nanostructured designs are quickly emerging. These designs necessitate comprehensive knowledge of the pharmacokinetic characteristics of nanotherapeutics, which are defined by their specific compositions and physical characteristics, and present additional regulatory challenges. It is anticipated that there will be a growing number of medicines that use nanotechnology in the next years thanks to all the advancements in science and technology. The construction of guidelines for the production and evaluation of nanomedicines is essential for a collective understanding among stakeholders for authorizing novel and cutting-edge nanomedicines for the drugs market. To encourage logical choices on regulatory and scientific issues, funding, and access to market, this approach must also be implemented in conjunction with interagency harmonization initiatives (Soares et al., 2018).

REFERENCES

Adabi, M., Naghibzadeh, M., Adabi, M., Zarrinfard, M. A., Esnaashari, S. S., Seifalian, A. M., Faridi-Majidi, R., Tanimowo Aiyelabegan, H., & Ghanbari, H. (2017). Biocompatibility and nanostructured materials: Applications in nanomedicine. *Artificial Cells, Nanomedicine, and Biotechnology*, *45*(4), 833–842. doi:10.1080/21691401.2016.1178134 PMID:27247194

Ai, J., Biazar, E., Jafarpour, M., Montazeri, M., Majdi, A., Aminifard, S., ... Rad, H. G. (2011). Nanotoxicology and nanoparticle safety in biomedical designs. *International Journal of Nanomedicine*, *6*, 1117. PMID:21698080

Albanese, A., Tang, P. S., & Chan, W. C. (2012). The effect of nanoparticle size, shape, and surface chemistry on biological systems. *Annual Review of Biomedical Engineering*, *14*(1), 1–16. doi:10.1146/annurev-bioeng-071811-150124 PMID:22524388

Almeida, L., Felzenszwalb, I., Marques, M., & Cruz, C. (2020). Nanotechnology activities: Environmental protection regulatory issues data. *Heliyon*, *6*(10), e05303. doi:10.1016/j.heliyon.2020.e05303 PMID:33102878

Arnaldi, S., & Muratorio, A. (2013). *Nanotechnology, uncertainty and regulation. A guest editorial* (Vol. 7). Springer.

Bawa, R. (2007). *Patents and nanomedicine*. Academic Press.

Bawa, R. (2008). *Nanoparticle-based therapeutics in humans: A survey*. Academic Press.

Bhardwaj, N., & Kundu, S. (2012). *Chondrogenic differentiation of rat MSCs on porous scaffolds of silk fibroin/chitosan blends*. Academic Press.

Bowman, D. M., & Hodge, G. A. (2006a). Nanotechnology: Mapping the wild regulatory frontier. *Futures, 38*(9), 1060–1073. doi:10.1016/j.futures.2006.02.017

Bowman, D. M., & Hodge, G. A. (2006b). A small matter of regulation: An international review of nanotechnology regulation. *Colum. Sci. & Tech. L. Rev., 8*, 1.

Buzea, C., Pacheco, I. I., & Robbie, K. (2007). *Nanomaterials and nanoparticles: Sources and toxicity*. Academic Press.

Cedervall, T., Lynch, I., Foy, M., Berggård, T., Donnelly, S. C., Cagney, G., Linse, S., & Dawson, K. A. (2007). Detailed identification of plasma proteins adsorbed on copolymer nanoparticles. *Angewandte Chemie International Edition in English, 46*(30), 5754–5756. doi:10.1002/anie.200700465 PMID:17591736

Chen, L., McCrate, J. M., Lee, J. C., & Li, H. (2011). The role of surface charge on the uptake and biocompatibility of hydroxyapatite nanoparticles with osteoblast cells. *Nanotechnology, 22*(10), 105708. doi:10.1088/0957-4484/22/10/105708 PMID:21289408

Cheng, C. J., Tietjen, G. T., Saucier-Sawyer, J. K., & Saltzman, W. M. (2015). A holistic approach to targeting disease with polymeric nanoparticles. *Nature Reviews. Drug Discovery, 14*(4), 239–247. doi:10.1038/nrd4503 PMID:25598505

Cheon, J., Chan, W., & Zuhorn, I. (2019). The Future of Nanotechnology: Cross-disciplined Progress to Improve Health and Medicine. *Accounts of Chemical Research, 52*(9), 2405–2405. doi:10.1021/acs.accounts.9b00423 PMID:31525881

Coles, D., & Frewer, L. J. (2013). Nanotechnology applied to European food production–A review of ethical and regulatory issues. *Trends in Food Science & Technology, 34*(1), 32–43. doi:10.1016/j.tifs.2013.08.006

Committee for Medicinal Products for Human Use. (2013). *Reflection paper on surface coatings: General issues for consideration regarding parenteral administration of coated nanomedicine products*. Author.

Cusato, J., Allegra, S., Nicolò, A. D., Calcagno, A., & D'Avolio, A. (2018). *Precision medicine for HIV: where are we?* Academic Press.

Das, S., Sen, B., & Debnath, N. (2015). *Recent trends in nanomaterials applications in environmental monitoring and remediation*. Academic Press.

De Jong, W. H., Hagens, W. I., Krystek, P., Burger, M. C., Sips, A. J., & Geertsma, R. E. (2008). Particle size-dependent organ distribution of gold nanoparticles after intravenous administration. *Biomaterials, 29*(12), 1912–1919. doi:10.1016/j.biomaterials.2007.12.037 PMID:18242692

De Villiers, M. M., Aramwit, P., & Kwon, G. S. (2008). *Nanotechnology in drug delivery*. Springer Science & Business Media.

de Vlieger, J. S., Crommelin, D. J., Tyner, K., Drummond, D. C., Jiang, W., McNeil, S. E., & Shah, V. P. (2019). *Report of the AAPS guidance forum on the FDA draft guidance for industry: "Drug products, including biological products, that contain nanomaterials"*. Springer.

Decuzzi, P., Pasqualini, R., Arap, W., & Ferrari, M. (2009). Intravascular delivery of particulate systems: Does geometry really matter? *Pharmaceutical Research*, *26*(1), 235–243. doi:10.100711095-008-9697-x PMID:18712584

Delgado-Ramos, G. C. (2014). Nanotechnology in Mexico: Global trends and national implications for policy and regulatory issues. *Technology in Society*, *37*, 4–15. doi:10.1016/j.techsoc.2013.09.005

Dhawan, A., & Sharma, V. (2010). Toxicity assessment of nanomaterials: Methods and challenges. *Analytical and Bioanalytical Chemistry*, *398*(2), 589–605. doi:10.100700216-010-3996-x PMID:20652549

El-Sayed, A., & Kamel, M. (2020). *Advances in nanomedical applications: diagnostic, therapeutic, immunization, and vaccine production*. Academic Press.

Elijah, V. T., & Odiyo, J. (2019). Perception of environmental spillovers across scale in climate change adaptation planning: The case of small-scale farmers' irrigation strategies. Academic Press.

Ema, M., Gamo, M., & Honda, K. (2016). *A review of toxicity studies of single-walled carbon nanotubes in laboratory animals*. Academic Press.

Fathi-Achachelouei, M., Knopf-Marques, H., Ribeiro da Silva, C. E., Barthès, J., Bat, E., Tezcaner, A., & Vrana, N. E. (2019). Use of Nanoparticles in Tissue Engineering and Regenerative Medicine. *Frontiers in Bioengineering and Biotechnology*, *7*, 113. doi:10.3389/fbioe.2019.00113 PMID:31179276

FDA. (2014). *Guidance for Industry: Considering Whether an FDA-Regulated Product Involves the Application of Nanotechnology*. FDA.

Foldvari, M., & Bagonluri, M. (2008). Carbon nanotubes as functional excipients for nanomedicines: II. Drug delivery and biocompatibility issues. *Nanomedicine (London)*, *4*(3), 183–200. doi:10.1016/j.nano.2008.04.003 PMID:18550450

Ganguly, J., Saha, S., Bera, A., & Ghosh, M. (2017). Exploring electro-optic effect and third-order nonlinear optical susceptibility of impurity doped quantum dots: Interplay between hydrostatic pressure, temperature and noise. *Optics Communications*, *387*, 166–173. doi:10.1016/j.optcom.2016.11.040

Gessner, A., Lieske, A., Paulke, B., & Müller, R. (2002). Influence of surface charge density on protein adsorption on polymeric nanoparticles: Analysis by two-dimensional electrophoresis. *European Journal of Pharmaceutics and Biopharmaceutics*, *54*(2), 165–170. doi:10.1016/S0939-6411(02)00081-4 PMID:12191688

Giunipero, L. C., Hooker, R. E., Joseph-Matthews, S., Yoon, T. E., & Brudvig, S. (2008). A decade of SCM literature: Past, present and future implications. *The Journal of Supply Chain Management*, *44*(4), 66–86. doi:10.1111/j.1745-493X.2008.00073.x

Grimshaw, J., Eccles, M., & Tetroe, J. (2004). Implementing clinical guidelines: Current evidence and future implications. *The Journal of Continuing Education in the Health Professions*, *24*(S1), S31–S37. doi:10.1002/chp.1340240506 PMID:15712775

Hafner, A., Lovrić, J., Lakoš, G. P., & Pepić, I. (2014). Nanotherapeutics in the EU: An overview on current state and future directions. *International Journal of Nanomedicine*, *9*, 1005–1023. doi:10.2147/IJN.S55359 PMID:24600222

Hsiao, I. L., & Huang, Y.-J. (2011). Effects of various physicochemical characteristics on the toxicities of ZnO and TiO2 nanoparticles toward human lung epithelial cells. *The Science of the Total Environment*, *409*(7), 1219–1228. doi:10.1016/j.scitotenv.2010.12.033 PMID:21255821

Hulla, J., Sahu, S., & Hayes, A. (2015). *Nanotechnology: History and future*. Academic Press.

Hussain, S., Hess, K., Gearhart, J., Geiss, K., & Schlager, J. (2005). *In vitro toxicity of nanoparticles in BRL 3A rat liver cells*. Academic Press.

Hussain, S. M., Warheit, D. B., Ng, S. P., Comfort, K. K., Grabinski, C. M., & Braydich-Stolle, L. K. (2015). At the Crossroads of Nanotoxicology in vitro: Past Achievements and Current Challenges. *Toxicological Sciences*, *147*(1), 5–16. doi:10.1093/toxsci/kfv106 PMID:26310852

ICH. (1997). Guideline: Preclinical Safety Evaluation of Biotechnology-Derived Pharmaceuticals. Academic Press.

Jain, N., Jain, R., Thakur, N., Gupta, B. P., Jain, D. K., Banveer, J., & Jain, S. (2010). Nanotechnology: A safe and effective drug delivery system. *Asian Journal of Pharmaceutical and Clinical Research*, *3*(3), 159–165.

Jones, D., Caballero, S., & Davidov-Pardo, G. (2019). Bioavailability of nanotechnology-based bioactives and nutraceuticals. *Advances in Food and Nutrition Research*, *88*, 235–273. doi:10.1016/bs.afnr.2019.02.014 PMID:31151725

Kahru, A., & Dubourguier, H.-C. (2010). *From ecotoxicology to nanoecotoxicology*. Academic Press.

Kaspar, A. A., & Reichert, J. M. (2013). Future directions for peptide therapeutics development. *Drug Discovery Today*, *18*(17), 807–817. doi:10.1016/j.drudis.2013.05.011 PMID:23726889

Ke, W., Crist, R. M., Clogston, J. D., Stern, S. T., Dobrovolskaia, M. A., Grodzinski, P., & Jensen, M. (2022). *Trends and Patterns in Cancer Nanotechnology Research: A Survey of NCI's caNanoLab and Nanotechnology Characterization Laboratory*. Academic Press.

Keck, C. M., Jansch, M., & Müller, R. H. (2012). Protein Adsorption Patterns and Analysis on IV Nanoemulsions-The Key Factor Determining the Organ Distribution. *Pharmaceutics*, *5*(1), 36–68. doi:10.3390/pharmaceutics5010036 PMID:24300396

Keller, K. H. (2007). Nanotechnology and society. *Journal of Nanoparticle Research*, *9*(1), 5–10. doi:10.100711051-006-9193-3

Kim, B. Y., Rutka, J. T., & Chan, W. C. (2010). Nanomedicine. *The New England Journal of Medicine*, *363*(25), 2434–2443. doi:10.1056/NEJMra0912273 PMID:21158659

Kim, E. M., & Jeong, H. J. (2017). Current Status and Future Direction of Nanomedicine: Focus on Advanced Biological and Medical Applications. *Nuclear Medicine and Molecular Imaging*, *51*(2), 106–117. doi:10.100713139-016-0435-8 PMID:28559935

Klimov, V. I. (2017). *Nanocrystal quantum dots*. CRC Press. doi:10.1201/9781420079272

Kothamasu, P., Kanumur, H., Ravur, N., Maddu, C., Parasuramrajam, R., & Thangavel, S. (2012). Nanocapsules: The weapons for novel drug delivery systems. *BioImpacts*, *2*(2), 71–81. doi:10.5681/bi.2012.011 PMID:23678444

Lee, H., Lee, B. P., & Messersmith, P. (2007). *A reversible wet/dry adhesive inspired by mussels and geckos*. Academic Press.

Lee, H.-J., Lee, S. J., Uthaman, S., Thomas, R. G., Hyun, H., Jeong, Y. Y., . . . Park, I.-K. (2015). *Biomedical applications of magnetically functionalized organic/inorganic hybrid nanofibers*. Academic Press.

Li, X., Zhang, X. N., Li, X. D., & Chang, J. (2016). Multimodality imaging in nanomedicine and nanotheranostics. *Cancer Biology & Medicine*, *13*(3), 339–348. doi:10.20892/j.issn.2095-3941.2016.0055 PMID:27807501

Lin, P. C., Lin, S., Wang, P. C., & Sridhar, R. (2014). Techniques for physicochemical characterization of nanomaterials. *Biotechnology Advances*, *32*(4), 711–726. doi:10.1016/j.biotechadv.2013.11.006 PMID:24252561

Liu, S., Xu, L., Zhang, T., Ren, G., & Yang, Z. (2010). *Oxidative stress and apoptosis induced by nanosized titanium dioxide in PC12 cells*. Academic Press.

Lloyd-Parry, O., Downing, C., Aleisaei, E., Jones, C., & Coward, K. (2018). Nanomedicine applications in women's health: State of the art. *International Journal of Nanomedicine*, *13*, 1963–1983. doi:10.2147/IJN.S97572 PMID:29636611

Lombardo, D., Kiselev, M. A., & Caccamo, M. T. (2019). Smart nanoparticles for drug delivery application: Development of versatile nanocarrier platforms in biotechnology and nanomedicine. *Journal of Nanomaterials*. doi:10.1155/2019/3702518

Louro, H. (2018). Relevance of Physicochemical Characterization of Nanomaterials for Understanding Nano-cellular Interactions. *Advances in Experimental Medicine and Biology*, *1048*, 123–142. doi:10.1007/978-3-319-72041-8_8 PMID:29453536

Maffre, P., Nienhaus, K., Amin, F., Parak, W. J., & Nienhaus, G. U. (2011). Characterization of protein adsorption onto FePt nanoparticles using dual-focus fluorescence correlation spectroscopy. *Beilstein Journal of Nanotechnology*, *2*, 374–383. doi:10.3762/bjnano.2.43 PMID:22003445

Massaro, S., & Lorenzoni, G. (2021). *Nanomedicine: A socio-technical system*. Academic Press.

Mast, M. P., Modh, H., Champanhac, C., Wang, J. W., Storm, G., Krämer, J., Mailänder, V., Pastorin, G., & Wacker, M. G. (2021). Nanomedicine at the crossroads - A quick guide for IVIVC. *Advanced Drug Delivery Reviews*, *179*, 113829. doi:10.1016/j.addr.2021.113829 PMID:34174332

McGee, E. (2009). *Nanomedicine: Ethical concerns beyond diagnostics, drugs, and techniques*. Academic Press.

Medina, C., Santos-Martinez, M., Radomski, A., Corrigan, O., & Radomski, M. (2007). Nanoparticles: Pharmacological and toxicological significance. *British Journal of Pharmacology*, *150*(5), 552–558. doi:10.1038j.bjp.0707130 PMID:17245366

Morato, Y. L., Marciello, M., Chamizo, L. L., Paredes, K. O., & Filice, M. (2021). Hybrid magnetic nanoparticles for multimodal molecular imaging of cancer. In *Magnetic Nanoparticle-Based Hybrid Materials* (pp. 343–386). Elsevier. doi:10.1016/B978-0-12-823688-8.00008-9

Nayak, P., Nanda, R., & Mishra, M. (2021). Role of Nanoparticles in Medical Imaging. In *Medical Imaging Methods* (pp. 107–144). CRC Press. doi:10.1201/9781003112068-5

Nordmann, A. (2007). If and then: A critique of speculative nanoethics. *NanoEthics*, *1*(1), 31–46. doi:10.100711569-007-0007-6

Ochekpe, N. A., Olorunfemi, P. O., & Ngwuluka, N. C. (2009). Nanotechnology and drug delivery part 1: Background and applications. *Tropical Journal of Pharmaceutical Research*, *8*(3). Advance online publication. doi:10.4314/tjpr.v8i3.44546

Ong, W.-J., Zheng, N., & Antonietti, M. (2021). *Advanced nanomaterials for energy conversion and storage: current status and future opportunities*. Academic Press.

Owens, D. E. III, & Peppas, N. A. (2006). Opsonization, biodistribution, and pharmacokinetics of polymeric nanoparticles. *International Journal of Pharmaceutics*, *307*(1), 93–102. doi:10.1016/j.ijpharm.2005.10.010 PMID:16303268

Patra, J. K., Das, G., Fraceto, L. F., Campos, E. V. R., Rodriguez-Torres, M. P., Acosta-Torres, L. S., Diaz-Torres, L. A., Grillo, R., Swamy, M. K., Sharma, S., Habtemariam, S., & Shin, H.-S. (2018). Nano based drug delivery systems: Recent developments and future prospects. *Journal of Nanobiotechnology*, *16*(1), 71. doi:10.118612951-018-0392-8 PMID:30231877

Petros, R. A., & DeSimone, J. M. (2010). Strategies in the design of nanoparticles for therapeutic applications. *Nature Reviews. Drug Discovery*, *9*(8), 615–627. doi:10.1038/nrd2591 PMID:20616808

Pino, P., Pelaz, B., Zhang, Q., Maffre, P., Nienhaus, G. U., & Parak, W. J. (2014). Protein corona formation around nanoparticles – from the past to the future. *Materials Horizons*, *1*(3), 301–313. doi:10.1039/C3MH00106G

Pita, R., Ehmann, F., & Papaluca, M. (2016). Nanomedicines in the EU-Regulatory Overview. *The AAPS Journal*, *18*(6), 1576–1582. doi:10.120812248-016-9967-1 PMID:27527889

Prasad, M., Lambe, U. P., Brar, B., Shah, I., J, M., Ranjan, K., Rao, R., Kumar, S., Mahant, S., Khurana, S. K., Iqbal, H. M. N., Dhama, K., Misri, J., & Prasad, G. (2018). Nanotherapeutics: An insight into healthcare and multi-dimensional applications in medical sector of the modern world. *Biomedicine and Pharmacotherapy*, *97*, 1521–1537. doi:10.1016/j.biopha.2017.11.026 PMID:29793315

Prasad, M., Lambe, U. P., Brar, B., Shah, I., J, M., Ranjan, K., Rao, R., Kumar, S., Mahant, S., Khurana, S. K., Iqbal, H. M. N., Dhama, K., Misri, J., & Prasad, G. (2018). Nanotherapeutics: An insight into healthcare and multi-dimensional applications in medical sector of the modern world. *Biomedicine and Pharmacotherapy*, *97*, 1521–1537. doi:10.1016/j.biopha.2017.11.026 PMID:29793315

Ramalingam, M., & Rana, D. (2015). Impact of Nanotechnology in Induced Pluripotent Stem Cells-driven Tissue Engineering and Regenerative Medicine. *Journal of Bionanoscience*, *9*(1), 13–21. doi:10.1166/jbns.2015.1277

Rickerby, D., & Morrison, M. (2007). Nanotechnology and the environment: A European perspective. *Science and Technology of Advanced Materials*, *8*(1-2), 19–24. doi:10.1016/j.stam.2006.10.002

Roco, M. C., Mirkin, C. A., & Hersam, M. C. (2011). *Nanotechnology research directions for societal needs in 2020: Retrospective and outlook*. Academic Press.

Rycroft, T., Trump, B., Poinsatte-Jones, K., & Linkov, I. (2018). *Nanotoxicology and nanomedicine: making development decisions in an evolving governance environment*. Academic Press.

Sadaf, A., Zeshan, B., Wang, Z., Zhang, R., Xu, S., Wang, C., & Cui, Y. (2012). Toxicity evaluation of hydrophilic CdTe quantum dots and CdTe@SiO2 nanoparticles in mice. *Journal of Nanoscience and Nanotechnology*, *12*(11), 8287–8292. doi:10.1166/jnn.2012.6667 PMID:23421208

Sahaym, U., & Norton, M. G. (2008). Advances in the application of nanotechnology in enabling a 'hydrogen economy'. *Journal of Materials Science*, *43*(16), 5395–5429. doi:10.100710853-008-2749-0

Sakiyama-Elbert, S., & Hubbell, J. (2001). Functional biomaterials: Design of novel biomaterials. *Annual Review of Materials Research*, *31*(1), 183–201. doi:10.1146/annurev.matsci.31.1.183

Sim, S., & Wong, N. K. (2021). Nanotechnology and its use in imaging and drug delivery [Review]. *Biomedical Reports*, *14*(5), 42. doi:10.3892/br.2021.1418 PMID:33728048

Singh, K. K., Singh, A., & Rai, S. (2022). *A study on nanomaterials for water purification*. Academic Press.

Singh, S., Kumar, V., Anil, A. G., Kapoor, D., Khasnabis, S., Shekar, S., . . . Singh, J. (2021). *Adsorption and detoxification of pharmaceutical compounds from wastewater using nanomaterials: A review on mechanism, kinetics, valorization and circular economy*. Academic Press.

Singh, T., Shukla, S., Kumar, P., Wahla, V., Bajpai, V. K., & Rather, I. A. (2017). Application of nanotechnology in food science: Perception and overview. *Frontiers in Microbiology*, *8*, 1501. doi:10.3389/fmicb.2017.01501 PMID:28824605

Siri, J. G., Fernando, C. A., & De Silva, S. (2020). Nanotechnology and Protection of Intellectual Property. *Emerging Trends.*, *14*(4), 307–327. PMID:32532198

Soares, S., Sousa, J., Pais, A., & Vitorino, C. (2018). Nanomedicine: Principles, Properties, and Regulatory Issues. *Frontiers in Chemistry*, *6*, 360. doi:10.3389/fchem.2018.00360 PMID:30177965

Sousa, F., Ferreira, D., Reis, S., & Costa, P. (2020). *Current insights on antifungal therapy: Novel nanotechnology approaches for drug delivery systems and new drugs from natural sources*. Academic Press.

Spagnolo, A. G., & Daloiso, V. (2009). *Outlining ethical issues in nanotechnologies*. Academic Press.

Suravi Pandit, Dewan, & Ahmed. (2016). Nanotechnology based biosensors and its application. *The Pharma Innovation*, *6*(5), 18–25.

Tinkle, S., McNeil, S. E., Mühlebach, S., Bawa, R., Borchard, G., Barenholz, Y. C., Tamarkin, L., & Desai, N. (2014). Nanomedicines: Addressing the scientific and regulatory gap. *Annals of the New York Academy of Sciences*, *1313*(1), 35–56. doi:10.1111/nyas.12403 PMID:24673240

Tiwari, S. B., & Amiji, M. M. (2006). A review of nanocarrier-based CNS delivery systems. *Current Drug Delivery*, *3*(2), 219–232. doi:10.2174/156720106776359230 PMID:16611008

Tonndorf, R., Aibibu, D., & Cherif, C. (2021). *Isotropic and anisotropic scaffolds for tissue engineering: Collagen, conventional, and textile fabrication technologies and properties.* Academic Press.

Trotta, F., Zanetti, M., & Cavalli, R. (2012). Cyclodextrin-based nanosponges as drug carriers. *Beilstein Journal of Organic Chemistry*, *8*, 2091–2099. doi:10.3762/bjoc.8.235 PMID:23243470

Xue, C., Zhu, H., Tan, D., Ren, H., Gu, X., Zhao, Y., Zhang, P., Sun, Z., Yang, Y., Gu, J., Gu, Y., & Gu, J. (2018). Electrospun silk fibroin-based neural scaffold for bridging a long sciatic nerve gap in dogs. *Journal of Tissue Engineering and Regenerative Medicine*, *12*(2), e1143–e1153. doi:10.1002/term.2449 PMID:28485084

Yang, W., Gong, Y., & Li, W. (2021). *A review: Electrode and packaging materials for neurophysiology recording implants*. Academic Press.

Zheng, G., Patolsky, F., Cui, Y., Wang, W. U., & Lieber, C. M. (2005). Multiplexed electrical detection of cancer markers with nanowire sensor arrays. *Nature Biotechnology*, *23*(10), 1294–1301. doi:10.1038/nbt1138 PMID:16170313

Zheng, X., Zhang, P., Fu, Z., Meng, S., & Dai, L. (2021). *Applications of nanomaterials in tissue engineering*. doi:10.1039/D1RA01849C

Zor, F., Selek, F. N., Orlando, G., & Williams, D. F. (2019). Biocompatibility in regenerative nanomedicine. *Nanomedicine (London)*, *14*(20), 2763–2775. doi:10.2217/nnm-2019-0140 PMID:31612774

Chapter 7
Comparative Study of Conventional Techniques and Functional Nanomaterials for PMI

Rahat Bukhari
University of Health Sciences, Pakistan

Rahat Abdul Rehman
University of Health Sciences, Pakistan

ABSTRACT

The most vital part of a medico-legal investigation is the estimation of the period since death. The procedures now in use for determining this time period depend on a wide range of variables that are frequently out of the forensic examiner's control. Therefore, using nanosensors to analyse changes in biological components has provided a special benefit to precisely estimate the PMI. The evaluation of post-mortem intervals using novel functional nanomaterials techniques is compared in detail to conventional methods in this chapter. It also discusses the difficulties in making an accurate assessment due to changing environmental conditions as well as the benefits of using nanomaterials.

INTRODUCTION

A significant challenge in forensic investigations is determining the post-mortem interval (PMI). Forensic case investigations are greatly influenced by early, accurate PMI estimates. There are many acknowledged techniques that can be used to study cadaver phenomena in order to identify early PMI, such as monitoring rectal temperature, post-mortem lividity, muscle relaxation, and stomach content. Due to significant variances in these methods and a variety of environmental and individual factors, a significant inaccuracy in PMI assessment may occur (Liao et al., 2020).

DOI: 10.4018/978-1-6684-8325-1.ch007

Definition of Postmortem Interval

The time between death and finding a body is called the post-mortem interval (PMI). In order to identify the time of death, it is necessary to find human remains. The time between a person passing away and a body being found is known as the post-mortem interval (PMI) (Gelderman, Boer, Naujocks, IJzermans, & Duijst, 2018).

Importance of Postmortem Interval

The PMI gives a temporal range that can be used to identify the human remains and help in the study of potential causes of death, therefore its understanding is crucial. To pinpoint the exact time of death, two techniques have been developed: the Henssge-nomogram and electrical or mechanical activation of skeletal muscles. Because the body ceases cooling once it reaches the ambient temperature, these three approaches are only useful when the PMI is transient. Up to 13 hours after a person has passed away, both mechanical and electric muscle stimulation are performable. it is challenging to pinpoint the precise time of death in case of longer PMI (Gelderman et al., 2018).

Introduction of Nanomaterials

Nanomaterials have over time found utility in catalytic applications such as environmental remediation, nanomedicine, and other fields due to their tiny size, large surface area, and enhanced reactivity. These nanoparticles can be functionalized with wide variety of functional groups, which both enhance and add to the nanomaterials' already existing characteristics. These functionalized nanomaterials have a much wider range of uses and also perfect for use in forensic research. Recent advances in the detection of explosives, illegal substances, chemical and biological warfare agents, and fingerprints identification have been made possible by these functionalized nanoscale materials (Rawtani, Tharmavaram, Pandey, & Hussain, 2019).

Estimation of Postmortem Interval

Death investigations are quite tough since it is impossible to measure the postmortem interval (PMI). PMIs are frequently the result of postmortem alterations such as hypostasis, rigor mortis, rectal temperature, or various supravital reactions in forensic practice. Even though numerous studies have been done to develop more accurate techniques, there is no such technique that can be used to estimate PMIs accurately, especially after prolonged PMIs and in more thoroughly degraded remains. There are currently no suitable or workable methods that can be applied as quick and accurate PMI estimators at crime scenes due to the inherent difficulties of keeping living materials (Xia, Zhai, Liu, & Mo, 2015).

Since autopsies provide easy access to biofluids, the PMI has been assessed using a number of biofluids over the years. However, the environment has a significant impact on the precision and reliability of biofluid measurements. For this purpose, vitreous humor gain attention because of its benefits over cerebrospinal fluid and blood, it is recognized as a suitable biological matter for post-mortem biochemical research. VH is the sole physically isolated fluid due to its anatomical location, resistance to pollutants, and ability to remain intact even in situations of open craniocerebral trauma. PMI has been calculated by using a variety of biochemical components of VH, including potassium, magnesium, and hypoxanthine

(HxBut research suggests that integrating Hx concentration with ambient temperature might result in a more precise assessment of PMI. Hx is produced as a byproduct of the metabolism of adenosine triphosphate (ATP). It undergoes additional breakdown to become uric acid (UA), which is then transformed into xanthine (XA) by xanthine oxidase (XOD). Spectrophotometry, electrophoresis, chemiluminescence, and HPLC are a few conventional methods that have been used to find Hx. However, the costs, the time required, and the complexity of these procedures are high (Liao et al., 2020).

An electrochemical biosensor for vitreous humor hypoxanthine detection is a useful tool in forensic scenarios for figuring out how long after the death the body was last examined. The biological components of VH, such as salt, chloride, calcium, magnesium, phosphate, and potassium, have been suggested for use in calculating PMI. Although studies on the vitreous sodium-potassium ratio have been published recently, no connection has been found between variations in the ratio and the postmortem time. Baniak found no indication of a relationship between vitreous Na+ and Cl ion concentrations and PMI, and also Fang didn't find any evidential relationship between vitreous Na+ and Cl ion concentrations and PMI. Measurement of vitreous potassium has been the profound parameter for PMI prediction that has received the most research out of all the vitreous constituents (Ding et al., 2017).

Figure 1. Representation of PMI by vitreous humor
Source: Zięba, Wiergowski, Cieślik, Anand, and Krzyżanowska (2023)

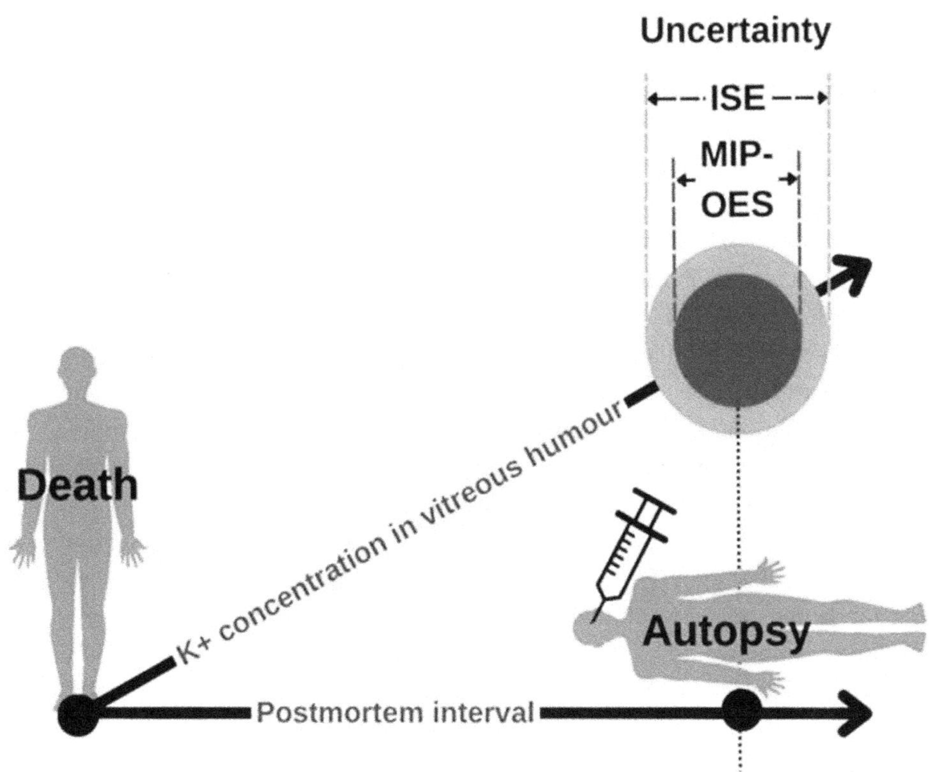

Techniques for Postmortem Intervals

In order to create before-and-after comparison, all of the samples are passes through post-mortem analyses. With the help of thermal desorption spectroscopy and ion beam analyses, different elements' concentrations and depth distributions can be discovered. Morphology and porosity can be determined using electron microscopy, and the form and variety of devices found in the substance can be determined using X-ray diffraction (XRD). Raman microscopy is also useful for this purpose. This method relies on an inelastic light scattering mechanism, which makes It is extremely sensitive to morphology, structure, and chemical content. Depending on the object's optical properties which are being probed, the surrounding surface can be detected using Raman microscopy, with a lateral resolution of a few micrometers (in the case of metals, a few tens of nanometers). This makes it possible to keep an eye on spatial inhomogeneities. The Raman microscopy technique can disclose the chemical bonds that hold materials together as well as the bonds with other impurities and hydrogen isotopes with metallic atoms and electrons. This can be done by providing information on atomic vibrations (Pardanaud et al., 2020).

Several techniques have been put out for determining skeletal remains at this time, including luminol chemiluminescence reaction, radioisotope studies, X-ray diffraction, and fluorescence tests. Additionally, vibrational spectroscopy methods like Fournier transform infrared (FTIR) and Raman spectroscopy may swiftly, directly, and economically acquire the spectrum information of molecules with little to no sample preparation. In light of this, it would seem that vibrational spectroscopy is the most effective technique for examining bone tissue for forensic purposes, and considerable effort has been expended on the study of changes to bones following death.

Fournier Transform Infrared (FTIR) Spectroscopy

Using FTIR spectra, the crystallinity index (C.I.) and carbonate-phosphate index may determine modern human bone fragments from prehistoric bone fragments. The carbonate, organic, and C. I. contents of pig bone can be measured using FTIR spectroscopy to reveal the time-dependent changes connected to soil environment. Researchers were able to identify the local distribution of organic and mineral components in forensic and archaeological bones by creating chemical maps utilizing mid-infrared microscopic imaging from the surface of the bones. Similar sensitivity to daily variations in the composition of human ribs and leg bones was found in Raman spectroscopy research. However, chemometric techniques were not fully utilized in these investigations, which prevent them from providing more thorough spectral data. In order to analyze bone components for clinical or forensic purposes, chemometrics has also been employed in conjunction with vibrational spectroscopic technology as a potent analytical tool. Our previous investigations using blood samples, spleen, liver, kidney, brain, and other human and animal tissues have shown that FTIR can unmistakably identify postmortem alterations. By using FTIR to estimate the PMI, the infrared spectrum demonstrates its astounding superiority. The objectives of this work were to discover sensitive spectral areas that would be able to provide crucial dating information for future research and to date human skull remains using FTIR in conjunction with chemometrics (Wang et al., 2017).

Raman Spectroscopy

In the case of crystal-triggered AKI, the undetermined stones in the renal tissue of a young male victim of a severe burn were determined as calcium oxalate dihydrate by Raman spectroscopy via forensic diag-

Figure 2. FTIR spectroscopy for estimating the postmortem interval

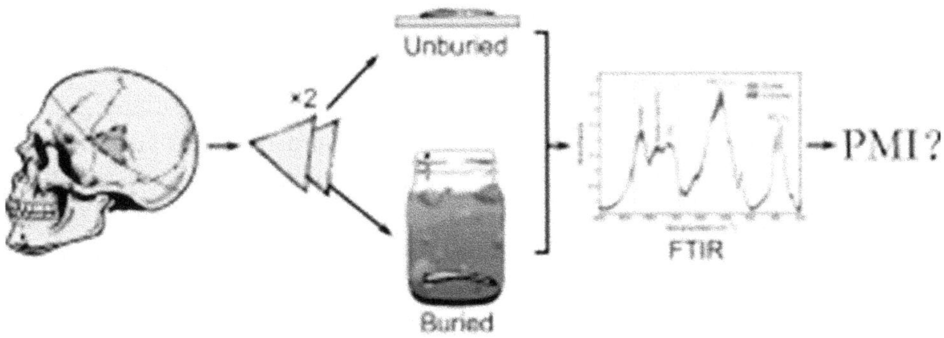

nostic tests. Additionally, in order to confirm the viability of using Raman spectroscopy in pathological biopsy that closely resembled the biopsy we created a mouse model of calcium oxalate dihydrate crystal nephropathy. We are the first to report using Raman spectroscopy in forensics to identify unknown crystalline chemicals in tissue slides (Yu et al., 2022).

Figure 3. Raman spectroscopy for estimating the postmortem interval

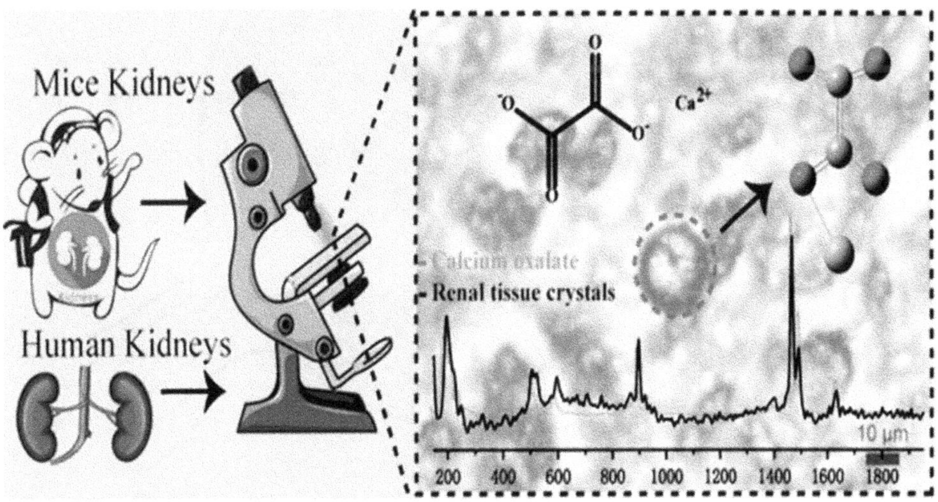

Method for Estimating Postmortem Interval by Nanomaterials

A unique method that combines electron paramagnetic resonance (EPR) and SQUID magnetometry has been used to examine the physical characteristics of ferric ions and iron-oxide nanoparticles in postmortem human brain tissue. We can calculate the volume of the low molecular weight iron pool and its electron spin relaxation rates using EPR. We also recognize iron mineralization products linked to a magnetite/maghemite phase and a ferrihydrite (ferritin) phase using SQUID magnetometry. Further calculations are made to determine the concentration of ferritin and magnetite/maghemite nanoparticles.

We examined brain tissue from an Alzheimer's patient and a healthy control in order to evaluate the new combination technology. Finally, we show that the magnetite/maghemite nanoparticles, whose sizes surpass 40–50 nm and whose magnetic moments are blocked at ambient temperature, are incompatible with the ferritin protein, whose core is typically 6–8 nm in size. Given that aberrant iron accumulation in the brain is a defining feature of numerous illnesses, including Alzheimer's Disease, we think that this approach could be useful in the research of neurodegenerative disorders. Blood, cerebrospinal fluid, and vitreous humor (VH), among other body fluids, are used to determine a variety of metabolic, physiological, and biochemical changes that are utilized in estimating the PMI through death. It has been asserted that PMI can be measured using the biological components of VH, such as sodium, chloride, calcium, magnesium, phosphate, and potassium. The vitreous sodium-potassium ratio has been studied recently, but no connections between these changes and the postmortem period have been found.

Figure 4. Vitreous humor

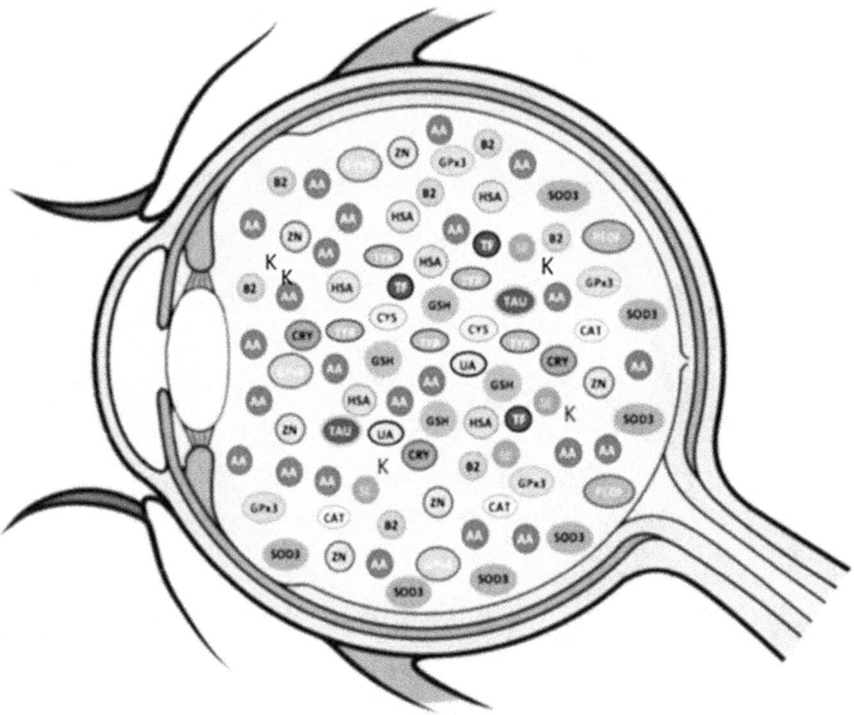

An innovative fluorescence sensing method can be determined by using silver nanoclusters (AgNCs), a new family of luminous materials. This is partly brought about by its distinctive characteristics, such as biomaterials, water solubility, and low toxicity. Over time, there have been reports of AgNCs being made utilizing a variety of templates. Single stranded DNA (ssDNA) as a green chemical template for the synthesis and characterization of AgNCs is the subject of extensive investigation at the moment. Single-stranded nucleic acid ligands known as aptamers provide a specialized epitope with high specificity and affinity for a certain target molecule. Low toxicity, ecologically friendly production, and biocompatibility are advantages of such an aptamer-based sensor, or apt sensor. It was customary to use the distinctive aptamer/AgNCs assembly as a fluorescent probe and a specific binding ligand, respectively. To the best of our knowledge, there aren't many research about the application of fluorescence sensing methods for PMI assessment and vitreous potassium detection in the forensic-related literature. In this study, fluorescent AgNCs encased in DNA were used as a tool to assess PMI and to establish a selective fluorescence sensing approach for vitreous potassium detection. These ground-breaking scientific papers served as our guide. Here, we create a unique DNA sequence that will allow AgNCs to self-assemble and precisely detect K+ ions in VH. In view of the findings from the recommended fluorescence assay's study of the VH sample, a novel formula explaining the relationship between PMI and [K+] has been estimated. The established fluorescent apt sensor could show the distinctive benefits of detecting efficacy, high sensitivity, selectivity, and simple preprocessing (Ding et al., 2017).

Figure 5. AgCN encased in DNA use for K+ detection vitreous humor
Source: Song et al. (2019)

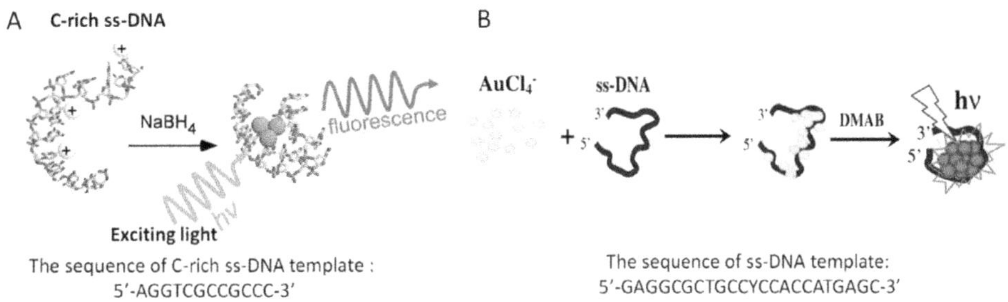

Biochemical Changes in Postmortem Intervals

Research on VH for calculating PMI has been conducted throughout the last few decades. In VH, a large number of biochemicals estimate the PMI, and some of these substances show a favorable correlation. There are many other sorts of components that can be categorized, including electrolytes, amino acids, hypoxanthine (Hx), creatinine, urea nitrogen, and uric acid. Because of these prior investigations, forensic investigators are able to compute PMI exactly. More study on the relationships between VH and PMI changes is required in light of this situation.

Potassium

The intracellular potassium levels are maintained at high concentrations throughout life while extracellular (VH) potassium levels are maintained at low levels as a result of the Na/K-action. pump's Cellular hypoxia, which causes ATP to be depleted and selective membrane permeability for ions to be lost after death, alters postmortem vitreous potassium levels. Postmortem vitreous potassium levels rise as a result of passive diffusion of intracellular potassium into the vitreous body. Due to alterations brought on by this process and cell autolysis, postmortem VH levels diverge from antemortem biochemistry, which has been researched in PMI estimation studies (Li et al., 2018).

The extracellular fluid is the location where sodium ions are present in the highest concentration. The total body sodium content determines the extracellular fluid compartment's volume, and the serum sodium concentration has the biggest impact on the crystal osmotic pressure. Chloride, the primary anion in extracellular fluid, is also closely related to alterations in the body's acid-base equilibrium and salt metabolism. Following this, vitreous sodium and chloride have generated interest, and some researchers have suggested that estimating PMI can be done by detecting sodium and chloride levels in the VH. However, numerous studies have not found a connection between PMI and vitreous sodium and chlorides. Currently, it is thought that the salt and chloride concentrations in vitreous matter have little bearing on calculating PMI. The salt and chloride concentrations in the VH steadily decreased by around 2.2 mM/day as the PMI increased. This might be due to the study's small sample size and the shifting decrease in vitreous sodium concentrations with a limited range of PMI. Thus, rather than PMI assessment, vitreous sodium and chloride concentrations may be more useful for establishing the cause of death (Li et al., 2018).

Magnesium

Magnesium is an important marker for the analysis of postmortem metabolic alterations. During the antemortem period, a small amount of magnesium diffuses from the retina to the lens and into the VH. Due to ATP consumption and the lack of selective cell membrane permeability after death, magnesium ions in postmortem samples are reallocated between the extracellular fluid and intracellular fluid. Therefore, the process producing the increase in postmortem magnesium can be used in PMI computation. It is still debatable if vitreous magnesium may be used as a predictor for PMI calculation. Previously, the PMI was computed using vitreous magnesium levels immediately after death (Li et al., 2018).

Hypoxanthine

Hx is created during the hypoxic breakdown of adenosine monophosphate, and antemortem hypoxia may cause it to increase. Postmortem Hx levels in the VH continue to rise once metabolism ceases, and they might even rise over time as postmortem duration lengthens. Since postmortem potassium changes are thought to be easily impacted by the environment, Hx is projected to replace vitreous potassium in the computation of PMI. According to past study, a number of factors, such as the way of death, the environment, and different analytical procedures, may have an effect on the precision of PMI calculation and the results of Hx detection (Li et al., 2018).

Nitrogenous Compounds

In postmortem serum samples, urea nitrogen, uric acid, and creatinine are thought to be rather stable byproducts of protein breakdown. Numerous researchers have examined the levels of nitrogenous substances in various biological samples obtained during autopsies, such as serum, pericardium, brain fluids, synovial fluid, and VH. Several authors investigated vitreous nitrogenous compounds to determine postmortem changes in nitrogenous compounds and study postmortem vitreous nitrogenous compound levels in relation to various causes of death. The concentrations of vitreous creatinine, uric acid, or urea did not correlate with the PMI. These researchers found that urea and creatinine are strikingly constant in postmortem values, demonstrating their remarkable stability in biological fluids (Li et al., 2018).

Amino Acids

Amino acids, the building blocks of enormous molecular proteins, are used in a variety of biological reaction pathways, including biosynthesis and catabolism. Currently, over 300 more amino acids that are used in diverse ways by people have been discovered in cells. The blood-vitreous barrier is permeable to amino acids in their free form. Numerous studies have focused on postmortem changes in amino acids in VH because these molecules are regarded to be useful indicators for identifying the PMI. The amount of free amino acids in the vitreous tissue was measured using HPLC, and it was found that the majority of these levels increased linearly with rising PMI. Amino concentrations are being used increasingly frequently in forensic studies to pinpoint the moment of death. Despite the fact that the VH amino acid determination for PMI estimate is less useful than vitreous potassium, studies have indicated that it has the potential to grow (Li et al., 2018).

CONCLUSION

We looked into how sensitive the Raman spectroscopy was to the dense, porous, and nanoparticle forms of tungsten oxide and nitride. This method has the benefit of quickly providing a sizable amount of data based on semi-qualitative spectroscopic observations from the perspective of an understudies' post-mortem campaign. This could help identify the samples that need time-consuming quantitative investigations utilizing methods like TOF-ERDA or electron microscopy (Pardanaud et al., 2020).

Significant progress has been made in the assessment of PMI using vitreous humor. Vitreous potassium remains the most reliable indicator of PMI. Many scientists have focused on creating cutting-edge statistical and analytical methods to precisely measure PMI. Future studies on PMI estimate could look into multiple constituents at once using cutting-edge detection techniques like fluorescent-sensing analytical methods. These techniques can examine ions, proteins, amino acids, glucose, and a wide range of other biological components. Similar to the microarray approach, different fluorescent probes may be created for a matrix in order to simultaneously detect the majority of biological components in the VH. It can be useful to estimate PMI and improve the accuracy of the equations used to calculate it by setting up an assessment instrument for multiple constituent simultaneous analysis to evaluate the outcomes of each individual component and take into account various causes of death, states of illness, and influencing factors (Li et al., 2018).

REFERENCES

Ding, Y., Li, X., Guo, Y., Duan, W., Ling, J., Zha, L., Yan, J., Zou, Y., & Cai, J. (2017). Estimation of postmortem interval by vitreous potassium evaluation with a novel fluorescence aptasensor. *Scientific Reports*, *7*(1), 1–9. doi:10.103841598-017-02027-1 PMID:28500353

Donaldson, A. E., & Lamont, I. L. (2014). Estimation of post-mortem interval using biochemical markers. *The Australian Journal of Forensic Sciences*, *46*(1), 8–26. doi:10.1080/00450618.2013.784356

Estracanholli, É. S., Kurachi, C., Vicente, J. R., de Menezes, P. F. C., & Bagnato, V. S. (2009). Determination of post-mortem interval using in situ tissue optical fluorescence. *Optics Express*, *17*(10), 8185–8192. doi:10.1364/OE.17.008185 PMID:19434150

Gelderman, H., Boer, L., Naujocks, T., IJzermans, A., & Duijst, W. (2018). The development of a post-mortem interval estimation for human remains found on land in the Netherlands. *International Journal of Legal Medicine*, *132*(3), 863–873. doi:10.100700414-017-1700-9 PMID:29110084

Kim, B. J., Han, C., Moon, H., Kwon, J., Jang, I.-S., Lim, S.-K., Park, K.-W., Choi, J.-S., & An, H. J. (2018). Monitoring of post-mortem changes of saliva N-glycosylation by nano LC/MS. *Analytical and Bioanalytical Chemistry*, *410*(1), 45–56. doi:10.100700216-017-0702-2 PMID:29164281

Li, W., Chang, Y., Cheng, Z., Ling, J., Han, L., Li, X., & Ding, Y. (2018). Vitreous humor: A review of biochemical constituents in postmortem interval estimation. *Journal of Forensic Science and Medicine*, *4*(2), 85. doi:10.4103/jfsm.jfsm_13_18

Liao, L., Xing, Y., Xiong, X., Gan, L., Hu, L., Zhao, F., Tong, Y., & Deng, S. (2020). An electrochemical biosensor for hypoxanthine detection in vitreous humor: A potential tool for estimating the post-mortem interval in forensic cases. *Microchemical Journal*, *155*, 104760. doi:10.1016/j.microc.2020.104760

Munoz Barus, J. I., Febrero-Bande, M., & Cadarso-Suárez, C. (2008). Flexible regression models for estimating postmortem interval (PMI) in forensic medicine. *Statistics in Medicine*, *27*(24), 5026–5038. doi:10.1002im.3319 PMID:18618426

Oliveira-Costa, J., & Mello-Patiu, C. A. d. (2004). Application of forensic entomology to estimate of the postmortem interval (PMI) in homicide investigations by the Rio de Janeiro Police Department in Brazil. *Anil Aggrawal's Internet Journal of Forensic Medicine and Toxicology*, *5*(1), 40–44.

Pardanaud, C., Dellasega, D., Passoni, M., Martin, C., Roubin, P., Addab, Y., ... Salomon, E. (2020). Post-mortem analysis of tungsten plasma facing components in tokamaks: Raman microscopy measurements on compact, porous oxide and nitride films and nanoparticles. *Nuclear Fusion*, *60*(8), 086004. doi:10.1088/1741-4326/ab9347

Rawtani, D., Tharmavaram, M., Pandey, G., & Hussain, C. M. (2019). Functionalized nanomaterial for forensic sample analysis. *Trends in Analytical Chemistry*, *120*, 115661. doi:10.1016/j.trac.2019.115661

Sato, T., Zaitsu, K., Tsuboi, K., Nomura, M., Kusano, M., Shima, N., Abe, S., Ishii, A., Tsuchihashi, H., & Suzuki, K. (2015). A preliminary study on postmortem interval estimation of suffocated rats by GC-MS/MS-based plasma metabolic profiling. *Analytical and Bioanalytical Chemistry*, *407*(13), 3659–3665. doi:10.100700216-015-8584-7 PMID:25749795

Song, C., Xu, J., Chen, Y., Zhang, L., Lu, Y., & Qing, Z. (2019). DNA-templated fluorescent nanoclusters for metal ions detection. *Molecules (Basel, Switzerland)*, *24*(22), 4189. doi:10.3390/molecules24224189 PMID:31752270

van Landeghem, F. K., Maier-Hauff, K., Jordan, A., Hoffmann, K.-T., Gneveckow, U., Scholz, R., Thiesen, B., Brück, W., & von Deimling, A. (2009). Post-mortem studies in glioblastoma patients treated with thermotherapy using magnetic nanoparticles. *Biomaterials*, *30*(1), 52–57. doi:10.1016/j.biomaterials.2008.09.044 PMID:18848723

Wang, Q., Ishikawa, T., Michiue, T., Zhu, B.-L., Guan, D.-W., & Maeda, H. (2012). Stability of endogenous reference genes in postmortem human brains for normalization of quantitative real-time PCR data: Comprehensive evaluation using geNorm, NormFinder, and BestKeeper. *International Journal of Legal Medicine*, *126*(6), 943–952. doi:10.100700414-012-0774-7 PMID:23010907

Wang, Q., Zhang, Y., Lin, H., Zha, S., Fang, R., Wei, X., Fan, S., & Wang, Z. (2017). Estimation of the late postmortem interval using FTIR spectroscopy and chemometrics in human skeletal remains. *Forensic Science International*, *281*, 113–120. doi:10.1016/j.forsciint.2017.10.033 PMID:29127892

Xia, Z., Zhai, X., Liu, B., & Mo, Y. (2015). Determination of Electrical Conductivity of Cadaver Skeletal Muscle: A promising method for the estimation of late postmortem interval. *Journal of Forensic Science and Medicine*, *1*(1), 16. doi:10.4103/2349-5014.155554

Yu, Y., Jiang, Q., Wan, H., Li, R., Sun, Y., Zhang, Z., Mao, Z., Cao, Y., & Chen, F. (2022). In Situ Identification of Unknown Crystals in Acute Kidney Injury Using Raman Spectroscopy. *Nanomaterials (Basel, Switzerland)*, *12*(14), 2395. doi:10.3390/nano12142395 PMID:35889619

Yuan, W., Tang, Y., Yang, X., & Wan, Z. (2012). Porous metal materials for polymer electrolyte membrane fuel cells–A review. *Applied Energy*, *94*, 309–329. doi:10.1016/j.apenergy.2012.01.073

Zhang, Y., Liu, L., & Ren, L. (2020). Liquid chromatography-tandem mass spectrometry (LC-MS/MS) determination of cantharidin in biological specimens and application to postmortem interval estimation in cantharidin poisoning. *Scientific Reports*, *10*(1), 10438. doi:10.103841598-020-67278-x PMID:32591575

Zięba, S., Wiergowski, M., Cieślik, B. M., Anand, J. S., & Krzyżanowska, M. (2023). Uncertainty of Postmortem Time Estimation Based on Potassium Ion Determination in Vitreous Humor Using Potentiometric Ion-Selective Electrode and Microwave-Induced Plasma with Optical Emission Spectrometry Methods. *Separations*, *10*(3), 201. doi:10.3390eparations10030201

Chapter 8
Employing Nano-Sensors for Explosives Detection

Mubeen Khalid
University of Health Sciences, Pakistan

Muhammad Farhan Khan
University of Health Sciences, Pakistan

Shahid Nazir
University of Health Sciences, Pakistan

ABSTRACT

This chapter presents the potential of the application of nano-sensors for trace explosive detection. A major challenge faced in detecting trace explosives is the detection of ultra-low quantities of trace compounds. This chapter discusses the selective and sensitive detection of trace explosives by employing nano-sensors, limitations of detecting trace explosives, and reevaluates the current technologies that are inefficient and inaccurate for trace explosive detection. Application of nano-sensors for the exposure of low concentration explosive compounds is an effective platform in the advancement of trace explosive detection.

INTRODUCTION

A reactive substance containing significant amount of potential energy is called as explosive, when suddenly released, rapidly and spontaneously decays with the progression of substantial amounts of heat and gas. Explosions caused by explosives are typically complemented by the assembly of light, sound, pressure, and heat.

The potential energy stowed in the explosive substances can be:

- Grain dust or nitroglycerin as chemical energy
- BLEVE or aerosol can and gas cylinder as compressed gas
- Isotopes of plutonium-239 and uranium -235 as nuclear energy

DOI: 10.4018/978-1-6684-8325-1.ch008

The pace at which explosive materials expand can be used to classify them. Low and high explosives are terms used to describe materials that detonate (faster than the speed of sound the chemical reaction mechanism passes through the material). Explosives can also be divided into groups based on how sensitive they are. Primary explosives are materials that can be ignited by relatively low levels of pressure and heat, whereas secondary and tertiary explosives are constituents that are comparatively insensitive.

Many different substances have the potential to detonate, albeit fewer are produced with the intention of being used as explosives. The remainder is too risky, expensive, unstable, decomposition prone, hazardous, delicate, or degradation quickly.

On the other hand, some substances are flammable or combustible only if they burn without detonating. The change is not significant, though. Under normal circumstances, some substances—powders, volatile organic liquids, dusts, powders, or gases —may only be inflammable or combustible, but in certain circumstances or under certain conditions—such as confinement or quick release—they may turn explosive (Meixner, Fleischer, Kern, Sheremet, & McMillan, 2022).

Classification of Explosives

Low Explosives

Low explosives typically deflagrate as opposed to detonating. 2000 < 3000 feet per second is the range of their reaction times. A nice illustration is black powder. These materials often have a enlivening influence during blasting, high flammability, low water resistance, and sensitivity to #6 blasting cover. Typically, low explosives are not as effective in breaking rock as high explosives (Viswanath, Ghosh, & Boddu, 2018). When triggered, low explosives like black powder and nitrocellulose self-ignite very slowly, releasing massive volumes of gas in a clear and controllable manner. Explosives of numerous varieties are produced (Shreve & Brink Jr, 1977).

High Explosives

A high explosive is any compound blend that explodes with a counter speed in more than 5,000 feet per second. Primary or initiation and secondary explosives are further categories for explosives. Due to their high sensitivity, initiating explosives are typically utilized in small quantities in percussion caps and detonators to set off greater amounts of secondary explosives. Nitramines, nitro compounds, and nitrates, in particular are far less sensitive to thermal or mechanical shock but detonate with a powerful explosion when ignited by the starting explosive. TNT (2,4,6-trinitrotoluene) and ammonium nitrate explosives are the primary secondary explosives created for both industrial and military applications (Shreve & Brink Jr, 1977). The blasting cap no. 8 is used for commencing the reaction. High explosives are:

1. **Straight Dynamite:** Nitroglycerin travels at speeds of between ten and twenties of thousands of feet per second and is absorbent. This dynamite is the most sensitive explosive ever used in trade. The filling's nitroglycerin content can be precisely determined by weight strength. This explosive has poor fumes, good water resistance, and poor cohesiveness.
2. **Ammonia Dynamite:** The solitary difference between ammonia and regular dynamite is that precise amount of nitroglycerin has been replaced by ammonium or sodium nitrate as well as different carbonaceous fuels. Three types of ammonia dynamite exist: High Density: The detonation veloc-

ity of this product ranges from 8,000 to 13,000 feet per second, and it has good water resistance and passable odors. Low density: This product offers good to good vapor resistance, good to poor water resistance, and detonation velocities between 7,000 and 11,000 feet per second. Acceptable kinds with the exception of the addition of cooling salts such sodium chloride, these products are comparable to low-density ammonia dynamite. Under specific usage restrictions, the US Bureau of Mines must authorize the permit. The typical resistance to vapor and water for this material is good to poor.

3. **Gelatin Dynamite:** Contains nitrocellulose and different absorbent filler materials are used to gel nitroglycerin. It condenses into a rubbery to soupy substance that is waterproof. Straight gelatin (a): 13,000 to 23,000 feet per second explosion velocity. Variety strengths greater than 60% have poor vapor characteristics. The substance is extremely cohesive and highly water resistant. (b) Special or Ammonia Gelatins - These gelatins have a similar structure to direct gelatin, but instead of nitroglycerin, they contain sodium and ammonium nitrates along with carbonaceous fuels. The range series of detonation velocity is between 10,000 to 23,000 feet per second. And they have good water resistance.
4. **Semi-Gelatin Dynamite:** A substance that combines dynamite and gelatin of ammonia and is water resistant but less powerful than gelatin amongst 10,000 and 15,000 feet per second in speed. The smokiness score is favorable.

Two-Component Explosives

Each component of these packages is non-explosive; thus, they can be carried in any manner. But conjoining outcomes must be managed and stored accordingly in a Class A explosive (Viswanath et al., 2018).

History of Explosive Discovery and Use

Mankind's history is also the history of war. Since ancestry these types of chemicals had been used by humans for different construction destruction purpose. Unplanned blasts during their revelation or creation have obviously exhibited the "enchanted power" of these materials. In spite of the fact that they were just used as firecrackers at strict services for quite a while, it was likewise understood that such materials could be utilized for "unusual distinction". The pre-arranged and foundational utilization of the properties of dangerous materials brought about the making of hazardous gadgets. Unstable gadgets can be utilized both in cautious and hostile tasks. Such kind of their tactical application changed the idea of fighting (Horváth & Szatai, 2020). The earliest man-made unstable explosive material was dark or smoky explosive, which is generally alluded to as black powder, as it was utilized for quite a while for the most part to fire shots from guns.

During the 19th and 20th centuries, the development of modern explosives such as dynamite and nitroglycerin led to their use in military conflicts around the world. During World War I, both sides used explosives extensively, from artillery shells and grenades to land mines and improvised explosive devices (IEDs). The bombing of civilian targets during this conflict, including the use of air raids and gas attacks, caused widespread devastation and civilian casualties. In World War II, the development of even more powerful explosives such as the atomic bomb and other incendiary devices led to even greater destruction and loss of life. The bombings of Hiroshima and Nagasaki in 1945, which killed over 200,000 people, remain some of the most devastating examples of explosive use for human destruction in history

Explosives have been used for military purposes for centuries. They have been used in various forms, such as gunpowder, dynamite, and plastic explosives, to name a few. The primary purpose of explosives in military applications is to cause destruction and damage to enemy targets, such as buildings, vehicles, and infrastructure (D. Kumar & Elias, 2019).

Explosives are also used to create barriers, clear obstacles, and make pathways through difficult terrain, such as mountains or forests. They can be used to construct tunnels, dig trenches, and blast away rocks.

But here we discuss how to detect the explosions using different methods especially nano sensors to limit the negative impact of these explosive materials.

Why Is Explosion Detection Necessary?

Explosion detection is required for several reasons, including:

- **Safety:** Explosions can be extremely dangerous and can cause serious injuries or even death. Detecting explosions as early as possible can help reduce the risk of harm to people and property.
- **Prevention:** Early detection of an explosion can help prevent further damage and loss of property or equipment. In some cases, detecting an explosion may even prevent an explosion from occurring in the first place by identifying potential hazards and addressing them before they become a problem.
- **Regulatory Compliance:** Many industries, such as oil and gas, chemical processing, and mining, are required by law to have explosion detection systems in place to ensure compliance with safety regulations.
- **Insurance Requirements:** Many insurance policies require that explosion detection systems be installed and maintained in order for coverage to be valid.

Conventional Method of Detecting Explosives

In terms of national security, finding hidden bombs quickly and accurately is of utmost importance. Small quantities of explosives in checked baggage or landmines have the potential to destroy property and injure or kill people. Explosive materials have been identified using a variety of detection techniques from many scientific disciplines, such as vapor detection and X-ray screening.

Active Neutron Interrogation in Explosive Detection

Neutron interrogation has just recently—roughly in the last 25 years—become a practical way to find potentially explosive elements. Then, a number of methods, such as fast neutron analysis, pulsed fast /thermal analysis, fast neutron analysis, thermal neutron analysis, fast neutron radiography, employ neutrons, or fast neutron scattering analysis, neutron elastic scattering and fast neutron radiography to identify explosives.

All neutron-free explosive investigation techniques have shortcomings that reduce their usefulness. Chemiluminescence, canines and ion mobility spectrometry are just a few techniques used to look for traces of explosives in the air (Furton & Myers, 2001). Unfortunately, these methods won't be able to distinguish the substance if the explosives are handled and packaged appropriately. There are also detection techniques that include probing targets with photon and X-ray transmission scattering (Moore

& Journal, 2007). High atomic number materials, on the other hand, can hide these photons, making it possible for anything concealed behind the shielding to remain undiscovered. Additionally, using these techniques atomic number and densities can be broadly calculated, but not so precisely. Confusion can result when looking at objects because of this. A false alarm will be triggered by a non-explosive substance with explosive-like physical characteristics. More crucially, it is still feasible for the explosive element to be mislabeled as innocuous depending on how the ingredients are oriented inside the object.

Neutron source are used with numerous unique ways to look for explosives, but these systems typically employ the same active interrogation strategies. When conducting a passive examination, the system does not speak to subject and merely using non-intrusive methods to gather data, including having a canine unit smell in air near subject's exterior. With an active the explosive and interrogation detection system actually engages with the target to obtain information more quickly, for instance by directing neutrons at it. Each of the systems covered in this article depends on neutrons interacting with relevant substances at a particular target. Depending on the cross section concentration of the neutron and interactions with the atoms in the target, neutrons are either more or less likely to interact and produce radiation (Lehnert & Kearfott, 2010). Gamma rays, characteristic X-rays, elastically and in-elastically scattered neutrons are among radiation produced by these neutron interactions (Buffler & Tickner, 2010). One or more secondary radiation types can be recorded by the detectors in active neutron interrogation systems. The equipment then processes the signal and either a person or a computer analyses it. This aids the operator in deciding whether more investigation is necessary to ascertain whether the items are illicit or harmful.

The fundamental benefit of neutron-based explosive detection systems is that neutrons, which penetrate over iron and lead with precise attenuation, are challenging to protect. This distinguishes it from many other systems, particularly those that use X-ray and gamma-ray beams that are directed at targets that can be protected by dense materials with a medium to high atomic number. The explosive might not be noticed if such materials are positioned in-between explosive and source. However, neutrons can penetrate materials with a high atomic number because they are better protected by light nuclei, and they eventually extent to explosive. An additional drawback of gamma and X-rays is their limited ability to differentiate between inorganic and organic materials, whereas neutrons are effective at doing so based on the chemical makeup of various organic molecules (Y Liu, Sowerby, Tickner, & Isotopes, 2008). They are established technologies that are reliable and mature. Any new strategy must offer a definite

Figure 1. Neutron detector assembly
Source: Marques et al. (2023)

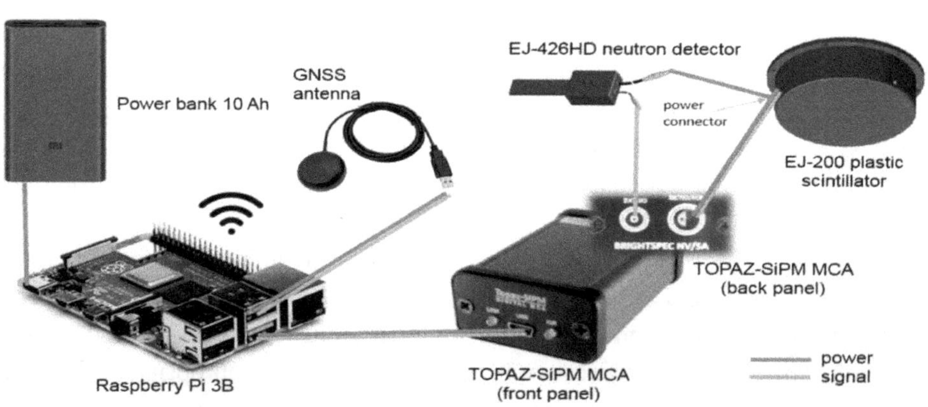

advantage or meet a specific need before it can be widely adopted. Other explosive detection methods may be replaced by active neutron interrogation, but its benefits and drawbacks must be carefully assessed against those of competing technologies (Whetstone, Kearfott, & Chemistry, 2014).

Explosives Detection System for Aviation

According to S. Singh and Singh (2003) there are four basic methods for screening checked baggage: manual hand searches (often known as "pat-downs"), sniffer canines i.e. dogs, trace explosive detection (TED), and inspection of automated X-ray with accredited explosives detection systems. The density and shape of the contents of the luggage are taken into account by X-ray imaging conventional method, by which bulk bombs are detected. Objects with a density that is comparable to that of recognized explosives are marked as potential risks. Although this could provide significant amount of obstacle to further traditional X-ray procedures, as we will discover later, there are a variety of less traditional methods that boost sensitivity to a particular media.

Several technical options for the bulk screening of baggage (both checked and carry-on) became commercially accessible decades ago, and some of them have been practically used no matter what. These systems are comprised on a variety of physical theories, including X-ray diffraction, dual- or multi-energy X-rays in transmission, Raman spectroscopy, and trace analysis. The device must have an extraordinary likelihood of detection in order to truly meet security requirements. In other words, it needs great sensitivity (like capacity of recognizing a threat when one exists) and high specificity, and low false alarm rate (capability to stain benign objects as null extortions). This chapter tries to discuss particularly high-performance technologies, namely those that utilize cutting-edge technology, considering the variations in the civil aviation security threats, the different approaches and tools that have been created to deal with them, and the science behind x-ray interrogation technologies. Since competing neutron interrogation techniques currently have far less developed technology, they are viewed as outside the scope of this paper (Wells, Bradley, & Isotopes, 2012).

Detection of Explosives by Conventional Electronic Noses

Electronic odor detection has garnered a lot of attention during the past 20 years, both for basic and complicated scents. As a result, so-called "electronic noses" have become commercially available. These devices, which typically consist of a number of semi-selective sensors and the obligatory pattern recognition software, have been used to solve a variety of industrial olfactory problems, ranging from food quality assertion to medical diagnosis. Here, the characteristics of electronic noses are examined, and a number of suggestions for improving performances in relation to exposure specificity and limit are made. This chapter discusses the general issue of using an electronic sensor nose to detect explosive materials as well as the unique requirement for compounds very low vapor pressure detection i.e., a detection perimeter of minimum 109 ppm in air and in some cases down to 1012 parts per trillion in air. Even while there is a great deal of potential to raise detection threshold of electronic noses which are sensor-based, it is dubious that this technology will soon overtake the ion mobility spectroscopy's "gold standard" for explosives detection. As a substitute, electronic noses can be thought of as an additional technique employed to look for unidentified substances (Gardner, 2004).

Figure 2. Working principle of e-nose
Source: Wu et al. (2020)

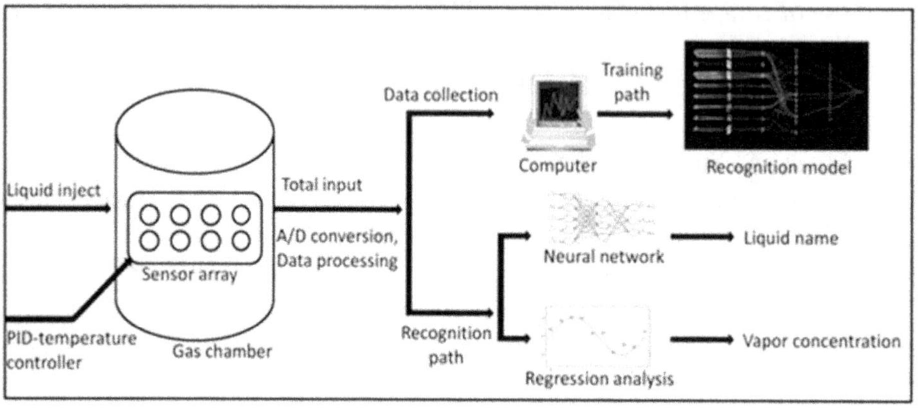

INTRODUCTION TO NANO-SENSORS

Nano-scale instruments known as nano-sensors are used to calculate physical quantities and transform them into signals for analysis and detection. "Nano-sensors are mechanical or chemical sensors that can be employed to either monitor physical parameters like temperature at the nano-scale or to detect the presence of chemicals and nanoparticles."

The same way a sensor works, nano-sensors are platforms with a distinguishing dimension of on the nanometer scale; they can detect either minute particles or minute amounts of something.

Today, a number of approaches to producing nano-sensors are put forth; these include molecular self-assembly, bottom-up assembly, and top-down lithography. Different kinds of nano-sensors are available on the market and are being developed for a variety of applications, particularly in the defense, environmental, and health sectors. The core procedure for these sensors is the same: the analyte is selectively bound, the nano-sensor interacts with the bio element to produce a signal, and the signal is then processed to produce meaningful metrics (Gorup et al., 2020).

Basic Necessities for Designing Nano Sensors

Some basic requirements are listed for modulating desire type of nano sensors:

- **Choose the Appropriate Sensing Mechanism:** The first step in designing a nano sensor is to choose the sensing mechanism that will be used to detect the target property. This could be based on electrical, optical, or mechanical properties, for example.
- **Choose the Appropriate Nanomaterial:** The next step is to choose the appropriate nanomaterial that will be used to construct the sensor. This could be a material with unique electrical or optical properties that will enable the detection of the target property. Carbon nanotubes, graphene, and metal oxide nanoparticles are some examples of materials that are commonly used in nano sensors.

- **Design the Sensor Structure:** Once the sensing mechanism and nanomaterial have been chosen, the next step is to design the structure of the sensor. This will involve determining the size and shape of the sensor, as well as the way in which the nanomaterial will be incorporated into the sensor.
- **Functionalize the Sensor:** In many cases, the sensor will need to be functionalized in order to increase its selectivity and sensitivity. This could involve adding a coating or modifying the surface of the sensor to enable specific interactions with the target property.
- **Test and Optimize the Sensor:** Finally, the sensor will need to be tested and optimized to ensure that it is able to detect the target property with sufficient sensitivity and selectivity. This may involve testing the sensor under different conditions and making adjustments to the design as needed (Nocerino et al., 2022; Javaid, Haleem, Singh, Rab, & Suman, 2021).

HOW DOES NANO SENSORS WORK

Mechanical and chemical nano sensors are two separate types of nano sensors, and they both have unique sensing methods.

When an analyte is discovered, nano sensors that can detect chemicals measure by the minute alternations in electrical conductivity of nanomaterial. A molecule's binding or adsorption reduces the electrical conductivity of many nanomaterials, which is high in general. The change that can be seen is what is measured. One-dimensional (1D) materials like nanotubes and nanowires make upright instances of chemical nano-sensors because, once an analyte is detected, their electrically restricted structure can function as both electronic and transducer wires (Javaid et al., 2021).

A change in a material's electrical conductivity is another way that mechanical nano-sensors function. But this works through an entirely different process. When a material is physically changed, such as when it is used to make mechanical nano-sensors, the material's electrical conductivity changes, and this physical change results in a detectable reaction. A linked capacitor that experiences a detectable change in capacitance as a result of the physical change can also be used to quantify this response (Azman, 2020).

Applications of Nano-Sensors

The greater sensitivity of nano-sensors is advantageous in a wide range of applications. Below, we'll go into greater detail on a few of them.

Optical Nanosensin

In biological applications, optical nano-sensors have proven to be a good sensing technology. This is particularly true in bioimaging, wherever a nano-sensor can accurately measure a molecule's fluorescence. Nanomaterials with non-toxic receptors attached to their surfaces are found in optical nano-sensors, which enable the detection of environmental optical changes. Optical nano-sensors can be utilized for fluorescence detection as well as ion concentration monitoring, measuring hazardous levels in the tracking any undesired or interfering species, and cellular environment.

Figure 3. Nano sensors applications
Source: Ahangaran and Navarchian (2020)

Nano-Sensors for Explosives

Explosive detection that is selective and sensitive is crucial in the battle in contradiction of terrorist extortions. Due to a variety of factors, including the extensive range of explosive materials, the lack of readily recognizable signatures, sheer number of possible delivery methods for these weapons, and the dearth highly selective and sensitive low-cost sensors, the exposure of trace explosives has turned into a actual expensive and complex endeavor. The ability to lower sensor organization costs over mass manufacture, high sensitivity, and selectivity are necessary for success in the fight against terrorism using explosives. Nano-sensors have the ability to meet all the criteria for a platform that is effective for explosive trace detection (Senesac & Thundat, 2008).

Given below is an example of illustration of SERS based three nitro-based explosives detection by employment of gold nanogap.

Environmental Monitoring

Microorganisms and hazardous chemical compounds can be monitored and analyzed using nano-sensors when they are present in environmental samples. Ion-selective electrodes (ISEs) and electrochemical

Figure 4. SERS-based explosion detection
Source: Adhikari et al. (2021)

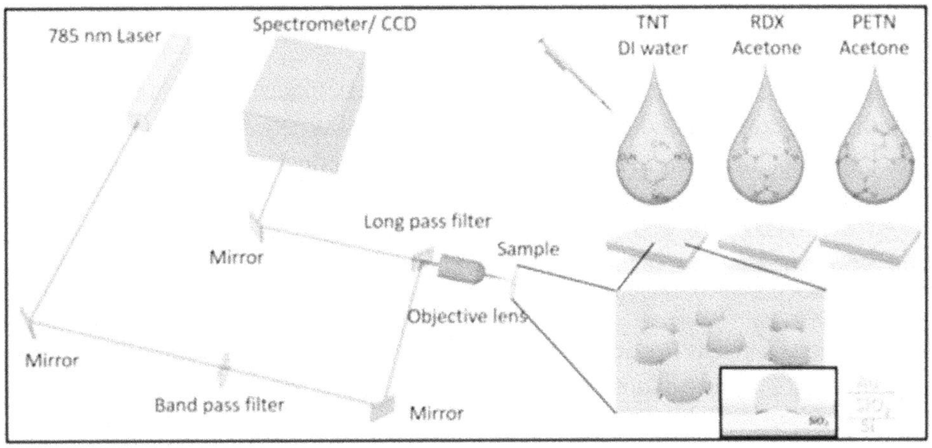

sensors, which are common methods for detecting small levels of metals, nitrates, phosphates, and pesticides in water samples, can be made more sensitive by adding nanomaterials to them. Real-time measurement capabilities, which are highly prized for environmental monitoring applications, are another advantage of nano-sensors.

Sensing Light

Numerous applications concentrate on detecting various chemicals in a certain environment. However, electromagnetic radiation can also be detected using nano-sensors. One illustration is the use of nanowires or nanorods of zinc oxide to sense low-level UV light. Because of alteration of their resistive nature and by providing a detectable response to electromagnetic radiation, nanowires are frequently used for electromagnetic radiation sensing applications. In order to provide a quick and effective reaction, utilization of nanowires can also be paralleled. In this case, electrons cascade over all of the nanowires.

Combining Nano-Sensors With Other Technologies

Conjoining nano sensors with other down to earth innovations, for example, microfluidic gadgets and Microelectromechanical Frameworks (MEMS) gadgets is turning out to be increasingly well known. This has been helpful for various applications, for example, the testimony of nanoparticles on silicon substrates for additional powerful gas and synthetic detecting applications, the utilization of gold nanowires in microfluidic gadgets to recognize blood tests cholesterol level, the utilization of carbon nanotubes (CNTs) on silicon to distinguish hurtful hints of alkali, and the utilization of fluid-based MEMS gadgets to identify minute measures of microorganisms in liquid sample.

Achievements of Nano-Sensors

Carbon-Based Advanced Polymeric Nanocomposites for Forensic Investigations

Due to their high surface area, nanomaterials, particularly those made of carbon have nanoscale dimensions, great strength, flexibility, and electrical conductivity, offer extraordinary physical and chemical traits. Their size, shape, and structure have an impact on their optical characteristics and responsiveness (N. Kumar & Kumbhat, 2016).

Detecting Illegal Drugs

Initial colorimetric tests, such as the Scott, Simon, Marquis, Duquenois-Levine, and Zimmermann tests are frequently followed by confirmatory tests using a variety of chromatographic techniques, including HPLC, GC/MS, FTIR, Raman spectroscopy, etc. These methods for nano matters enable a more effective, sensitive, and sensible strategy (Laurila, Sainio, & Caro, 2017).

Due to its ability to behave as an aptamer fluorescence quencher, GO is utilized to identify illicit substances. While quenching includes the physical adsorption of the aptamer on GO, limiting the sensitivity of detection and perhaps producing fallacious signals, there are other associated concerns. Many common nanomaterials may highly adsorb poly-cytosine (poly-C) DNA, which is DNA sequences with cytosines in the range of 0 to 30, and this method can be utilized to further improve sensitivity in the examination of illegal preparations. In order to recognize cocaine (H. Shi et al., 2018) on the base panel of GO employed an aptamer containing poly-C adsorbed on the top. A nonionic surfactant (Tween 20) was employed to further enhance the characteristics and inhibit the binding of other medications. Tween 20's lipophilic chains are probably how it interacts with GO. By the use of this specific method, the LOD (limit of detection) was reduced to 2.45 pM. And the outcomes demonstrate the potential of our method for analytical applications.

Nafion is a polymer made of perflurosulfonic acid (PFSA) and polytetrafluoroethylene (PTFE), which is frequently used in electrochemistry. An electrochemical sensor with a Nafion membrane and a mesh composed of SWCNTs was created by (Wester et al., 2019) for the targeted exposure of codeine (CO) and morphine (MO). Dry transfer was used to create the SWCNT network electrodes, which were then coated by dipping them in a Nafion membrane. The effective simultaneous exposure of CO and MO in a solution containing significant surfeit of ascorbic acid (AA) and uric acid served as proof of the sensor's efficiency. Nanoscale channels covered with negatively charged sulfone groups are present in the Nafion membrane. Only cationic compounds, like opioids, can thus pass through it. Given that on neutral pH the majority of interferents are negatively charged, such as UA and AA, this is the main benefit of using this type of sensor while working with biological medium. In multicomponent biological media, the Nafion-coated SWCNT electrode demonstrated high sensitivity and selectivity for analyte detecting and was capable of detecting MO in the range of 0.05-10 M and CO in the range of 0.1-50 M. This promising electrode also allowed measurements in human plasma that had not been treated or even mildly diluted without protein precipitation.

The alkaloid tramadol (C16H25NO2) reduces pain by preventing the uptake of norepinephrine and serotonin. In many countries around the world, this potent palliative is utilized for postoperative care.

Even though this alkaloid is thought to be safe to consume, an overdose can have fatal consequences. A electrochemical sensor for tramadol was created by Arabali, Malekmohammadi, and Karimi (2020) using pencil graphite electrode (PGE) that has been altered with PPy and CuO nanoparticles. Using AFM and SEM pictures, the morphology of the produced sensor was examined. With a 4.6-fold increase in the oxidation current compared to the bare PGE electrode, electrocatalytic activity of PGE/CuO-NPs/Ppy for the measurement of tramadol oxidation via the square wave voltammetry (SWV) approach was outstanding. The LOD was close to 1.0 nM, while the linearity range was 0.38–5.0 nM. Additionally, the created electrode was effectively used to detect tramadol in drug samples.

One of the most popular illegal narcotics supplied worldwide is cocaine. Therefore, it is highly desired to identify and measure this medication using quick and sensitive methods. On the surface of Pd-decorated G, Florea, Cowen, Piletsky, and De Wael (2019) created an electrode for the discerning detection of cocaine using the MIP assembly. The CV method to produce the MIP used the medicine as the template and aminobenzoic acid as the monomer, respectively. They looked into and made adjustments to the experimental Pd NP deposition parameters such as pH, electro polymerization mixture composition, etc. The amount of drug was measured using the SWV method, with a linearity range of 0.1-0.5 M and a LOD of 0.05 M. Analysis was done on actual drug samples, saliva, and drug-laced river water. The electrode was capable of measuring cocaine levels close to additional levels.

Xylazine, a phenothiazine derived known as “tranq,” is a potent sedative that was first created to treat hypertension, but it also results in bradycardia and brief elevations in blood sugar levels. The FDA (Food and Drug Administration*)* has restricted its use due to its side effects (Ruiz-Colón, Chavez-Arias, Díaz-Alcalá, & Martínez, 2014). Xylazine has gained popularity as a recreational drug recently and is frequently used to adulterate illegal substances including heroin, cocaine, and their mixes. Due to limitations on its administration to people, the toxicity of this medication and others is still unknown. Saisahas et al. (2022) constructed an electrochemical method with paper usage as a substrate in order to detect xylazine. They are extremely helpful in forensics because of their low cost, light weight, versatility, portability, sustainability, and appropriateness for mass production. PANI, which has a significant surface area and encourages electron transmission, was used to functionalize the ink G after it had been applied to paper. The benzene rings of G and xylazine interacted to improve analyte adsorption. This sensor was tested directly on xylazine using differential pulse voltammetry (DPV), and a linear response was discovered from 0.18 to 5 g ml-1 and from 5 to 100 g ml-1, with a LOD of 0.06 g ml-1. Recoveries from the sensor’s use to find this substance in samples of fortified beverages ranged from 84 to 106%. Illegal substances can be found using a portable system in forensic investigations.

Nano Sensors for Doping Substances

Doping is the use of unlawful, prohibited substances to improve athletic performance, preparation, or operation. For many years, professionals and elite athletes have heavily resorted to drugs to aid in their athletic endeavors. In addition to illegal narcotics like amphetamine, cocaine, ephedrine, and other narcotic analgesics, the most popular types of doping substances also include medicines marketed as nutritional supplements like peptide hormones, anabolic steroids, and diuretics, among others (Mazzei, Antiochia, Botrè, Favero, & Tortolini, 2014).

A crucial protein utilized in blood doping is erythropoietin (EPO). Because it promotes the production of new red blood cells, athletes use EPO illegally to boost oxygen delivery to tissues and hence improve performance. Han, Zhuo, Chai, Xiang, and Yuan (2015) created a sandwich-type immunosensor for detecting EPO in human serum that has C60 functionalized with PAMAM and AuNPs. PAMAM was mixed with ethanol to form PAMAM-C60 NPs, which were then covered in gold nanoparticles. PAMAM was also modified using fullerene. The antibody was coated with a film of protein A and nanodendrites to immobilize it on GCE. The electrochemical response was captured by CV after adding a surfactant (TOA+), which functions as a supportive agent to boost the inherent redox activity of C60. The acquired immunosensor responded linearly between 0.01 and 80 mU/mL.

Nano Sensors for Toxic Gases

Since polymer-functionalized graphene has a permeable design that improves the movement of synthetics into the sensor layers, it is utilized for destructive gas detecting. Furthermore, macromolecules and a few gases can collaborate through H-securities or - stacking to further develop responsiveness.

A transparent and flexible sensor made by Yoon et al. (2018) contains CVD graphene that had been functionalized with PPy. Graphene/polyethylene terephthalate (SLG/PET) screen on a single-layer, the modification was accomplished by electrochemical polymerization of monomers, which were treated to up to 70 cycles and shown excellent cyclability. The manufactured sensor was utilized o measure a variety of gases, including ammonia and NO2, with these two compounds having the lowest LOD values (0.03 and 0.04 ppb). After coming into contact with NO2, the sensor's electrical resistance dropped, whereas coming into contact with NH3 caused it to increase. The ordered PPy/graphene structure of the nanocomposite, which increases charge mobility, was credited with the outstanding sensing capability of the material.

By electropolymerizing PPy on CVD-G, NH3 sensors were created using a similar technique, and the sensors' performance was shown to be better than those without polymer layers. When the sensors were subjected to 5 ppm NH3 at 25 °C in the presence of the interference-causing gases CH2O and NO2, very good selectivity was discovered. Electron transfer from NH3 to PPy is thought to be the main cause of the greater sensitivity.

A gas sensor for NH3 was likewise made utilizing PPy/rGO crossovers, in which PPy was connected to rGO by an in-situ substance oxidation polymerization strategy. The collaborations among nanomaterial and polymer, which increment sensor opposition, were approved by Raman and FTIR tests. A straight reaction was found in the reaction of rGO/PPy, rGO, and plain PPy, nanocomposite versus smelling salts fixations in the scope of 1-10 ppm, showing great selectivity while using alcohols or different gases.

Nano Sensors for Explosives

Due to safety and human health concerns, artifacts detection, specifically 2,4,6-trinitrotoluene (TNT), is major curiosity on a global scale. These explosives actually pollute the environment, especially natural waters. In the field of forensic science, it is crucial to develop novel, quick, affordable, and reliable tests for the exposure of explosives in both gaseous and aqueous environments. In this regard, an apta sensor for extremely sensitive TNT detection was created by Shahdost-Fard and colleagues (Shahdost-Fard, Roushani, & Bioelectronics, 2017). Following covalent attachment of the TNT/aptamer mixture to the AuNPs@C60-modified GCE, DOP polymerization was carried out. After the TNT is removed, the MIP works in concert along added aptamer for creating nanohybrid receptor (aptamer-MIP) with superior acknowledgeable capabilities compared to aptamer alone. This method produced very low LOD of 3.5

amol L1, a large range of linearity from 0.01 mol L1 to 1.5 mol L1, and a wide range of linearity. This method can be applied to other types of molecules, including proteins, peptides, and antibodies.

Through covalent attachment of nanocomposite of AgNPs/thiol-GQD and the use of Ru as electrochemical indicator, the same authors created a second TNT aptamer sensor (Shahdost-fard & Roushani, 2017). The aptamer structure folded as a result of TNT attaching to the aptamer-modified electrode surface, which decreased the DPV signal of Ru. Thanks to the enhanced performance of the synthesized nanocomposite, surface-immobilized aptasensor's high concentration, and chosen redox system specificity, this new aptasensor was successfully employed to calculate the amount of this explosive in soils and water.

Figure 5. A scheme showing electrochemical aptasensor for TNT quantification synthesis

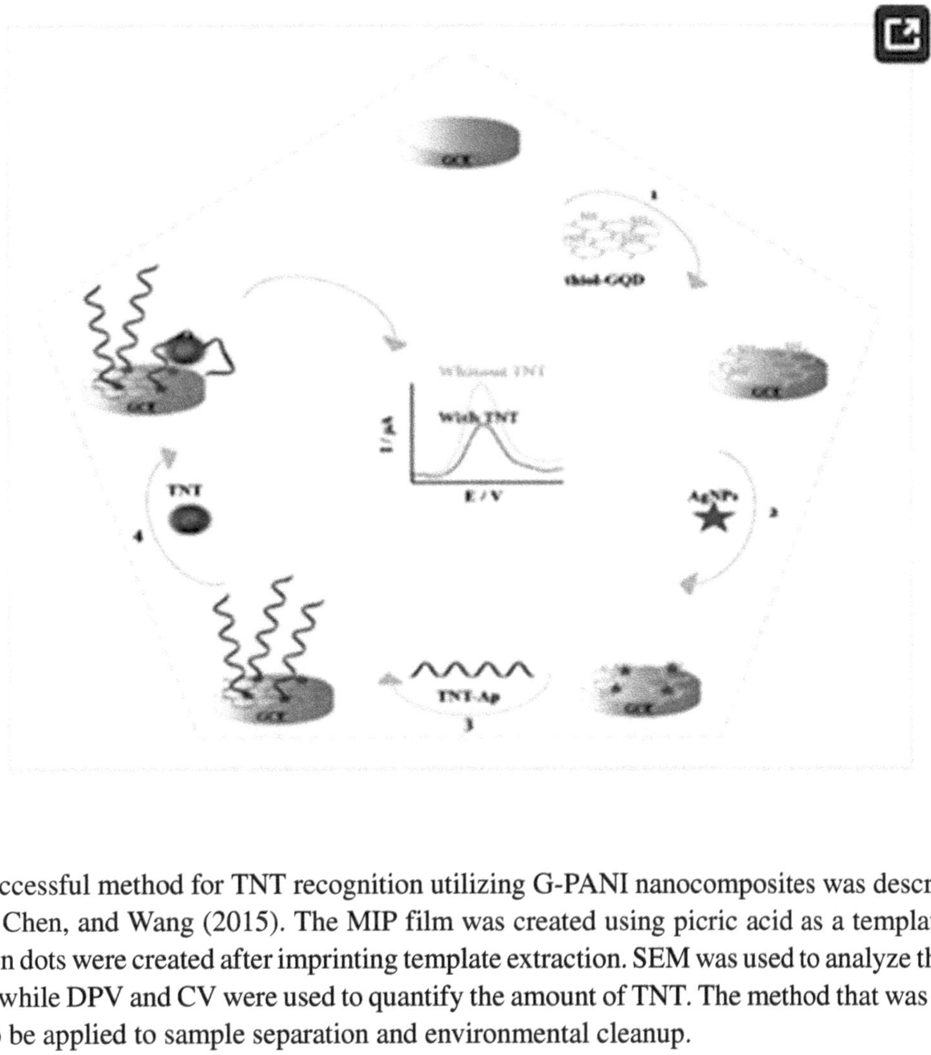

The successful method for TNT recognition utilizing G-PANI nanocomposites was described by L. Shi, Hou, Chen, and Wang (2015). The MIP film was created using picric acid as a template, and the recognition dots were created after imprinting template extraction. SEM was used to analyze the shape of the films, while DPV and CV were used to quantify the amount of TNT. The method that was developed could also be applied to sample separation and environmental cleanup.

Shamsipur, Tabrizi, Mahkam, and Aboudi (2015) created a sensor which is electrochemical for identifying the same explosive that contained a hybrid of rGO and poly(amidoamine) (ErGO-PAMAM).

With a linearity range of 0.05 to 1.3 ppm and a modest limit of detection of 0.002 ppm, the produced electrode performed better than other modified G-based electrodes in tests using AFM, FTIR, SEM, EIS, and EDXS.

NANO EXPLOSIVES CHANGED THE ERA OF EXPLOSIVES

Nano-EM preparation has proven to be a very effective method for creating new EMs, such as nano explosives, which have the potential to be widely used for both military and civilian uses. Numerous articles are available that review and explore various facets of nano-EM (Sarawadekar et al., 2008). Additionally, porous inorganic silicon nano explosive energetic composites that incorporate a metal-metal oxidizing agent (MIC) as well as certain energetic ionic liquids have both been reported (R. P. Singh, Verma, Meshri, & Shreeve, 2006; Zhang, Rossi, Ardila Rodriguez, Tenailleau, & Alphonse, 2007). In particular, reviews on several significant nanometer-sized explosives that concentrate on the production of these materials and the examination of their properties have been published. The molecular makeup of some of the explosives under discussion as shown in Figure 6.

Figure 6. Molecular structures of explosives used from last 20 years
Source: Gao et al. (2019)

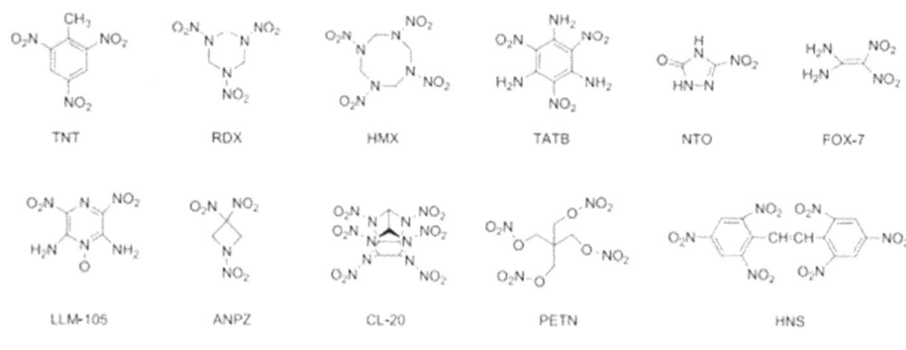

In the development of nano-EM, studies on nano explosives in the areas of synthesis, preparation, characterization, application, and assessment are crucial. The key ideas for nano explosives that we give in this chapter ought to help readers gain a basic comprehension of the development phases that are listed.

There are several benefits of using nano explosives as EM, including greater insensitivity, a lesser temperature of thermal decomposition and a nano "melting point" influence, and the potential for somewhat higher detonation velocities than those of non-nanometer sized explosives. In conclusion, there are nano-advantages in enhancing characteristics like low sensitivity to nano-explosives, including distinct explosives like HMX/GO, CL-20, and HMX composites, CL-20/b-CD, along with unique nano-co-crysts. To enhance the overall effectiveness of original explosives, it is now possible to create a nano explosive form with brand-new type of EM (Han et al., 2015).

Figure 7. Illustration of nano-EMs from preparation to applications
Source: Gao et al. (2019)

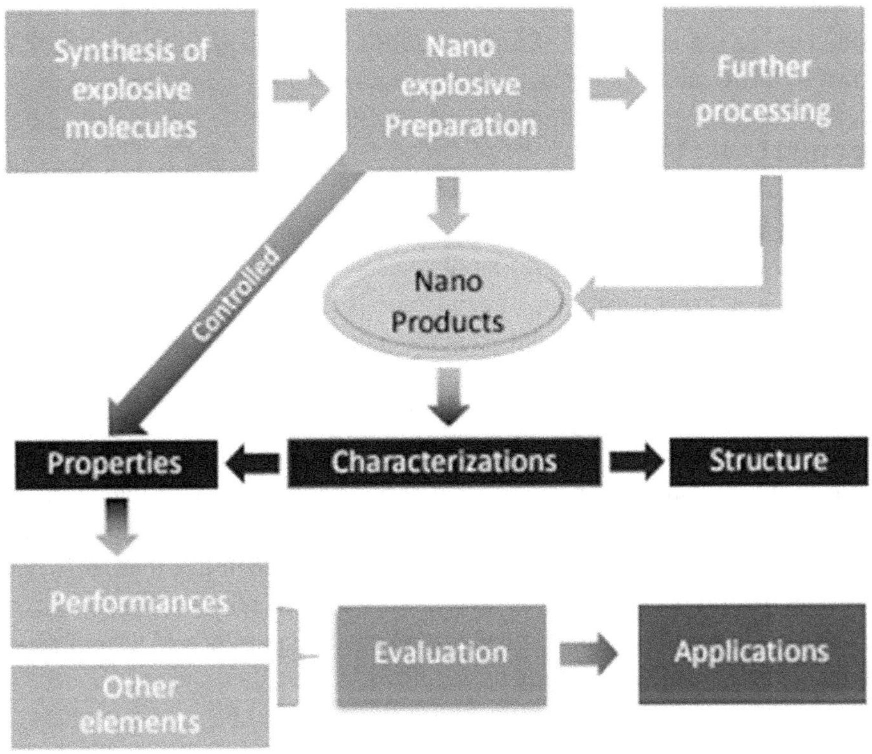

NANO SENSORS: A HELPING HAND FOR COMBATING RISING CRIME AND TERRORISM

Terrorism is today's greatest crime of humankind causing mass destruction of humanity. Nano sensors have the potential to play a role in combating terrorism and criminal activities by improving the capabilities of laboratory and law enforcement agencies to detect and prevent terrorist and criminalist activities. One potential use of nano sensors is in the detection of explosives. Nano sensors can be designed to detect the chemical signatures of explosive materials, even in very small quantities. This could help law enforcement agencies to detect hidden explosives and prevent terrorist attacks before they occur. In recent studies it was discovered that curcumin based nps can be used for detecting TNT (Trinitrotoluene) even the quantity of 1nM can be detected. Silver and gold nanoparticles are picked ones as their SPR range is 400 - 800 nm, these are steady, idle areas of strength for and as well as lessening specialists. Nanoparticles, connected with cysteine, structure Mesienheimer buildings explosives with nitro sweet-smelling compound and location of RDX (imperial destruction hazardous) up to 0.15 mg/L in water tests is additionally conceivable. QDs has recently referenced can likewise be utilized for the recognition of DNT (dinitrotoulene), dynamite, RDX, PETN and tetryl (Daeid, Hackman, Adegoke, & Nic Daeid, 2021; Nazari, 2020).

Nano sensor framework for identification is based on nanotubes, nanowires have high surface region to volume proportion which are high assimilation capacities. The framework comprises of exhibit of sensors creating electrical sign for fume being consumed by the detecting chamber. The fume created by explosives when come in contact with these sensors gets assimilated into them furthermore, produce a substance signature or a print one of a kind to a substance compound which can be matched to the library of known standard mixtures (Senesac & Thundat, 2008).

Warfare Agents' Detection

Nano sensors are helpful in detection of warfare biological and chemical based elements help in preventing warfare terrorism. These compounds are profoundly poisonous, touchy, have quick activity potential and are exceptionally diligent. Fighting specialists exist in fluid, gas and spray structure which gets scattered into the climate quickly causing mass annihilation on openness. Nano based sensors here end up being of extraordinary worth in making a gigantic commitment by perceiving these specialists through nano-structured destinations where they bind, respond and show color variation. Nanoparticles have high likely applications in planning of sensors for fighting specialists, their annihilation, labeling, recognition and tracking purposes (Yu Liu et al., 2012). Microprocessors based chips having nanoparticles well defined for a substance or natural specialist are implanted in a gel covered by a layer of polymer and a layer of substrate is put on top of it. These central processors are tiny and their thickness is in nanometer range. These nanomaterial microprocessors are inserted on to the garments, structures, arms, and transports and so forth. These microprocessors color changes indicate the presence of fighting specialist when it interacts with the particular specialist (Song et al., 2015).

Nano Trackers

Nano trackers are being infused into the assemblage of crooks so that if they even attempt and prevail with regards to getting away from the penitentiaries, could once more be gotten through utilizing this global positioning framework. Hoodlums in the wake of being let out of the prison or remedial focuses could not necessarily follow the cultural standards and show degenerate way of behaving which should be observed. This is conceivable by infusing Nano trackers into their body which will empower the observing system of such delivered convicts a less dreary interaction. Chronic wrongdoers can be followed down assuming that they engage in violations again (Pandya & Shukla, 2018). The articles can likewise be labeled with peptide based taggants which truly mark them. These can be applied on the outer layer of things and when recuperated, swabs of such taggants on investigation show color variation or fluorescence change. Additionally, according to the Locard's trade standard, these taggants can likewise be recuperated from the individual who is engaged with the wrongdoing related with the labeled article (Roopachandrika & Reddy, 2023).

Nano Sensors for Forgeries

Barcode tags, notwithstanding Nano trackers, additionally help in preventive legal sciences. These are explicit examples drawn on articles which can decide the credibility of an item which helps in forestalling the falsification or duplication of different products. Nano innovation has presented other extraordinary strategies, for example, nano strands and Nano specks with different variety which is utilized for the discovery and confirmation of items. The Nano filaments and nano specks can be breaking down utilizing different legal light sources like UV (bright), IR (infrared), and so forth that will distinguish and translate these nano fibers what's more, nano dots.

Figure 8. General barcode system
Source: Wang, Zheng, You, and Ju (2023)

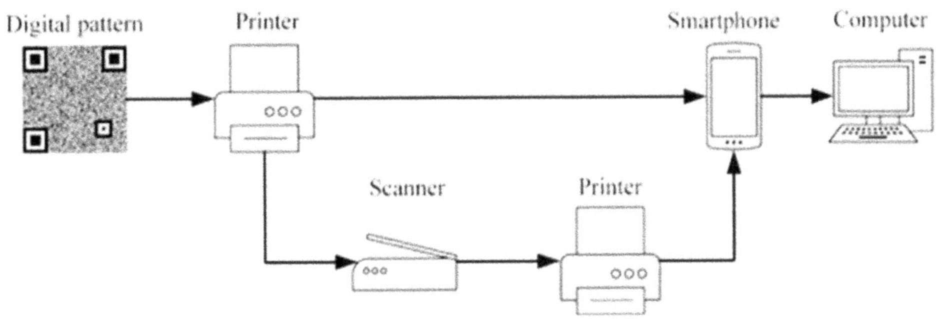

Nano sensors can also be used in the monitoring of critical infrastructure, such as banks and government buildings, to detect any potential tampering or security breaches. Another potential application of nano sensors is in the area of forensic science. Nano sensors can be used to detect and identify trace amounts of DNA, chemicals, or other materials at crime scenes, which can help investigators to identify suspects and solve crimes. Overall, the use of nano sensors in the fight against crime has the potential to greatly improve the ability of law enforcement agencies to detect and prevent criminal activities. However, as with any technology, it is important to carefully consider the potential ethical and privacy implications of their use.

CHALLENGES INVOLVED

Due to their excellent performance, the manufacture of nano explosives on a large scale is one of the essential measures for more application. Top-down approaches have their peculiar constraints in relations with particle size homogeneity, reduction and efficacy, despite being more adaptable and commercially viable than other procedures. The particle size reduction process for top down frequently takes a lengthy period. Therefore, it is not feasible to shrink the particle size below a particular threshold of 100 nm using conventional top-down techniques. A typical particle size ranges from 150 to 300 nm. In addition, changes in micron product of solid state and chemical deterioration can also take place during grinding or mechanical grinding. It is necessary to talk about the restrictions on nano explosives' synthesis (Li & Sun, 2022).

On the other hand, the majority of the development information that have been published in relation to bottom-up methodologies have generally concentrated on the cost effectiveness, size-controlling, and continuous nano explosives manufacturing. Although several synergistic combinations of recrystallization, freezing and spraying, vacuum co-deposition ultrasound, electrospray and oil/water microemulsion, etc. have been carried out in the last few years, a methodical methodology for preparation methodology of nanoexplosives is still unknown. Added to it the other concern is evaluating the high performance of nanoexplosives; when particles are smaller than 100 nm, especially at the nanoscale, properties of nanomaterials differ dramatically from those of volumes (Bae et al., 2022).

Selecting an energetically adequate combination that might offer advantages is one of the primary issues in designing or creating nano-cocrystals. When commencing with the two powerful explosives combination, like acetone CL-20 and HMX, investigating solubility combinations would be one of the important issues to handle. When examining combinations with highly soluble explosives, variant solubility might potentially can be an concern and cause the component units to crystallize.

As a result, by anticipating material results, the studying of supramolecular interactions can be accountable for physicochemical features of nano-explosives in the creation of computational techniques. By addressing present issues and creating these predictive models, it should be possible to create nano explosives based on current goods more quickly. Computational science should present security implications and prospective applications.

REFERENCES

Adhikari, S., Ampadu, E. K., Kim, M., Noh, D., Oh, E., & Lee, D. (2021). Detection of explosives by SERS platform using metal nanogap substrates. *Sensors (Basel)*, *21*(16), 5567. doi:10.339021165567 PMID:34451009

Ahangaran, F., & Navarchian, A. H. (2020). Recent advances in chemical surface modification of metal oxide nanoparticles with silane coupling agents: A review. *Advances in Colloid and Interface Science*, *286*, 102298. doi:10.1016/j.cis.2020.102298 PMID:33171357

Ansari, S., & Masoum, S. (2021). *A hybrid imprinted polymer based on magnetic graphene oxide and carbon dots for ultrasonic assisted dispersive solid-phase microextraction of oxycodone*. Academic Press.

Arabali, V., Malekmohammadi, S., & Karimi, F. (2020). *Surface amplification of pencil graphite electrode using CuO nanoparticle/polypyrrole nanocomposite; a powerful electrochemical strategy for determination of tramadol*. Academic Press.

Azman, A. M. (2020). Fast, easy, reproducible method for planting fingerprints for ninhydrin, iodine development. *Journal of Chemical Education*, *97*(2), 571–573. doi:10.1021/acs.jchemed.9b00357

Bae, G., Kim, M., Jang, M., Kwon, Y. M., Yim, S., Song, W., Myung, S., Lee, S. S., Hasan, T., & An, K. S. (2022). Photon-Pen Writing for Metal Oxide-Based Versatile Nanosensors. *Advanced Functional Materials*, *32*(38), 2204821. doi:10.1002/adfm.202204821

Buffler, A., & Tickner, J. (2010). *Detecting contraband using neutrons: challenges and future directions*. Academic Press.

Daeid, N. N., Hackman, L., Adegoke, O., & Nic Daeid, N. (2021). Colorimetric optical nanosensors for trace explosive detection using metal nanoparticles: Advances, pitfalls, and future perspective. *Emerging Topics in Life Sciences*, *5*(3), 367–379. doi:10.1042/ETLS20200281 PMID:33960382

Florea, A., Cowen, T., Piletsky, S., & De Wael, K. (2019). *Electrochemical sensing of cocaine in real samples based on electrodeposited biomimetic affinity ligands*. Academic Press.

Furton, K. G., & Myers, L. (2001). *The scientific foundation and efficacy of the use of canines as chemical detectors for explosives*. Academic Press.

Gardner, J. W. (2004). *Review of conventional electronic noses and their possible application to the detection of explosives. In Electronic Noses & Sensors for the Detection of Explosives*. Springer. doi:10.1007/1-4020-2319-7

Gorup, L. F., Amorin, L. H., Camargo, E. R., Sequinel, T., Cincotto, F. H., Biasotto, G., & La Porta, F. A. (2020). *Methods for design and fabrication of nanosensors: the case of ZnO-based nanosensor. In Nanosensors for smart cities*. Elsevier.

Han, J., Zhuo, Y., Chai, Y.-Q., Xiang, Y., & Yuan, R. (2015). *New type of redox nanoprobe: C60-based nanomaterial and its application in electrochemical immunoassay for doping detection*. Academic Press.

Horváth, T., & Szatai, J. Z. (2020). A History of Detection of Explosive Devices 1. *Land Forces Academy Review*, *25*(3), 189–200. doi:10.2478/raft-2020-0023

Javaid, M., Haleem, A., Singh, R. P., Rab, S., & Suman, R. (2021). Exploring the potential of nanosensors: A brief overview. *Sensors International*, *2*, 100130. doi:10.1016/j.sintl.2021.100130

Kumar, D., & Elias, A. J. (2019). The Explosive Chemistry of Nitrogen: A Fascinating Journey From 9th Century to the Present. *Resonance*, *24*(11), 1253–1271. doi:10.100712045-019-0893-2

Kumar, N., & Kumbhat, S. (2016). *Essentials in nanoscience and nanotechnology*. John Wiley & Sons. doi:10.1002/9781119096122

Laurila, T., Sainio, S., & Caro, M. (2017). *Hybrid carbon based nanomaterials for electrochemical detection of biomolecules*. Academic Press.

Lehnert, A. L., & Kearfott, K. (2010). *The detection of explosive materials: review of considerations and methods*. Academic Press.

Li, J., & Sun, D. (2022). Molecularly imprinted ratiometric fluorescence nanosensors. *Langmuir*, *38*(44), 13305–13312. doi:10.1021/acs.langmuir.2c01925 PMID:36278924

Liu, Y., Chen, C.-L., Zhang, Y., Sonkusale, S. R., Wang, M. L., & Dokmeci, M. R. (2012). SWNT based nanosensors for wireless detection of explosives and chemical warfare agents. *IEEE Sensors Journal*, *13*(1), 202–210. doi:10.1109/JSEN.2012.2211007

Liu, Y., Sowerby, B., & Tickner, J. (2008). *Comparison of neutron and high-energy X-ray dual-beam radiography for air cargo inspection*. Academic Press.

Marques, L., Félix, L., Cruz, G., Coelho, V., Caetano, J., Vale, A., Cruz, C., Alves, L., & Vaz, P. (2023). Neutron and Gamma-Ray Detection System Coupled to a Multirotor for Screening of Shipping Container Cargo. *Sensors (Basel)*, *23*(1), 329. doi:10.339023010329 PMID:36616926

Mazzei, F., Antiochia, R., Botrè, F., Favero, G., & Tortolini, C. (2014). *Affinity-based biosensors in sport medicine and doping control analysis*. Academic Press.

Meixner, A. J., Fleischer, M., Kern, D. P., Sheremet, E., & McMillan, N. (2022). *Applications*. Walter de Gruyter GmbH & Co KG. doi:10.1515/9783110442908

Moore, D. S. (2007). *Recent advances in trace explosives detection instrumentation*. Academic Press.

Nazari, A. (2020). *Nanosensors for smart cities: An introduction. In Nanosensors for smart cities*. Elsevier. doi:10.1016/B978-0-12-819870-4.00001-3

Nocerino, V., Miranda, B., Tramontano, C., Chianese, G., Dardano, P., Rea, I., & De Stefano, L. (2022). Plasmonic nanosensors: Design, fabrication, and applications in biomedicine. *Chemosensors (Basel, Switzerland), 10*(5), 150. doi:10.3390/chemosensors10050150

Pandya, A., & Shukla, R. K. (2018). New perspective of nanotechnology: Role in preventive forensic. *Egyptian Journal of Forensic Sciences, 8*(1), 1–11. doi:10.118641935-018-0088-0

Roopachandrika, C., & Reddy, D. K. A. (2023). (Forthcoming). Alerting and Tracking System of Dislocated Object. *International Journal for Innovative Engineering & Management Research.*

Ruiz-Colón, K., Chavez-Arias, C., Díaz-Alcalá, J. E., & Martínez, M. (2014). *Xylazine intoxication in humans and its importance as an emerging adulterant in abused drugs: A comprehensive review of the literature*. Academic Press.

Saisahas, K., Soleh, A., Promsuwan, K., Saichanapan, J., Phonchai, A., Sadiq, N. S. M., ... Limbut, W. J. A. o. (2022). *Nanocoral-like Polyaniline-Modified Graphene-Based Electrochemical Paper-Based Analytical Device for a Portable Electrochemical Sensor for Xylazine Detection*. Academic Press.

Senesac, L., & Thundat, T. G. (2008). Nanosensors for trace explosive detection. *Materials Today, 11*(3), 28-36.

Senesac, L., & Thundat, T. (2008). *Nanosensors for trace explosive detection*. Academic Press.

Shahdost-Fard, F., & Roushani, M. (2017). *Designing an ultra-sensitive aptasensor based on an AgNPs/thiol-GQD nanocomposite for TNT detection at femtomolar levels using the electrochemical oxidation of Rutin as a redox probe*. Academic Press.

Shahdost-fard, F., & Roushani, M. (2017). *Impedimetric detection of trinitrotoluene by using a glassy carbon electrode modified with a gold nanoparticle@ fullerene composite and an aptamer-imprinted polydopamine.* Academic Press.

Shamsipur, M., Tabrizi, M. A., Mahkam, M., & Aboudi, J. J. E. (2015). A High Sensitive TNT Sensor Based on Electrochemically Reduced Graphene Oxide-Poly (amidoamine). *Electroanalysis, 27*(6), 1466–1472. doi:10.1002/elan.201400634

Shi, H., Xiang, W., Liu, C., Shi, H., Zhou, Y., & Gao, L. (2018). *Highly sensitive detection for cocaine using graphene oxide-aptamer based sensors in combination with tween 20*. Academic Press.

Shi, L., Hou, A., Chen, L., & Wang, Z. (2015). *Electrochemical sensor prepared from molecularly imprinted polymer for recognition of TNT*. Academic Press.

Shreve, R. N., & Brink, J. A. Jr. (1977). *Chemical Process Industries*. McGraw-Hill Book Co.

Singh, R. P., Verma, R. D., Meshri, D. T., & Shreeve, J. (2006). *Energetic nitrogen-rich salts and ionic liquids.* Academic Press.

Singh, S., & Singh, M. (2003). Explosives detection systems (EDS) for aviation security. Academic Press.

Song, Y., Chen, J., Hu, D., Liu, F., Li, P., Li, H., Chen, S., Tan, H., & Wang, L. (2015). Ratiometric fluorescent detection of biomakers for biological warfare agents with carbon dots chelated europium-based nanoscale coordination polymers. *Sensors and Actuators. B, Chemical*, *221*, 586–592. doi:10.1016/j.snb.2015.07.008

Viswanath, D. S., Ghosh, T. K., & Boddu, V. M. (2018). *Emerging energetic materials: Synthesis, physicochemical, and detonation properties*. Springer. doi:10.1007/978-94-024-1201-7

Wang, T., Zheng, H., You, C., & Ju, J. (2023). A Texture-Hidden Anti-Counterfeiting QR Code and Authentication Method. *Sensors (Basel)*, *23*(2), 795. doi:10.339023020795 PMID:36679589

Wells, K., & Bradley, D. (2012). *A review of X-ray explosives detection techniques for checked baggage*. Academic Press.

Wester, N., Mynttinen, E., Etula, J., Lilius, T., Kalso, E., Kauppinen, E. I., . . . Koskinen, J. (2019). *Simultaneous detection of morphine and codeine in the presence of ascorbic acid and uric acid and in human plasma at nafion single-walled carbon nanotube thin-film electrode*. Academic Press.

Whetstone, Z., & Kearfott, K. (2014). *A review of conventional explosives detection using active neutron interrogation*. Academic Press.

Wu, Z., Wang, H., Wang, X., Zheng, H., Chen, Z., & Meng, C. (2020). Development of electronic nose for qualitative and quantitative monitoring of volatile flammable liquids. *Sensors (Basel)*, *20*(7), 1817. doi:10.339020071817 PMID:32218148

Yoon, T., Jun, J., Kim, D. Y., Pourasad, S., Shin, T. J., Yu, S. U., . . . Kim, K. (2018). *An ultra-sensitive, flexible and transparent gas detection film based on well-ordered flat polypyrrole on single-layered graphene*. Academic Press.

Zhang, K., Rossi, C., Ardila Rodriguez, G. A., Tenailleau, C., & Alphonse, P. (2007). *Development of a nano-Al/ Cu O based energetic material on silicon substrate*. Academic Press.

Chapter 9
Fluorescent Nanomaterials and Its Application in Biomedical Engineering

Sudipto Datta
Department of Materials Engineering, Indian Institute of Science, Bengaluru, India

Ranjit Barua
Centre for Healthcare Science and Technology, Indian Institute of Engineering Science and Technology, India

ABSTRACT

Currently, in the field of biomedical engineering and biological applications, the use of soft florescent nanomaterials has increased because of their excellent biocompatibility, easy biofunctionalization, and high brightness properties. This chapter summarizes the current developments of nano-sized fluorescent soft biological imaging agents. Many fluorescent soft nanoparticles like biomaterial-based NPs, vesicles, micelles, nanogels, small-molecule organic NPs, semiconducting polymer NPs, and dye-doped polymer NPs are mentioned briefly starting from the preparation methods, their structures, their optical properties, as well as their functionalization. Depending upon the nano-sized imaging agents' functional as well as optical properties, their uses are briefly described in relation to Vivo imaging, cellular process imaging, and in vitro imaging by using nonspecific and specific targeting.

INTRODUCTION

Nanotechnology is becoming popular in the present time. It is being used in many fields. It has become a part of our life. It is being used in various fields like nanomedicines, biomaterials, consumer products, and nanoelectronics. It involves interdisciplinary fields like mechanical engineering, electrical engineering, chemical engineering, material science, physics, biology, and chemistry. In biomedical domain, nanomaterials find various applications because of: good biocompatibility, highly loadable surface, optical, chemical, and physical properties, and their small size (Wilczewska et al., 2012). Because nanomaterials

DOI: 10.4018/978-1-6684-8325-1.ch009

match sizes with various biomolecules like viruses (size variety of 1 to 100 nm), proteins, glucose, oligonucleotides, antibodies, DNA, and numerous in vivo and in vitro reports connecting biomaterials and nanomaterials are presented (Narayan., 2017). In various biomedical bids, the main use the nano-materials are in cells and tissues bio-imaging, it has held the huge interest on various researchers worldwide. This technology has now turn out to be an important tool for exploring many biological methods happening at the subcellular and cellular stages and achieving status in both clinical diagnostics and essential research. Although myriad inorganic and organic imaging probes are now accessible, nanomaterials applications as an imaging probe are now multiplying. Nanomaterials application in the bioimaging field covers point of care (POC) testing to in-depth imaging of tissues (Syedmoradi et al., 2017). In the medical field, there are many imaging modalities for accessing images of most tissues and cells. Among them includes (LA-ICP-MS) laser ablation-inductively coupled plasma-mass spectroscopy, electrochemical imaging, CT imaging, resonance imaging (MRI), radio imaging, and many more. Mentioned few methods above are destructive, need considerable sample preparation, and hold limited resolution. Because of this factors, the applications are limited for subcellular and cellular structures imaging inside the human body. This has flagged the technique for the progress of imaging probes that will have great determination, and can deliver pictures in a noninvasive way, and gives images of multinodal for gaining in-depth data. Through the expansion of novel functional biomaterials, nanomaterials bioimaging interests considerable attention in biomedical research because of the high-resolution properties, rich in contrast, sensitivity, and versatility available in this technology. This also integrates orthodox imaging techniques with a new advanced probe to record events at the cellular level having high clarity and contrast. Previously, the resolution of optical imaging was only in a limited range which is ~200 nm. With the improvement of imaging methods using fluorescence techniques, like stimulated emission depletion (STED) microscopy and photo-activated localization microscopy (PALM), the resolution is improved (Betzig et al., 2006, Willig et al., 2006). Though, a few limitations are still there for studying animal models which need in-depth imaging and high resolution. This crackdown can be limited by utilizing newly developed fluorescent nanomaterials imaging probes. Nanomaterials bioimaging mainly accepts the three main methods – (a) for making tissues and cells fluorescent, injection of fluorescent nanomaterials is done on tissues and cells; (b) targeted bioimaging; and (c) employing sensors nanomaterials for sensing the cellular biochemical types that are non-fluorescent inherently. For bioimaging of targeted tissues or cells, the precise target location is imaged by fluorescent nanomaterials. For those cases, the functional groups can be ligands, receptors, oligomers, and many more for identifying a specific location. Figure 1 shows the application of fluorescent nanomaterial in biomedical engineering area. Imaging agents of nanosizes are categorized into two types depending upon their biological environment behavior and chemical nature. They are soft nanomaterials (e.g., micelles, nanogels, polymeric NPs) and hard nanomaterials (e.g. metal nanoclusters [Sharma et al., 2011, Yu et al., 2008], carbon NPs [Liu et al., 2013, Li et al., 2013], lanthanide-doped NPs and quantum dots, dye-doped silica).

Hard nanomaterials are developed using inorganic components, especially under strict environments, by various methods. Though they are extensively used in fluorescent imaging for in vitro by surface alteration, the potential bioincompatibility and inorganic nature obstruct their real-world uses for imaging in vivo imaging (Swierczewska et al., 2011, Tantra et al., 2011, Cho et al., 2010). Presently, only one nano-sized hard imaging agent are accepted by the Drug Administration (FDA) and U.S. Food for clinical trial phase I (Cheng et al., 2013). Moreover, the fabrication of soft optical nano-sized imaging agents under mild conditions and less toxic organic molecules is done. Additionally, numerous FDA-accepted biocompatible molecules are accessible for the expansion of nano-sized imaging products, like

Figure 1. Application of fluorescent nanomaterial in biomedical engineering area

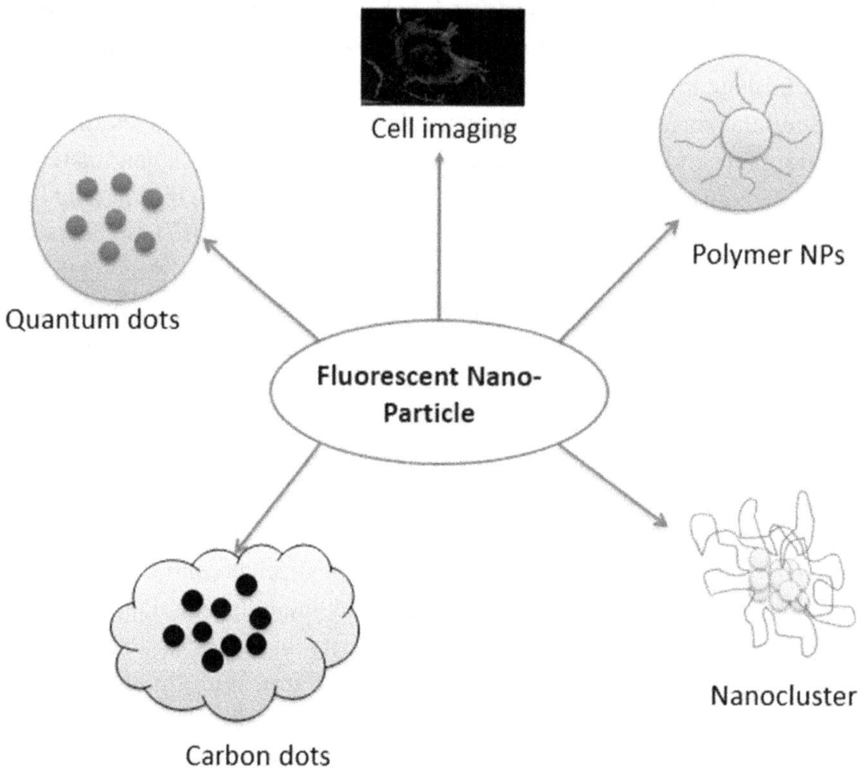

the biodegradable polymers poly(methyl methacrylate) (PMMA), poly(lactic-co-glycolic acid) (PLGA) derivatives, poly(lactic acid) (PLA), poly(ethylene glycol) (PEG) and fluorescent dye indocyanine green (ICG) (Benya et al., 1989). The review highlights a wider variety of soft fluorescent nano-biomaterials NPs, small-molecule organic NPs, and semiconducting polymers, including dye-doped polymer NPs, vesicles, micelles, and hydrogels. The development methods, as well as the design rules, are covered from the viewpoint of surface functionalization, optical properties, and particle structure. The applications are covered in view point of in vivo imaging, in vitro imaging and cellular process imaging. We think that the readers of this review paper will achieve some information for the soft fluorescent nano-sized imaging agents construction and choosing the needed fluorescent organic NPs having accurate targeting groups, fluorophores, nanostructure for biological molecules of attention. We also would like to inform you that the review paper is restricted to only the biological imaging applications, the fluorescent organic nano-biomaterials NPs in the situation of fluorescence-based multi-functional organic NPs, and fluorescent nanocomposites (e.g. composition of organic shell and inorganic core) are left off. We also restrain our debates to fluorescent organic nano-biomaterials NPs which have consistent shape, manageable size of the particle its surface chemistry, hence we neglect debates on fluorescent nano-aggregates which dearth of distinct morphological info or the surface chemistry.

FLUORESCENCE IMAGING CELLS

In living organisms, the organelles and tissues are of so small size that their clear image capturing is very difficult. This uses contrasting imaging agents for imaging localized area tissues and organelles for seeing them under microscopes to study the various biologicals processes happing there. The issues of the contrasting structure of a localized area and overall structure are solved by using fluorescence. In fluorescence which is one kind of luminescence, there is a process of radiation absorption and chased by light emission of a definite wavelength through the object. Fluorescence is the ability of a molecule to emit light upon absorption. This imaging technique is a non-invasive technology that allows many biological processes to happen in a living environment. It covers an extensive variety of annotations, which contains interactions occurring in tissues and cells, gene expression, and tissues protein expression. This method mostly includes fluorescent dyes and proteins picturization to cognize numerous molecular processes happening in the cells of living being, thus attending to the reason of having a possible tool for uses biochemical applications. In the case of fluorescence microscope, the models are branded using a probe of the fluorescent instrument, and finally, the model is lightened using an intense light. The absorption and emission of a fixed wavelength are done by a labeled fluorophore. For this reason, on a dark background mainly backgrounds of black color are used and the image is obtained having high contrast. For studying a biological process, mostly used imaging technique is fluorescence. Mainly, two kinds of fluorescence imaging, they are extrinsic fluorescence imaging and intrinsic imaging. The first kind of imaging is related to extrinsic fluorescence that is shown by probes and synthetic dyes, sensors, nanomaterials, and so on. This one is usually used for the identification of species that are not seen directly by fluorescence imaging. The another kind of imaging is mainly depend on fluorescence emission using intrinsic fluorophoric species like NADH in tissues, tryptophan residues in proteins, and so on. Many imaging techniques using fluorescence imaging are currently present for studying the tissues and cells structures in the body. This involves super-resolution microscopy, total internal reflection fluorescence (TIRF) microscopy, light sheet microscopy, two-photon microscopy, parallel confocal microscopy, point scanning confocal microscopy, and fluorescent wide-field microscopy.

SYNTHESIS OF FLUORESCENT NANOMATERIALS

Carbon Dots

New emerging nanomaterials having sizes 10 nm in the nanocarbon family are Carbon dots (CDs) that were primarily obtained during the single-walled carbon nanotubes (SWCNTs) purification using electrophoresis in 2004 (Xu et al., 2004). This should be noted that carbon dots CDs generally replace semiconductor quantum dots in the field of better biocompatibility, excellent cell permeability, and preferable elasticity in the case of surface alteration, excitation-dependent multicolor emission, high photostability, low cytotoxicity and high solubility in water (Li et al., 2015, Zuo et al., 2016). Usually, CDs mostly contain graphene quantum dots (GQDs) and carbon quantum dots (CQDs). CDs with tunable sizes having a mass of synthetic methods can be differentiated into two major types: physical methods and chemical methods (Baker et al., 2010). Figure 2 shows the different types of carbon nanoparticles (CNPs).

Figure 2. Different types of carbon nanoparticles (CNPs)

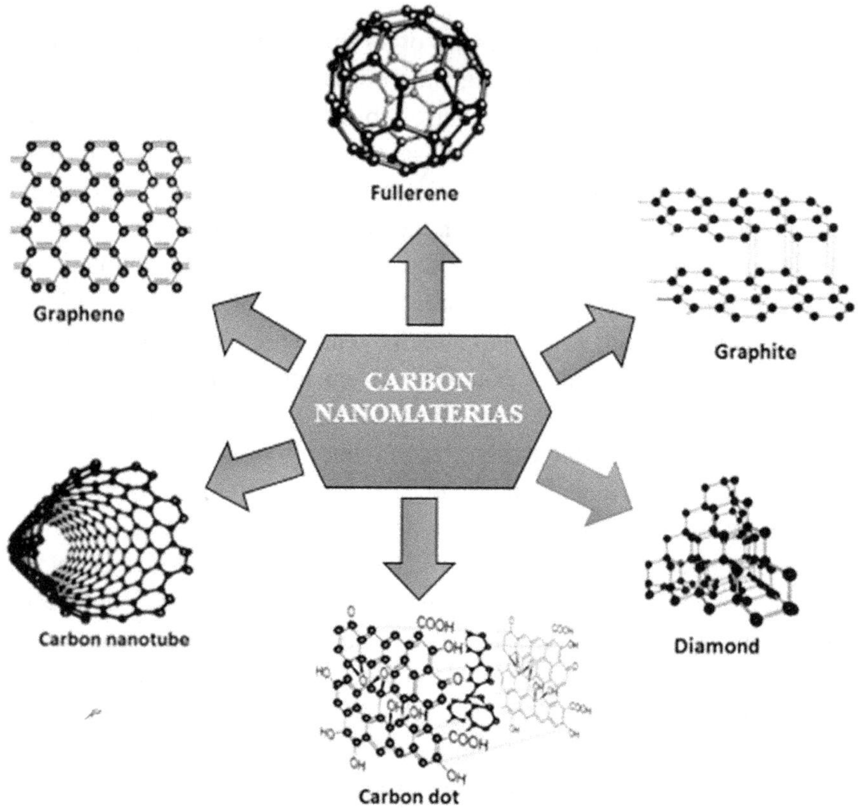

Quantum Dots (Semiconductor Crystals)

In the past decade, the research spot was mainly quantum dots (QDs) having huge symmetric photoluminescence and absorption spectra, large effective Stokes shifts, high molar extinction coefficients, high resistance to photobleaching, and high quantum yield (Jiang et al. 2010). The mechanism of formation of the QDs, when holes and electrons charged carriers are controlled by a potential barrier for a specific location, then the semiconductor shows a huge quantum size effect which results in shifting of the fluorescence and absorption spectrum. Small locations are smaller than the de Broglie wavelength for a charged carrier, or equally, the diameter of the nanocrystal is smaller than two times the Bohr atomic radius of excitation for bulk solid material (Nozik et al., 2010). During the confinement of the charge carriers is done by potential obstructions for 3D spatial dimensions, then QDs are produced which generally contains atoms of IV-VI (PbS, PbSe), III-V (GaAs, InP) or II-VI (CdSe, ZnS). The first production for QD was in 1982 (Efros et al., 1982, Ekimov et al., 1982). In glass matrices, microcrystals of semiconductors and nanocrystals are grown. Through the advancement of fluorescent materials, quantum dots i.e. QDs are produced by other methods like the mixture of preceding preparation methods, situ methods, linker-assisted adsorption methods, and direct adsorption methods. The previous methods include a combination of the produced semiconductor with QD precursors as well as a combination of semiconductor precursors with previously prepared QDs, for which the QDs or the semiconductor are produced separately (Bajorowicz et al., 2018). For the QDs synthesis after a research series, most of the researchers reported the QDs fluorescent property studies. The synthesis of the QDs having narrow size distributions by bringing into picture semiconductor precursors like cadmium telluride (CdTe), cadmium selenide (CdSe), and cadmium sulfide (CdS) for investigating the optical size-dependent properties of the QDs (Dabbousi et al., 1997). From that time, the most similar chemical composition of the QDs became CdSe with numerous protective inorganic shells (Hines et al., 1996) and surface modifications (Gerion et al., 2001, Guo et al., 2003, Yang et al., 2019) having been applied to reduce colloidal stability.

Chemical Synthetic Methods

For the preparation of carbon dots, the most commonly used is Chemical synthetic methods because the output carbon dots produced have some outstanding properties like low toxicity, resistance to photobleaching, ease of functionalization, chemical inertness, and superior water solubility. It includes supported synthesis (Zong et al., 2011), solution chemistry methods (Yan et al., 2010), microwave-assisted/ultrasonic treatment (Xiao., et al., 2013, Tang et al., 2012, DaSilva., 2011), hydrothermal carbonization (Yang et al., 2012), acidic oxidation (Tao et al., 2012, Liu et al., 2007), electrochemical synthesis (Li et al., 2011, Zhao et al., 2008) etc. In the past decade, among various methods of synthesis, the most commonly used method is electrochemical synthesis. A new method of electrooxidation synthesis by (Zhao et al. 2008) produced CDs having low cytotoxicity. The produced CDs were synthesized by graphitic column electrode oxidation counter to a saturated calomel electrode having Pt wire counter electrode aqueous solution of NaH_2PO_4 (Zhao et al., 2008). The ultrafiltration of the supernatant was done by a device having a centrifugal filter for obtaining the carbon dots consisting of yellow and blue fluorescence correspondingly. Qu et al. reported an approach of direct electrochemical synthesis for GQDs having a uniform size ranging from 3 to 5 nm by graphene electrode electrochemical oxidation in a solution of phosphate buffer (Li et al., 2011). Green was the photoluminescent (PL) color for the particles. In 2007, Mao et al. finished the oxidation and combustion production of carbon dots using mixing an oxidant with candle soot, following refluxing, centrifugation, and finally dialysis for getting purified carbon dots. The produced carbon dots spectra of photoluminescence have a huge range of colors having wavelengths of varying emission-peak which ranges from 415 nm to 615 nm i.e. (violet to red-orange). Finally, the produced carbon dots CDs were more exposed to electrophoresis gel of polyacrylamide to distinguish the carbon dots with dissimilar optical categorization. For the synthesis of carbon dots which are stable nanoparticles, acidic oxidation is broadly used. After the carbon dots/graphite as well as refluxing were treated with acid, the total CDs obtained were 3 to 4 nm which showed a transparent liquid that discharged fluorescence yellow under UV light having good stability in saline. This produced carbon dots having long wave (yellow/orange/red) fluorescence holding good penetration. The preservation of the carbon dots should be done at normal room temperature for a certain time and no precipitations were produced which could cause fluorescence loss (Tao et al., 2012). For the synthesis process, ultrasonic/microwave synthesis has mainly and gradually become a secondary technology for CDs (Simoes et al., 2016). Xiao et al. produced fluorescent CDs, with a diameter of 3 to 5 nm using a rapid, green, and economical microwave-assisted method (Xiao et al., 2013). The greatest outstanding property of this one-step method was that both the functionalization and formation of carbon dots were finished instantaneously via microwave pyrolysis resulting first time from the ionic liquids (Xu et al., 2017). In the reactions occurring in a microwave oven, the carbon source used is the cheap ionic liquids, the solution color changed from no color i.e. colorless to deep brown as the reaction progressed with time (Esteves et al., 2011). Using the ultrasonic method, Tang et al. developed monodispersed water-soluble carbon dots where the carbon source was active carbon or glucose. The emitted fluorescence was colorful and bright (Tang et al., 2012). Likewise, Vanesa Romero et al. produced carbon dots that were co-doped with sulfur (S) and nitrogen (N) having good fluorescence after vegetable carbohydrates' photochemical oxidation. For increasing the surface of carbon dots, co-doping of S and N was done, which enhanced the performance of luminescence (Romero et al., 2019). For the determination of doxycycline, a fluorescent probe, and nitrogen-doped carbon quantum dots (NCQDs) were used (Feng et al., 2018). Using the microwave hydrothermal method, co-doped carbon dots with sulfur and nitrogen i.e. (NSCDs) was prepared by Pathak et al. by tris–acetate-ethylenediamine buffer and thiourea. Because of the multicolour fluorometry of NSCDs, they were used in the imaging of human buccal epithelial cells and various pathogenic bacteria (Pathak et al., 2019). Seeing all of the stated artificial approaches required for tough acid, numerous complex investigational process stages, and additional changes with new compounds were done for improving the solubility of water of the carbon dots and improving their photoluminescence. Many research teams used the carbohydrates photoluminescence hydrothermal carbonization like citric acid, glucose, and chitosan for avoiding the complicated and time taking functionalization and purification methods (Parvin et al., 2017). A one-step method of synthesis for carbon dots having extremely amino-functionalized fluorescent and quantum yield (QY) of nearly 7.8% using chitosan hydrothermal carbonization at slight temperature was reported by Yang et al. Neither surface passivation reagent nor strong acid solvent are needed for this method. Keeping aside the carbon dots' functional groups, the water solubility, as well as the bio toxicity, were improved by surface improvement (Yang et al., 2012). The synthesis of the multi-doped carbon dots (MCDs), having color-tunable and bright release was done by method of one-pot technique deprived of having any additional surface passivation. The produced MCDs were mixed with plentiful components which are biogenic like (P, N, O) and ultimately shows good excitation-wavelength-dependent characteristics, fluorescent emission, ion-stability, high optical stability, and good aqueous solubility. MCDs can measure the intracellular Fe^{3+} using multi-color fluorescence imaging, it can also delicately identify Fe^{3+} in blue light at 15.9 nm (Zhang et al., 2018).

Physical Synthetic Methods

The physical synthetic methods available currently are plasma treatment, laser ablation/passivation, and arc discharge. (Xu et al., 2004) oxidized the arc-discharge soot by HNO_3 then divided the suspension using the process of gel electrophoresis into the SWCNTs. Ultimately, they separated the bands of the highly fluorescent carbon dots nanoparticles which were moving fast (Xu et al., 2004). By using the approach of mild laser ablation, Li et al. produced CDs with nano-carbon materials using the predecessor and green solvent liquid medium (Li et al., 2011). Moreover, Gokus et al. showed that oxygen plasma can initiate strong fluorescence for single-layer graphene (Gokus et al., 2009).

Carbon Nanotubes

In the biomedical domain, 1-dimensional of 1D carbon nanotubes have got huge attention because of their excellent optical and electronic properties. Depending upon the layers of cylindrical graphene, the carbon nanotubes can be classified into two types namely multi-walled carbon nanotubes (MWCNTs) and single-walled carbon nanotubes (SWCNTs). SWCNTs have only a layer of graphene sheet which is rolled into a cylinder whereas, for MWCNTs, many concentric graphene sheet layers are present. The

Figure 3. Different types of silica nanoparticles

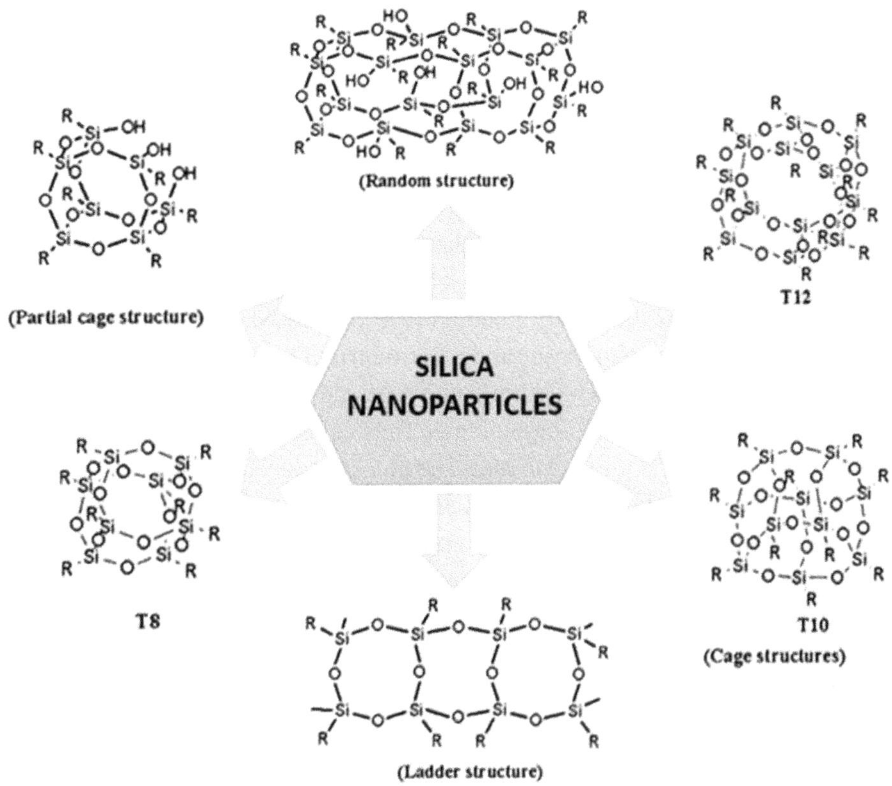

length of CNTs is a few or several millimeters while the outer diameter stays within 100 nm, so it leads to a large surface area and a very high aspect ratio (Prajapati et al., 2019). Furthermore, the exclusive organization of the carbon atoms inside carbon nanotubes produces a highly π-electron conjugation outward to the nanotube (Karimi et al., 2016). Moreover, the CNTs are gifted with fluorescence in the NIR region and strong absorption (Nagai et al., 2019). Because of all these properties CNTs along with their good interaction with various biomolecules, makes them a suitable material for various biomedical uses. Synthetic methods affect the quality, chirality, structure, length, and diameter of the CNTs. It must be checked that this process is responsible for huge-scale production. Most preferred methods are chemical vapor deposition (Zhao et al., 2021), laser ablation (Mwafy et al., 2019) and arc-discharge (Roslan et al., 2019). On top of the list, it is most important to functionalize the CNTs for improving the solubility as well as stop them from collecting in biological media and solvents. In CNTs, the defects can be introduced by covalent functionalization, which causes a huge reduction in the total damage of NIR fluorescence. Moreover, noncovalent functionalization having molecules which are amphiphilic like polymers can reserve the fluorescent properties and structure of the CNTs but reduces the QY of CNTs. For overcoming all these limitations, new methods for the functionalization and fabrication of CNTs have been stated. (Lee et al.2011) stated that the addition of reducing agent dithiothreitol can improve the SWCNTs' fluorescent QY for the initial time, causing fluorophores containing brightness the same as of QDs (Lee et al., 2011). Hou et al. examined the dithiothreitol adding to SWCNTs functionalized along with various surfactants. For SDS and DNA-covered SWCNTs, the fluorescent QY contained in them improved expressively, though, for other surfactants at different extents, fluorescence quenching was seen (Lee et al., 2020). This results in dithiothreitol addition to SDS and DNA-covered SWCNTs are achievable solutions for getting the carbon nanotube applications in biomedicine.

Organic Frameworks

Novel crystalline porous materials having outstanding stability, low toxicity, and adsorption are Covalent-organic frameworks (COFs). The combination of the fluorescence method of determination with the fluorescent small organic molecules design may be used for constructing more effective nanoprobes [60]. For the determination of suitable 2,4,6-trinitrophenol (TNP), a new NaphthalimideBenzothiazole compound was produced as a fluorescent nanoprobe and colorimetric. The receptor's Tefuorescence emission peaks were quenched selectively using TNP along with a detection boundary as less as 1.613×10^{-10} M. A new group of multifunctional inorganic-organic materials called metal-organic frameworks (MOFs) having many functionalized 3D crystalline structures and holes produced by linkers and metal ions. MOFs gives potential applications in catalysis, separation, and various features because of their solo qualities like the confinement of the pores, specific surface area, and excellent chemical tenability. Few MOFs are luminescent and have light intensity and quantum yield which can be influenced using excitation wavelength and temperature (Shing et al., 2020). By doxycycline addition, (Yu et al. 2020) fabricated a novel useful metal-organic structure of europium and pyromellitic acid, that shows amazing fluorescence improvement at 617 nm as well as 526 nm. From the data, it is seen that both the intensities of the fluorescence were connected with the concentration of doxycycline. The exclusive fluorescence reaction of the structure can distinguish doxycycline from various tetracycline antibiotics having great selectivity.

Phosphors

In biomedicine applications, phosphors are widely used because of their advantage in decreasing light-scattering interference and autofluorescence from tissues. Generally, phosphors consist of doped ions and host materials (Qin et al., 2020). Through the phosphors, the most commonly used and promising phosphor is yttrium oxide (Y2O3) because of its minimum phonon energy and photo durability. In phosphors, the abundant doped materials are Lanthanides because of their huge energy transfer channels and electron levels. Various processes are reported for the synthesis of phosphors like co-precipitation processes (Perhaita et al., 2020), sol-gel (Rocha et al., 2018), fame spray pyrolysis (Khan et al., 2020), and hydrothermal (Yu et al., 2020). The most economical and efficient process for the development of phosphors is hydrothermal synthesis. With sodium citrate by hydrothermal method, Yu et al. produced Y2O3:Eu3+ phosphors (Yu et al., 2020). The produced phosphors properties are decided by the hydrothermal process where the sodium citrate concentration, as well as the amount of addition of Eu and NaOH, are used. For oxide-based phosphors, Flame spray pyrolysis is the suitable method for quick and successive fabrication of oxide-based phosphors. This process gives phosphors with huge homogenous dopant distribution and high crystallinity compared to other conventional methods. By using fame spray pyrolysis (Qin et al., 2020, Khan et al., 2020). produced Y_2O_3 phosphors doped with Tb^{3+} which were having more or less a diameter of 100 nm with constricted size distribution. Here in this method, with other metal nitrate precursors, mixing of alkali salt was done which successfully controlled the distribution of sizes in the constricted range. The method of sol-gel fabrication gives many pros like narrow particle

Figure 4. Various types of graphene-based nanoparticles

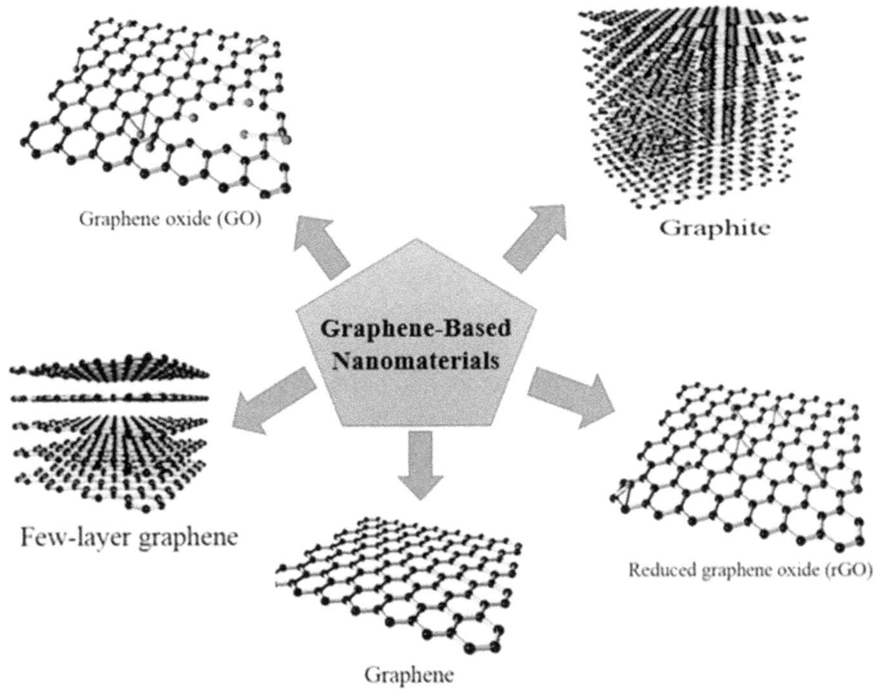

size distribution, uniform particle morphology, reduced synthesis time, purity, and high homogeneity (Khan et al., 2020). Using this method of sol-gel, Leonardo et al. achieved SiO_2-Gd_2O_3 phosphors doped with Sm^{3+} (Perhaita et al., 2020).

Silica Nanoparticles

Taking the advantage of the properties of silica nanoparticles like stabilization of the embedded fluorophores, robustness, mechanical stability, and transparency it is mostly been used in bio-organic applications. Currently, silica core or shell NPs are used for the intracellular identification of H_2PO^{4-} and Zn2+ in existing cells and were produced as "of/on" nanosensors having fluorescent. In current years, Nanoparticles dye doping of organic type is been used widely and produced for various biodetection applications (Tummala et al., 2020). The commonly used production method for the silica nanoparticles are reverse microemulsion technique also the Stöber technique which was primarily defined in late 1960s (Stober et al., 1968). The method of TeStöber was first introduced in the 1960s (Stober et al., 1968), and includes alkyl silicate hydrolysis and following silicic acid condensation in alcoholic solutions which are catalyzed by ammonia addition. Another method of NP formation of silica is the method of reverse microemulsion that includes an alkyl silicates reaction, usually TEOS, within the droplets of water for microemulsion of water-in-oil (Finnie et al., 2007). (He et al., 2010) produced three types of nanoparticles having dye doping silica with the reverse microemulsion method and the Stöber method by implanting dye inside the particle core. Molecules that are responsive functionally to the Zn^{2+} are placed on the surface of the particle. In HeLa cells, Tefuorescent silica nanoparticles were applied for intracellular Zn^{2+} ($H2PO^{4-}$) fluorescent imaging. Addition of Zn^{2+} when done to the Zn^{2+} nanosensor, the NPs showed the capability of $H2PO^{4-}$ detection ratiometrically. Furthermore, nanocomposites of silica having good biocompatibility and monodispersity can be simply rectified with the help of the functional groups (Wang et al., 2015, Gawande et al., 2015, Liu et al., 2015). Doping of the magnetic nanoparticles into silica nanoparticles having fluorescent dyes was reported by Lee et al. For drug delivery in anti-cancers, fluorescence imaging, and magnetic resonance (MR), the use of these silica nanocomposites can be done (Lee et al., 2011). In short, silica particles might be an investigation spot in such situations having significantly prolonged uses. Figure 3 shows different types of silica nanoparticles.

Metal Nanomaterial

An infield of biomedical engineering, the gold nanoparticles quantum mechanical effects, for example, photoluminescence release or AuNPs is perfect for another in-vivo nanosensor (Kuhn et al., 2006, Sonnichen et al., 2005). AuNPs have involved extensive technical awareness by the benefit of their comfort of synthesis and exclusive characteristics, and varied synthetic approaches have been described. The most significant procedures, chemical procedures are usually completed by considering a water solvent solution of chloroauric along with the decreasing products in the incidence of a soothing product. Citric acid is commonly utilized, which can perform as together a dropping and stabilizer agent (Karimadom et al., 2021). Though, citric acid with AuNPs stabilized can go via an irretrievable increase throughout functionalization improvement along with the thiolate ligands. The difficulty is defeated through creating the response that occurs in the existence of surfactants, water-soluble polymers, or covering products which benefit to deliver greater steadiness also avoid combination of nanoparticles. The shape and size of AuNPs is measured by varying the surface-modifying products or gold-citrate proportion. Though AuNPs are non-toxic in certain investigational settings, their harmfulness and side effects are essential to be carefully observed (Chansuvarn et al., 2015). Ag fluorescent nanoclusters are more attracted because of their exclusive chemical as well as physical properties. The production procedure of these nanoclusters are categorized by the steadying scaffold into DNA proteins, polymers, oligonucleotides, and DNA peptides. Also, widespread research has verified few green production, for example, the use of water solvent stem extracts of S. Alba and D. trifoliate to enhance the training circumstances (Li et al., 2014). Cu NCs or Copper nanoclusters are comparatively commonly applied as metal materials of noble type, however, their production is quiet threatened because of their liability to oxidation. In recent times, stable copper nanoclusters were successfully prepared by (Kawasaki et al. 2011), where DNA can be applied as the pattern for the production of fluorescent. A method of double-strand DNA was offered by Mohir et al in solution to get copper nanoclusters with high selectivity (Rotaru et al., 2010). Applying the properties of fluorescent of the copper nanoclusters, it stood efficiently exploited as a suitable fluorescent initiate signal sign for the selective purpose of RDX (Aparna et al., 2019).

GRAPHENE-BASED NANOMATERIALS

As graphene with its products are extensively investigated for various biomedical uses for example drug delivery and bioimaging. Graphene nanomaterials contain graphene oxide (GO), graphene nanosheets, etc. They contain great surface capacities and exclusive surface characteristics which permit non-covalent connections with biomolecules, water-insoluble drugs, and dye molecules. Figure 4 shows several types of Graphene based nanoparticles. Most investigators have described dissimilar graphene production approaches. Synthetic approaches of grapheme nanomaterials should be categorized in two groups, i.e. bottom-up, and top-down. The top-down approaches contain the separation from arranged graphite sheets to procedure graphene sheets, containing solvent-based exfoliation (Ma et al., 2020), electrochemical exfoliation (Lee et al., 2020) and mechanical exfoliation (Sinclair et al., 2019). Bottom-up methods contain the identification of carbon atoms by alternate carbon sources. CVD (chemical vapor deposition) (Wang et al., 2020), and Epitaxial growth (Liu et al., 2020) is the most usually applied bottom-up synthesis procedures.

BIOLOGICAL AND BIOMEDICAL APPLICATIONS

Application of "In-Vivo" Imaging

The fluorescence imaging transformations from in-vitro to in-vivo are still an interesting job. The main major problems need severe concern. The optical signals' penetration depth, is significantly reduced by the scattering, auto-fluorescence, photon absorption, scattering in biological materials. For dealing with this difficulty, nanoparticles of fluorescent organic type having emission of longer-wavelength, generally in the (Near-infrared) NIR series, are greatly favored due to the concentrated interface inside the 600–1300 nm (tissue optical window) (Pansare et al., 2012). The additional concern is the bio-organic obstacles that the body utilizes to hit and eliminate exogenous materials, for example, viruses, bacteria, drugs, and medical implants. The difficulties inflicted through the body can be mostly distinguish as either physiological or cellular. Physiological difficulties consist of the liver, kidneys, blood, spleen, immune system, and guards which can avert extravasations of outer materials from the blood (Kievit et al., 2011). Consequently, some nano-sized element surface imaging composition must be cautiously planned so that it can avoid biological difficulty and get to the object's place before it is detached through the body. The barriers of cells contain the cell membrane, intracellular trafficking, and the endosome/lysosome trafficking system. Figure 5 shows the in vivo imaging of the mice treated with Cy5.5-labeled F-Bcl-2 ASO, Cy5.5-labeled SNA (F-Bcl-2 ASO), and free Cy5.5 at altered time intervals after IV injection. The red-dashed areas shows the tumor positions (Mou et al., 2019).

Nonspecific Imaging

Numerous fluorescent nanoparticles with NIR (Near-infrared) or red emission is being arranged for unfocused imaging in-vivo. These elements are steady by any surfactants (Mou et al., 2019, Kilin et al., 2014), amphiphilic polymers (Cova et al., 2013) or PEG (Reul et al., 2012, Ding et al., 2013). Matters of importance can be imaged in-vivo by tracing the fluorescent trail of organic nanoparticles. For instance, in the healthy mice brain ventricles, the human amniotic fluid cells are labeled with polymer RhB-doped nanoparticles that were implanted, and with the help of fluorescence imaging, the human amniotic fluid cells longitudinal bio-distribution in altered areas of brain was defined (Kilin et al., 2014). The blood vessels are perfect for common imaging in-vivo as they are the chief portion for element transference next to (intravenous) IV injection. A two-photon fluorescence imaging method was used for actual vasculature imaging in mice (live), and in the brain, ear, and bone marrow, three models were analyzed. To execute in-vivo micro-angiography imaging on live zebrafish embryos, DiI-TPB (a bright fluorescent

Figure 5. In vivo imaging of the mice treated with Cy5.5-labeled F-Bcl-2 ASO, Cy5.5-labeled SNA (F-Bcl-2 ASO), and free Cy5.5 at altered time intervals after IV injection. The red-dashed areas shows the tumor positions.
Source: Mou et al. (2019)

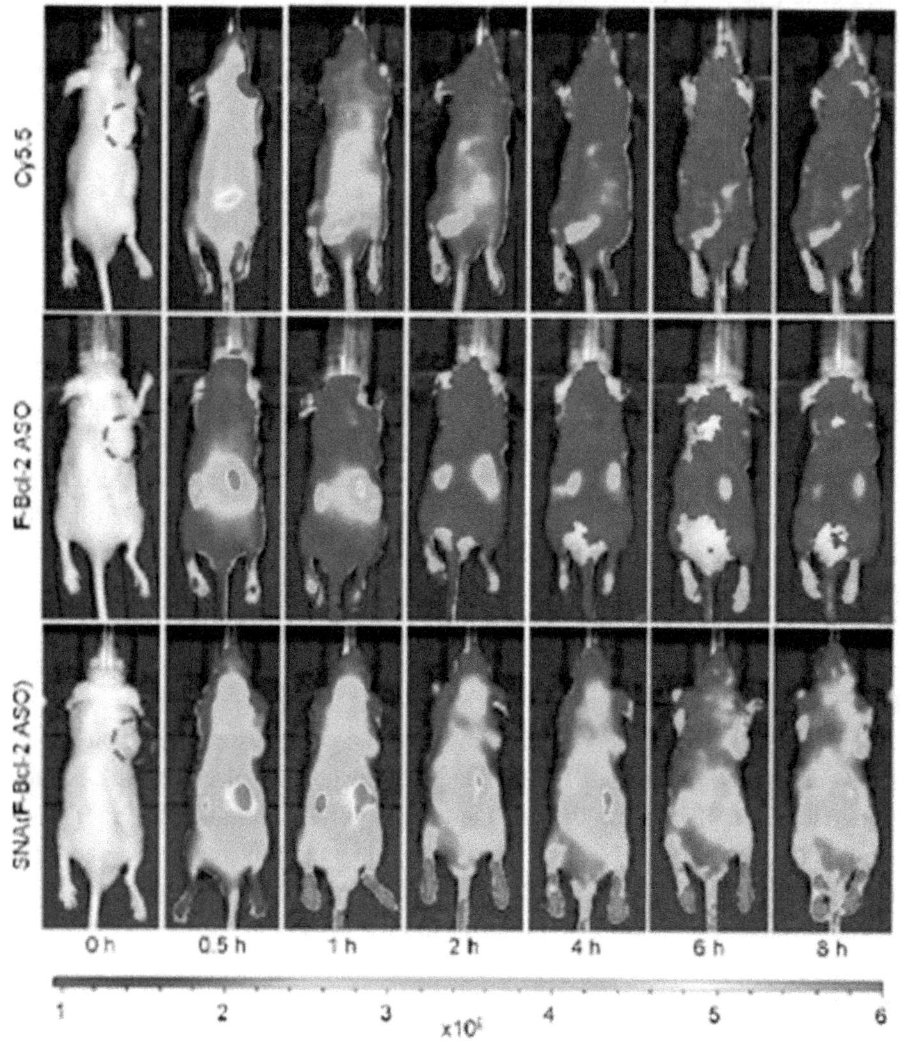

nano-sized droplet) was designed by (Kilin et al., 2014). (Mou et al., 2019) the injection of the droplets done in the zebrafish embryo's sinus venosus.

Specific Imaging

The EPR (enhanced permeability and retention) consequence is active for fluorescent nanoparticles of inertly aimed tumors without ligands. With the help of this effect, drug-loaded micelles (fluorescent PEO stabilized) were utilized to drug delivery imaging through circulation of blood (Tao et al., 2013). The fluorescent imaging real-time has shown that in the tumor area micelles is grown until they touch an extreme piont at 24 hours after the injection. Though, passive targeting depending on EPR, is smooth to perform; active targeting displays improved specificity along with the recognized receptor-ligand interfaces. Maximum for the detailed imaging studies in-vivo were done targeting the fluorescent nanoparticles, for example, NIR (Near-infrared) polymer vesicles adapted with the antibodies (Zhang et al., 2012). (Near-infrared) NIR polymer nanoparticles join together with peptide-coated (Near-infrared) NIR nanoparticles (Ding et al., 2013) & folate groups (Li et al., 2012, Mini et al., 2010) give considerably improved recognition particularity and exactness because of the specificity between ligands and receptors. As well as ligands that impasse precisely for the object, remaining ligands can be also required to support the traverse special biological obstacles. Currently, the brain in-vivo fluorescent imaging is challenging, due to the existence of the (BBB) blood-brain barrier and the difficult dependency insertion of the probe for the obstruction on the properties of surface and size. In recent times, numerous efforts were prepared to precisely image brains with nanoparticles–ligand bio-conjugates. PLGA-based nanoparticles having a NIR (Near-infrared) color as well as the peptide for targeting the brain-specific location was described by (Tosi et al.,2011, Bunschoten et al., 2012). Studies of the fluorescent and confocal microscopic brain tissues approved the insertion of the NIR (Near-infrared) nanoparticles through the blood-brain barrier (BBB) and delivery inside the cerebral tissue cell.

FUTURE TRENDS

There are numerous possible uses for fluorescent nanoparticles in biomedical engineering. Future applications could include:

1. Fluorescent nanoparticles can be utilized to image biological tissues and cells by acting as contrast agents. They can also be utilized for in vivo tumor imaging, enabling a more accurate and precise diagnosis.
2. **Drug Delivery:** Drugs can be delivered to particular cells or tissues in the body using fluorescent nanoparticles. They can also be used to track the distribution of drugs in real time, enabling more individualized and effective treatment (Barua et al., 2018).
3. Fluorescent nanomaterials are functional as biosensors because they may be used to identify particular molecules or proteins in biological samples (Zhang et al., 2022). Early disease detection, treatment effectiveness monitoring, and medication discovery can all benefit from this.
4. **Tissue Engineering:** In tissue engineering applications, fluorescent nanoparticles can be utilized to track the development and differentiation of cells (Vieira et al., 2017). Additionally, they can be utilized to build 3D models of tissues for use in drug testing and individualized treatment.
5. Fluorescent nanoparticles can be utilized in photodynamic treatment as photosensitizers (Chen et al., 2021). Reactive oxygen species are created when they are activated by light, and they can specifically kill cancer cells while sparing healthy cells damage.

Generally, the diagnosis, treatment, and monitoring of diseases may be revolutionized by using fluorescent nanoparticles in biomedical engineering. But more investigation is required to understand and maximize their potential properly.

CONCLUSION

We have studied current advancements for the improvement of the soft fluorescent nanoparticles for biological & biomedical imaging processes. In contrast to fluorescent nanoparticles (solid), it has the capability of biomedical as well as biological imaging. Different fluorescent nanoparticles (organic) are expanded as imaging agent materials, as well as nanoparticles of the semiconducting polymer, small organic molecule, dye-doped polymer nanoparticles, nano-gels, vesicles, and polymeric micelles. These organic nanoparticles are prepared by post-polymerization (reprecipitation and miniemulsion) procedures, direct polymerization, or self-assembly procedures. The mentioned processes are considered as the feature of preliminary materials, which decides the contact forces among the molecules as well as the steadiness of the particles, and the accumulation status of the fluorophores within the nanoparticles that affect the properties of fluorescence. In the case of soft fluorescent nanoparticles display varied imaging abilities and optical properties. Moderately, AIE-type small-molecule organic nanoparticles and fluorescent semiconducting polymer nanoparticles are recognized as super brightness that is enviable for imaging high-contrast materials. Polymeric nanoparticles having dye-doped could have low brightness frail point, however, a huge collection of dyes are obtainable for the making of dye-doped nanoparticles having multipurpose imaging capabilities. Also, these types of nanoparticles are derived from biodegradable biomaterials which are appropriate for imaging in-vivo due to their small poisonousness. For the case of micelles polymer, vesicles, as well as the nanogels, the unattached nanostructures allow the bio-analytes or incitements to influence the inner fluorophores or particle matrix and cellular procedures could be imaged consequently. The imaging agents are broadly described for in-vitro imaging; though still, it is needed for advance, for example, custom-made size of the particle and as well as the surface chemistry for aiming and imagining organelles. Concerning in-vivo imaging, nanoparticles having fluorophores of long-wavelength are usually implemented. However, the fluorescence imaging are restricted for the smaller tissues, although organic nanoparticles NIR-II, is improved to enhance the access depth. For undertaking this limitation, one probable approach is that to increase nanoprobes multimodality with new contrast products which are combined along with the fluorescent nanoparticles. For instance, 64Cu–DOTA can be accepted to construct Gd-DTPA MRI (optical agents), and Positron emission tomography (PET) (optical dual-modality imaging agents). Then again, the soft fluorescent imaging agents (nano-sized) could get further useful functions in the surgical process.

REFERENCES

Aparna, R. S., Anjali Devi, J. S., Anjana, R. R., Nebu, J., & George, S. (2019). Zn(II) ion modulated red emitting copper nanocluster probe for the fluorescence turn on sensing of RDX. *Sensors and Actuators. B, Chemical*, *291*, 298–305. doi:10.1016/j.snb.2019.04.051

Bajorowicz, B., Kobylanski, M. P., Golabiewska, A., Nadolna, J., Zaleska-Medynska, A., & Malankowska, A. (2018). Quantum dot-decorated semiconductor micro- and nanoparticles: A review of their synthesis, characterization and application in photocatalysis. *Advances in Colloid and Interface Science*, *256*, 352–372. doi:10.1016/j.cis.2018.02.003 PMID:29544654

Baker, S. N., & Baker, G. A. (2010). Luminescent carbon nanodots: emergent nanolights. Angewandte Chemie-Int, 49(38), 6726–6744.

Barua, R., Datta, S., Chowdhury, A. R., & Datta, P. (2018). Advances in MEMS and Micro-Scale technologies for application in controlled Drug-Dosing systems: MEMS-Based drug delivery systems. In *Design and Development of Affordable Healthcare Technologies* (pp. 165–179). IGI Global. doi:10.4018/978-1-5225-4969-7.ch007

Benya, R., Quintana, J., & Brundage, B. (1989). Adverse reactions to indocyanine green: A case report and a review of the literature. *Catheterization and Cardiovascular Diagnosis, 17*(4), 231–233. doi:10.1002/ccd.1810170410 PMID:2670244

Betzig, E., Patterson, G. H., Sougrat, R., Lindwasser, O. W., Olenych, S., Bonifacino, J. S., Davidson, M. W., Lippincott-Schwartz, J., & Hess, H. F. (2006). Imaging intracellular fluorescent proteins at nanometer resolution. *Science, 313*(5793), 1642–1645. doi:10.1126cience.1127344 PMID:16902090

Bunschoten, A., Buckle, T., Kuil, J., Luker, G. D., Luker, K. E., Nieweg, O. E., & van Leeuwen, F. W. B. (2012). Targeted non-covalent self-assembled nanoparticles based on human serum albumin. *Biomaterials, 33*(3), 867–875. doi:10.1016/j.biomaterials.2011.10.005 PMID:22024362

Chansuvarn, W., Tuntulani, T., & Imyim, A. (2015). Colorimetric detection of mercury(II) based on gold nanoparticles, fluorescent gold nanoclusters and other gold-based nanomaterials. *Trends in Analytical Chemistry, 65*, 83–96. doi:10.1016/j.trac.2014.10.013

Chen, C., Song, M., Du, Y., Yu, Y., Li, C., Han, Y., Yan, F., Shi, Z., & Feng, S. (2021). Tumor-Associated-Macrophage-Membrane-Coated Nanoparticles for Improved Photodynamic Immunotherapy. *Nano Letters, 21*(13), 5522–5531. doi:10.1021/acs.nanolett.1c00818 PMID:34133181

Cheng, L. T. J. (2013)... *Nano Today, 8*, 290–312. doi:10.1016/j.nantod.2013.04.007 PMID:23997809

Cho, E. C., Glaus, C., Chen, J. Y., Welch, M. J., & Xia, Y. N. (2010). Inorganic nanoparticle-based contrast agents for molecular imaging. *Trends in Molecular Medicine, 16*(12), 561–573. doi:10.1016/j.molmed.2010.09.004 PMID:21074494

Cova, L., Bigini, P., Diana, V., Sitia, L., Ferrari, R., Pesce, R. M., Khalaf, R., Bossolasco, P., Ubezio, P., Lupi, M., Tortarolo, M., Colombo, L., Giardino, D., Silani, V., Morbidelli, M., Salmona, M., & Moscatelli, D. (2013). Biocompatible fluorescent nanoparticles for *in vivo* stem cell tracking. *Nanotechnology, 24*(24), 245603. doi:10.1088/0957-4484/24/24/245603 PMID:23690139

da Silva, J., & Goncalves, H. M. R. (2011). Analytical and bioanalytical applications of carbon dots. *Trends in Analytical Chemistry, 30*(8), 1327–1336. doi:10.1016/j.trac.2011.04.009

Dabbousi, B., Rodriguez-Viejo, J., Mikulec, F. V., Heine, J., Mattoussi, H., Ober, R., Jensen, K., & Bawendi, M. (1997). (CdSe) ZnS core-shell quantum dots: Synthesis and characterization of a size series of highly luminescent nanocrystallites. *The Journal of Physical Chemistry B, 101*(46), 9463–9475. doi:10.1021/jp971091y

Ding, D., Goh, C. C., Feng, G. X., Zhao, Z. J., Liu, J., Liu, R. R., Tomczak, N., Geng, J. L., Tang, B. Z., Ng, L. G., & Liu, B. (2013). Ultrabright Organic Dots with Aggregation-Induced Emission Characteristics for Real-Time Two-Photon Intravital Vasculature Imaging. *Advanced Materials, 25*(42), 6083–6088. doi:10.1002/adma.201301938 PMID:24038281

Ding, D., Liu, J., Feng, G. X., Li, K., Hu, Y., & Liu, B. (2013). Bright Far-Red/Near-Infrared Conjugated Polymer Nanoparticles for In Vivo Bioimaging. *Small, 9*(18), 3093–3102. doi:10.1002mll.201300171 PMID:23625815

Efros, A. L., & Efros, A. L. (1982). Interband absorption of light in a semiconductor sphere. *Soviet Phys Semicond USSR, 16*(7), 772–775.

Ekimov, A., & Onushchenko, A. (1982). Quantum size effect in the opticalspectra of semiconductor micro-crystals. *Soviet Phys Semicond USSR, 16*(7), 775–778.

Esteves da Silva, J. C. G., & Goncalves, H. M. R. (2011). Analytical and bioanalytical applications of carbon dots. *Trends in Analytical Chemistry, 30*(8), 1327–1336. doi:10.1016/j.trac.2011.04.009

Feng, X. T., Ashley, J., Zhou, T. C., & Sun, Y. (2018). Fluorometric determination of doxycycline based on the use of carbon quantum dots incorporated into a molecularly imprinted polymer. *Mikrochimica Acta, 185*(11), 9. doi:10.100700604-018-2999-8 PMID:30293175

Finnie, K. S., Bartlett, J. R., Barbé, C. J. A., & Kong, L. (2007). Formation of silica nanoparticles in microemulsions. *Langmuir, 23*(6), 3017–3024. doi:10.1021/la0624283 PMID:17300209

Gawande, M. B., Monga, Y., Zboril, R., & Sharma, R. K. (2015). Silica-decorated magnetic nanocomposites for catalytic applications. *Coordination Chemistry Reviews, 288*, 118–143. doi:10.1016/j.ccr.2015.01.001

Gerion, D., Pinaud, F., Williams, S. C., Parak, W. J., Zanchet, D., Weiss, S., & Alivisatos, A. P. (2001). Synthesis and properties of biocompatible water-soluble silica-coated CdSe/ZnS semiconductor quantum dots. *The Journal of Physical Chemistry B, 105*(37), 8861–8871. doi:10.1021/jp0105488

Gokus, T., Nair, R. R., Bonetti, A., Boehmler, M., Lombardo, A., Novoselov, K. S., Geim, A. K., Ferrari, A. C., & Hartschuh, A. (2009). Making Graphene Luminescent by Oxygen Plasma Treatment. *ACS Nano, 3*(12), 3963–3968. doi:10.1021/nn9012753 PMID:19925014

Guo, W. H., Li, J. J., Wang, Y. A., & Peng, X. G. (2003). Luminescent CdSe/CdS core/shell nanocrystals in dendron boxes: Superior chemical, photochemical and thermal stability. *Journal of the American Chemical Society, 125*(13), 3901–3909. doi:10.1021/ja028469c PMID:12656625

He, C., Zhu, W., Xu, Y., Zhong, Y., Zhou, J., & Qian, X. (2010). Ratiometric and reusable fluorescent nanoparticles for Zn2+ and H 2 PO 4− detection in aqueous solution and living cells. *Journal of Materials Chemistry, 20*(47), 10755–10764. doi:10.1039/c0jm01925a

Hines, M. A., & Guyot-Sionnest, P. (1996). Synthesis and characterization of strongly luminescing ZnS-capped CdSe nanocrystals. *Journal of Physical Chemistry, 100*(2), 468–471. doi:10.1021/jp9530562

Jiang, X., Röcker, C., Hafner, M., Brandholt, S., Dörlich, R. M., & Nienhaus, G. U. (2010). Endo-and exocytosis of zwitterionic quantum dot nanoparticles by live HeLa cells. *ACS Nano, 4*(11), 6787–6797. doi:10.1021/nn101277w PMID:21028844

Karimadom, B. R., & Kornweitz, H. (2021). Mechanism of producing metallic nanoparticles, with an emphasis on silver and gold nanoparticles, using bottom-up methods. *Molecules (Basel, Switzerland), 26*(10), 2968. doi:10.3390/molecules26102968 PMID:34067624

Karimi, P. (2016). Effects of structure and partially localization of the pi electron clouds of single-walled carbon nanotubes on the cation-pi Interactions. Iran J Chem Chem Eng Int, 35(3), 35–43.

Kawasaki, H., Kosaka, Y., Myoujin, Y., Narushima, T., Yonezawa, T., & Arakawa, R. (2011). Microwave-assisted polyol synthesis of copper nanocrystals without using additional protective agents. *Chemical Communications (Cambridge)*, *47*(27), 7740–7742. doi:10.1039/c1cc12346g PMID:21647527

Khan, S., Choi, Y., Ahn, H. Y., Han, J. H., Ju, B. K., Chung, J., & Cho, S. H. (2020). Control of particle size in flame spray pyrolysis of Tb-doped Y(2)O(3)for bioimaging. *Materials (Basel)*, *13*(13), 2987. doi:10.3390/ma13132987 PMID:32635558

Kievit, F. M., & Zhang, M. Q. (2011). Cancer Nanotheranostics: Improving Imaging and Therapy by Targeted Delivery Across Biological Barriers. *Advanced Materials*, *23*(36), H217–H247. doi:10.1002/adma.201102313 PMID:21842473

Kilin, V. N., Anton, H., Anton, N., Steed, E., Vermot, J., Vandamme, T. E., Mely, Y., & Klymchenko, A. S. (2014). Counterion-enhanced cyanine dye loading into lipid nano-droplets for single-particle tracking in zebrafish. *Biomaterials*, *35*(18), 4950–4957. doi:10.1016/j.biomaterials.2014.02.053 PMID:24661553

Kuhn, S., Hakanson, U., Rogobete, L., & Sandoghdar, V. (2006). Enhancement of single-molecule fluorescence using a gold nanoparticle as an optical nanoantenna. *Physical Review Letters*, *97*(1), 4. doi:10.1103/PhysRevLett.97.017402 PMID:16907406

Lee, A. J., Wang, X. Y., Carlson, L. J., Smyder, J. A., Loesch, B., Tu, X. M., Zheng, M., & Krauss, T. D. (2011). Bright fluorescence from individual single-walled carbon nanotubes. *Nano Letters*, *11*(4), 1636–1640. doi:10.1021/nl200077t PMID:21417364

Lee, H., Choi, J. I., Park, J., Jang, S. S., & Lee, S. W. (2020). Role of anions on electrochemical exfoliation of graphite into graphene in aqueous acids. *Carbon*, *167*, 816–825. doi:10.1016/j.carbon.2020.06.044

Lee, J. E., Lee, N., Kim, T., Kim, J., & Hyeon, T. (2011). Multifunctional mesoporous silica nanocomposite nanoparticles for theranostic applications. *Accounts of Chemical Research*, *44*(10), 893–902. doi:10.1021/ar2000259 PMID:21848274

Li, J., Zhu, J.-J., & Xu, K. (2014). Fluorescent metal nanoclusters: From synthesis to applications. *Trends in Analytical Chemistry*, *58*, 90–98. doi:10.1016/j.trac.2014.02.011

Li, J. L., Tang, B., Yuan, B., Sun, L., & Wang, X. G. (2013). A review of optical imaging and therapy using nanosized graphene and graphene oxide. *Biomaterials*, *34*(37), 9519–9534. doi:10.1016/j.biomaterials.2013.08.066 PMID:24034502

Li, S. Q., Pan, R. F., Mehdi, Y. A., Xiao, D. L., & He, H. (2015). One-step spontaneous synthesis of fluorescent carbon nanoparticles with thermosensitivity from polyethylene glycol. *New Journal of Chemistry*, *39*(9), 7033–7039. doi:10.1039/C5NJ01129A

Li, X., Wang, H., Shimizu, Y., Pyatenko, A., Kawaguchi, K., & Koshizaki, N. (2011). Preparation of carbon quantum dots with tunable photoluminescence by rapid laser passivation in ordinary organic solvents. *Chemical Communications (Cambridge)*, *47*(3), 932–934. doi:10.1039/C0CC03552A PMID:21079826

Li, Y., Hu, Y., Zhao, Y., Shi, G. Q., Deng, L. E., Hou, Y. B., & Qu, L. T. (2011). An electrochemical avenue to green-luminescent graphene quantum dots as potential electron-acceptors for photovoltaics. *Advanced Materials*, *23*(6), 776–780. doi:10.1002/adma.201003819 PMID:21287641

Li, Z. H., Wu, L. Y., Hu, P. R., Han, S. H., Zhang, T., Fan, H. L., Jin, W., Jin, Q. H., & Mu, Y. (2012). Soft nanomaterial-based targeting polymersomes for near-infrared fluorescence multispectral in vivo imaging. *Nanoscale*, *4*(22), 7097–7105. doi:10.1039/c2nr32047a PMID:23069779

Liu, H. P., Ye, T., & Mao, C. D. (2007). Fluorescent carbon nanoparticles derived from candle soot. Angewandte Chemie-Int Ed, 46(34), 6473–6475.

Liu, J. N., Bu, W. B., & Shi, J. L. (2015). Silica coated upconversion nanoparticles: A versatile platform for the development of efficient theranostics. *Accounts of Chemical Research*, *48*(7), 1797–1805. doi:10.1021/acs.accounts.5b00078 PMID:26057000

Liu, Q., & Guo, B. D. (2013). Article. *Nano Letters*, *13*, 2436–2441. doi:10.1021/nl400368v PMID:23675758

Liu, Z. Z., Xu, Q. F., Zhang, C. T. F., Sun, Q. Y., Wang, C. B., Dong, M. D., Wang, Z. G., Ohmori, H., Kosinova, M., Goto, T., Tu, R., & Zhang, S. (2020). Laser-induced growth of large-area epitaxial graphene with low sheet resistance on 4H-SiC(0001). *Applied Surface Science*, *514*, 145938. doi:10.1016/j.apsusc.2020.145938

Ma, H., & Shen, Z. G. (2020). Exfoliation of graphene nanosheets in aqueous media. *Ceramics International*, *46*(14), 21873–21887. doi:10.1016/j.ceramint.2020.05.314

Miki, K., Oride, K., Inoue, S., Kuramochi, Y., Nayak, R. R., & Matsuoka, H. (2010). H.Harada H, Hiraoka M and Ohe K. *Biomaterials*, *31*, 934–942. doi:10.1016/j.biomaterials.2009.10.005 PMID:19853909

Mou, Q., Ma, Y., Ding, F., Gao, X., Yan, D., Zhu, X., & Zhang, C. (2019). Two-in-One Chemogene Assembled from Drug-Integrated Antisense Oligonucleotides To Reverse Chemoresistance. *Journal of the American Chemical Society*, *141*(17), 6955–6966. doi:10.1021/jacs.8b13875 PMID:30964284

Mwafy, E. A., & Mostafa, A. M. (2019). Multi walled carbon nanotube decorated cadmium oxide nanoparticles via pulsed laser ablation in liquid media. *Optics & Laser Technology*, *111*, 249–254. doi:10.1016/j.optlastec.2018.09.055

Nagai, Y., Yudasaka, M., Kataura, H., & Fujigaya, T. (2019). Brighter near-IR emission of single-walled carbon nanotubes modified with a cross-linked polymer coating. *Chemical Communications (Cambridge)*, *55*(48), 6854–6857. doi:10.1039/C9CC02712B PMID:31123733

Narayan, R. (Ed.). (2017). *Nanobiomaterials: Nanostructured Materials for Biomedical Applications*. Elsevier.

Nozik, A. J., Beard, M. C., Luther, J. M., Law, M., Ellingson, R. J., & Johnson, J. C. (2010). Semiconductor quantum dots and quantum dot arrays and applications of multiple exciton generation to third-generation photovoltaic solar cells. *Chemical Reviews*, *110*(11), 6873–6890. doi:10.1021/cr900289f PMID:20945911

Pansare, V. J., Hejazi, S., Faenza, W. J., & Prud'homme, R. K. (2012). Review of Long-Wavelength Optical and NIR Imaging Materials: Contrast Agents, Fluorophores, and Multifunctional Nano Carriers. *Chemistry of Materials*, *24*(5), 812–827. doi:10.1021/cm2028367 PMID:22919122

Parvin, N., & Mandal, T. K. (2017). Dually emissive P, N-co-doped carbon dots for fluorescent and photoacoustic tissue imaging in living mice. *Mikrochimica Acta, 184*(4), 1117–1125. doi:10.100700604-017-2108-4

Pathak, A., Suneesh, P. V., Stanley, J., & Babu, T. G. S. (2019). Multicolor emitting N/S-doped carbon dots as a fluorescent probe for imaging pathogenic bacteria and human buccal epithelial cells. *Mikrochimica Acta, 186*(3), 10. doi:10.100700604-019-3270-7 PMID:30715615

Perhaita, I., Muresan, L. E., Tudoran, L. B., Silipas, D. T., & Borodi, G. (2020). Synthesis of silicate apatite phosphors with enhanced luminescence via optimized precipitation technique through pH control. *Journal of Sol-Gel Science and Technology, 96*(2), 498–510. doi:10.100710971-020-05400-1

Prajapati, S. K., Malaiya, A., Kesharwani, P., Soni, D., & Jain, A. (2019). Biomedical applications and toxicities of carbon nanotubes. *Drug and Chemical Toxicology, 17094*, 92. doi:10. 1080/ 01480 545 PMID:31908176

Qin, X. Y., Wang, J., & Yuan, Q. (2020). Synthesis and biomedical applications of lanthanides-doped persistent luminescence phosphors with NIR emissions. *Frontiers in Chemistry, 8*, 608578. doi:10.3389/fchem.2020.608578 PMID:33381494

Reul, R., Tsapis, N., Hillaireau, H., Sancey, L., Mura, S., Recher, M., Nicolas, J., Coll, J. L., & Fattal, E. (2012). Near infrared labeling of PLGA for in vivo imaging of nanoparticles. *Polymer Chemistry, 3*(3), 694–702. doi:10.1039/c2py00520d

Rocha, L. A., Campos-Junior, P. H. A., Esbenshade, J., Siqueira, R. L., Schiavon, M. A., & Ferrari, J. L. (2018). Biocompatibility and photoluminescence of Sm3+doped SiO2-Gd2O3: A promising non-toxic red phosphor to plasmatic membrane tracking. *Ceramics International, 44*(4), 3726–3733. doi:10.1016/j.ceramint.2017.11.153

Romero, V., Vila, V., de la Calle, I., Lavilla, I., & Bendicho, C. (2019). Turn–on fluorescent sensor for the detection of periodate anion following photochemical synthesis of nitrogen and sulphur co–doped carbon dots from vegetables. *Sensors and Actuators. B, Chemical, 280*, 290–297. doi:10.1016/j.snb.2018.10.064

Roslan, M. S., Chaudhary, K. T., Doylend, N., Agam, A., Kamarulzaman, R., Haider, Z., Mazalan, E., & Ali, J. (2019). Growth of wall-controlled MWCNTs by magnetic field assisted arc discharge plasma. *Journal of Saudi Chemical Society, 23*(2), 171–181. doi:10.1016/j.jscs.2018.06.003

Rotaru, A., Dutta, S., Jentzsch, E., Gothelf, K., & Mokhir, A. (2010). Selective dsDNA-templated formation of copper nanoparticles in solution. *Angewandte Chemie International Edition in English, 49*(33), 5665–5667. doi:10.1002/anie.200907256 PMID:20629055

Sharma, J., Yeh, H. C., Yoo, H., Werner, J. H., & Martinez, J. S. (2011). Silver nanocluster aptamers: In situ generation of intrinsically fluorescent recognition ligands for protein detection. *Chemical Communications (Cambridge), 47*(8), 2294–2296. doi:10.1039/C0CC03711G

Simões, E. F. C., Leitão, J. M. M., & da Silva, J. C. G. E. (2016). Carbon dots prepared from citric acid and urea as fluorescent probes for hypochlorite and peroxynitrite. *Mikrochimica Acta, 183*(5), 1769–1777. doi:10.100700604-016-1807-6

Sinclair, R. C., Suter, J. L., & Coveney, P. V. (2019). Micromechanical exfoliation of graphene on the atomistic scale. *Physical Chemistry Chemical Physics*, *21*(10), 5716–5722. doi:10.1039/C8CP07796G PMID:30801077

Singh, V., & Tiwari, M. K. (2020). UV emitting Pb2+ doped Ca2La8(SiO4)(6)O-2 phosphors prepared by sol-gel procedure. *Optik (Stuttgart)*, *206*, 163600. doi:10.1016/j.ijleo.2019.163600

So, M. K., Xu, C. J., Loening, A. M., Gambhir, S. S., & Rao, J. H. (2006). Self-illuminating quantum dot conjugates for in vivo imaging. *Nature Biotechnology*, *24*(3), 339–343. doi:10.1038/nbt1188 PMID:16501578

Sonnichsen, C., Reinhard, B. M., Liphardt, J., & Alivisatos, A. P. (2005). A molecular ruler based on plasmon coupling of single gold and silver nanoparticles. *Nature Biotechnology*, *23*(6), 741–745. doi:10.1038/nbt1100 PMID:15908940

Stöber, W., Fink, A., & Bohn, E. (1968). Controlled growth of monodisperse silica spheres in the micron size range. *Journal of Colloid and Interface Science*, *26*(1), 62–69. doi:10.1016/0021-9797(68)90272-5

Swierczewska, M., Lee, S., & Chen, X. Y. (2011). Inorganic Nanoparticles for Multimodal Molecular Imaging. *Molecular Imaging*, *10*(1), 3–16. doi:10.2310/7290.2011.00001 PMID:21303611

Syedmoradi, L., Daneshpour, M., Alvandipour, M., Gomez, F. A., Hajghaseem, H., & Omidfar, K. (2017). Point of care testing: The impact of nanotechnology. *Biosensors & Bioelectronics*, *87*, 373–387. doi:10.1016/j.bios.2016.08.084 PMID:27589400

Tang, L. B., Ji, R. B., Cao, X. K., Lin, J. Y., Jiang, H. X., Li, X. M., Teng, K. S., Luk, C. M., Zeng, S. J., Hao, J. H., & Lau, S. P. (2012). Deep ultraviolet photoluminescence of water-soluble self-passivated graphene quantum dots. *ACS Nano*, *6*(6), 5102–5110. doi:10.1021/nn300760g PMID:22559247

Tantra, R., & Knight, A. (2011). Cellular uptake and intracellular fate of engineered nanoparticles: A review on the application of imaging techniques. *Nanotoxicology*, *5*(3), 381–392. doi:10.3109/17435390.2010.512987 PMID:20846020

Tao, H. Q., Yang, K., Ma, Z., Wan, J. M., Zhang, Y. J., Kang, Z. H., & Liu, Z. (2012). In vivo NIR fluorescence imaging, biodistribution, and toxicology of photoluminescent carbon dots produced from carbon nanotubes and graphite. *Small*, *8*(2), 281–290. doi:10.1002mll.201101706 PMID:22095931

Tao, Z. M., Hong, G. S., Shinji, C., Chen, C. X., Diao, S., Antaris, A. L., Zhang, B., Zou, Y. P., & Dai, H. J. (2013). Biological Imaging Using Nanoparticles of Small Organic Molecules with Fluorescence Emission at Wavelengths Longer than 1000 nm. *Angewandte Chemie International Edition*, *52*(49), 13002–13006. doi:10.1002/anie.201307346 PMID:24174264

Tosi, G., Bondioli, L., Ruozi, B., Badiali, L., Severini, G. M., Biffi, S., De Vita, A., Bortot, B., Dolcetta, D., Forni, F., & Vandelli, M. A. (2011). NIR-labeled nanoparticles engineered for brain targeting: In vivo optical imaging application and fluorescent microscopy evidences. *Journal of Neural Transmission (Vienna, Austria)*, *118*(1), 145–153. doi:10.100700702-010-0497-1

Tummala, S., Huang, W. A., Wu, B. H., Chang, K. C., & Ho, Y. P. (2020). Fluorescent mesoporous nanoparticles for beta-lactamase screening assays. *ChemistryOpen*, *9*(10), 1074–1081. doi:10.1002/open.202000221 PMID:33117628

Vieira, S., Vial, S., Reis, R. L., & Oliveira, J. M. (2017). Nanoparticles for bone tissue engineering. *Biotechnology Progress*, *33*(3), 590–611. doi:10.1002/btpr.2469 PMID:28371447

Wang, J., & Yan, B. (2019). Improving covalent organic frameworks fluorescence by triethylamine pinpoint surgery as selective biomarker sensor for diabetes mellitus diagnosis. *Analytical Chemistry*, *91*(20), 13183–13190. doi:10.1021/acs.analchem.9b03534 PMID:31529947

Wang, J. B., Ren, Z., Hou, Y., Yan, X. L., Liu, P. Z., Zhang, H., Zhang, H. X., & Guo, J. J. (2020). A review of graphene synthesis at low temperatures by CVD methods. *New Carbon Materials*, *35*(3), 193–208. doi:10.1016/S1872-5805(20)60484-X

Wang, Y., & Gu, H. C. (2015). Core-shell-type magnetic mesoporous silica nanocomposites for bioimaging and therapeutic agent delivery. *Advanced Materials*, *27*(3), 576–585. doi:10.1002/adma.201401124 PMID:25238634

Warnement, M. R., Tomlinson, I. D., & Rosenthal, S. J. (2007). Fluorescent Imaging Applications of Quantum Dot Probes. *Current Nanoscience*, *3*(4), 273–284. doi:10.2174/157341307782418595

Wilczewska, A. Z., Niemirowicz, K., Markiewicz, K. H., & Car, H. (2012). Nanoparticles as drug delivery systems. *Pharmacological Reports*, *64*(5), 1020–1037. doi:10.1016/S1734-1140(12)70901-5 PMID:23238461

Willig, K. I., Rizzoli, S. O., Westphal, V., Jahn, R., & Hell, S. W. (2006). STED microscopy reveals that synaptotagmin remains clustered after synaptic vesicle exocytosis. *Nature*, *440*(7086), 935–939. doi:10.1038/nature04592 PMID:16612384

Xiao, D., Yuan, D., He, H., & Gao, M. (2013). Microwave assisted one-step green synthesis of fluorescent carbon nanoparticles from ionic liquids and their application as novel fluorescence probe for quercetin determination. *Journal of Luminescence*, *140*, 120–125. doi:10.1016/j.jlumin.2013.02.032

Xu, H., Zhang, K., Liu, Q., Liu, Y., & Xie, M. (2017). Visual and fluorescent detection of mercury ions by using a dually emissive ratiometric nanohybrid containing carbon dots and CdTe quantum dots. *Mikrochimica Acta*, *184*(4), 1199–1206. doi:10.100700604-017-2099-1

Xu, X., Ray, R., Gu, Y., Ploehn, H. J., Gearheart, L., Raker, K., & Scrivens, W. A. (2004). Electrophoretic analysis and purification of fluorescent single-walled carbon nanotube fragments. *Journal of the American Chemical Society*, *126*(40), 12736–12737. doi:10.1021/ja040082h PMID:15469243

Yan, X., Cui, X., & Li, L. (2010). Synthesis of large, stable colloidal grapheme quantum dots with tunable size. *Journal of the American Chemical Society*, *132*(17), 5944–5945. doi:10.1021/ja1009376 PMID:20377260

Yang, Y., Cui, J., Zheng, M., Hu, C., Tan, S., Xiao, Y., Yang, Q., & Liu, Y. (2012). One-step synthesis of amino-functionalized fluorescent carbon nanoparticles by hydrothermal carbonization of chitosan. *Chemical Communications (Cambridge)*, *48*(3), 380–382. doi:10.1039/C1CC15678K PMID:22080285

Yang, Y. K., Chang, Y. Y., Guo, Y. Y., Yu, L. G., Zhang, G. H., Zhai, D. D., Wang, X. M., & Sun, X. T. (2019). Fluorometric microplate-based dimethoate assay using CdSe/ZnS quantum dots coated with a molecularly imprinted polymer. *Mikrochimica Acta, 186*(8), 10. doi:10.100700604-019-3649-5 PMID:31367777

Yu, J. H., Choi, S. M., Richards, C. I., Antoku, Y., & Dickson, R. M. (2008). Live Cell Surface Labeling with Fluorescent Ag Nanocluster Conjugates †. *Photochemistry and Photobiology, 84*(6), 1435–1439. doi:10.1111/j.1751-1097.2008.00434.x PMID:18764887

Yu, L., Chen, H. X., Yue, J., Chen, X. F., Sun, M. T., Hou, J., Alamry, K. A., Marwani, H. M., Wang, X. K., & Wang, S. H. (2020). Europium metal-organic framework for selective and sensitive detection of doxycycline based on fluorescence enhancement. *Talanta, 207*, 8. doi:10.1016/j.talanta.2019.120297 PMID:31594583

Zhang, L. Z., Lin, Y., Zhang, Y. J., Chen, R., Zhu, Z. S., Wu, W., & Jiang, X. Q. (2012). Fluorescent Micelles Based on Star Amphiphilic Copolymer with a Porphyrin Core for Bioimaging and Drug Delivery. *Macromolecular Bioscience, 12*(1), 83–92. doi:10.1002/mabi.201100197 PMID:22052617

Zhang, Y., Chan, K. F., Wang, B., Chiu, P. W. Y., & Zhang, L. (2018). Spore-derived color-tunable multi-doped carbon nanodots as sensitive nanosensors and intracellular imaging agents. *Sensors and Actuators. B, Chemical, 271*, 128–136. doi:10.1016/j.snb.2018.05.112

Zhang, Y., Tang, H., Chen, W., & Zhang, J. (2022). Nanomaterials Used in Fluorescence Polarization Based Biosensors. *International Journal of Molecular Sciences, 23*(15), 8625. doi:10.3390/ijms23158625 PMID:35955779

Zhao, H., Tu, N., Zhang, W. B., Zhang, M., & Wang, J. (2021). Novel synthesis of Silicon/Carbon nanotubes microspheres as anode additives through chemical vapor deposition in fluidized bed reactors. *Scripta Materialia, 192*, 49–54. doi:10.1016/j.scriptamat.2020.09.048

Zhao, Q. L., Zhang, Z. L., Huang, B. H., Peng, J., Zhang, M., & Pang, D. W. (2008). Facile preparation of low cytotoxicity fluorescent carbon nanocrystals by electrooxidation of graphite. *Chemical Communications (Cambridge), 41*(41), 5116–5118. doi:10.1039/b812420e PMID:18956040

Zong, J., Zhu, Y., Yang, X., Shen, J., & Li, C. (2011). Synthesis of photoluminescent carbogenic dots using mesoporous silica spheres as nanoreactors. *Chemical Communications (Cambridge), 47*(2), 764–766. doi:10.1039/C0CC03092A PMID:21069221

Zuo, P. L., Lu, X. H., Sun, Z. G., Guo, Y. H., & He, H. (2016). A review on syntheses, properties, characterization and bioanalytical applications of fluorescent carbon dots. *Mikrochimica Acta, 183*(2), 519–542. doi:10.100700604-015-1705-3

Chapter 10
Nanomaterial-Based Bio-Detection

Iqra Zareef
University of Health Sciences, Pakistan

Rahat Rehman
University of Health Sciences, Pakistan

Shahid Nazir
University of Health Sciences, Pakistan

Ahsan Riaz
University of Health Sciences, Pakistan

ABSTRACT

Analysis of body fluids is an important aspect of forensic investigation; therefore, rapid and sensitive detection is required for their analysis as well as characterization. This chapter presents an overview of the applications of different nanomaterials with respect to the biological material detection. If we talk about forensic biological samples, we especially need to handle them with care, because such samples from crime scenes are always in traces. This chapter briefly discusses the use of functional nanoparticles for the detection of biological materials. Genetic material obtained from crime scenes is one of the most important pieces of evidence in forensic investigations, and the chapter focuses on the application of functional nanomaterials for their detection.

INTRODUCTION

Nanomaterials are in the midst of rapid development owing to their potential applications in almost all scientific fields (Cao & Wang, 2004). With daily technological advancements, novel properties of nanomaterials and nanotechnology are being discovered, studied, and utilized in medicine, diagnostic, forensic, pharmaceutical, clinical, industrial, agricultural, environmental monitoring and safety purposes (Perfézou et al., 2012; Zhang et al., 2009). Several nanomaterial types have already been synthesized

DOI: 10.4018/978-1-6684-8325-1.ch010

Figure 1. Application of nanotechnology in forensics

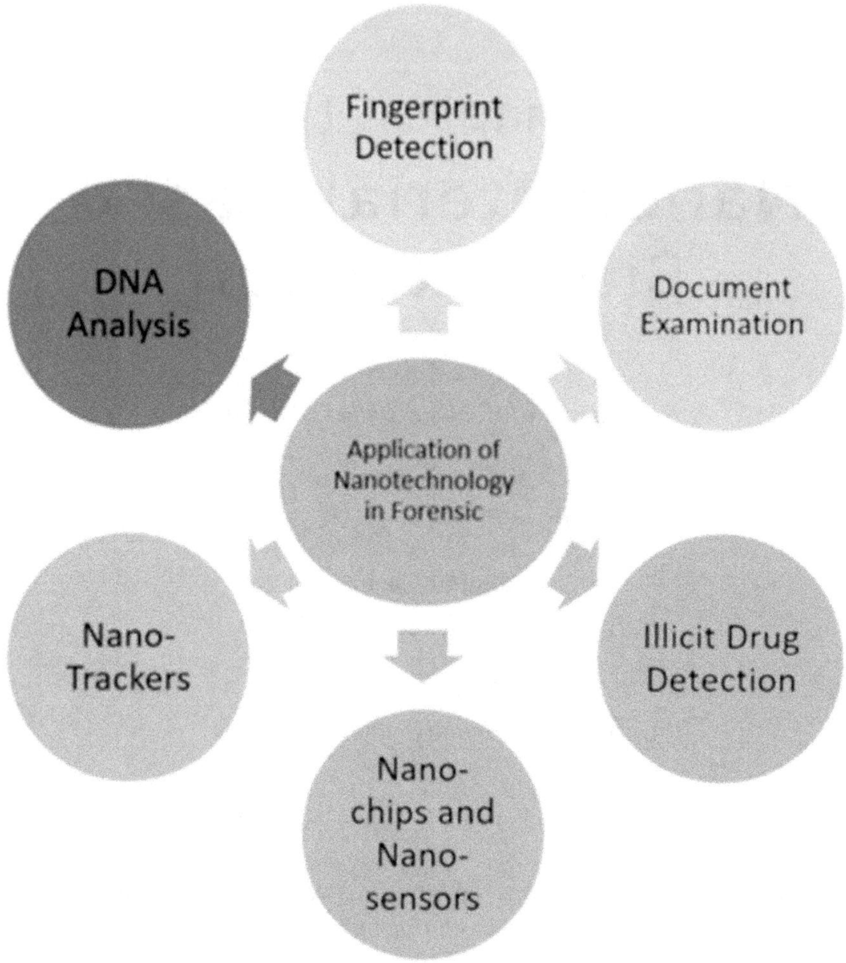

and utilized for such purposes; among which carbon nanoparticles, magnetic nanoparticles, silver nanoparticles, carbon nanotubes, quantum dots, graphene, and hybrid nanocomposites are widely used because of their optical, thermal and structural properties (Chen & Chatterjee, 2013; Yang et al., 2015).

Owing to the complexity of forensic analysis procedures, it is necessary to develop highly efficient and rapid technological methods that can provide accurate results with limited available evidence. Most of the time, the evidence secured from a crime scene is either partially degraded, impure, or in trace amounts (C. Liu et al., 2019). Furthermore, securing and preserving this evidence is always at risk of compromise as it requires a great deal of accuracy. In this regard, nanotechnology, illustrated in Figure 1, has gained a high level of interest in the forensic field because it offers rapid results and ease in complex analytical procedures (Hagarová, 2020).

Many conventional methods that are widely used in forensic applications have significant downfalls. For instance, manual extraction and isolation of DNA from biological samples leads to denaturation (Abdulrahman & Ghanem, 2018). The latent fingerprinting powders used in fingerprint development most often lead to overlapping of granules, resulting in distorted results. However, the detection and isolation

Figure 2. Utilization of nanomaterial in forensics

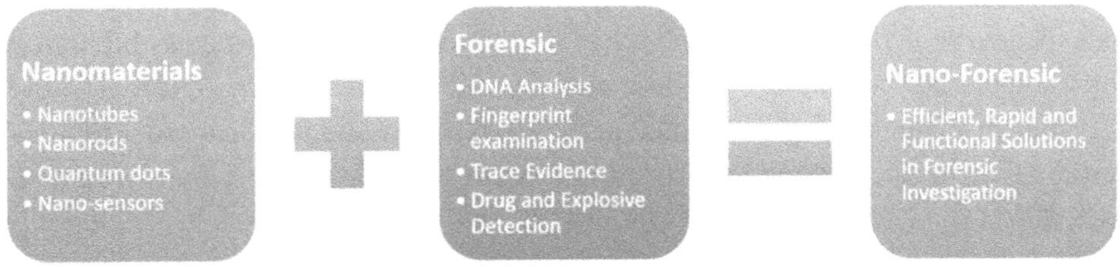

of toxic substances and illicit drugs are complex and expensive (Chan et al., 2020; Sanli et al., 2020). Conventional methods for the detection of explosives and chemical warfare substances are inadequate because they lack in terms of sensitivity and selectivity (A. S. Lodha et al., 2016; Rawtani et al., 2019). Aside from this, forensic field heavily lacks in rapid and accurate detection of biological, chemical, and toxic substances for on-site crime scene analysis (Nadar et al., 2021). To overcome these shortcomings, nanoparticles have been utilized in several forensic analytical techniques because of their modifiable properties. The utilization of nanomaterials in the field of forensics, giving rise to nano-forensics, see Figure 2, has helped enhance the investigation process and provide accurate and time-efficient results (Anuradha & Bhatia, 2022).

The latest advancements in nanotechnology have enabled the detection of trace evidence such as DNA from gunshot residues, fingerprints, trace explosives, and heavy metals (Ganesh, 2016). The nano dimensions of various nanoparticles, such as nanotubes, quantum dots, nanorods, nanofibers, and supramolecules, provide them with a high surface area that increases molecular interactions and exhibits high sensitivity and selectivity (El-Ansary & Faddah, 2010; Huang et al., 2010; Xu et al., 2009).

One of the crucial aspects of any criminal investigation is the analysis of body fluids, which leads to identification. However, many body fluids that are found at a crime scene are hardly visible to the naked eye and often times they are present in the form of mixtures. As such, the most important factor is to determine the presence of a specific body fluid in the stain. The most frequently found body fluids at a crime scene are saliva, blood, semen, sweat, vaginal fluid or urea. Depending on the type of crime scene, any such body fluid may be present which provides valuable DNA evidence. Often times, these fluids are present in trace amounts but provide significant evidence for identification (Virkler & Lednev, 2009). Conventional techniques for identification and analysis of biological evidence present at crime scenes include alternative light sources (ALS), crystal tests, electrophoretic techniques, chemiluminescence, spectroscopic, immunologic, and microscopic techniques (Rawtani et al., 2019). These techniques fall short because of their low sensitivity and selectivity, destruction of valuable evidence, and time consumption, which hinders case investigation. The use of nanotechnology and the application of functional nanoparticles for the analysis of biological fluids provide novel solutions to the abovementioned conventional problems.

BIOLOGICAL EVIDENCE

The materials located at a crime scene which are of biological origin are termed as biological evidence. These biological materials contain deoxyribose nucleic acid (DNA) in them. In forensic investigations, DNA is confirmatory evidence that is accepted in the court of law and juridical proceedings. These biological materials help in establishing the crucial link between the crime scene and a person or persons (Safarik et al., 2002).

Classification of Biological Evidence

There are two major categories of biological evidence that are found at a crime scene, which are further divided based on their areas of origin, 1) Body Fluids 2) Non-Body Fluids. Further classification of the biological evidence that may be recovered from crime scenes is provided in the Figure 3.

Figure 3. Classification of biological evidence

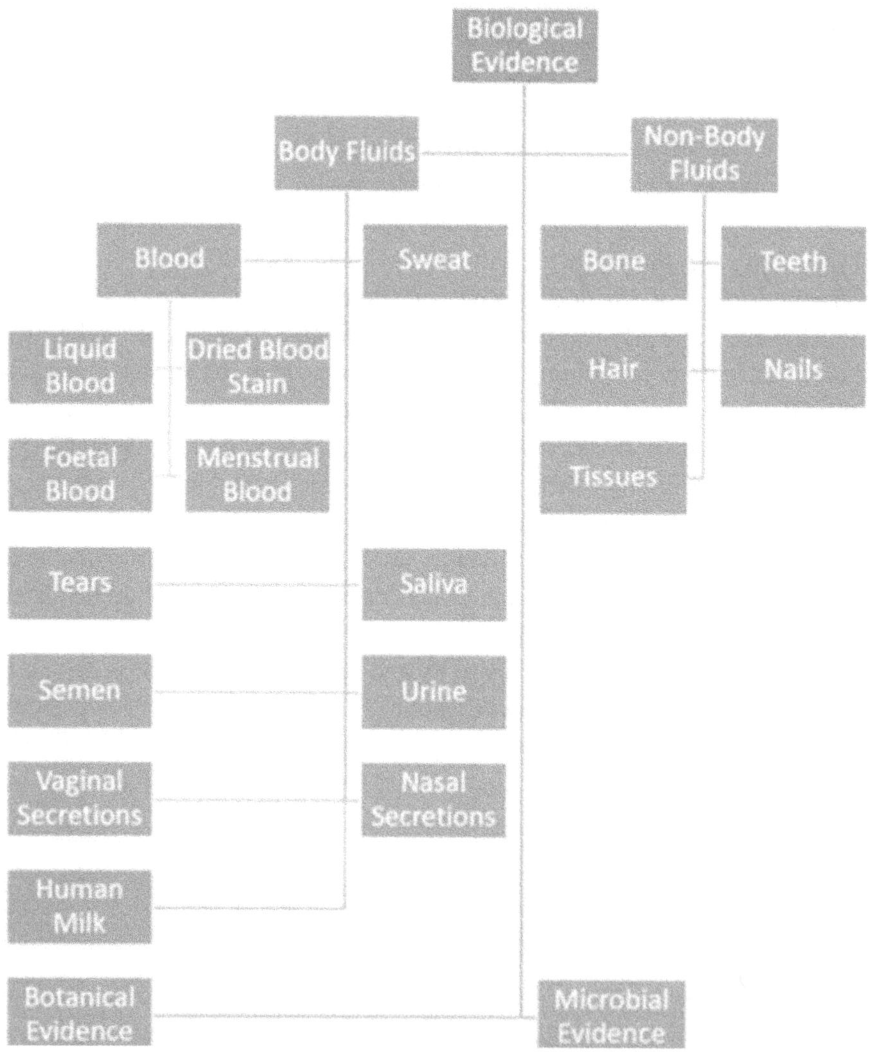

The biological evidence that is most commonly found and analyzed from crime scenes includes saliva, blood, semen, vaginal secretion, and urine. DNA can be isolated from all biological materials. The type of evidence that can be found and recovered from a crime scene depends on the nature of the crime. In Table 1, several possible locations from which different types of biological fluids can be recovered during crime scene investigation are listed, as well as the possible amount of DNA that can be isolated from the respective biological fluid (Rao et al., 2020). However, the amount of DNA extracted may vary depending on the condition of the crime scene, its collection, as well as preservation, and the applied analytical techniques.

Table 1. Location where biological fluids are recovered from

Biological Fluids	Possible Location of Recovery	Amount of DNA
Blood	i. Pool of blood next to body of crime ii. Weapon of Offense Fingertips Wall Splatter and Trail in Floor iii. Gloves iv. Bullet Casings and Projectile	Liquid Blood: 20000–40000ng/ml Blood Stain: 250–500ng/cm^2
Semen/vaginal fluid	i. Used Condom ii. Clothing iii. Floor and Bed of crime scene	Seminal fluid: 150000–300000ng/ml Post-intercourse vaginal swab: 10–3000ng/swab
Saliva	i. Food ii. Rim of Glass, Bottle, Cans iii. Toothpick iv. Fingernails v. End of Pen vi. Corner of Currency	Liquid saliva: 1000–10000ng/ml Oral swab; 100–1500ng/ swab
Sweat	i. Clothing ii. Footwear iii. Place of Seating (Chair, backrest, handle) iv. Personal Belongings (Wallet, Watches, Jewelry, Phone Cover)	-

Source: Rao et al. (2020)

Markers of Biological Fluids

Depending on whether the individuals secrete a separate gene in their biological fluids or not, they are classified as secretors and non-secretors, respectively. The individuals that are termed secretors secrete their blood type antigens into their body fluids (mucus and saliva), whereas, the non-secretors do not secrete their blood type antigen in the body fluids. This separate gene is different from the one that specifies the individual's blood type (R. Li, 2008). As the biological fluids vary due to their different area of origin, the biological markers specific to that biological fluid also vary from each other. Biosensors for the detection and analysis of different biological fluids are based on these specific biological markers (Yáñez-Sedeño et al., 2019). In Table 2, different biological markers specific to each biological fluid are listed.

Table 2. Specific Markers of biological fluids

Biological Fluids	Specific Markers	Reference
Blood	Glycophorin A	(de Beijer et al., 2018)
Menstrual Blood	MMP7 MMP10 MMP11	(Jakubowska et al., 2014; Moawad et al., 2018)
Semen	Prostate specific antigen (PSA) Semenogelin	(de Beijer et al., 2018; Matsuzawa et al., 1994; Tsutsumi et al., 1984)
Saliva	Statherin (STATH) gene	(Tsai et al., 2018)
Sweat	Dermicidin	(de Beijer et al., 2018)

Commercially Available Biosensor Kits

Conventional analytical techniques and biosensor kits currently used in forensic investigations are neither highly accurate nor highly specific. Additionally, these are time-consuming and expensive techniques. In Figure 4, some of the commercially available biosensor kits utilized in forensic investigations are listed (Dube et al., 2020).

Figure 4. Commercially available biosensor kits

Biological Fluids	Target Biomarker	Commercially available biosensor kit
Blood	Hemoglobin	Hemastix® test (Bayer AG, Leverkusen, Germany)
		ABAcard® Hematrace® test (Abacus Diagnostics®, West Hills, CA, USA)
		SERATEC® HemDirect test (SERATEC® GmbH, Göttingen, Germany)
	Glycophorin A	RSID™-Blood test (Independent Forensics, Hillside, IL, USA)
Saliva	Salivary amylase	The Phadebas® test (Magel Life Sciences, Lund, Sweden
		RSID™-Saliva test (Independent Forensics, Hillside, IL, USA)
Semen	PSA	SERATEC® PSA Semiquant test (SERATEC® GmbH, Göttingen, Germany)
		ABAcard® p30 test (Abacus Diagnostics®, West Hills, CA, USA)
		BioSign® PSA test (Princeton BioMeditech Corporation, Princeton, NJ, USA)
	Semenogelin	RSID™-Semen test (Independent Forensics, Hillside, IL, USA)
Urine	Tamm–Horsfall glycoprotein	RSID™-Urine test (Independent Forensics, Hillside, IL, USA)

NANOMATERIAL-BASED BIO DETECTION

With the use of nanomaterials of different types, such as nanochips, nanoprobes, and nano-bio sensors, the use of nanotechnology for forensic investigation ranges from the detection of illicit drugs, toxins, and poisons to the detection of fingerprints, analysis of forensic documents, and examination of DNA and body fluids (Lad et al., 2016). For DNA-based applications of nanotechnology, analytical techniques used in forensic sample analyses, such as polymerase chain reaction (PCR), are optimized by incorporating various nanoparticles to increase the sensitivity of the technique. This resulted in nanomaterial-assisted PCR (nano-PCR). Different nanoparticles, such as AuNPs, are used in nano-PCR to increase the sensitivity, specificity, and efficiency of analytical reactions (Pan et al., 2012). For fingerprint analysis, various nanoparticles such as AgNPs, AuNPs, Al2O3-NPs, ZnO-NPs, and fluorescent carbon-based nanoparticles have been exploited to develop and detect fingerprints (Prasad et al., 2020). Quantum dots of carbon, graphene, and silicon have been utilized for screening and sensing illicit drugs to aid forensic investigations (Ganesan & Nagaraaj, 2020). For example, graphene-based quantum dots have been reported to detect cyanide ions (Hou et al., 2013) and Mn-doped ZnS quantum dots have been reported to successfully detect cocaine in saliva (Chantada–Vázquez et al., 2018). On the other hand, luminescent quantum dots have been utilized to detect counterfeit inks in forensic examination of documents(Kesarwani et al., 2020) whereas; quantum dot-based nanohybrid have been developed to detect trinitrotoluene (TNT) (Wu et al., 2015).

Carbon Nanodots for Bio Detection

The carbon dots (CDs) are nanomaterials with a size of less than 10 nm (Bae et al., 2010). Carbon nanodots (CNDs) are a new class of carbon dots with high luminescence and water solubility. These CNDs have been utilized in electrochemical sensors because of their DNA-sensing ability (Loo et al., 2016; Mulpur et al., 2015; Verhagen & Kelarakis, 2020).

Carbon Black Nanomaterials for Bio Detection

Carbon-based nanomaterials have been studied for quite a long time because of their highly customizable properties and cost-effectiveness. A new type of carbon nanomaterial called Carbon Black has recently piqued the interest of researchers, which is different from carbon nanotube, nanopore and nanorod because of its exploitative functionalization (Arduini et al., 2020). To increase the specificity and sensitivity for the analysis of DNA sequences, atomic force microscopy techniques have been optimized by utilizing carbon nanotubes. Carbon black nanomaterials are used in biosensors to detect various biological materials, including DNA, enzymes, and antibodies (Aydın et al., 2018).

Hybrid Nanocomposites Bio Detection

Biosensors developed using hybrid nanocomposites have the potential to detect biological materials during forensic investigations (Gao et al., 2017). Hybrid sensing probes for hybrid nanocomposites detect biological materials through thermal, optical, or electrical signals (Stewart et al., 2015; Takahashi et al., 2018).

DNA Detection

Conventional techniques for DNA detection and analysis are destructive, and in forensic investigations, evidence in the form of DNA is most often obtained in minute quantities. Nanomaterials are useful for trace amounts of DNA found at crime scenes, as they provide rapid, selective, and efficient results (F. Li et al., 2013). For this purpose, several platforms have already been developed, in which AuNPs are exploited for the sensing of DNA. SiNPs are reported to have been used for the same purpose owing to their unique geometric properties and higher stablilty (Wang & Liu, 2007). Dendrimers, nanowires, and quantum dots have been utilized for DNA detection (Rawtani et al., 2019).

For the optimization of analytical techniques employed during the analysis of DNA, mainly colorimetric techniques, AuNPs, and AgNPs have been used to functionalize Halloysite Nanotubes (HNT) for detection of DNA (Vikrant et al., 2019).

Semen Detection

One piece of evidence found at a sexual crime site is semen or vaginal fluid. On-site examination and detection techniques for such body fluids are in great demand in forensics. To provide such a platform for forensic analytical techniques, a study published in 2015 probed the potential of using a hand-held biosensor device for the detection of semen using surface plasmon coupled emission (SPCE) with Ag-C60 substrate, a new composite fullerene (Mulpur et al., 2015). The basic principle behind the simplistic GLAG/FLAG-C60 SPCE platform for rapid on-site detection of semen is illustrated in Figure 5. The signals from the SPCE sensor are then detected in smartphone which acts as a detector. It provides rapid detection of semen on-site at crime scenes and reduces the chances of contamination and destruction

Figure 5. The basic principle behind the simplistic GLAG/FLAG-C60 SPCE platform for rapid on-site detection of semen
Source: Mulpur et al. (2015)

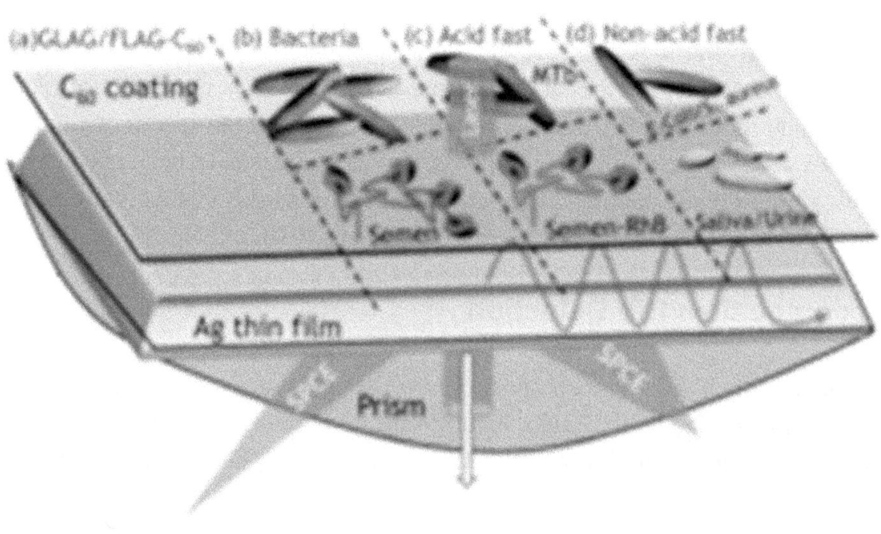

of the evidence. This study probes the potential of handheld sensing devices that can replace expensive and time-consuming spectrometry procedures.

OPTICAL BIOSENSOR FOR TRACE BIOLOGICAL DETECTION

Body fluids frequently found in trace amounts at crime scenes include blood, semen, and saliva. On-site tests routinely performed for the detection of trace biological evidence are presumptive tests that are highly destructive in nature and thus hinder forensic investigation. Majority of nanoparticles incorporated biosensors that are used to detect biological evidence found in literature function on electrochemical properties. Such functional nanoparticle based biosensors are required that will function on optical transduction mechanism (Costanzo et al., 2023). Currently, very limited literature data is present that targets the optical properties of biosensors and nanoparticles for the detection of biological materials.

Biosensor for Blood Detection

CdSe/ZnS Quantum Dots were used as nanomaterials conjugated with an anti-glycophorin A antibody that targets the glycophorin A marker present in the human blood. It was observed that treatment with quantum dots in the biosensor did not have a degrading effect on DNA and full DNA profiles were obtained after detection (Frascione et al., 2012).

Biosensor for Semen Detection

For the detection of human semen found at crime scenes, various nanomaterials have been utilized in biosensors that target different markers present in the semen. Zirconium metal-organic frameworks and gold nanoparticles have been utilized in colorimetric biosensors for semen detection, yielding positive results with 90% accuracy (Sun et al., 2019). Another marker present in semen is prostate-specific antigen (PSA), which has been targeted using polyamine nanoparticles functionalized with magnetic iron oxide cores (Nam et al., 2003). The principle of PSA detection is based on the mechanism designed by Frascione et al. for PSA biosensors, as illustrated in Figure 6.

Figure 6. The principle of PSA detection using quantum dot
Source: Frascione et al. (2012)

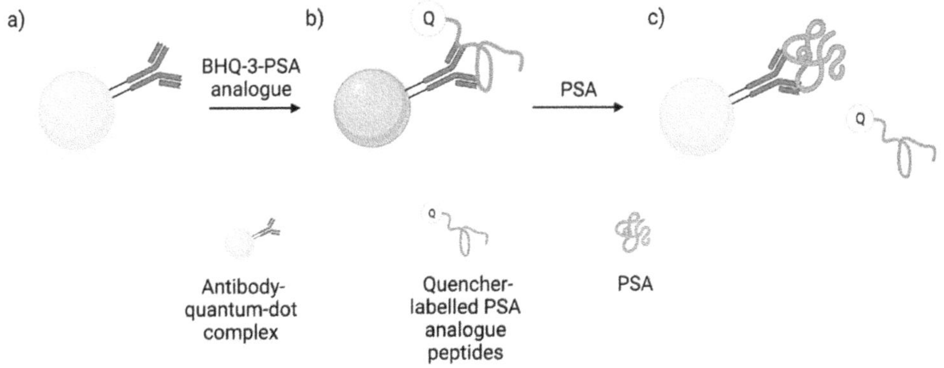

Acid phosphatase present in semen has been targeted using CuInS2 Quantum Dots based on fluorescent "turn off-on" biosensor application. The presence of acid phosphatase does not allow the aggregation of quantum dots of CuInS2 on the biosensor, and the quantum dots emit fluorescence, which indicates positive detection (Z. Liu et al., 2015).

Biosensor for Saliva Detection

Instead of targeting a specific marker present in saliva, a study targeted the specific bacteria unique to saliva fluid (X. Li et al., 2019). The optical biosensor developed using bacteria as a specific target for saliva detection used silicon carbide nanoparticles, which were stabilized using bovine serum albumin and functionalized using the antibacterial peptide GH12. Streptococcus salivarius is the targeted bacteria unique to the saliva fluid for its detection. A schematic diagram of the basic principle of biosensor operation is shown in Figure 7.

Figure 7. Detection of bacteria unique to saliva
Source: X. Li et al. (2019)

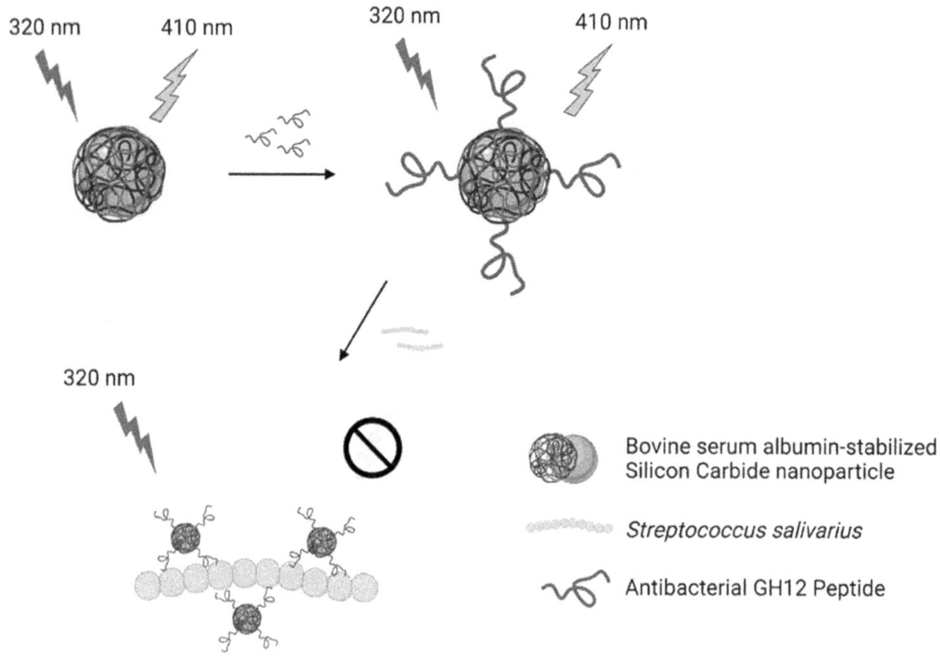

LABS ON A CHIP

Nanotechnology has enabled new platforms that integrate the analytical procedures of capillary electrophoresis on a simple film and allow on-site detection and analysis of biological samples in a matter of minutes. Micro-chip based real-time PCR have been developed that have greatly reduced the time and cost of analytical procedures. The integrated DNA chip system has inbuilt small chambers for extraction and purification of DNA and separation and amplification PCR chambers that allow the complete processing of extraction for genotyping of a blood sample (McCord, 2006). Schematics of the integrated DNA chip are illustrated in Figure 7.

Figure 8. Integrated DNA chip. This system is developed by James Landers at the University of Virginia. It has the potential to integrate all current processes of the forensic laboratory in one device.
Source: McCord (2006)

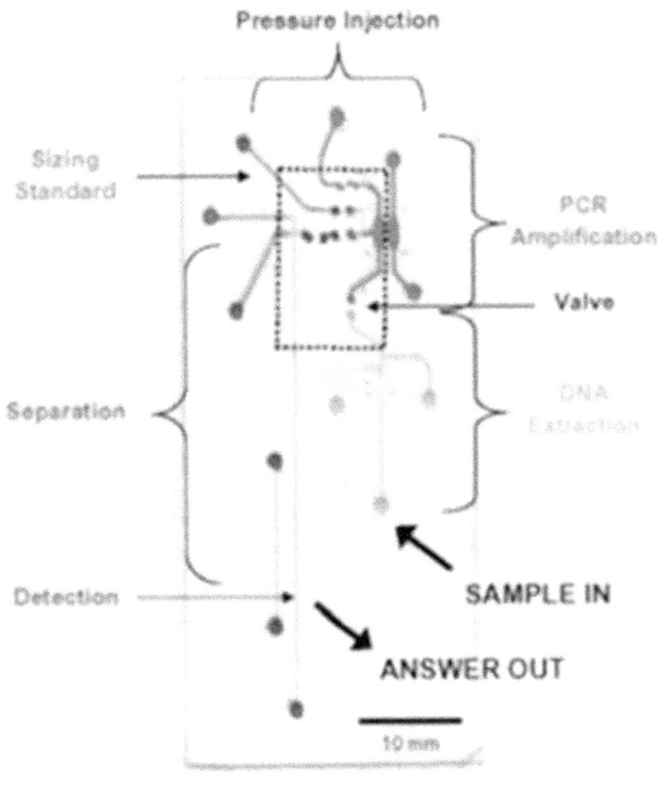

FORENSIC ANALYSIS OF CELLULAR COMPONENTS USING MNPS

Magnetic nanoparticles have been extensively studied because of their sensitivity and absorption capacity (Kudr et al., 2017). Applications of magnetic nanoparticles in forensic investigations have brought forth elaborate solutions to the shortcomings of conventional methods. Functionalized magnetic nanoparticles have been used for the optimization of solid-phase extraction (SPE) of hair samples for the detection of drug abuse cases (Buszewski & Szultka, 2012). In addition, magnetic nanoparticles have been utilized for forensic fingerprint analysis and forensic genetics (A. Lodha, 2018). Most notably, magnetic nanoparticles have been used for the detection and separation of cellular components, including nucleic acid purification and protein isolation, which have also been implemented in forensic investigations.

DNA Extraction

Magnetic nanoparticles are effectively used for the isolation of DNA from biological samples recovered from crime scenes. The magnetic nanoparticles were chemically modified using silica, amino groups, and polyethylenimine to craft functional nanoparticles. These functional magnetic nanoparticles are introduced in conventional DNA extraction procedures, where they bind to DNA and hold it down to reduce the rate of contamination and wastage during the isolation process (Haddad et al., 2017). An illustration of DNA extraction using functionalized magnetic nanoparticles is shown in Figure 8.

Figure 9. DNA extraction using functionalized magnetic nanoparticle
Source: Haddad et al. (2017)

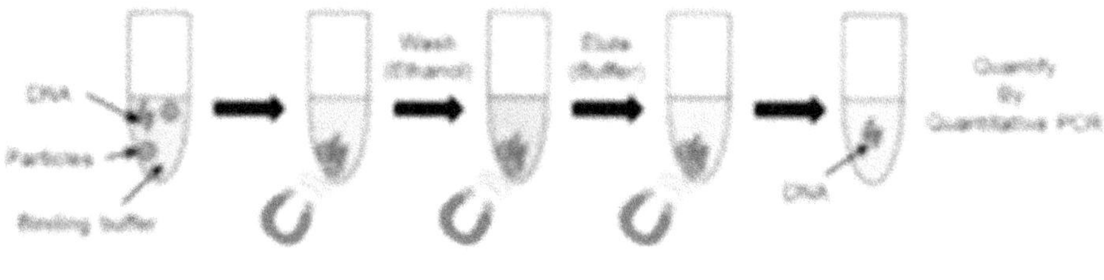

Similarly, magnetic nanoparticles have also been used to extract DNA from dried blood samples (Bagban et al., 2018). Blood samples recovered from crime scenes were either in liquid form or in dried stain. It is challenging to analyze and extract DNA from dried stains of blood because there is a high risk of degradation. In conclusion, magnetic nanoparticles have facilitated the extraction and isolation of DNA from samples recovered from crime scenes (Nadar et al., 2021).

Seminal Fluid Detection and Isolation

Magnetic nanoparticles have been used to isolate and extract sperm cells from mixtures of cells (Anslinger et al., 2008). The use of magnetic nanoparticles for sperm cell separation did not interfere with or disrupt the DNA recovered from sperm, and an accurate single-source STR profile was generated (Zhao et al., 2016). Therefore, extensive studies have established that functional magnetic nanoparticles are increasingly efficient, selective, and specific in their behavior. Functional MNPs significantly reduce the destruction, wastage, and contamination of important biological evidence recovered from crime scenes during their detection, isolation, and separation.

LIMITATIONS

Although extensive research is being conducted on the use of nanoparticles and the application of nanotechnology for forensic investigation, there are certain limitations that influence the implementation of newly developed nanotechnology for daily use. Biosensors integrated with nanoparticles are still in the experimental phase and need to be optimized according to each analytical procedure before they can be mass-produced and made available for public use.

Biological fluids are susceptible to degradation and thus require special handling and sensitive analytical techniques to limit the rate of degradation, as biological fluids are recovered in trace amounts from the scene of crime. Mass production of nanotechnology-based biosensors requires a large amount of capital for their optimization and specificity before they can be made available for use in forensic investigations. They need to be optimized to maximize accuracy, as the results directly affect court proceedings.

REFERENCES

Abdulrahman, A., & Ghanem, A. (2018). Recent advances in chromatographic purification of plasmid DNA for gene therapy and DNA vaccines: A review. *Analytica Chimica Acta, 1025*, 41–57. doi:10.1016/j.aca.2018.04.001 PMID:29801607

Anslinger, K., Bayer, B., Danilov, S. M., & Metzger, R. (2008). Application of sperm-specific antibodies for the separation of sperm from cell mixtures. *Forensic Science International. Genetics Supplement Series, 1*(1), 394–395. doi:10.1016/j.fsigss.2007.10.140

Anuradha & Bhatia, T. (2022). Novel nanomaterials in forensic investigations: A review. *Materials Today: Proceedings, 50*, 1071–1079. https://doi.org/https://doi.org/10.1016/j.matpr.2021.07.466

Arduini, F., Cinti, S., Mazzaracchio, V., Scognamiglio, V., Amine, A., & Moscone, D. (2020). Carbon black as an outstanding and affordable nanomaterial for electrochemical (bio) sensor design. *Biosensors & Bioelectronics, 156*, 112033. doi:10.1016/j.bios.2020.112033 PMID:32174547

Aydın, M., Aydın, E. B., & Sezgintürk, M. K. (2018). A highly selective electrochemical immunosensor based on conductive carbon black and star PGMA polymer composite material for IL-8 biomarker detection in human serum and saliva. *Biosensors & Bioelectronics, 117*, 720–728. doi:10.1016/j.bios.2018.07.010 PMID:30014946

Bae, S., Kim, H., Lee, Y., Xu, X., Park, J.-S., Zheng, Y., Balakrishnan, J., Lei, T., Ri Kim, H., Song, Y., Kim, Y.-J., Kim, K. S., Özyilmaz, B., Ahn, J.-H., Hong, B. H., & Iijima, S. (2010). Roll-to-roll production of 30-inch graphene films for transparent electrodes. *Nature Nanotechnology, 5*(8), 574–578. doi:10.1038/nnano.2010.132 PMID:20562870

Bagban, M. A., Mansuri, R., & Jain, N. (2018). A novel nanoparticle-based method to isolate DNA from dried saliva and semen samples. *Journal of Advanced Medical Sciences and Applied Technologies, 4*(1), 13–20.

Buszewski, B., & Szultka, M. (2012). Past, Present, and Future of Solid Phase Extraction: A Review. *Critical Reviews in Analytical Chemistry, 42*(3), 198–213. doi:10.1080/07373937.2011.645413

Cao, G., & Wang, Y. (2004). Nanostructures and nanomaterials: Synthesis. *Properties and Applications, 2*, 94.

Chan, W., Wong, G. F., Hung, C., Wong, Y., Fung, K., Lee, W., Dao, K.-L., Leung, C., Lo, K., Lee, W., & Cheung, B. K. (2020). Interpol review of toxicology 2016–2019. *Forensic Science International. Synergy, 2*, 563–607. doi:10.1016/j.fsisyn.2020.01.018 PMID:33385147

Chantada–Vázquez, M. P., de–Becerra–Sánchez, C., Fernández–del–Río, A., Sánchez–González, J., Bermejo, A. M., Bermejo–Barrera, P., & Moreda–Piñeiro, A. (2018). Development and application of molecularly imprinted polymer – Mn-doped ZnS quantum dot fluorescent optosensing for cocaine screening in oral fluid and serum. *Talanta*, *181*, 232–238. doi:10.1016/j.talanta.2018.01.017 PMID:29426506

Chen, A., & Chatterjee, S. (2013). Nanomaterials based electrochemical sensors for biomedical applications. *Chemical Society Reviews*, *42*(12), 5425–5438. doi:10.1039/c3cs35518g PMID:23508125

Costanzo, H., Gooch, J., & Frascione, N. (2023). Nanomaterials for optical biosensors in forensic analysis. *Talanta*, *253*, 123945. doi:10.1016/j.talanta.2022.123945 PMID:36191514

de Beijer, R. P., de Graaf, C., van Weert, A., van Leeuwen, T. G., Aalders, M. C. G., & van Dam, A. (2018). Identification and detection of protein markers to differentiate between forensically relevant body fluids. *Forensic Science International*, *290*, 196–206. doi:10.1016/j.forsciint.2018.07.013 PMID:30071450

Dube, S., Tharmavaram, M., Pandey, G., Rawtani, D., & Mustansar Hussain, C. (2020). Sensors for the Detection of Biological Fluids. *Technology in Forensic Science: Sampling, Analysis, Data and Regulations*, 239–258.

El-Ansary, A., & Faddah, L. M. (2010). Nanoparticles as biochemical sensors. *Nanotechnology, Science and Applications*, *3*, 65–76. doi:10.2147/NSA.S8199 PMID:24198472

Frascione, N., Pinto, V., & Daniel, B. (2012). Development of a biosensor for human blood: New routes to body fluid identification. *Analytical and Bioanalytical Chemistry*, *404*(1), 23–28. doi:10.100700216-012-6111-7 PMID:22644154

Ganesan, M., & Nagaraaj, P. (2020). Quantum dots as nanosensors for detection of toxics: A literature review. *Analytical Methods*, *12*(35), 4254–4275. doi:10.1039/D0AY01293A PMID:32940270

Ganesh, E. N. (2016). Application of nanotechnology in forensic science. *Environment*, *1*(1).

Gao, W., Wang, C., Muzyka, K., Kitte, S. A., Li, J., Zhang, W., & Xu, G. (2017). Artemisinin-luminol chemiluminescence for forensic bloodstain detection using a smart phone as a detector. *Analytical Chemistry*, *89*(11), 6160–6165. doi:10.1021/acs.analchem.7b01000 PMID:28492080

Haddad, Y., Dostalova, S., Kudr, J., Zitka, O., Heger, Z., & Adam, V. (2017). DNA-magnetic particle binding analysis by dynamic and electrophoretic light scattering. *Journal of Visualized Experiments*, 129. PMID:29155773

Hagarová, I. (2020). Magnetic Solid Phase Extraction as a Promising Technique for Fast Separation of Metallic Nanoparticles and Their Ionic Species: A Review of Recent Advances. *Journal of Analytical Methods in Chemistry*, *8847565*, 1–12. Advance online publication. doi:10.1155/2020/8847565 PMID:32963882

Hou, X., Zeng, F., Du, F., & Wu, S. (2013). Carbon-dot-based fluorescent turn-on sensor for selectively detecting sulfide anions in totally aqueous media and imaging inside live cells. *Nanotechnology*, *24*(33), 335502. doi:10.1088/0957-4484/24/33/335502 PMID:23892368

Huang, L., Peng, Z., Guo, Y., & Porter, A. L. (2010). Identifying the emerging roles of nanoparticles in biosensors. *Journal of Business Chemistry*, *7*(1).

Jakubowska, J., Maciejewska, A., Bielawski, K. P., & Pawłowski, R. (2014). mRNA heptaplex protocol for distinguishing between menstrual and peripheral blood. *Forensic Science International. Genetics*, *13*, 53–60. doi:10.1016/j.fsigen.2014.07.006 PMID:25082137

Kesarwani, S., Parihar, K., Sankhla, M. S., & Kumar, R. (2020). Nano-forensic: New perspective and extensive applications in solving crimes. *Latent in Applied Nanobioscience*, *10*(1), 1792–1798. doi:10.33263/LIANBS101.17921798

Kudr, J., Haddad, Y., Richtera, L., Heger, Z., Cernak, M., Adam, V., & Zitka, O. (2017). Magnetic Nanoparticles: From Design and Synthesis to Real World Applications. In Nanomaterials (Vol. 7, Issue 9). doi:10.3390/nano7090243

Lad, A. N., Pandya, A., & Agrawal, Y. K. (2016). Overview of nano-enabled screening of drug-facilitated crime: A promising tool in forensic investigation. *Trends in Analytical Chemistry*, *80*, 458–470. doi:10.1016/j.trac.2015.07.016

Li, F., Pei, H., Wang, L., Lu, J., Gao, J., Jiang, B., Zhao, X., & Fan, C. (2013). Nanomaterial-based fluorescent DNA analysis: A comparative study of the quenching effects of graphene oxide, carbon nanotubes, and gold nanoparticles. *Advanced Functional Materials*, *23*(33), 4140–4148. doi:10.1002/adfm.201203816

Li, R. (2008). *Forensic biology: identification and DNA analysis of biological evidence*. CRC Press. doi:10.4324/9781420043440

Li, X., Ding, Y., Ling, J., Yao, W., Zha, L., Li, N., Chang, Y., Wang, Y., & Cai, J. (2019). Bacteria-targeting BSA-stabilized SiC nanoparticles as a fluorescent nanoprobe for forensic identification of saliva. *Mikrochimica Acta*, *186*(12), 1–10. doi:10.100700604-019-3890-y PMID:31707552

Liu, C., Wu, S., Yan, Y., Dong, Y., Shen, X., & Huang, C. (2019). Application of magnetic particles in forensic science. *Trends in Analytical Chemistry*, *121*, 115674. doi:10.1016/j.trac.2019.115674

Liu, Z., Lin, Z., Liu, L., & Su, X. (2015). A convenient and label-free fluorescence "turn off–on" nanosensor with high sensitivity and selectivity for acid phosphatase. *Analytica Chimica Acta*, *876*, 83–90. doi:10.1016/j.aca.2015.03.027 PMID:25998462

Lodha, A. (2018). Forensic nanotechnology in forensic genetics. *Peer Rev J Forensic Genet Sci*, *1*(2), 19–22.

Lodha, A. S., Pandya, A., & Shukla, R. K. (2016). Nanotechnology: An applied and robust approach for forensic investigation. *Forensic Res Criminol Int J*, *2*(1), 44.

Loo, A. H., Sofer, Z., Bouša, D., Ulbrich, P., Bonanni, A., & Pumera, M. (2016). Carboxylic carbon quantum dots as a fluorescent sensing platform for DNA detection. *ACS Applied Materials & Interfaces*, *8*(3), 1951–1957. doi:10.1021/acsami.5b10160 PMID:26762211

Matsuzawa, S., Itoh, Y., Kimura, H., Kobayashi, R., Nakagawa, T., & Ohno, S. (1994). Rapid detection of human seminal plasma proteins by membrane aspiration test (MAT). *Forensic Science International*, *64*(2), 119–124. doi:10.1016/0379-0738(94)90221-6 PMID:8175082

McCord, B. (2006). Nanotechnology and its potential in forensic DNA analysis. *Profiles in DNA*, *9*(2), 7–9.

Moawad, A. M., zaghlol, H., Abdelsalam, M. H., Abdelfattah, A., Sabry, D., & Atef, A. (2018). Differential genes expression biomarkers for menstrual and peripheral blood stains analysis. *Egyptian Journal of Forensic Sciences*, *8*(1), 41. doi:10.118641935-018-0072-8

Mulpur, P., Yadavilli, S., Mulpur, P., Kondiparthi, N., Sengupta, B., Rao, A. M., Podila, R., & Kamisetti, V. (2015). Flexible Ag–C 60 nano-biosensors based on surface plasmon coupled emission for clinical and forensic applications. *Physical Chemistry Chemical Physics*, *17*(38), 25049–25054. doi:10.1039/C5CP04268B PMID:26345678

Nadar, S. S., Kelkar, R. K., Pise, P. V., Patil, N. P., Patil, S. P., Chaubal-Durve, N. S., Bhange, V. P., Tiwari, M. S., & Patil, P. D. (2021). The untapped potential of magnetic nanoparticles for forensic investigations: A comprehensive review. *Talanta*, *230*, 122297. doi:10.1016/j.talanta.2021.122297 PMID:33934767

Nam, J.-M., Thaxton, C. S., & Mirkin, C. A. (2003). Nanoparticle-based bio-bar codes for the ultrasensitive detection of proteins. *Science*, *301*(5641), 1884–1886. doi:10.1126cience.1088755 PMID:14512622

Pan, D., Mi, L., Huang, Q., Hu, J., & Fan, C. (2012). Genetic analysis with nanoPCR. *Integrative Biology*, *4*(10), 1155–1163. doi:10.1039/c2ib20076g PMID:22907590

Perfézou, M., Turner, A., & Merkoçi, A. (2012). Cancer detection using nanoparticle-based sensors. *Chemical Society Reviews*, *41*(7), 2606–2622. doi:10.1039/C1CS15134G PMID:21796315

Prasad, V., Lukose, S., Agarwal, P., & Prasad, L. (2020). Role of Nanomaterials for Forensic Investigation and Latent Fingerprinting—A Review. *Journal of Forensic Sciences*, *65*(1), 26–36. doi:10.1111/1556-4029.14172 PMID:31454084

Rao, P. K., Pandey, G., & Tharmavaram, M. (2020). Biological Evidence and Their Handling. In *Technology in Forensic Science* (pp. 35–53). https://doi.org/https://doi.org/10.1002/9783527827688.ch3

Rawtani, D., Tharmavaram, M., Pandey, G., & Hussain, C. M. (2019). Functionalized nanomaterial for forensic sample analysis. *Trends in Analytical Chemistry*, *120*, 115661. doi:10.1016/j.trac.2019.115661

Safarik, M. E., Jarvis, J. P., & Nussbaum, K. E. (2002). Sexual Homicide of Elderly Females: Linking Offender Characteristics to Victim and Crime Scene Attributes. *Journal of Interpersonal Violence*, *17*(5), 500–525. doi:10.1177/0886260502017005002

Sanli, S., Ghorbani-Zamani, F., Moulahoum, H., Gumus, Z. P., Coskunol, H., Odaci Demirkol, D., & Timur, S. (2020). Application of Biofunctionalized Magnetic Nanoparticles Based-Sensing in Abused Drugs Diagnostics. *Analytical Chemistry*, *92*(1), 1033–1040. doi:10.1021/acs.analchem.9b04025 PMID:31800231

Stewart, A. J., Hendry, J., & Dennany, L. (2015). Whole blood electrochemiluminescent detection of dopamine. *Analytical Chemistry*, *87*(23), 11847–11853. doi:10.1021/acs.analchem.5b03345 PMID:26530000

Sun, Z., Wu, S., Ma, J., Shi, H., Wang, L., Sheng, A., Yin, T., Sun, L., & Li, G. (2019). Colorimetric Sensor Array for Human Semen Identification Designed by Coupling Zirconium Metal–Organic Frameworks with DNA-Modified Gold Nanoparticles. *ACS Applied Materials & Interfaces*, *11*(40), 36316–36323. doi:10.1021/acsami.9b10729 PMID:31522499

Takahashi, F., Nitta, S., Shimizu, R., & Jin, J. (2018). Electrochemiluminescence and voltammetry of tris (2, 2′-bipyridine) ruthenium (II) with amphetamine-type stimulants as coreactants: An application to the discrimination of methamphetamine. *Forensic Toxicology*, *36*(1), 185–191. doi:10.100711419-017-0388-3

Tsai, L.-C., Su, C.-W., Lee, J. C.-I., Lu, Y.-S., Chen, H.-C., Lin, Y.-C., Linacre, A., & Hsieh, H.-M. (2018). The detection and identification of saliva in forensic samples by RT-LAMP. *Forensic Science, Medicine, and Pathology*, *14*(4), 469–477. doi:10.100712024-018-0008-5 PMID:30058014

Tsutsumi, A., Yamamoto, Y., Okamura, K., Oki, Y., Hamada, K., & Ishizu, H. (1984). LDH (lactate dehydrogenase) isoenzyme in human semen and seminal stains. *Hanzaigaku Zassi*, *50*, 41–48.

Verhagen, A., & Kelarakis, A. (2020). Carbon dots for forensic applications: A critical review. *Nanomaterials (Basel, Switzerland)*, *10*(8), 1535. doi:10.3390/nano10081535 PMID:32764380

Vikrant, K., Bhardwaj, N., Bhardwaj, S. K., Kim, K.-H., & Deep, A. (2019). Nanomaterials as efficient platforms for sensing DNA. *Biomaterials*, *214*, 119215. doi:10.1016/j.biomaterials.2019.05.026 PMID:31146176

Virkler, K., & Lednev, I. K. (2009). Analysis of body fluids for forensic purposes: From laboratory testing to non-destructive rapid confirmatory identification at a crime scene. *Forensic Science International*, *188*(1), 1–17. doi:10.1016/j.forsciint.2009.02.013 PMID:19328638

Wang, Y., & Liu, B. (2007). Silica nanoparticle assisted DNA assays for optical signal amplification of conjugated polymer based fluorescent sensors. *Chemical Communications*, *34*(34), 3553–3555. doi:10.1039/b705936a PMID:18080542

Wu, P., Xu, C., Hou, X., Xu, J.-J., & Chen, H.-Y. (2015). Dual-emitting quantum dot nanohybrid for imaging of latent fingerprints: Simultaneous identification of individuals and traffic light-type visualization of TNT. *Chemical Science (Cambridge)*, *6*(8), 4445–4450. doi:10.1039/C5SC01497B PMID:30155001

Xu, K., Huang, J., Ye, Z., Ying, Y., & Li, Y. (2009). Recent Development of Nano-Materials Used in DNA Biosensors. In Sensors (Vol. 9, Issue 7, pp. 5534–5557). doi:10.339090705534

Yáñez-Sedeño, P., Agüí, L., Campuzano, S., & Pingarrón, J. M. (2019). What Electrochemical Biosensors Can Do for Forensic Science? Unique Features and Applications. In Biosensors (Vol. 9, Issue 4). doi:10.3390/bios9040127

Yang, N., Chen, X., Ren, T., Zhang, P., & Yang, D. (2015). Carbon nanotube based biosensors. *Sensors and Actuators. B, Chemical*, *207*, 690–715. doi:10.1016/j.snb.2014.10.040

Zhang, X., Guo, Q., & Cui, D. (2009). Recent Advances in Nanotechnology Applied to Biosensors. In Sensors (Vol. 9, Issue 2, pp. 1033–1053). doi:10.339090201033

Zhao, X.-C., Wang, L., Sun, J., Jiang, B.-W., Zhang, E.-L., & Ye, J. (2016). Isolating Sperm from Cell Mixtures Using Magnetic Beads Coupled with an Anti-PH-20 Antibody for Forensic DNA Analysis. *PLoS One*, *11*(7), e0159401. doi:10.1371/journal.pone.0159401 PMID:27442128

Chapter 11
Nano Forensic Testing of Illicit Drugs

Ahsan Riaz
University of Health Sciences, Pakistan

Iqra Zareef
University of Health Sciences, Pakistan

Anam Munawar
University of Health Sciences, Pakistan

Allah Rakha
University of Health Sciences, Pakistan

Naveed A. Shad
National Institute for Biotechnology and Genetic Engineering, Faisalabad, Pakistan

ABSTRACT

Nano forensics utilizes nano-based sensors in matters of crime or where the legality of the process is challenged. There exist many conventional techniques, but considering the approach of criminals and the modern-day standard of investigation, there is a need for nanotechnology-based sensors for their sensitive, selective, and rapid detection of illicit drugs. Illicit drugs are psychoactive in nature. They act upon the CNS that results in different perceptions and long-lasting responses. This chapter covers the drugs that can be the potential source of causing abuse and those commonly detected in forensic lab testing. The conventional techniques have a few drawbacks. They are time-consuming, highly expensive, and some of these techniques are also destructive. Nanosensors are nanoparticle-based probing techniques that constitute a sensing platform for the detection of the biological substance present in the evidence sample collected from the crime scene. These nanosensors can work with limited quantities, and the quality of the sample yields valuable information for preparing the scientific report.

DOI: 10.4018/978-1-6684-8325-1.ch011

INTRODUCTION

From the 20th Century onwards there has been an era of advancement in every field of science. Specifically, in the last few decades, science and its associated fields have progressed to new heights. Of all the scientific disciplines one of the most fascinating ones is the field of forensic science which also has progressed in every field of life (Nikkel, 2020). Forensic science is a science of its own having many sub-disciplines that work together to produce a scientific report that helps both the criminal investigation and judiciary towards the resolution of the case. Under the sky of the forensic world, one of its disciplines nano-forensic testing has gained its utmost attention in recent times (Tully et al., 2020).

Nano-forensic is the modern approach that focuses on the applicability of nano-based sensors in matters of crime or where the legality of the process is challenged (Paikrao, Tajane, Patil, & Dipole, 2022). There exist many conventional techniques that are being applied in forensic testing, but considering the approach of criminals and the modern-day standard of investigation. There exists a deficiency in producing the scientific report that is very common (Graziano, Anzillotti, Mannocchi, Pichini, & Busardò, 2019). Whenever a crime happens anywhere in the world some sort of evidence is always left. The evidence is most often in the fragile form left that is not only contaminated but also very compromised in its composition. Even though the existing evidence is being collected properly from the crime scene, the conventional techniques are sometimes unable to detect meaningful information and the production of any conclusive results (Walters, 2018).

Nano-forensic testing covers every bit that is applied by using scientific techniques based upon the preparations from the nano-scale materials that are equal to the millionth part of a meter. An illustration is given in Figure 1.

Figure 1. The modern sensing device based upon an electrochemical sensing platform used for illicit drug testing work at the scene of the crime. (A) shows the individual components of the sensing platform 1-potentiometer, 2-buffer solution, 3-SPE, 4- spreader, 5-Pasteur pipette and 6-evidence sample. (B) shows sample preparation. In (C), the sample is loaded in the designated area of the sensing SPE platform, and (D) shows the interface having the results of electrochemical sensing.
Source: Parrilla et al. (2022)

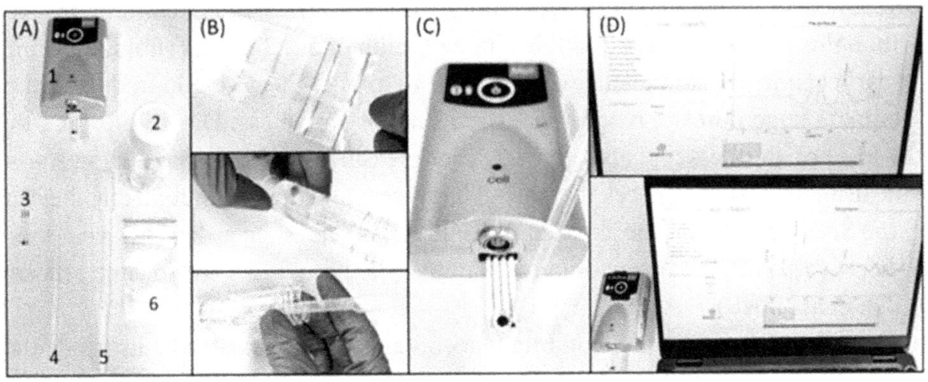

Forensic science itself faced many limitations in its process that can be eliminated through the induction of nanotechnology in it. The modern-day forensic issues and the scientific limitations can both be overcome by using specific nanomaterials, ultimately helping the judiciary and law enforcement in coming closer to the conclusion of the case. Nanotechnology is a very broad field of study; yet to discover many new things in the scope of forensic science (Singh & Samal, 2021). The nanosensors that produced the unparalleled results are quantum dots, nano-crystals, nanotubes, nanoparticles and nanocomposites for the biological and physical evidence such as in isolating the DNA, detecting different body fluids, visualizing fingerprints, in the detection of explosives, fibers and illicit drugs. In this chapter, we are going to study the nano-forensic testing of illicit drugs (Adegoke, Zolotovskaya, Abdolvand, & Daeid, 2022).

Illicit Drugs in the Field of Forensic Science

Psychotropic Substance

Illicit Drugs are chemical substances that are psychoactive in nature they directly act upon the CNS that results in different perceptions, emotions and long-lasting responses. Every drug is not an illicit drug and even the drugs used for medicinal purposes can alter the level of consciousness producing severe responses. Examples of psychotropic substances include legal drugs given only over prescription such as medicinal drugs which are used for pharmaceutical purposes including pain killers like morphine and codeine and many other illegal drugs that are used for their psychoactive purposes like cocaine, amphetamines, cannabinoids and marijuana (Göl & Çok, 2019; Hondebrink et al., 2020) (Hondebrink et al., 2020).

Illicit drugs can be classified by many different categories like their site of occurrence, their mode of action, availability of their active ingredients and into the popular five schedules classified on their ability to cause the abuse. Here we will focus on the drugs that can be the potential source of causing abuse and the drugs most commonly detected in forensic lab testing.

Stimulants: Stimulants include those drugs which are given to activate the sympathetic nervous system (SNS). These drugs show their mode of action by binding with adrenergic receptors that in turn increase cardiac and respiratory activity. Knowing the nature of responses produced by stimulants they are given in patients having depressing mental and body status. The medicinal role of the stimulant drug includes the treatment of attention deficit hyperactivity disorder (ADHD) in children that are unable to control their thoughts (Howard et al., 2020; Mejia de Grubb et al., 2020) (Taubin, Wilson, & Wilens, 2022). In the earlier time, the stimulants were chewed to extract the active ingredient but now people are looking for extra large doses to reach the desired state of happiness. Due to this fact, the stimulant intensifies the level of alertness and also gives its user pleasuring effects. The person starts taking the overdose on the drugs to maximally seek the state of happiness and the satisfaction that the drug produces by activating the SNS. As a result, the abuse effect of drugs is seen such as loss of sleep, loss of appetite and irregular heart rhythms. The most common examples of these drugs are cocaine and amphetamine (Mejia de Grubb et al., 2020).

Cocaine: Cocaine is the CNS stimulant drug that contains its active ingredients inside the leaf of the Erythroxylum coca plant. The plant is widely spread in the amazon forest of South America. The second most common abuse drug in America after opiates exists in two forms: crack is the free base that is used for smoking and hydrochloride salt can be snorted or taken intravenously and orally. In countries where cocaine is locally produced people usually extract the cocaine by chewing rather than then the other look for chemical methods to extract the cocaine. Cocaine produces its mechanism of action by stimulating the CNS to increase the production of many stimulating hormones like serotonin and dopamine that results in euphoria-like signs and symptoms and also causes an increase in blood pressure, respiration rate and motor activity as well (Zibbell et al., 2022).

Amphetamines: Amphetamines are a strong stimulant of CNS and are used for many recreational purposes in the past. They are used to speed the process of sending impulses between the brain and the body. Most commonly amphetamines are converted into methamphetamines and their active ingredient is an herb that is known as ephedrine. Due to the fact ephedrine can cross the blood-brain barrier it can affect sleeping patterns and is sometimes used for inducing sleep in many disorders like narcolepsy and asthma. The drug can be introduced into the body in two ways orally and intravenously by both routes different effects are produced. The oral administration of the drug is recommended by pharmaceutical companies because it can only produce drug dependence whereas the intravenous mode of administration can also cause the person addicted to the drug (Zibbell et al., 2022; Ganson, Murray, & Nagata, 2021).

Depressants: Depressants work in opposition to the action of stimulants they tend to slow down the working of neurotransmitters in the brain. The psychoactive substances found in the depressant drug tend to make a person more relaxed by slowing the rate of the vital organ by sending impulses from the brain to the body. The most common depressant is barbiturates (Smith, 2019).

Barbiturates: Barbiturates are group depressants and are also called sedative drugs; they are used to induce sleep treating seizures and many anxiety-based disorders. They all have a common effect, which is to decrease brain activity. Their dose is directly proportional to the effect that they produce in the body if they are taken in small doses, they will cause sedation and if they are taken in larger doses, they will cause the effect termed as hypnotics. Barbiturates are called sedative-hypnotics due to the very reason they are utilized in the body as sleeping pills and for inducing the effect of sedation (Purohit, Pandey, Tharmavaram, Rawtani, & Mustansar Hussain, 2020).

Narcotics: At the beginning of the last Century term narcotics were referred for any kind of drug that is highly dangerous. Narcotics were only obtained from plant sources in the past but nowadays they are mostly produced synthetically through organic sources. They are utilized for various pharmaceutical purposes due to their medicinal effect, which they produce by binding with their specific receptors (Priyonugroho et al., 2018). One of their uses is as a potent drug of choice in relieving severe pain, by binding with opioid receptors in the brain. Due to their psychoactive nature, they can cause addiction to drugs if they are used in excess amounts. Narcotics are not completely illicit considering their benefits as strong pain relievers they are legally used for severe pain and during surgeries. Example of the legal drug under narcotics includes morphine and in some part of the world heroin (which is even more potent than morphine) can be used (Priyonugroho et al., 2018) (Purohit et al., 2020). Example of synthetic opioid includes Fentanyl.

Fentanyl and Its Analogues: Fentanyl is a synthetic opioid that is prescribed to treat severe pain, such as that experienced by cancer patients. However, it is also produced illegally and used as a recreational drug, often mixed with other drugs, which have led to many overdoses and deaths. Fentanyl analogues, also known as fentalogs or fentanyl derivatives, are synthetic drugs that are chemically similar to fentanyl but have slight structural differences. Some analogues are more potent than fentanyl and can be even more dangerous, with smaller doses needed to cause an overdose. These analogues are often produced in clandestine laboratories and sold on the black market, often mixed with other drugs such as heroin, cocaine, or benzodiazepines (Pergolizzi Jr et al., 2021).

The use of fentanyl and its analogues has become a significant public health crisis in many countries, including the United States and Canada. Some commonly known fentanyl analogues include carfentanil, acetylfentanyl, and furanylfentanyl. It is important to note that these drugs are illegal and highly dangerous, and using them can result in severe health consequences, including respiratory depression, coma, and death (Crepeault et al., 2023).

Hallucinogens: Hallucinogens are psychoactive substances that are known to cause hallucinations. They affect the sensation and perception of the brain in such a way that the user enters an unreal dreamy kind of world. Example of the drugs includes lysergic acid diethylamide (LSD) and marijuana. They can be obtained through natural sources as well as synthetic means. When taken in an overdose the drug can produce serious effects on the body that can lead to death of the person (Henningfield et al., 2022).

Lysergic Acid Diethyl Amide: LSD is one of the hallucinogenic substances that doesn't occur in nature they are produced synthetically. LSD is an illicit drug commonly used among street drug users as a white powder. LSD is obtained in large amount from many fungus species, they produce effects that remains for a long period. Though they are known to disrupt the perception and the working of the brain in many ways but are not addictive (Kopra et al., 2022).

Marijuana: Marijuana is an illicit drug that is obtained from the cannabis plant. The active ingredient is tetra hydro Cannabinol (THC), which is the product obtained after drying the various parts of the plant like leaves and flowers. Depending upon its users and the way of using the drug has many names such as ganja and joint. The drug is also used for pharmaceutical practices and as a taste enhancer in many food products. The drug is known as a hallucinogenic agent as its active ingredient (THC) is passed directly from the respiratory tract to the circulation and then reaches the brain. Its antiemetic effects are utilized in the treatment of cancer patients (van Dongen, Alderliefste, Onderwater, Ferrari, & Terwindt, 2021).

New Trends in Ongoing Microdosing of LSD and Psilocybin: Microdosing is the practice of taking very small amounts of psychedelic substances like LSD or psilocybin (the active ingredient in magic mushrooms) for an extended period, typically every few days or once a week, with the aim of improving mood, focus, creativity, and overall well-being. While microdosing is not a new practice, it has gained popularity in recent years as anecdotal evidence suggests that it can have positive effects on mental health and cognitive function. One new trend in ongoing microdosing of LSD and psilocybin is the use of "stacking." This involves taking small amounts of multiple substances, including psychedelics and other nootropics, with the aim of enhancing their effects. For example, some microdosers may combine LSD or psilocybin with caffeine, L-theanine, or other supplements to improve focus and concentration (Adegoke et al., 2019).

Another trend is the use of personalized dosing regimens, where individuals adjust their microdose amounts and frequency based on their individual needs and experiences. This allows for a more tailored approach to microdosing and may help individuals achieve their desired effects more effectively. Finally, there is also growing interest in using microdosing as a form of self-therapy, where individuals use psychedelics in combination with mindfulness practices and other therapeutic techniques to address specific mental health issues or improve overall well-being. While the research in this area is still limited, there is growing evidence to suggest that microdosing may have therapeutic potential for certain conditions, such as anxiety and depression. However, it is important to note that the use of psychedelics for therapeutic purposes should only be done under the guidance of a trained healthcare professional (Anderson et al., 2019).

Opioid Epidemic in USA

The opioid epidemic in the United States is a public health crisis that has been ongoing for several years. It refers to the dramatic increase in the misuse and abuse of opioid drugs, including prescription painkillers, heroin, and synthetic opioids such as fentanyl. According to the Centers for Disease Control and Prevention (CDC), over 500,000 people have died from opioid-related overdoses in the United States between 1999 and 2019. The epidemic has had a significant impact on individuals, families, communities, and the healthcare system (Marshall et al., 2019).

The opioid epidemic is the result of several factors, including the over-prescription of opioid painkillers by healthcare providers, the illegal diversion and sale of prescription opioids, the increasing availability of illicit opioids such as heroin and fentanyl, and a lack of access to effective addiction treatment and recovery services. To address the opioid epidemic, policymakers and healthcare professionals have taken a range of actions, including expanding access to addiction treatment and recovery services, improving prescription drug monitoring programs, increasing access to the overdose-reversing drug naloxone, and reducing the over-prescription of opioids. However, much work remains to be done to prevent new cases of opioid misuse and abuse and to help those struggling with opioid addiction to recover and lead healthy lives (Drake et al., 2020).

Rise of Darknet Marketplaces

Darknet marketplaces have become more prevalent in recent years due to the anonymity and privacy they provide to both buyers and sellers. These marketplaces operate on the dark web, which is a part of the internet that is not indexed by traditional search engines and is accessible only through specialized software, such as Tor. Darknet marketplaces allow buyers and sellers to conduct transactions anonymously using cryptocurrencies, such as Bitcoin, which further enhances the privacy and security of these transactions. They offer a wide range of illegal goods and services, including drugs, weapons, stolen data, and counterfeit items (Omar and Ibrahim, 2020).

The rise of darknet marketplaces is also linked to the increased adoption of cryptocurrencies and the growing demand for privacy and security in online transactions. However, it is important to note that the use of darknet marketplaces for illegal activities has significant legal and personal consequences, and individuals engaging in such activities risk facing criminal charges and potential harm to their health and safety (Kethineni and Cao, 2020).

Conventional Methods for Illicit Drug Testing

The Conventional testing techniques for illicit drug detection in the piece of evidence collected from a crime scene includes chromatography and spectroscopy. Sometimes both of these techniques are combined to produce the most accurate results in Figure 2.

Mass Spectrometry

Mass spectrometry utilizes the different masses of particles for finding the abundance patterns of the target analytes. The sample containing the target analyte is passed through the heated chamber that converts it into the vapor form and further moves the particles through a magnetic field. The particles

Figure 2. Processing an unknown sample collected from a crime scene is processed. (A) shows the physical evaluation of an unknown sample. (B) shows the processing of the unknown sample by the electrochemical sensing platform, and (C) shows the detection of specific illicit drugs such as cocaine and amphetamines.
Source: Parrilla et al. (2022)

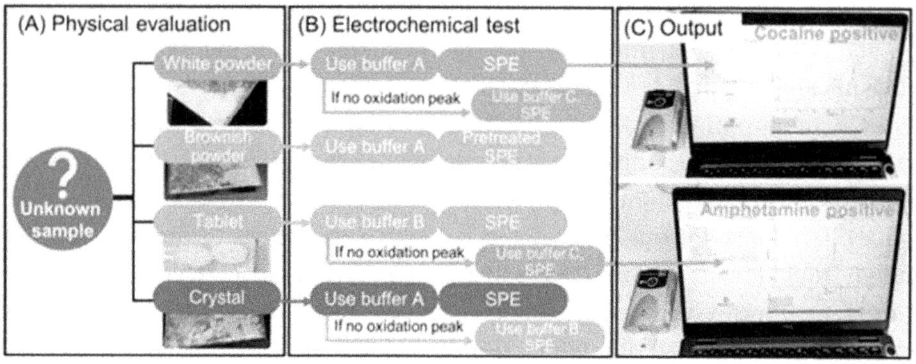

of different magnetic fields strike the detector and their specific intensities are calculated. The specific mass-to-charge ratio of different drugs is calculated by using this technique. The analysis of drug samples submitted in forensic laboratories is mostly done by this technique. Mass spectrometry in combination with chromatographic techniques like gas chromatography/mass spectrometry (GC/MS) is utilized for illicit drug testing with increased accuracy (Moore, Foss, Juhascik, Botch-Jones, & Kero, 2019). There are a few drawbacks associated with these techniques, they are destructive techniques sample used for processing is being destroyed which make them less suitable for forensic testing and also, they are very expensive techniques as their reagent are costly (Chiang et al., 2019).

Ion Mobility Spectrometry

Ion Mobility Spectrometry (IMS) is another spectrometry-based analytical method that is used for illicit drug analysis. The drug analysis in IMS is dependent upon ionizing the particles and moving them through the gaseous phase containing respective buffers the speed of the ions results in the separation and identification. The IMS focuses on three characteristic properties of the ion for their drug analysis which includes their charge, mass and the collisions in between them. IMS works with only those drug sample that is heat stable, the sample separation and identification is done by first ionizing the particles and then applying the electric field that separates the ions based on their respective charges. Ions have to maintain their trajectory for going through the filter screen reaching the detector and getting separated; the trajectory is only maintained if the ions are having specific mobility.

Raman Spectroscopy

Raman spectroscopy is the spectroscopic technique used for the forensic detection of illicit drug samples. The optical techniques look for the Raman scattering that is produced by directing the laser beam over the sample and causing the change in the vibrational energy of particles due to the change in their rotations. The scattering particles have molecular interaction with each other producing different energies and striking the detector and getting their detection done (Azimi & Docoslis, 2022). Raman spectroscopy can be utilized for very similar pharmaceutical substances those having a similar number of molecules (molecular formulae) and different arrangements of their molecules (molecular formulae) can also be separated. Raman Spectroscopy has limited application in the processing of illicit drugs that exhibit fluorescence such as heroin (Liu, He, Xu, & Hua, 2021).

Nano Sensors-Based Detection of Drug

Nanosensor-based detection of illicit drugs is the modern-day approach that is utilized to strengthen the criminal justice system that aims to exonerate the innocent and give punishment to the guilty. Nanosensor is a nanoparticle-based probing technique that constitutes a specific element and a transducer for the detection of the biological substance present in the evidence sample collected from the crime scene (Kesarwani, Parihar, Sankhla, & Kumar, 2020).

Figure 3. Sample containing the target analyte is introduced into the sensor system through the receptor channel. The receptors are tagged with specific recognition sites that selectively bind with the target analyte; the remaining interfering substances are unable to bind with receptors. The labelled molecules travel through the transducer that converts them into a specific signal molecule. The further processing of the sensor system depends upon the type of sensing platforms such as an optical sensor or the electrochemical sensor that perform their specific signal processing.
Source: Gavrilă, Stoica, Iordache, and Sârbu (2022)

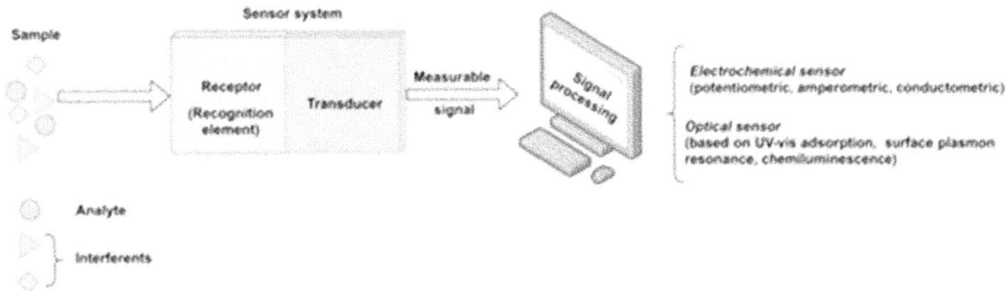

The nanosensors are synthesized in two steps first step is a modification of the probe with the specific nanomaterials like nanowires, nanotubes, nanorods and nanoshells. The second step is the attachment of the respective analyte with the respective nanostructure which leads to the development of the nano biosensor. The attachment surface can be further modified with physical and chemical methods that ultimately give the large surface area and excellent sensing properties (Solaimuthu, Vijayan, Murali, & Korrapati, 2020; Shahbazi & Zare-Dorabei, 2019).

Surface-Enhanced Raman Scattering for Illicit Drug Detection (SERS)

SERS is a powerful detection technique that can yield information from the very low concentration of the analyte present in the sample. The working of the SERS is highly dependent upon the specific substrate and the type of nanomaterials being chosen. Considering the choice of substrate uniformity is highly important so that reproducibility in the detection process is retained (Azimi & Docoslis, 2022). The scheme for working SERS is given in Figure 4.

Figure 4. The typical illustration of the surface enhanced raman spectroscopy. The sensing platform is optimized with the induction of internal standards for detecting illicit drugs. The specific nanoparticle used in this SERS sensor is Au NPs respectively.
Source: Yan et al. (2017)

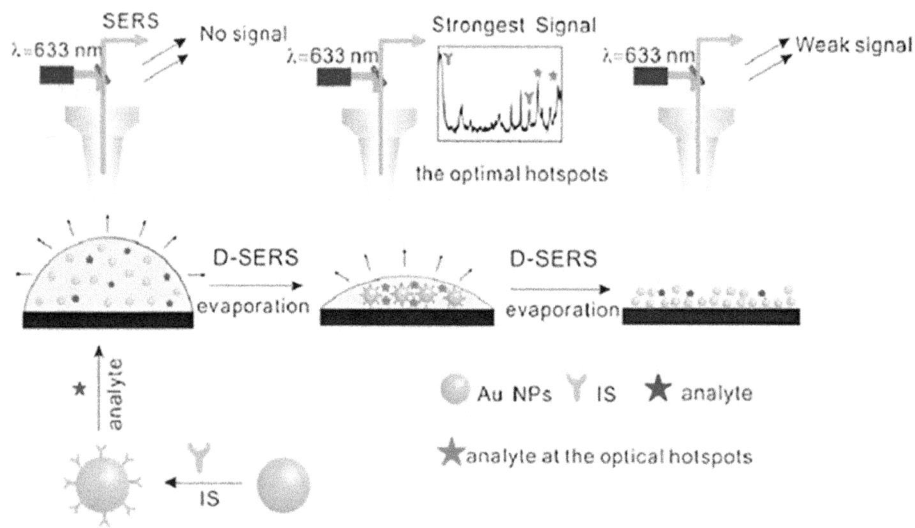

There are two ways for substrate molecules attachment that can be used for the detection of illicit drugs: (1) nanostructures on solid supports and (2) and nanostructures with colloidal surfaces.

Nanostructures on Solid Supports: In this method, nanomaterials are attached to the solid support substrate materials that give good analytical applications. The nanomaterial is fixed to the substrate surfaces in such a way that is in a close location with each other. The configuration of nanoparticles is also modified to give them different orientations so that the SERS is better utilized. The crucial indicator for the SERS analytical work includes the inter-particle distance, rough surfaces and the size of the films (Sarfo, Izake, O'Mullane, & Ayoko, 2019).

Nanostructures With Self-Assemblies Surfaces: In this method, nanomaterials are attached to a large substrate surface that gives the nanomaterials a large surface area. The nanomaterials exist in layers and electrostatic interactions exist between the nanomaterials in different layers. Uniformity of the layer is very important that is achieved by using the different capping agents, the distance of the particle from one another must be 1 nm to achieve the homogeneity of the layers. Multiple layers of nanomaterials can also be synthesized by adding subsequent layers, for adding multiple layers specific linker molecules are added and the uniformity of layer is difficult to achieve with many layers and associated clustering of nanoparticles (Song et al., 2022).

Different types of nanomaterial have been attached to the substrate surface with this self-assembling method. A few examples of this approach include (a) assembling and attachment of gold nanoparticles with the 3D substrate structure of halogen ions and (b) assembling of octahedral gold nanoparticles in layer structure with strong electrostatic interactions between different layers. The important consideration in all of the substrate attachment methods is to keep the inter-particle distance very small so that uniformity is obtained which is crucial for producing reproducible SERS measurements (Mitra & Basak, 2022).

Nano Forensic Testing of Common Illicit Drugs

Codeine Detection

Codeine is included in the family of opioids used as a prescribed drug for pain relief and is extracted from the poppy plant. Due to its psychoactive properties, it can cause drug addiction and serious abuse effects if the drug is not taken within the prescribed, leading to serious health and social issues (Kesarwani et al., 2020).

Multiwall Carbon Nanotubes/Glassy Carbon (MWCNT/ GCE) Electrode-based nanosensors have shown great sensing properties for codeine. The limit of detection of the biosensor was recorded as 0.19 μM showing its specificity to analyze different body fluids like urine for the codeine analysis. The selectivity of the method was checked by analyzing codeine with glucose and uric acid which also shows no interferences (Mousaabadi, Ensafi, & Rezaei, 2022).

Morphine Detection

Morphine is also a subclass of heroin and codeine included in the Opioid family of drugs, Due to its potent analgesic effect, it is given in severe pain to the heart patient and those undergoing surgical procedures. It works by acting upon CNS depressing the brain centers for pain management, due to its direct effect on brain receptors its medicinal role can be turned into abuse and addiction to drugs. Its benefit with associated risk for abuse suggests the requirement for the use of its medicinal purposes within prescribed limits (Feng et al., 2019). The sensor working selectively for the detection of morphine and codeine is given in Figure 5.

Many analytical methods exist for the detection of morphine such as below. Carbon paste electrode (CPE) along with gold nanoparticles (AuNP) is used for the determination of morphine by Voltammetry analysis. AuNP is particularly utilized for increasing the available surface area that increases the electrostatic forces of interaction leading to the sharp peaks of morphine leading to increase sensing properties of the biosensor. The electrode is set up in such a way that morphine is present at the positive electrode, so the cation exchange polymer has been added that moves towards the cation. The morphine detection in the urine sample was 87.2 nM as compared with the LOD of standard solution 1.33 nM. This relationship proves the non-selectivity of the biosensor and indicates the presence of interfering substances (Atta, Galal, El-Ads, & Hassan, 2019).

Figure 5. Top: The electrode setup based upon SWCNT and Nafion for the selective detection of illicit drugs such as codeine and morphine. Down: The cyclic voltammograms show the individual blue peaks for the pure SWCNT and the red peaks for SWCNT coated with Nafion.
Source: Wester et al. (2019)

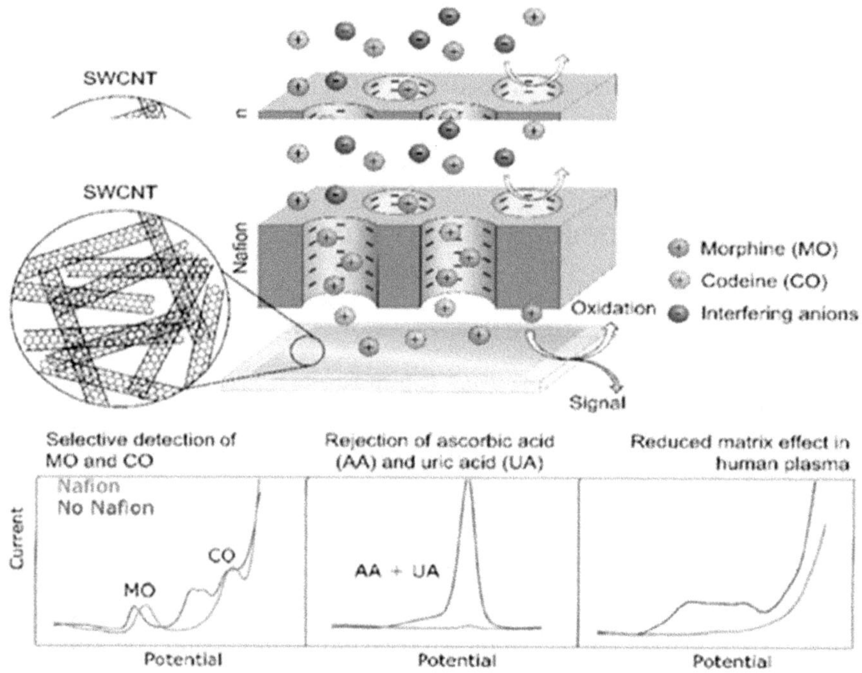

Cocaine Detection

The fluorescence-based Quantum dots (QDs) have shown numerous applications in the analysis of biological samples such as the one being collected from the crime scene. QDs utilize fluorescence resonance energy transfer (FRET) based nanosensors for the detection of illicit drugs such as cocaine. The main principle of detection is the labeling of the specific receptor molecule with the QDs that shows the affinity for the particular drug of the target (Atta et al., 2019). These nanosensors have the advantage that they can work and cause the detection of a single molecule with both sensing mechanisms like by switching the signal on and off. A nanosensor detecting cocaine through molecularly impregnated polymer (MIP) is given in Figure 6.

These sensors work through anti-cocaine receptors that directly bind with cocaine through the hydrophobic junctions. The binding of the nanosensor with the target results in a change of absorbance that results in the visible color change in the cocaine-containing sample. These nanosensors show good sensing properties with little sample preparation their detection limit is 0.5 μM showing maximum sensitivity with only a little consumption of the sample which makes them ideal for forensic testing of illicit drugs (Kesarwani et al., 2020).

Figure 6. Represent nanoparticle-based sensing platform that is modified with MIP for the detection of cocaine
Source: Florea et al. (2019)

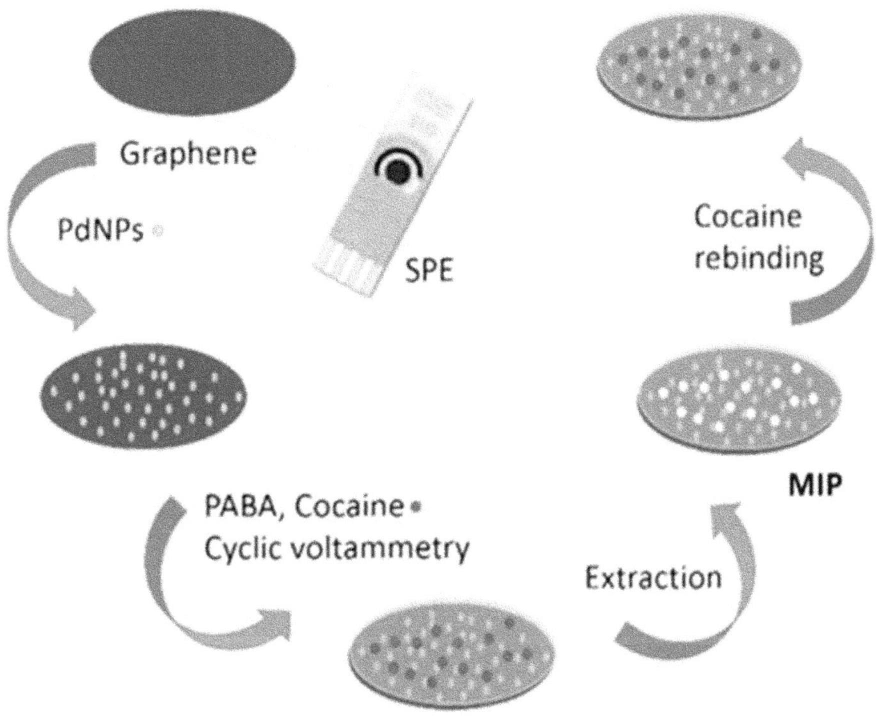

Methamphetamine Detection

Surface Plasmon Resonance (SPR) is a nano-biosensor highly specific for methamphetamine (MA) detection. The basic synthesis of the sensor involves the formation of a thin gold (Au) film by adsorbing the Au nanoparticles and the second step involves conjugating the SPR sensor over the thin gold film. The developed nano-biosensor based on SPR shows the maximum sensitivity toward methamphetamine and shows the detection of MA within the range from 0.1 to 1000 ng per ml (Peng et al., 2021). An example of an electrochemical sensing platform for the detection of MDMA is given in Figure 7.

Clonazepam Detection

The drug-facilitated crime mainly involving physical assaults and robbery cases is most commonly done through the use of the illicit drug clonazepam. Looking at the rate of this drug-facilitated crime only a few detection techniques are available. The nanotechnology-based sensor of gold nanoparticles (AuNPs) shows the maximum sensitivity, selectivity and the lowest limit of detection which is the prerequisite of the evidence sample processing obtained from the scene of the crime (Lodha, Pandya, Sutariya, &

Figure 7. An illustration of a typical electrochemical sensing platform for the detection of MDMA
Source: Soni et al. (2022)

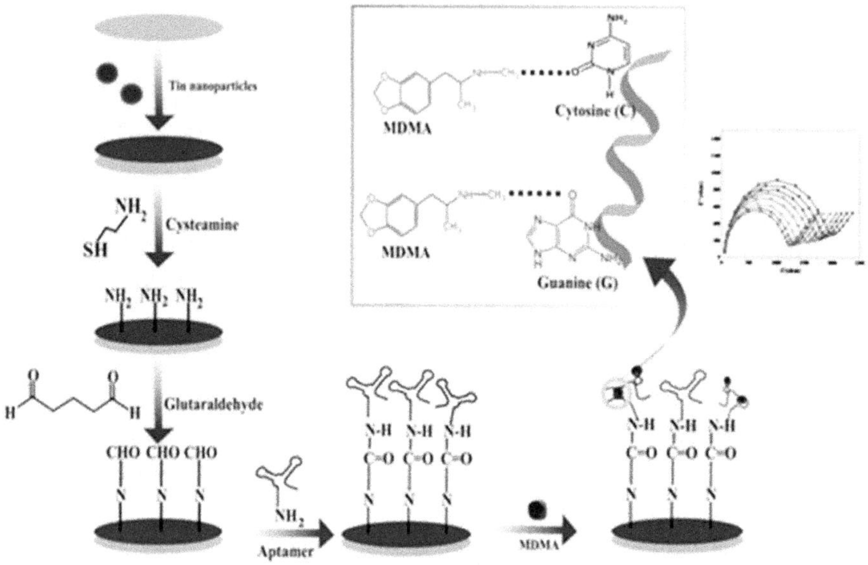

Menon, 2013). The colorimetric technique works by showing the visible change in the color of AuNPs from red to blue in response to the reaction between the AuNPs and the clonazepam. The nanosensor developed in this way shows the least amount of LOD as compared with all the existing techniques. Other closely related substances like diazepam, lorazepam and alprazolam did not show any significant amount of interference. The on-spot testing of the crime scene for clonazepam detection can be done through this nanosensor. The nanosensor has also shown significant analytical and sensing abilities for the bone remains sample that is crucial evidence in many large-scale crimes (Lodha et al., 2013).

Illicit Drug Detection on the Basis of Nano Chips

The development of nanochips is a new approach; the idea gets motivation from the previously developed biochips for the detection of pathogens and diagnostic purposes. These biochips developed for different routine and special assays like ELISA (Kong, Chong, Squire, & Wang, 2018), genotyping assays (Munir, Ahmed, Ibrahim, Khalid, & Ojha, 2020), transfusion assays and immunoassays have resulted in speeding up the diagnostic and research processes at large. Now, the nanochip work with the same scope for the detection of illicit drugs resulting in the quick processing of crime scene samples. By using this modern-day nano chip approach, we can save both time and money. The principle of nano chip detection involves in the very first step the fixation of the targeted drug on the nanomaterial structure. For the attachment of the drug to the nanomaterial surface, different techniques are used like nanocontact printing and lithography on nanoimprinted particles (Riehemann et al., 2009).

The different recognition of nanomaterials is present that is specific to the illicit drug of choice. The nanochip constructed in this way is highly sensitive and selective for the detection of illicit drugs with less processing time and cost (Lad, Pandya, & Agrawal, 2016).

Future Aspects of Nano Forensic in Toxicology

Illicit drugs have posed great threats to human societies because of the basic nature of the substance that is psychoactive in nature. The ability of these drugs to alter the CNS and cause serious health hazards has prompted the local authorities towards the need for their careful production and transportation. In the past, many conventional techniques have been used for the detection of illicit drugs, but there is certain limitation attached to these existing techniques (Lad et al., 2016; Mao et al., 2020). These instruments require proper laboratory settings for their functioning, require professionals with a great deal of expertise, long processing time and the highly costly as well. All these limitations are very easily overcome by the nanotechnology-based nanosensors that can work with the same efficiency in terms of sensitivity and selectivity but also, but they can work with small samples that are mostly the case with crime scene samples (Sharma et al., 2020).

Many types of nanosensors can be used more frequently in the coming future, the main types include fluorescence-based nanosensors, electrochemical conduction-based nanosensors and one including different chromogens as their sensing platforms. These nanosensors can work with limited quantities and the quality of the sample yields valuable information for preparing the scientific report. Due to their nanoscale sensing approach, they can work with the sample in the nm range, so these samples have the lowest LOD as compared with the conventional approaches. There is no doubt that the researchers and the scientific community in the modern area are looking for these nanosensors to be used more continuously. Especially the smoothness with which nanosensors work and the speed with which the results are produced by them are surely going to replace the conventional techniques in the future for the detection of illicit drugs. All these approaches are going to help the scientific community and forensic science at large (Dagar, Yadav, Sai, Satija, & Bhatia, 2022; Rawtani, Tharmavaram, Pandey, & Hussain, 2019).

REFERENCES

Adegoke, O., Mckenzie, C., & Daeid, N. N. (2019). Multi-shaped cationic gold nanoparticle-L-cysteine-ZnSeS quantum dots hybrid nanozyme as an intrinsic peroxidase mimic for the rapid colorimetric detection of cocaine. *Sensors and Actuators. B, Chemical*, *287*, 416–427. doi:10.1016/j.snb.2019.02.074

Adegoke, O., Zolotovskaya, S., Abdolvand, A., & Daeid, N. N. (2022). Fabrication of a near-infrared fluorescence-emitting SiO2-AuZnFeSeS quantum dots-molecularly imprinted polymer nanocomposite for the ultrasensitive fluorescence detection of levamisole. *Colloids and Surfaces. A, Physicochemical and Engineering Aspects*, *646*, 129013. doi:10.1016/j.colsurfa.2022.129013

Anderson, T., Petranker, R., Rosenbaum, D., Weissman, C. R., Dinh-Williams, L.-A., Hui, K., Hapke, E., & Farb, N. A. S. (2019). Microdosing psychedelics: Personality, mental health, and creativity differences in microdosers. *Psychopharmacology*, *236*(2), 731–740. doi:10.100700213-018-5106-2 PMID:30604183

Atta, N. F., Galal, A., El-Ads, E. H., & Hassan, S. H. (2019). Cobalt oxide nanoparticles/graphene/ionic liquid crystal modified carbon paste electrochemical sensor for ultra-sensitive determination of a narcotic drug. *Advanced Pharmaceutical Bulletin*, *9*(1), 110–121. doi:10.15171/apb.2019.014 PMID:31011565

Azimi, S., & Docoslis, A. (2022). Recent Advances in the Use of Surface-Enhanced Raman Scattering for Illicit Drug Detection. *Sensors (Basel)*, *22*(10), 3877. doi:10.339022103877 PMID:35632286

Chiang, C.-H., Lee, H.-H., Chen, B.-H., Lin, Y.-C., Chao, Y.-Y., & Huang, Y.-L. (2019). Using ambient mass spectrometry and LC–MS/MS for the rapid detection and identification of multiple illicit street drugs. *Journal of Food and Drug Analysis, 27*(2), 439-450.

Crepeault, H., Socias, M. E., Tobias, S., Lysyshyn, M., Custance, A., Shapiro, A., & Ti, L. (2023). Examining fentanyl and its analogues in the unregulated drug supply of British Columbia, Canada using drug checking technologies. *Drug and Alcohol Review*, *42*(3), 538–543. doi:10.1111/dar.13580 PMID:36423900

Dagar, M., Yadav, S., Sai, V., Satija, J., & Bhatia, H. (2022). Emerging trends in point-of-care sensors for illicit drugs analysis. *Talanta*, *238*, 123048. doi:10.1016/j.talanta.2021.123048 PMID:34801905

Drake, J., Charles, C., Bourgeois, J. W., Daniel, E. S., & Kwend, E. M. (2020). Exploring the impact of the opioid epidemic in Black and Hispanic communities in the United States. *Drug Science, Policy and Law*, *6*, 2050324520940428. doi:10.1177/2050324520940428

Feng, J., Lepetre-Mouelhi, S., Gautier, A., Mura, S., Cailleau, C., Coudore, F., Hamon, M., & Couvreur, P. (2019). A new painkiller nanomedicine to bypass the blood-brain barrier and the use of morphine. *Science Advances*, *5*(2), eaau5148. doi:10.1126ciadv.aau5148 PMID:30788432

Florea, A., Cowen, T., Piletsky, S., & De Wael, K. (2019). Electrochemical sensing of cocaine in real samples based on electrodeposited biomimetic affinity ligands. *Analyst (London)*, *144*(15), 4639–4646. doi:10.1039/C9AN00618D PMID:31250860

Ganson, K. T., Murray, S. B., & Nagata, J. M. (2021). Associations between eating disorders and illicit drug use among college students. *International Journal of Eating Disorders*, *54*(7), 1127–1134. doi:10.1002/eat.23493 PMID:33638571

Gavrilă, A. M., Stoica, E. B., Iordache, T. V., & Sârbu, A. (2022). Modern and Dedicated Methods for Producing Molecularly Imprinted Polymer Layers in Sensing Applications. *Applied Sciences (Basel, Switzerland)*, *12*(6), 3080. doi:10.3390/app12063080

Göl, E., & Çok, I. (2019). New psychoactive substances in Turkey: Narcotics cases assessed by the Council of Forensic Medicine between 2016 and 2017 in Ankara, Turkey. *Forensic Science International*, *294*, 113–123. doi:10.1016/j.forsciint.2018.11.003 PMID:30502695

Graziano, S., Anzillotti, L., Mannocchi, G., Pichini, S., & Busardò, F. P. (2019). Screening methods for rapid determination of new psychoactive substances (NPS) in conventional and non-conventional biological matrices. *Journal of Pharmaceutical and Biomedical Analysis*, *163*, 170–179. doi:10.1016/j.jpba.2018.10.011 PMID:30316062

Henningfield, J. E., Coe, M. A., Griffiths, R. R., Belouin, S. J., Berger, A., Coker, A. R., ... Nichols, C. D. (2022). Psychedelic drug abuse potential assessment research for new drug applications and Controlled Substances Act scheduling. *Neuropharmacology*, *218*, 109220. doi:10.1016/j.neuropharm.2022.109220 PMID:35987353

Hondebrink, L., Nugteren-van Lonkhuyzen, J. J., Hunault, C. C., van den Berg, J., Van Der Gouwe, D., & van Riel, A. J. (2020). New psychoactive substances (NPS) in the Netherlands: Occurrence in forensic drug samples, consumer drug samples and poisons center exposures between 2013 and 2017. *Addiction (Abingdon, England)*, *115*(4), 716–725. doi:10.1111/add.14868 PMID:31656050

Howard, A. L., Kennedy, T. M., Mitchell, J. T., Sibley, M. H., Hinshaw, S. P., Arnold, L. E., Roy, A., Stehli, A., Swanson, J. M., & Molina, B. S. (2020). Early substance use in the pathway from childhood attention-deficit/hyperactivity disorder (ADHD) to young adult substance use: Evidence of statistical mediation and substance specificity. *Psychology of Addictive Behaviors*, *34*(2), 281–292. doi:10.1037/adb0000542 PMID:31886682

Kesarwani, S., Parihar, K., Sankhla, M. S., & Kumar, R. (2020). Nano-forensic: new perspective and extensive applications in solving crimes. *Latent in Applied Nanobioscience, 10*(1), 1792-1798.

Kethineni, S., & Cao, Y. (2020). The rise in popularity of cryptocurrency and associated criminal activity. *International Criminal Justice Review*, *30*(3), 325–344. doi:10.1177/1057567719827051

Kong, X., Chong, X., Squire, K., & Wang, A. X. (2018). Microfluidic diatomite analytical devices for illicit drug sensing with ppb-Level sensitivity. *Sensors and Actuators. B, Chemical*, *259*, 587–595. doi:10.1016/j.snb.2017.12.038 PMID:29755211

Kopra, E. I., Ferris, J. A., Rucker, J. J., McClure, B., Young, A. H., Copeland, C. S., & Winstock, A. R. (2022). Adverse experiences resulting in emergency medical treatment seeking following the use of lysergic acid diethylamide (LSD). *Journal of Psychopharmacology (Oxford, England)*, *36*(8). doi:10.1177/02698811221099650 PMID:35672900

Lad, A. N., Pandya, A., & Agrawal, Y. (2016). Overview of nano-enabled screening of drug-facilitated crime: A promising tool in forensic investigation. *Trends in Analytical Chemistry*, *80*, 458–470. doi:10.1016/j.trac.2015.07.016

Lea, T., Amada, N., & Jungaberl, E. H. (2020). Psychedelic microdosing: A subreddit analysis. *Journal of Psychoactive Drugs*, *52*(2), 101–112. doi:10.1080/02791072.2019.1683260 PMID:31648596

Liu, C. M., He, H. Y., Xu, L., & Hua, Z. D. (2021). New qualitative analysis strategy for illicit drugs using Raman spectroscopy and characteristic peaks method. *Drug Testing and Analysis*, *13*(3), 720–728. doi:10.1002/dta.2963 PMID:33142047

Lodha, A., Pandya, A., Sutariya, P. G., & Menon, S. K. (2013). Melamine modified gold nanoprobe for "on-spot" colorimetric recognition of clonazepam from biological specimens. *Analyst (London)*, *138*(18), 5411–5416. doi:10.1039/c3an00184a PMID:23807945

Mao, K., Zhang, H., Pan, Y., Zhang, K., Cao, H., Li, X., & Yang, Z. (2020). Nanomaterial-based aptamer sensors for analysis of illicit drugs and evaluation of drugs consumption for wastewater-based epidemiology. *Trends in Analytical Chemistry*, *130*, 115975. doi:10.1016/j.trac.2020.115975 PMID:32834242

Marshall, B., Bland, M. K., Hulla, R., & Gatchel, R. J. (2019). Considerations in addressing the opioid epidemic and chronic pain within the USA. *Pain Management*, *9*(2), 131–138. doi:10.2217/pmt-2018-0070 PMID:30806566

Mejia de Grubb, M. C., Salemi, J. L., Gonzalez, S. J., Chima, C. C., Kowalchuk, A. A., & Zoorob, R. J. (2020). Opioid, cocaine, and amphetamine use disorders are associated with higher30-day inpatient readmission rates in the United States. *Substance Abuse*, *41*(3), 365–374. doi:10.1080/08897077.2019.1635964 PMID:31295052

Mitra, S., & Basak, M. (2022). Diverse bio-sensing and therapeutic applications of plasmon enhanced nanostructures. *Materials Today*, *57*, 225–261. doi:10.1016/j.mattod.2022.05.023

Moore, A., Foss, J., Juhascik, M., Botch-Jones, S., & Kero, F. (2019). Rapid screening of opioids in seized street drugs using ambient ionization high resolution time-of-flight mass spectrometry. *Forensic Chemistry (Amsterdam, Netherlands)*, *13*, 100149. doi:10.1016/j.forc.2019.100149

Mousaabadi, K. Z., Ensafi, A. A., & Rezaei, B. (2022). Simultaneous determination of some opioid drugs using Cu-hemin MOF@ MWCNTs as an electrochemical sensor. *Chemosphere*, *303*, 135149. doi:10.1016/j.chemosphere.2022.135149 PMID:35660395

Munir, S., Ahmed, S., Ibrahim, M., Khalid, M., & Ojha, S. C. (2020). A Spellbinding interplay between biological barcoding and nanotechnology. *Frontiers in Bioengineering and Biotechnology*, *8*, 883. doi:10.3389/fbioe.2020.00883 PMID:33014994

Nikkel, B. (2020). Fintech forensics: Criminal investigation and digital evidence in financial technologies. *Forensic Science International: Digital Investigation*, *33*, 200908.

Omar, Z. M., & Ibrahim, J. (2020). An overview of Darknet, rise and challenges and its assumptions. *International Journal of Computer Science and Information Technologies*, *8*, 110–116.

Paikrao, H. M., Tajane, D. S., Patil, A. S., & Dipale, A. D. (2022). *Applications of Nanotechnology in Forensic Science. In Engineered Nanomaterials for Innovative Therapies and Biomedicine*. Springer.

Parrilla, M., Slosse, A., Van Echelpoel, R., Felipe Montiel, N., Langley, A. R., Van Durme, F., & De Wael, K. (2022). Rapid on-site detection of illicit drugs in smuggled samples with a portable electrochemical device. *Chemosensors (Basel, Switzerland)*, *10*(3), 108. doi:10.3390/chemosensors10030108

Peng, X., Kotnala, A., Rajeeva, B. B., Wang, M., Yao, K., Bhatt, N., Penley, D., & Zheng, Y. (2021). Plasmonic Nanotweezers and Nanosensors for Point-of-Care Applications. *Advanced Optical Materials*, *9*(13), 2100050. doi:10.1002/adom.202100050 PMID:34434691

Pergolizzi Jr, J. V., Dahan, A., Ann Lequang, J., & Raff, A. (2021). Overdoses due to fentanyl and its analogues (F/FAs) push naloxone to the limit. *Journal of Clinical Pharmacy and Therapeutics*, *46*(6), 1501–1504. doi:10.1111/jcpt.13462 PMID:34111307

Priyonugroho, G., Zaini, J., Samoedro, E., Firmansyah, I., Nurwidya, F., Antariksa, B., ... Susanto, A. (2018). Correlation between urinary cotinine, exhaled carbon monoxide, and nicotine dependence among Indonesian individuals in the national narcotics and illicit drug rehabilitation center. *Journal of Natural Science, Biology, and Medicine*, *9*(2).

Purohit, S., Pandey, G., Tharmavaram, M., Rawtani, D., & Mustansar Hussain, C. (2020). Sensors for the Detection of Illicit Drugs. *Technology in Forensic Science: Sampling, Analysis, Data and Regulations*, 221-238.

Rawtani, D., Tharmavaram, M., Pandey, G., & Hussain, C. M. (2019). Functionalized nanomaterial for forensic sample analysis. *Trends in Analytical Chemistry*, *120*, 115661. doi:10.1016/j.trac.2019.115661

Riehemann, K., Schneider, S. W., Luger, T. A., Godin, B., Ferrari, M., & Fuchs, H. (2009). Nanomedicine—Challenge and perspectives. *Angewandte Chemie International Edition*, *48*(5), 872–897. doi:10.1002/anie.200802585 PMID:19142939

Sarfo, D. K., Izake, E. L., O'Mullane, A. P., & Ayoko, G. A. (2019). Fabrication of nanostructured SERS substrates on conductive solid platforms for environmental application. *Critical Reviews in Environmental Science and Technology*, *49*(14), 1294–1329. doi:10.1080/10643389.2019.1576468

Shahbazi, N., & Zare-Dorabei, R. (2019). A novel "off-on" fluorescence nanosensor for sensitive determination of sulfide ions based on carbon quantum dots and gold nanoparticles: Central composite design optimization. *Microchemical Journal*, *145*, 996–1002. doi:10.1016/j.microc.2018.12.022

Sharma, S., Shekhar, S., Gautam, S., Sharma, B., Kumar, A., & Jain, P. (2020). Carbon-based nanomaterials as novel nanosensors. *Nanofabrication for smart nanosensor applications*, 323-347.

Singh, S., & Samal, N. (2021). Nanotechnology: A Powerful Tool in Forensic Science for Solving Criminal Cases. *Arab Journal of Forensic Sciences & Forensic Medicine*, *3*(2), 273–296. doi:10.26735/YDDJ4516

Smith, C. (2019). *Neurotransmitters and the narcotic analgesics. In Chemical and Biological Aspects of Drug Dependence*. CRC Press.

Solaimuthu, A., Vijayan, A. N., Murali, P., & Korrapati, P. S. (2020). Nano-biosensors and their relevance in tissue engineering. *Current Opinion in Biomedical Engineering*, *13*, 84–93. doi:10.1016/j.cobme.2019.12.005

Song, C., Ye, B., Xu, J., Chen, J., Shi, W., Yu, C., An, C., Zhu, J., & Zhang, W. (2022). Large-Area Nanosphere Self-Assembly Monolayers for Periodic Surface Nanostructures with Ultrasensitive and Spatially Uniform SERS Sensing. *Small*, *18*(8), 2104202. doi:10.1002mll.202104202 PMID:34877766

Soni, S., Jain, U., Burke, D. H., & Chauhan, N. (2022). Development of Nanomaterial-Modified Impedimetric Aptasensor—A Single-Step Strategy for 3, 4-Methylenedioxymethylamphetamine Detection. *Biosensors (Basel)*, *12*(7), 538. doi:10.3390/bios12070538 PMID:35884341

Taubin, D., Wilson, J. C., & Wilens, T. E. (2022). ADHD and Substance Use Disorders in Young People: Considerations for Evaluation, Diagnosis, and Pharmacotherapy. *Child and Adolescent Psychiatric Clinics of North America*, *31*(3), 515–530. doi:10.1016/j.chc.2022.01.005 PMID:35697399

Tully, G., Cohen, N., Compton, D., Davies, G., Isbell, R., & Watson, T. (2020). Quality standards for digital forensics: Learning from experience in England & Wales. *Forensic Science International: Digital Investigation*, *32*, 200905.

van Dongen, R. M., Alderliefste, G. J., Onderwater, G. L., Ferrari, M. D., & Terwindt, G. M. (2021). Migraine prevalence in visual snow with prior illicit drug use (hallucinogen persisting perception disorder) versus without. *European Journal of Neurology*, *28*(8), 2631–2638. doi:10.1111/ene.14914 PMID:33979006

Walters, G. D. (2018). Resistance to peer influence and crime desistance in emerging adulthood: A moderated mediation analysis. *Law and Human Behavior*, *42*(6), 520–530. doi:10.1037/lhb0000293 PMID:29927278

Wester, N., Mynttinen, E., Etula, J., Lilius, T., Kalso, E., Kauppinen, E. I., Laurila, T., & Koskinen, J. (2019). Simultaneous detection of morphine and codeine in the presence of ascorbic acid and uric acid and in human plasma at nafion single-walled carbon nanotube thin-film electrode. *ACS Omega*, *4*(18), 17726–17734. doi:10.1021/acsomega.9b02147 PMID:31681878

Zibbell, J. E., Clarke, S. D., Kral, A. H., Richardson, N. J., Cauchon, D., & Aldridge, A. (2022). Association between law enforcement seizures of illicit drugs and drug overdose deaths involving cocaine and methamphetamine, Ohio, 2014–2019. *Drug and Alcohol Dependence*, *232*, 109341. doi:10.1016/j.drugalcdep.2022.109341 PMID:35134733

Chapter 12
Nanotech and Document Security

Salah Ud Din
University of Health Sciences, Pakistan

Shoaib Shafique
University of Punjab, Lahore, Pakistan

Faisal Khan
University of the Punjab, Lahore, Pakistan

ABSTRACT

Being a multidisciplinary field, forensic science has greatly evolved over the previous decades. The recent advances in nanotechnology have opened new windows for forensic sciences as it offers cutting-edge methods for simplifying the laborious forensic analysis of the questioned documents. In this regard, various nanoparticles are most widely exploited and coupled with the traditional methods for resolving cases marked with forgery. These methods are now emerging to be the extensively used strategies for the future analysis of questioned documents to obtain the analytical, quantitative, as well as qualitative profile of the ink on real samples. This chapter thus explores the trending (traditional as well as novel nano-based) methods used so far in the questioned documents analysis.

INTRODUCTION

The inks evaluation on questioned documents is decisive, especially when the artifices of scripts are involved. Ink analysis has remained challenging forensically due to the non-specificity between the inks. Almost, all the inks contain the same basic composition with trace differences. These differences could not be exploited to serve forensic needs by conventional techniques. In addition, most of the traditional methods for ink analysis are destructive further limiting its application in forensic testing. Therefore, the application of nanotechnology can be employed to uncover the potential of ink analysis and ink aging as corroborative evidence. Nanotechnology has gained popularity in the Forensic community due to its

DOI: 10.4018/978-1-6684-8325-1.ch012

ability to produce, easy, quick, economical, and reliable results. Nano-based testing offers the benefit of not needing any pre-treatment process, which ensures that the original document remains intact and unaltered during the analysis. This means that the inks present on the document can be directly examined without any risk of damaging or compromising the integrity of the document. Gold and platinum nano-composites can be utilized to analyze various sources of blue ink and estimate the aging parameter. This means that by using these nano-composites, it is possible to identify the origin of blue ink on a document and estimate how old the ink is. Nano composites are materials made by combining two or more types of nanoparticles. These composites are widely used in the analysis of inks due to their ability to enhance the sensitivity of detecting even minor differences between inks. By combining the unique properties of different nanoparticles, nano composites amplify the detecting properties of the inks and enable more accurate identification of the origin and properties of the inks present on a document.

It is common for forensic laboratories to encounter ink-related evidence that stems from the use of pens during their investigations (Calcerrada & García-Ruiz, 2015). The examination of inks is a very important forensic process that can describe beneficial facts about a questioned document. A ''questioned'' document can be any communication presented in document form whose source or legitimacy is unsure and it may include suicidal notes, wills, death certificates, threatening letters, insurance claims, checks, petitions, lottery voucher, etc. which can be found in different criminal inspections (Braz et al., 2013). Document analysis starts by identifying the device used to produce the document. Next, a thorough examination of the inks is conducted to detect any changes or additions to the original content. This process involves analyzing the chemical and physical properties of the inks, as well as their relative ages and contributions to the document. By examining the chronological sequence of the ink contributions, it becomes possible to determine the order in which different portions of the document were written. All of these steps are necessary to ensure accurate document analysis and interpretation (Jones et al., 2006). Table 1 defines and elaborates the various relevant terminologies.

Table 1. Definitions of important terminologies

Technical Terms	Definition
Question Documents	A questioned document is a type of document that is often linked to criminal or civil investigations and is the subject of a dispute regarding its authenticity or origin.
Forensic Ink Analysis	It is the scientific examination of ink to determine if a document is authentic or has been forged.
Ink Dating	Ink dating is the forensic analysis of ink to determine document age or authenticity.
Nanoparticles:	Nanoparticles are extremely small particles, ranging from 1 to 100 nanometers in size, with unique physical and chemical properties.
Nanohybrids	Materials with organic and inorganic components that are linked together either by noncovalent bonds or covalent bonds at the nanometer scale are called nanohybrids.
Potentiometer	A potentiometer is an instrument used to measure voltage or potential difference by comparing an unknown voltage to a known reference.

METHODS OF INK ANALYSIS

Ink analysis can be done by using both destructive and non-destructive methods. Capillary zone electrophoresis (Rohde et al., 1997), Gas chromatography (Bügler et al., 2005), Liquid chromatography (Kher et al., 2006), Mass spectrometry (Grim et al., 2001), Thin layer chromatography (Djozan et al., 2008), and Gas chromatography–mass spectrometry (Sun et al., 2016) are destructive, expensive and time-consuming techniques. All these conventional methods require sample preparation which may include cutting of strips, punching holes, and piercing of ink-beared fibers from the question document. Once dismantled, those pieces are put down into a solvent to remove the ink and are then further analyzed by these techniques (Calcerrada & García-Ruiz, 2015). Despite producing reliable outputs, these techniques are not suggested for forensic cases and for precious or historical documents where the integrity of the documents could not be compromised (Neumann & Margot, 2009). Furthermore, these methods need skilled sample preparation and are also quite time-consuming (Jones et al., 2006).

Therefore, non-destructive examination methods are preferred because they are simple to execute, require minimal or no alteration of the document, and provide detailed information without compromis-

Figure 1. XRF analysis of the map
Source: Kriznar et al. (2022)

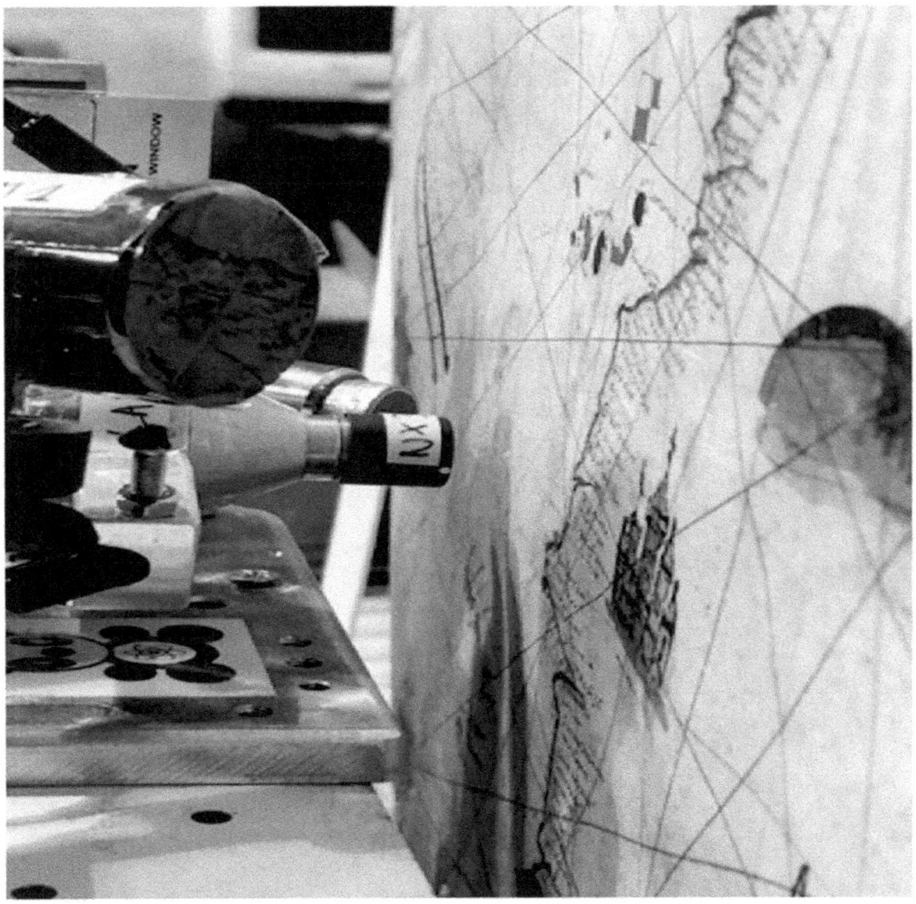

ing the evidential value of the documents (Kumar & Sharma, 2017). Non-destructive methods include microscopy and optic excitation. By integrating filters to intensify ink contrast, microscopy can be used for ink analysis. This allows for examination of ink properties, differentiation of different inks, and preservation of the document's integrit. Optical excitation can be employed to produce a specific molecular imprint of ink by using light of various wavelengths to generate a spectrum. Analyzing the peaks and dips in the spectrum provides insights into the ink's molecular structure and allows for its identification. In one study, a portolan chart from Juan Vespucci's collection in Spain was analyzed which is comprised of parchment and features images of the European and African beaches and islands (Kriznar et al., 2022). The map was studied using a variety of non-interfering methods, including infrared, ultraviolet light and X-ray fluorescence along with a digital microscope to determine the materials used and to look for

Figure 2. (A) visible representation of the scale bar together with the date as well as sign. The inscriptions are not evident under IR, confirming the employment of iron-gall ink. (B) UV photograph of the identical area with (C) IRR imaging of the exact same spot. Vermillion is represented by the white lines and bands; azurite is represented by the dark bands, which are likely carbon black underpaint; and thin black lines are a copper-based green pigment, as determined by XRF.
Source: Kriznar et al. (2022)

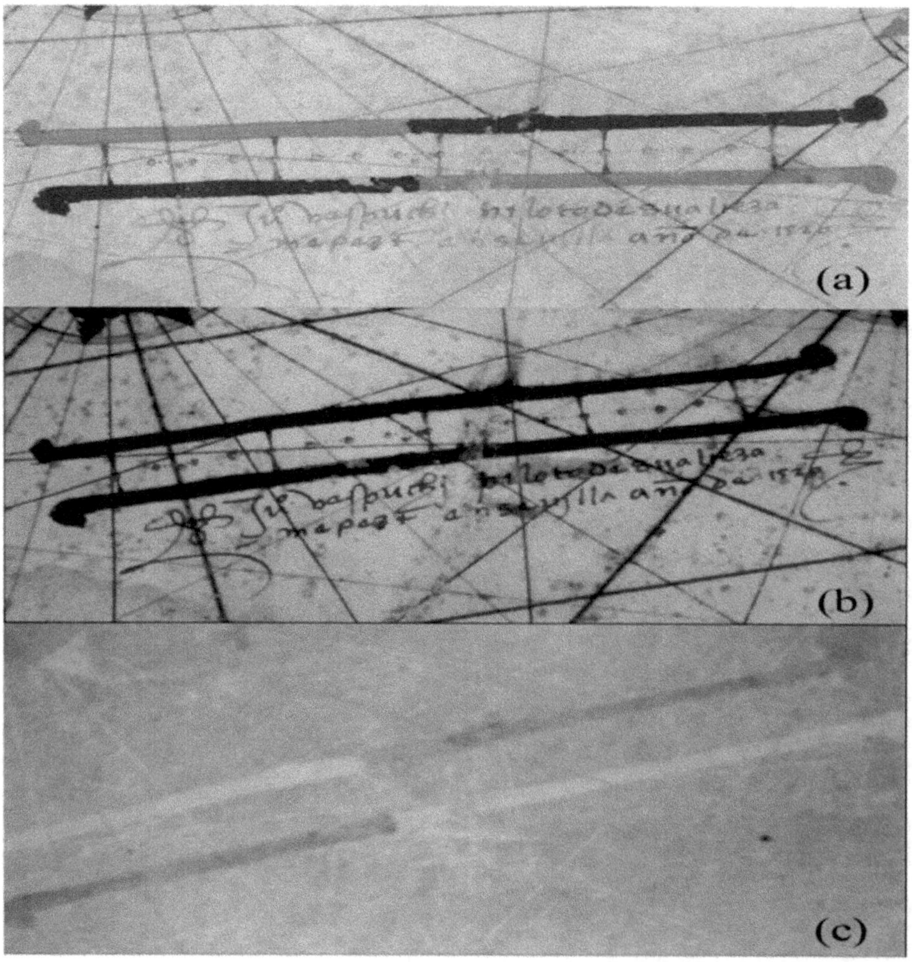

signs of retouching or renovation work. The following Figure 1 shows the XRF analysis of the map in the study (Kriznar et al., 2022).

The study revealed the following findings:

By using UV-Induced Visible Fluorescence technique, you can learn quickly about later interventions like overpaintings, renovations, fresh varnishes, and even fake signs. Similarly, Infrared reflectography (IRR) was used to examine the artwork for potential alterations, interventions, and pigment distinction. It is among the most popular image registration techniques in technical examination of art work because it can find infrared non-reflecting materials (like carbon black) behind painted layers that are invisible to the human eye. The map's elemental chemical makeup was examined using XRF. This method can detect chemical components heavier than that of Si, moreover, it is possible to identify the majority of the inorganic materials relevant to this study, including pigments, certain inks and preparation (Artioli, 2010). The following Figure 2 shows visible representation of the scale bar together with the date as well as sign and other information (Kriznar et al., 2022).

These methods need expertise of the investigator as they are less sensitive and require enough quantity of ink. Therefore, concocting more qualitative tools for question document examinations becomes imperative and concurrently, a challenge, as the modus operandi for counterfeiting documents grows more sophisticated. (Braz et al., 2013).

ROLE OF NANOTECHNOLOGY IN FORENSIC ANALYSIS OF INK

Nanotechnology provides promising success in this regard and offers new horizons for ink analysis forensically with minimal impairment in the quality of the document by improving resolution, enabling analysis of small/degraded samples, and enhancing sensitivity of techniques like SERS/fluorescence. It also allows flexibility in analyzing complex inks (Raza & Saha, 2013).

Nanotechnology involves exploring, designing, modeling, creating, manipulating, and utilizing functional materials, devices, and systems by managing matter at the nanometer scale(Y.-F. Chen, 2011). Nano-forensics is the latest innovation and has a consequential impact in the domain of forensic science. Incognito evidence which is worrisome to discern by conventional techniques could be easily spotted with nano-sensors and nano-devices (Prasad et al., 2020).

Coupling nano-technology with analytical techniques for analysis of ink chemistry in question documents is addressing roughly all the shortcomings encountered by conventional techniques (Y.-F. Chen, 2011; Huynh et al., 2016; Paikrao et al., 2022; Prasad et al., 2020; Raza & Saha, 2013).

FORENSIC ANALYSIS OF QUESTION DOCUMENTS

Forensic analysis of Question Documents is very crucial as they can be served as evidence in courts against crimes like question signatures, falsification, counterfeit currency, threatening letters, fraudulent travel documents, fake wills, bogus credit cards, or even terrorist attacks (Ahmed, 2022).

A wide range of techniques are employed to determine the authenticity, origin, and content of written materials. Among these techniques, dating studies, ink analysis, paper analysis, and studies of intersecting lines are some of the most commonly used methods for investigating questioned documents. As shown in

Figure 3. Fields of investigation in questioned document analysis: Dating, ink analysis, paper analysis, and intersecting lines studies

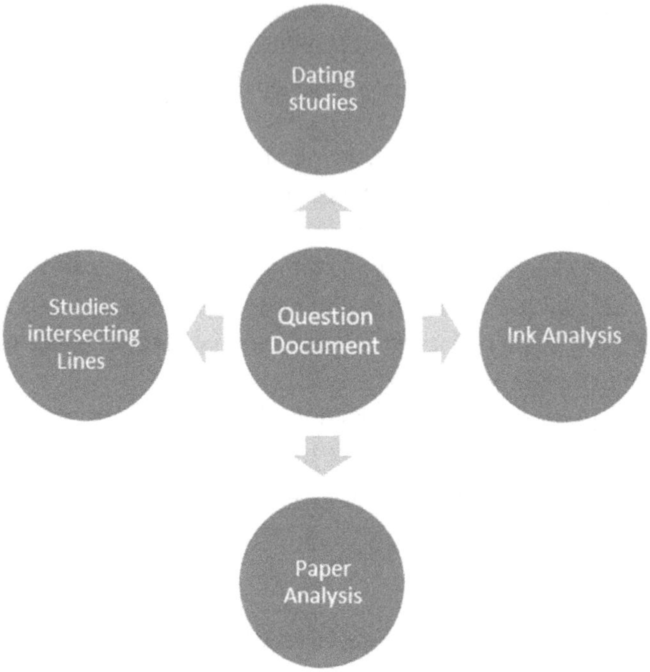

Figure 3, these domains of questioned document analysis are interconnected, with each field providing valuable insights and evidence for forensic document examiners.

Notwithstanding pseudoscientific studies like '*Graphology*', in which handwriting is analyzed, Physical, Optical, and Chemical methods for ink analysis are preferred as they are more validating (Dazzi & Pedrabissi, 2009). There are generally two aspects when examining a questioned document. The first aspect is the examination of the ink constitution to differentiate or contrast between the inks and ink strokes go after the study of compounds that debase with time to approximate the age of a questioned paper entry. Pen inks are a thick blend of various pigments and dyes that provide color, along with organic solvents that act as a medium, wax to add adhesive properties and viscosity, and additional substances such as driers, plasticizers, binders, antioxidants, waxes, preservatives, and other trace elements to enhance the quality of the ink. The ink vehicle consists of waxes, oils and solvents, and it sustains the flowing and drying property and imparts the right viscosity required to the ink. The precise formulation of pen inks is kept secret by the brand or manufacturer as it is their proprietary which further harbors the complexity of ink analysis in question documents. The application of three pen groups; Fountain pens, Ballpoint pens, and Gel pens, is very frequent both in official and non-official writings. The inks in fountain pens are less viscous and bore high water content which is conveyed to the paper through the nib whereas in Ballpoint inks, oil content is high and more viscous which leaves its prints on the paper by a rolling tiny hard sphere at the tip and the ink dries very quickly. Ballpoints have now replaced fountain pens as the most commonly used pen for everyday writing due to its reliability, cheapness, and handiness. The gel pen's ink viscosity is moderate due to an amalgam of both oil and water molecules in their composition. They dispense the ink to the paper by the application of a ball tip that rolls over the surface of the paper.

Viscous inks (oil based) are hard to be absorbed on paper as compared to less-viscous ink. However, the oil-based-inks components degrade slowly as compared to inks containing a high proportion of water (Calcerrada & García-Ruiz, 2015).

Ink dating is an exacting issue, as each compound/molecule starts to perish from the very moment, they land on the paper, through varied mechanisms. Dyes decompose, volatile compounds diffuse through the paper or evaporate and the polymerization of resins begins and the strokes become stiffer (Kapoor et al., 2021).

DATING STUDIES

Due to the significance of this topic, the aging of samples and pertinent studies on the subject are taken into consideration before analyzing the most updated literature on the examination of inks from different samples.

When examining bibliography, there are two distinct yet complimentary viewpoints to consider. Examining a contested document. The first approach tries to identify the age of entries present on a questioned document by using compounds that deteriorate with time. The second approach depends on the investigation of the ink constituents for the classification, identification, and comparison of inks or ink strokes.

Given that each chemical starts to disintegrate differently after the ink has been deposited on the paper, ink dating is a difficult problem to solve. In addition to volatile components (solvents) evaporating into the environment, dyes break down over time, and also travel through the fibers of the paper (Cantú, 2012); third resins begin to polymerize, making the strokes harder. In 2010, a comprehensive study was conducted on the methods used to date ballpoint pen inks over the previous three decades. The study demonstrated the broad spectrum of investigations that have been undertaken in this area (Ezcurra et al., 2010).

Here, we have made an effort to compile only the most recent studies on the analytical method-based aging of ink samples. Be aware that writing instruments other than ballpoint pens have undergone aging research and have been described in the section that follows. In comparison to pen ink, other samples like printer ink or office paper have not received nearly as much research (Ezcurra et al., 2010).

INK ANALYSIS

Chemical constituents of different inks sold on the market, including fountain, ballpoint and gel pen inks, has changed over time. Documents' dates can be improved by tracing these evolutions. The first to use optical and TLC qualities to distinguish between ink-based chemical compositions was the Zurich Police department, Switzerland (Cantu, 1995). Following this, numerous innovative methods were used for dating. Numerous researchers employed various analytical methods, including TLC, LIBS, HPLC, FTIR, GC, DRIFTS, LAICP-MS, HPLC-DAD, and TOF-SIMS are examples of advanced methods for differentiating inks. These methods can be used to find out the chemical constituents of the ink (Pachuta & Staral, 1994).

The chemical constituents of different other inks, such as typewriter printing and ribbon inks can also be analyzed using analytical techniques like Raman spectroscopy, infrared TLC and x-ray fluores-

Figure 4. Shows the different steps involved in the LIBS process
Source: Al-Ameri et al. (2022)

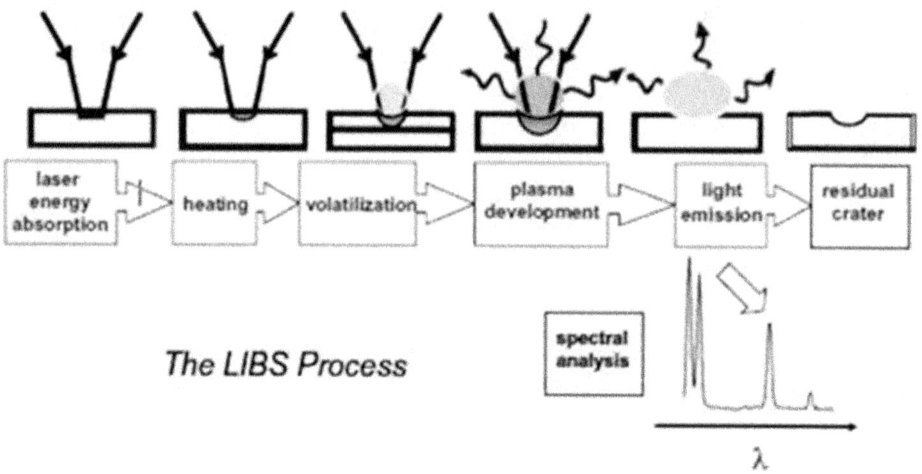

cence. The disadvantage, however, results from the fact that there are substantially fewer databases that include information about these inks. An instance depositions of a dating pencil is comparable. It can be accomplished to some kind of figuring out the makeup of pencil remnants by figuring out the makeup of the binder by using GC-MS spectroscopy. In contrast to typewriter ribbon ink dating and dating using pencil residue, these approaches may be used to date writing inks (Zięba-Palus & Trzcińska, 2011). Moreover, since the different agencies obtaining printed documents must verify the documents' validity as well as look for signs of forgery. One study has recommended a system for detecting unsupervised forgery that can determine if a document is fake depends on the spectroscopic examination of the ink used in that document (Al-Ameri et al., 2022). In this regards LIBS technique was used to acquire the spectra of the examined document inks (original as well as questioned). The results demonstrated that the suggested method offered higher accurateness in detecting forged documents as well as an increased rates of detection speed. The following Figure 4 illustrates the various steps involved in the LIBS Process (Al-Ameri et al., 2022).

PAPER ANALYSIS

To make paper, different synthetic and natural fibers are woven together with a few chemicals. Fibrous materials and covering ingredients like white pigments, fillers, colours and dyes are some of the components of the paper. In response to consumer demands and to improve paper properties, numerous changes have been made to the paper composition. For instance, in the paper formulation in 1940–1941, urea and formaldehyde waxy resins was included as sizing and covering components. Similar to this, optical whiteners made their debut in 1950 when they were first used in paper composition. The chemical makeup of the paper is increased by these basic components. Using the relevant data, the composition is determined (Koppenhaver, 2007). analytical methods include TLC, GC-MS, LAICP-MS, and LIBS, as well as UV or IR fluorescence.

Based on the year when these components were first introduced to the market, it can reduce the time frame. Examining watermarks is another approach to paper analysis.

They might be able to connect with the producers and offer details regarding the watermark's initial application. The watermark itself may have a date code in some conditions. Additionally, some books claim that certain databases, including lockwood-directory Post's of the Pulp, Paper, and Allied Trades, are accessible for watermark data. Greater effectiveness will be provided by the collective results of the chemical constituents and the watermark examination (Dirwono et al., 2010).

STUDIES OF INTERSECTING LINES

A variety of different samples used in contested documents that have been investigated for forensic purposes. Although, there may be more than one type of ink present on the paper in many documents, either as crossing lines or as a mix of ink. It's vital to note that recent research on sequencing of the intersecting lines is done since it can help with forensic casework. The development of techniques, which permits the accurate sequencing of junctions, also makes this trend possible. The analysis of the pattern of intersecting lines has demonstrated the potential of 3D micro-topography. A correct order sequence was found in the majority of cases. Additionally, intersections of different mixed inks of the same colour were also accomplished effectively (Spagnolo, 2006).

The intersections among intersecting strokes in printed form that are made using written implements have been studied using PIA-600 and Docu Centre expert softwares but this method was ineffective for accurately identify the sequence of intersecting lines (Kaur et al., 2013). Visible Soil Contact VSC was used to test junctions of the same type. The evaluation included black typewriter ink green, red and blue ballpoint pen, ballpoint pen ink, and blue, black, red, and green fountain pen strokes. Once more, signals from printers and typewriters outweighed those from pens, so results were equivocal, and it was recommended that this equipment not be used to analyze intersecting lines.

The order of writing lines between the intersections of a printer and a ballpoint pen was obtained using ATR-FTIR imaging. The increased absorption of toner colour in contrast to accepted papers ballpoint pens was demonstrated by the results. Blind tests were conducted, and 100% of the outcomes were accurate. For inkjet, gel pens and roller ball this practice was ineffective, but. The interaction among black ballpoint pens red-sealing inks and has been investigated using -ATR-FTIR. The proper order of deposition was effectively achieved by correlating the characteristic peaks of results of each sample to the signal of the intersection produced for the sequence of laser toner and seal ink, FTIR was additionally integrated with "SEM-EDS mapping" to improve the precision of the outcomes. The intersection's elemental distribution was determined, improving the reliability of the findings (Bojko et al., 2008). The following Figure 5 shows the SEM reference image of four cotton (Geminiani et al., 2022).

TECHNIQUES USED IN QUESTION DOCUMENT ANALYSIS

Chromatographic and electrophoretic techniques are in practice for question document analysis since the 90s (Zlotnick & Smith, 1999) followed by optical and physical methods (H.-S. Chen et al., 2002). The analysis of the inks and their dating started majorly with the chromatographic techniques in which the different components of the inks were separated and analyzed to compare and deduce useful information.

Figure 5. Showing the SEM reference image (A) cotton, (B) jute, (C) hemp, (D) flax, and (E) which describes the ATR-FTIR spectra of the reference materials of all these four cotton. The identification challenge is further aided by the red arrows that identify morphological characteristics
Source: Geminiani et al. (2022)

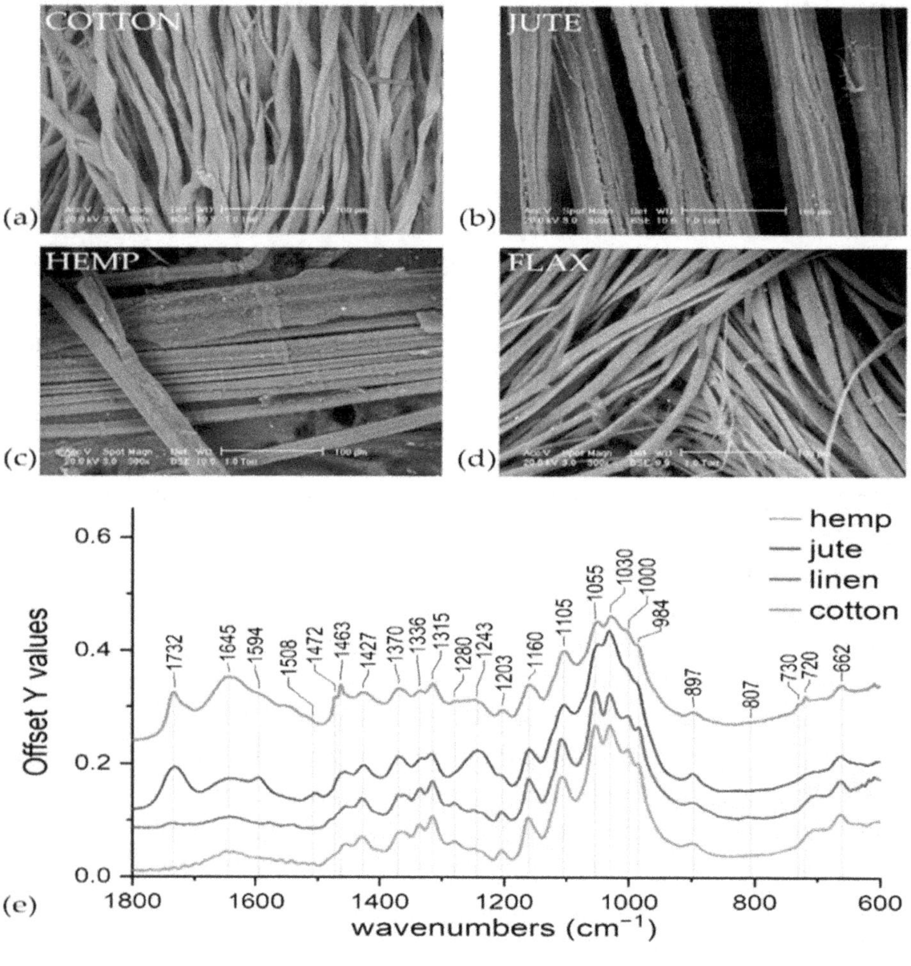

Thin Layer Chromatography is used as a classical method for the analysis of inks and their age estimation since the 60s (Brewer et al., 2005) and more recently HPTLC whereas the application of HPLC, GC Gas Chromatography, and Capillary Electrophoresis CE are least reported for ink analysis (Calcerrada & García-Ruiz, 2015). Due to high discriminating power, TLC has been used for the formation of the US international ink library (Neumann & Margot, 2009) and it is also being used as a positive control to confirm the results of other modern analytical techniques (Quinteros, 2022).

Despite producing qualitative output, these methods are not suitable for forensics analysis of inks where the document bores evidential value (Huynh et al., 2016).

Therefore, nowadays optical examinations/spectroscopic techniques are preferred for ink analysis because of their painless nature, and momentous, and easy in results interpretation (Ivanova et al., 2022).

Differentiation of inks is more objective and authentic as spectrometric techniques can locate major and trace elements (Ahmed, 2022). Atomic force microscope, scanning electron microscope, Transmis-

Figure 6. Shows the different signal types produced in TEM and SEM examination. The significant ones are shown in red circles.
Source: Díez-Pascual et al. (2022)

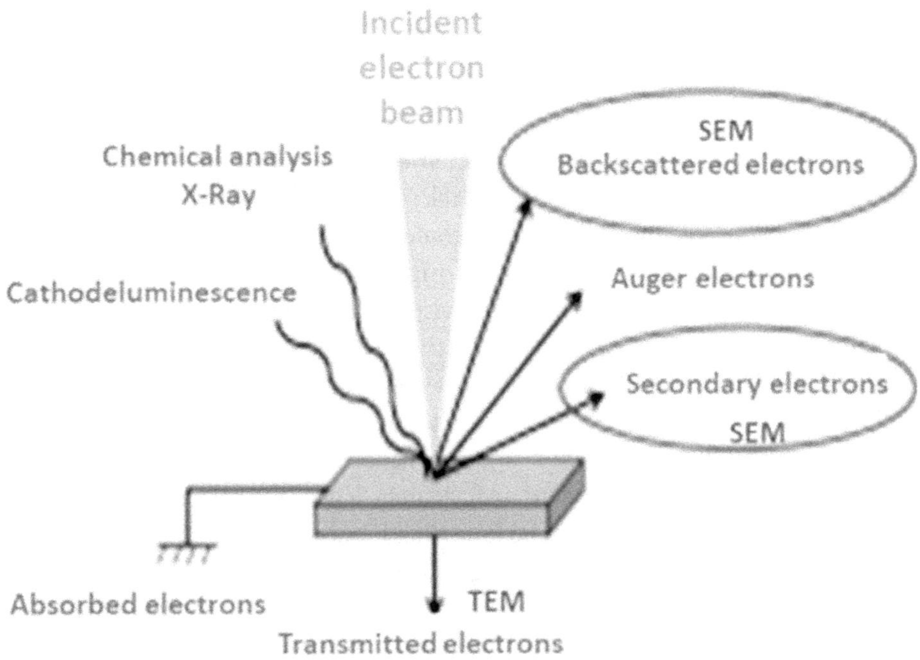

sion electron microscope, and spectrometric techniques like Raman Spectroscopy, Mass Spectroscopy, Infrared Spectroscopy, and UV-Vis Spectroscopy are on the rise in the field of question document analysis over the years (Sanjuan-Navarro et al., 2022), (Khofar et al., n.d.). SEM examination is frequently combined with an energy dispersive X-ray spectroscopy or EDXS detector, which provides data on composition of the sample. Given that it provides the quantity of each sample constituent, this analysis is a semiquantitative one. The source for TEM is usually an electron beam that travels across the specimen while being accelerated by a high voltage. Although the sample preparation is relatively time-consuming and necessitates very thin samples, it offers a greater resolution than SEM (Díez-Pascual et al., 2022).

Figure 6 shows the different signal types produced in TEM and SEM examination (Díez-Pascual et al., 2022).

Spectrometric techniques that furnish elemental composition gives beneficial results in the field of forensic science, for example, Fourier-transform infrared spectroscopy and Laser ablation inductively coupled plasma mass spectrometry (Gui et al., 2022).

Nonetheless, optical similarity among the inks has been reported which lessens the discriminating power of the optical examination. Furthermore, the results obtained from the optical methods were not reliable enough to present them as evidence in the court to serve justice. Therefore, a more comprehensive and robust intervention for ink analysis is required in which the integrity of the document is reserved along with producing reliable results. Nanotechnology and progressive techniques are very helpful in forensic investigations (Erdoğan, 2022).

Nanomaterials have played a developmental role in different facets of science and technology. Furthermore, to its active applications in many other disciplines, nanotechnology also has an indispensable role in forensic science. Nano forensics can be the possible solution for the complications encountered in ink analysis. The ability of nanoparticles to interact with molecules at the nano-scale can differentiate the trace difference between the inks which otherwise could not be detected by using conventional optical techniques like microscopy and spectroscopy (Raza & Saha, 2013).

Coupling the analytical optical techniques like UV-Vis Spectrometry, Raman Spectroscopy, SEM, TEM or AFM microscopy with nanoparticles can increase the limit of detection LOD of trace molecules up to many folds allowing sound identification of the ink source (Paikrao et al., 2022). Nanomaterials can be made by different methods like the hydrothermal route of synthesis. Nanomaterials can be characterized by state-of-the-art techniques such as Dynamic Light Scattering, Scanning Electron Microscope, X-ray Diffraction, Brunauer- EmmettTeller, and Fourier Transform Infra-Red (FTIR), etc. Designed nanomaterials can be optimized for Sensitivity, Selectivity, Replicability, and Aging based on selected transduction approaches like PG STAT and UV-Vis Spectrophotometer and many other methods. In one study, the cyclic voltammetry (CV) method was used where an innovative AuNP/CPE electrochemical was created through the electrodeposition of silver nanoparticles on carbon pasted electrode. The modified electrode served as a straightforward and accurate sensor for detecting CR dye in a variety of experimental samples (Ganash et al., 2023). The following Figure 7 illustrates the various steps involved in the detection of CR dye using nanoparticles (Ganash et al., 2023).

Figure 7. Shows the various steps involved in the detection of CR dye using nanoparticles
Source: Ganash et al. (2023)

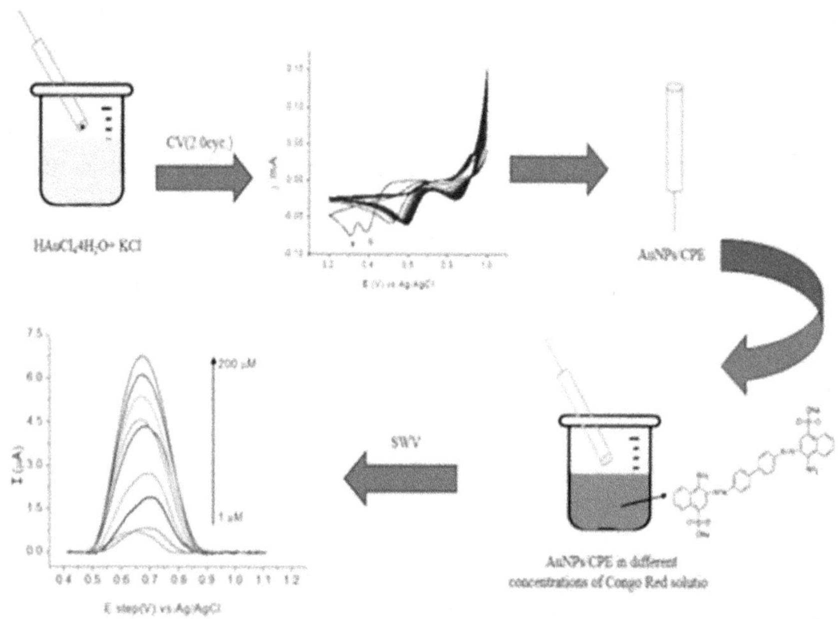

Figure 8. Shows high resolution TEM N-CDs (A), focused ion beam SEM of N-CDs/starch (B), N-CDs (C), and lastly CDs (D)
Source: Nugroho et al. (2022)

The nano-based technique can be used in forensic science for the investigation of cases related to question documents can ultimately lead to the conviction of criminals and solving crimes. In this regards, Carbon Dots (CDs) are of prime importance in forensic analysis as they have promising results in other aspects of crime scene investigations. These carbon dots have previously been used in fingerprint detection (Nugroho et al., 2022), and thus further research is needed for their validity in analyzing the questioned documents. Figure 8 and Figure 9 show high resolution TEM N-CDs and the two basic methods for the synthesis of Carbon quantum dots and carbon nanodots, respectively (Carbonaro et al., 2019; Nugroho et al., 2022).

Figure 9. Shows the two basic methods for the synthesis of carbon quantum dots and carbon nanodots
Source: Carbonaro et al. (2019)

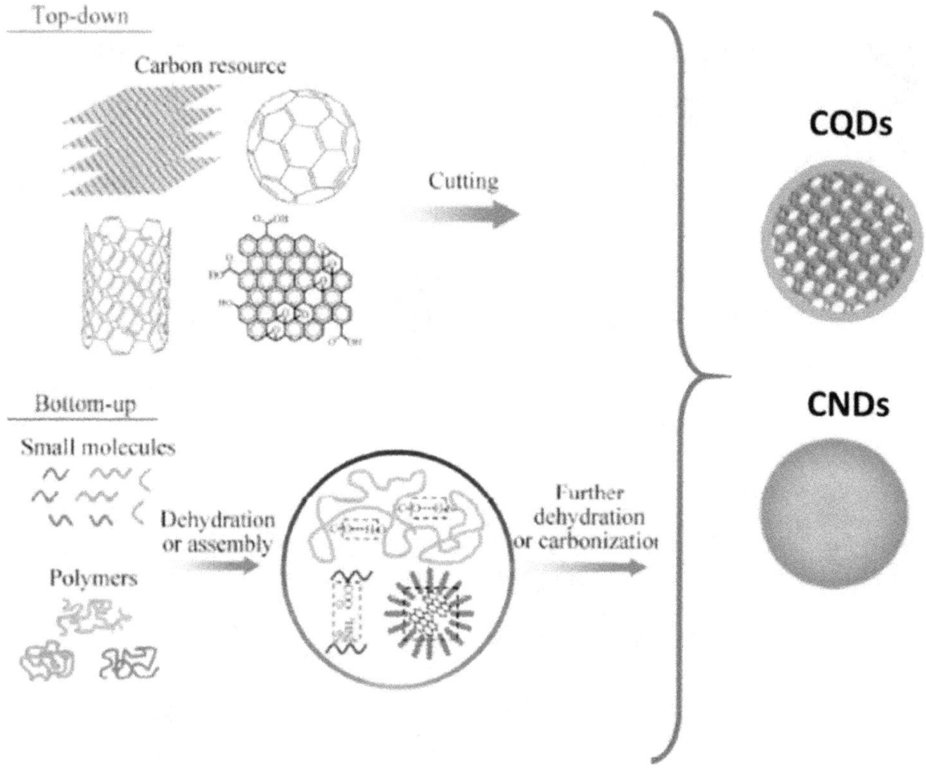

STATIC, DYNAMIC, AND SUPPLEMENTARY APPROACHES FOR DOCUMENT ANALYSIS

The identification and characterization of materials used in the creation of documents can provide valuable information about their origin, age, and authenticity. Table 2 presents a list of techniques commonly used for document analysis, including both static and dynamic approaches.

Static approaches involve the examination of the document's physical characteristics and composition without causing any significant changes to the sample. These techniques include HPLC, TLC, SEM, XRD, electrophotography, and LIBS-PCA. Static approaches are often used for the analysis of inks, papers, and other materials used in document creation.

Dynamic approaches involve the application of external stimuli to the sample to observe changes in its properties. These techniques include FTIR, GC-MS, IP-HPLC, micro spectrophotometry, and pyrolysis GC-MS. Dynamic approaches can provide information on the chemical composition of the document and the changes that may have occurred over time.

In addition to these techniques, the table also includes supplementary approaches, such as Raman spectroscopy, radio-carbon dating (C14), and nanotechnology investigation using SEM-EDX. These techniques can provide complementary information to static and dynamic approaches and can aid in the identification and characterization of materials used in document creation.

Table 2. Provides a list of approaches used for document analysis

Sr.No.	Technique	Approach	References
1	FTIR	Dynamic Approach	(Brunelle & Crawford, 2003)
2	GC–MS	Dynamic Approach	(Leal et al., 2019)
3	IP-HPLC	Dynamic Approach	(Menzyk et al., 2015)
4	Micro spectrophotometry	Dynamic Approach	(Ezcurra et al., 2010)
5	Pyrolysis GC–MS	Dynamic Approach	(Ortiz-Masiá et al., 2016)
6	Raman Spectroscopy	Supplementary Approach	(Gorshkova et al., 2020)
7	Radio-carbon Dating- (C14)	Supplementary Approach	(Fang et al., 2021)
8	HPLC	Static Approach	(Brunelle & Crawford, 2003)
9	TLC	Static Approach	(LaPorte, 2015)
10	SEM	Static Approach	(Area & Cheradame, 2011)
11	XRD	Static Approach	(Duran et al., 2009)
12	Electrophotography	Static Approach	(LaPorte, 2015)
13	LIBS-PCA	Static Approach	(Hui et al., 2019)
14	Nanotechnology investigation of nanomaterials using (SEM-EDX)	Supplementary Approach	(Pandya & Shukla, 2018)

Overall, the use of these techniques in document analysis can provide valuable insights into the material composition, age, and authenticity of documents, contributing to the fields of forensic science, history, and archaeology.

With its ability to manipulate materials at the nanoscale level, nanotechnology offers a promising approach for the detection, analysis, and characterization of forensic evidence. It can aid in the development of more advanced and sensitive analytical tools and techniques that can help solve complex forensic cases. Moreover, the integration of nanotechnology with other fields such as chemistry, physics, and biology can provide a multidisciplinary approach to forensic science, allowing for a more comprehensive understanding of the evidence. As such, it is expected that nanotechnology will continue to play a crucial role in the future of forensic science, contributing to the advancement and innovation of forensic methods and techniques.

Abbreviations: ESI-MS, Electrospray Sequential Ionization–Mass Spectrometry; GC–MS, Gas Chromatography-Mass Spectrometry; HPLC, High-Performance Liquid Chromatography; IP-HPLC, Ion Pair High-Performance Liquid Chromatography; SEM-EDX, Scanning Electron Microscopy- Energy Dispersive X-ray Spectroscopy; XRD, X-ray Diffraction; TLC, Thin Layer Chromatography; Py-GC–MS, Pyrolysis Gas Chromatography- Mass Spectrometry; LIBS, Laser-induced breakdown spectroscopy; FTIR, Fourier-Transform Infrared Spectroscopy.

LIMITATIONS/CHALLENGES WITH NANOPARTICLES

Despite its varied applications and several benefits, nanomaterials cannot be termed as flawless but still the need of time until outweighed by any better option in the future.

1. With a minor change in synthesis protocol variation in size and shape can lead to varied physical and chemical interactions
2. Au-Pt nanocomposites cannot measure the difference of the inks having a duration difference of fewer than six months approximately.
3. Unforeseen interactions of fabricated nanoparticles can occur with different interfering substances present in real samples which may lead to unpredictable results.

CONCLUSION

The chapter provides a comprehensive overview of the techniques employed in the analysis of questioned documents in forensic science. It highlights the significance of traditional methods for examining the physical characteristics and composition of documents, but also notes their limitations in the detection of forgeries. However, with the advent of nanotechnology, there has been a remarkable transformation in the process of document analysis. The incorporation of nanoparticles with conventional mass spectroscopy methods has significantly improved the sensitivity and selectivity of the analysis, particularly in the case of ink and dye analysis. This has made it easier to detect forgeries and has revolutionized the field of questioned document analysis. The chapter emphasizes the importance of incorporating nanotechnology in the analysis of questioned documents to enhance the accuracy and reliability of forensic investigations.

REFERENCES

Ahmed, Z. E. (2022). Forensic documents examination. *Asian Journal of Forensic Sciences, 1*(1).

Al-Ameri, M. A. A., Ciylan, B., & Mahmood, B. (2022). Spectral Data Analysis for Forgery Detection in Official Documents: A Network-Based Approach. *Electronics (Basel), 11*(23), 23. Advance online publication. doi:10.3390/electronics11234036

Area, M. C., & Cheradame, H. (2011). Paper aging and degradation: Recent findings and research methods. *BioResources, 6*, 5307–5337.

Artioli, G. (2010). *Scientific Methods and Cultural Heritage: An introduction to the application of materials science to archaeometry and conservation science*. OUP Oxford. doi:10.1093/acprof:oso/9780199548262.001.0001

Bojko, K., Roux, C., & Reedy, B. J. (2008). An examination of the sequence of intersecting lines using attenuated total reflectance–Fourier transform infrared spectral imaging. *Journal of Forensic Sciences, 53*(6), 1458–1467. PMID:18624894

Braz, A., Lopez-Lopez, M., & Garcia-Ruiz, C. (2013). Raman spectroscopy for forensic analysis of inks in questioned documents. *Forensic Science International, 232*(1–3), 206–212. doi:10.1016/j.forsciint.2013.07.017 PMID:24053882

Brewer, J. D., Hagan, K. A., & Egan, J. M. (2005). Forensic analysis of blue ballpoint pen inks using capillary electrophoresis. *Research and Technology, 7*(3).

Brunelle, R. L., & Crawford, K. R. (2003). *Advances in the Forensic Analysis and Dating of Writing Ink*. Charles C Thomas Publishers.

Bügler, J. H., Buchner, H., & Dallmayer, A. (2005). Characterization of ballpoint pen inks by thermal desorption and gas chromatography-mass spectrometry. *Journal of Forensic Sciences*, *50*(5), 2004486–6. doi:10.1520/JFS2004486 PMID:16225233

Calcerrada, M., & García-Ruiz, C. (2015). Analysis of questioned documents: A review. *Analytica Chimica Acta*, *853*, 143–166. doi:10.1016/j.aca.2014.10.057 PMID:25467455

Cantu, A. A. (1995). A sketch of analytical methods for document dating. Part I. The static approach: Determining age independent analytical profiles. *International Journal of Forensic Document Examiners*, *1*(1), 40–51.

Cantú, A. A. (2012). A study of the evaporation of a solvent from a solution—Application to writing ink aging. *Forensic Science International*, *219*(1–3), 119–128. doi:10.1016/j.forsciint.2011.12.008 PMID:22261554

Carbonaro, C. M., Corpino, R., Salis, M., Mocci, F., Thakkar, S. V., Olla, C., & Ricci, P. C. (2019). On the Emission Properties of Carbon Dots: Reviewing Data and Discussing Models. *C*, *5*(4), 4. Advance online publication. doi:10.3390/c5040060

Chen, H.-S., Meng, H.-H., & Cheng, K.-C. (2002). A survey of methods used for the identification and characterization of inks. *Forensic Science Journal*, *1*(1), 1–14.

Chen, Y.-F. (2011). Forensic Applications of Nanotechnology. *Journal of the Chinese Chemical Society (Taipei)*, *58*(6), 828–835. doi:10.1002/jccs.201190129

Dazzi, C., & Pedrabissi, L. (2009). Graphology and personality: An empirical study on validity of handwriting analysis. *Psychological Reports*, *105*(3), 1255–1268. doi:10.2466/PR0.105.F.1255-1268 PMID:20229925

Díez-Pascual, A. M., Cruz, D. L., & Redondo, A. L. (2022). Advanced Carbon-Based Polymeric Nanocomposites for Forensic Analysis. *Polymers*, *14*(17), 17. Advance online publication. doi:10.3390/polym14173598 PMID:36080673

Dirwono, W., Park, J. S., Agustin-Camacho, M. R., Kim, J., Park, H. M., Lee, Y., & Lee, K. B. (2010). Application of micro-attenuated total reflectance FTIR spectroscopy in the forensic study of questioned documents involving red seal inks. *Forensic Science International*, *199*(1–3), 6–8. doi:10.1016/j.forsciint.2010.02.009 PMID:20413232

Djozan, D., Baheri, T., Karimian, G., & Shahidi, M. (2008). Forensic discrimination of blue ballpoint pen inks based on thin layer chromatography and image analysis. *Forensic Science International*, *179*(2–3), 199–205. doi:10.1016/j.forsciint.2008.05.013 PMID:18639403

Duran, A., Perez-Rodriguez, J. L., Espejo, T., Franquelo, M. L., Castaing, J., & Walter, P. (2009). Characterization of illuminated manuscripts by laboratory-made portable XRD and micro-XRD systems. *Analytical and Bioanalytical Chemistry*, *395*(7), 1997–2004. doi:10.100700216-009-2992-5 PMID:19685044

Erdoğan, A. (2022). Analysis and Chemical Imaging of Blue Inks for the Investigation of Document Forgery by XPS. *Microchemical Journal, 183*, 108062. doi:10.1016/j.microc.2022.108062

Ezcurra, M., Góngora, J. M., Maguregui, I., & Alonso, R. (2010). Analytical methods for dating modern writing instrument inks on paper. *Forensic Science International, 197*(1–3), 1–20. doi:10.1016/j.forsciint.2009.11.013 PMID:20061099

Fang, L., Jenk, T. M., Singer, T., Hou, S., & Schwikowski, M. (2021). Radiocarbon dating of alpine ice cores with the dissolved organic carbon (DOC) fraction. *The Cryosphere, 15*(3), 1537–1550. doi:10.5194/tc-15-1537-2021

Ganash, A., Alshammari, S., & Ganash, E. (2023). Development of a Novel Electrochemical Sensor Based on Gold Nanoparticle-Modified Carbon-Paste Electrode for the Detection of Congo Red Dye. *Molecules (Basel, Switzerland), 28*(1), 1. Advance online publication. doi:10.3390/molecules28010019 PMID:36615215

Geminiani, L., Campione, F. P., Corti, C., Luraschi, M., Motella, S., Recchia, S., & Rampazzi, L. (2022). Differentiating between Natural and Modified Cellulosic Fibres Using ATR-FTIR Spectroscopy. *Heritage, 5*(4), 4. Advance online publication. doi:10.3390/heritage5040213

Gorshkova, K. O., Rossinskaya, E. R., Kirillova, N. P., Fogel, A. A., Kochemirovskaia, S. V., & Kochemirovsky, V. A. (2020). Investigation of the new possibility of mathematical processing of Raman spectra for dating documents. *Science & Justice, 60*(5), 451–465. doi:10.1016/j.scijus.2020.06.007 PMID:32873385

Grim, D. M., Siegel, J., & Allison, J. (2001). *Evaluation of desorption/ionization mass spectrometric*. Academic Press.

Gui, L., Chen, B., Zhou, Z., Liang, Y., He, M., & Hu, B. (2022). Phytic acid functionalized magnetic adsorbents for facile enrichment of trace rare earth elements in environmental water, digested atmospheric particulates and the extracts followed by inductively coupled plasma mass spectrometry detection. *Talanta, 244*, 244123426. doi:10.1016/j.talanta.2022.123426 PMID:35381498

Hui, Y. W., Mahat, N. A., Ismail, D., & Ibrahim, R. K. R. (2019). Laser-induced breakdown spectroscopy (LIBS) for printing ink analysis coupled with principle component analysis (PCA). AIP Conf. Proc. https://doi.org/ doi:10.1063

Huynh, V., Phelps, M. S., Golden, T. D., & Verbeck, G. F. (2016). Direct analyte-probed nanoextraction (DAPNe) coupled to matrix-assisted laser desorption ionization (MALDI) for examination of the ink chemistry on documents. *Forensic Chemistry (Amsterdam, Netherlands), 2*, 86–92. doi:10.1016/j.forc.2016.10.007

Ivanova, I., Spiridonov, I., & Lasheva, V. (2022). Spectroscopic technique for studying the characteristics of paper. Academic Press.

Jones, R. W., Cody, R. B., & Mcclelland, J. F. (2006). Differentiating writing inks using direct analysis in real time mass spectrometry. *Journal of Forensic Sciences, 51*(4), 915–918. doi:10.1111/j.1556-4029.2006.00162.x PMID:16882239

Kapoor, N., Sulke, P., Shukla, R. K., Kakad, R., Pardeshi, P., & Badiye, A. (2021). Forensic analytical approaches to the dating of documents: An overview. *Microchemical Journal, 170*, 170106722. doi:10.1016/j.microc.2021.106722

Kaur, R., Saini, K., & Sood, N. C. (2013). Sequencing the intersections of printed strokes with writing instrument strokes using DocuCentre expert (PIA 6000-EDF). *Science & Justice, 53*(2), 206–211. doi:10.1016/j.scijus.2012.08.005 PMID:23601730

Kher, A., Mulholland, M., Green, E., & Reedy, B. (2006). Forensic classification of ballpoint pen inks using high performance liquid chromatography and infrared spectroscopy with principal components analysis and linear discriminant analysis. *Vibrational Spectroscopy, 40*(2), 270–277. doi:10.1016/j.vibspec.2005.11.002

Khofar, P. N. A., Karim, U. K. A., Elias, E., Safian, M. F., & Halim, M. I. A. (n.d.). Trends of Forensic Analysis of Pen Ink Using Attenuated Total Reflectance Fourier Transform Infrared (ATR- FTIR) Spectroscopy. *Indonesian Journal of Chemistry*.

Koppenhaver, K. M. (2007). *Forensic document examination*. Humana Press Incorporated.

Kriznar, A., Ager, F., Macías, L. R., Ortega Feliu, I., Tubío, B. G., & Respaldiza, M. Á. (2022). Pigments and Inks Applied in Juan Vespucci's Portolan Chart (1520). *Colorants, 1*(4), 4. Advance online publication. doi:10.3390/colorants1040026

Kumar, R., & Sharma, V. (2017). A novel combined approach of diffuse reflectance UV-Vis-NIR spectroscopy and multivariate analysis for non-destructive examination of blue ballpoint pen inks in forensic application. *Spectrochimica Acta. Part A: Molecular and Biomolecular Spectroscopy, 175*, 17567–17575. doi:10.1016/j.saa.2016.12.008 PMID:28024250

LaPorte, G. M. (2015). Chemical analysis for the scientific examination of questioned documents. In Fundam. Appl (pp. 318–353). John Wiley & Sons, Ltd. doi:10.1002/9781118897768.ch8

Leal, T. A., Ferreira, C., Quintas, A., & Bernardo, A. (2019). Dating inks on paper through chromatographic analysis of volatile compounds: A mini-review. *Ann. Med, 51*(supl), 180. . doi:10.1080/07853890.2018.1562750

Menzyk, A., Zadora, G., & Sajewicz, M. (2015). Physicochemical analysis of ink – dating and establishing the sequence of intersecting lines of ink entries. *Z Zagadnien Nauk Sadowych, 104*, 279–302.

Neumann, C., & Margot, P. (2009). New perspectives in the use of ink evidence in forensic science: Part I. Development of a quality assurance process for forensic ink analysis by HPTLC. *Forensic Science International, 185*(1–3), 29–37. doi:10.1016/j.forsciint.2008.11.016 PMID:19147310

Nugroho, D., Keawprom, C., Chanthai, S., Oh, W.-C., & Benchawattananon, R. (2022). Highly Sensitive Fingerprint Detection under UV Light on Non-Porous Surface Using Starch-Powder Based Luminol-Doped Carbon Dots (N-CDs) from Tender Coconut Water as a Green Carbon Source. *Nanomaterials (Basel, Switzerland), 12*(3), 3. Advance online publication. doi:10.3390/nano12030400 PMID:35159745

Ortiz-Masiá, D., Cosín-Roger, J., Calatayud, S., Hernández, C., Alós, R., Hinojosa, J., Esplugues, J. V., & Barrachina, M. D. (2016). M1 Macrophages Activate Notch Signalling in Epithelial Cells: Relevance in Crohn's Disease. *Journal of Crohn's and Colitis, 10*(5), 582–592. doi:10.1093/ecco-jcc/jjw009 PMID:26802079

Pachuta, S. J., & Staral, J. S. (1994). Nondestructive analysis of colorants on paper by time-of-flight secondary ion mass spectrometry. *Analytical Chemistry, 66*(2), 276–284. doi:10.1021/ac00074a015

Paikrao, H. M., Tajane, D. S., Patil, A. S., & Dipale, A. D. (2022). Applications of Nanotechnology in Forensic Science. In Engineered Nanomaterials for Innovative Therapies and Biomedicine. Springer. doi:10.1007/978-3-030-82918-6_11

Pandya, A., & Shukla, R. K. (2018). New perspective of nanotechnology: Role in preventive forensic. *Egyptian Journal of Forensic Sciences, 8*(1), 57. doi:10.118641935-018-0088-0

Prasad, V., Lukose, S., Agarwal, P., & Prasad, L. (2020). Role of Nanomaterials for Forensic Investigation and Latent Fingerprinting-A Review. *Journal of Forensic Sciences, 65*(1), 26–36. doi:10.1111/1556-4029.14172 PMID:31454084

Quinteros, E. D. (2022). *Analysis of Blue Fountain Pen Ink by UV-Vis Spectroscopy, ATR-FTIR Spectroscopy, and Thin Layer Chromatography Enhanced with RGB Spot Evaluation to Construct a Database for Comparison and Identification*. Academic Press.

Raza, A., & Saha, B. (2013). Silver nanoparticles doped agarose disk: Highly sensitive surface- enhanced Raman scattering substrate for in situ analysis of ink dyes. *Forensic Science International, 233*(1–3), 21–27. doi:10.1016/j.forsciint.2013.08.004 PMID:24314497

Rohde, E., Mcmanus, A. C., Vogt, C., & Heineman, W. (1997). Separation and comparison of fountain pen inks by capillary zone electrophoresis. *Journal of Forensic Sciences, 42*(6), 14252J. doi:10.1520/JFS14252J

Sanjuan-Navarro, L., Moliner-Martínez, Y., & Campíns-Falcó, P. (2022). The state of art of nanocarbon black as analyte in a variety of matrices: A review. *Trends in Analytical Chemistry, 157*, 116769. doi:10.1016/j.trac.2022.116769

Spagnolo, G. S. (2006). Potentiality of 3D laser profilometry to determine the sequence of homogenous crossing lines on questioned documents. *Forensic Science International, 164*(2–3), 102–109. doi:10.1016/j.forsciint.2005.12.004 PMID:16431054

Sun, Q., Luo, Y., Zhang, Q., Yang, X., & Xu, C. (2016). How much can a forensic laboratory do to discriminate questioned ink entries? *Journal of Forensic Sciences, 61*(4), 1116–1121. doi:10.1111/1556-4029.13067 PMID:27364295

Zięba-Palus, J., & Trzcińska, B. M. (2011). Establishing of chemical composition of printing ink. *Journal of Forensic Sciences, 56*(3), 819–821. doi:10.1111/j.1556-4029.2011.01734.x PMID:21361954

Zlotnick, J., & Smith, F. (1999). *Chromatographic and electrophoretic approaches in ink analysis*. Academic Press.

Chapter 13
Role of Quantum Dots in Volatile Poison Detection

Shumaila Kiran
Government College University, Faisalabad, Pakistan

Saba Naz
Government College University, Faisalabad, Pakistan

Muniba Rahmat
Jiangsu University, Zhenjiang, China

Tahir Farooq
Government College University, Faisalabad, Pakistan

ABSTRACT

Volatile poisons are volatile organic carbon-based compounds that easily release from indoor surroundings (for example paints, furnishings, building materials) or industry productions because of their relatively low boiling points. The detection of volatile toxins can be done in a variety of ways. Various kinds of sensors have been created to analyze volatile organic toxins. Volatile poison sensors are generally easy to use, inexpensive, programmable, and portable. Future developments will include innovations or combinations of advanced technologies for volatile poisons sampling, detection, and analysis, as well as validation and standardization of these methods for practical clinical use. This chapter will provide an overview and critical analysis of the sampling, collecting, detection, and analytical techniques linked to volatile substances, with a focus on current advancements in nanomaterials such quantum dot-based sensors.

INTRODUCTION

Analytical environmental control needs to develop appropriate sensors which can make reliable and fast detection as well as measurement of environmentally important chemical compounds. From a broad range of pollutants, the control of high-vapor-pressure toxins in term of the volatile organic species is

DOI: 10.4018/978-1-6684-8325-1.ch013

very essential in work-related protection and health control (for example) in atmospheres of workplace) as they can generate the harmful effects on health of human beings.

Due to their desirable characteristics, such as strong visible-band absorption, superior mechanical, and high electrical conductivity features, developing nanomaterials can now be employed to customize optical and electrochemical (bio)sensors. In fact, a wide range of nanomaterials are used, including spheres and particles [metallic nanoparticles (NPs), quantum dots (QDs), and magnetic beads], nanorods, nanotubes, nanofibers, and nanowires, as well as nanocomposites made of nanopolymers, nanofilms, and nanoplates (Arduini et al., 2020; Wang et al., 2023). In this light, it is noteworthy that the configuration of nanomaterials and biomaterials plays a significant role in the design of hybrid nanostructured instruments to achieve "greater operational/storage stability, resistance, response time, ease of sampling, low sample volume and better selectivity with increased detection performance" (Arduini et al., 2016; Munteanu et al., 2023).

Volatile Poisons

The analytes known as volatile poisons are those whose chemical structures largely contain unstable carbon atoms that are unstable at room temperature. A volatile organic compound (VOC) is defined as "any organic compound possessing a 0.01 kPa or more vapour pressure at 293.15 K, or a resulting volatility under specified conditions of usage" in the EU Regulation on Emission Limitations for Volatile Organic Compounds. Since it has a low boiling point and a high vapour pressure, the organic component will evaporate at 20 °C in this case (room temperature). This feature allows the easy, quick, and imperceptible transmission of volatile poisons to humans through the air (Cumeras et al., 2019; Montero-Montoya et al., 2018).

Volatile poisons are the volatile organic carbon based compounds which easily release from indoor surroundings (for example paints, furnishings, building materials) or industry productions that are because of their relatively low boiling points. These compounds are discharged as gases from different liquids or solids. Organic molecules including alcohols, carbonyls, aromatics, alkanes, alkanes, esters, amides, and ethers can be released when you light a fire, cook, paint a house, drive a car, mow a lawn, use building materials, use pesticides, or simply breathe too much. Nearly all of a person's daily activities result in the release of biological materials into the environment. Anthropogenic pollutants, such as hazardous gases and volatile organic compounds (VOCs), have been released since the Industrial Revolution through a variety of processes, including the petroleum industry (Han et al., 2023). Exhaust gases, water separation technologies, industrial wastewater, batch operations, petroleum refining, paint manufacturing, and petrochemical processing are a few examples of specific sources and processes that frequently produce considerable volumes of volatile pollutants. Thus, rising VOC emissions and their impact on air quality are now seen as a major environmental problem (Yang et al., 2023).

Concentrations of numerous organic toxic species are consistently higher indoors than outdoors (up to 10 times greater) and the air in workplaces, houses, schools, and other business buildings is thought to contain between 50 and 300 distinct VOCs at any given moment. Volatile organic species are emitted by thousands of products, comprising solvent-based paints, printing inks, many consumer products, petroleum products and organic solvents, pesticides, cleaning products, and furniture and construction. Thus, VOC levels are employed as parameters to measure indoor air quality in addition to other indications like temperature and CO_2 levels (Almaie et al., 2022).

Volatile poisons comprise a wide range of chemicals; many of them may have temporary and permanent hazardous effects on environment and health of human beings, counting the potential cause of cancer. Furthermore, many exhaled volatile species are found as efficient biomarkers which can be employed to detect some diseases, comprising lung cancer (Yang et al., 2022; Timbrell & Barile, 2023). Therefore, there is a huge request on the selective and sensitive detection of volatile poisonous species in the gas phase for process control, medical diagnostics purposes and environmental monitoring.

Impact of Volatile Poisons on the Environment

Volatile poisonous compounds show a significant role in the production of ozone and fine particulates in the atmosphere. Volatile toxic species under sunlight reaction with NOx produced primarily from industrial activities, power plants and vehicles to make ozone, that in result support the development of fine particulates. Accumulation of fine particles, ozone and other gaseous pollutants causes smog that reduces visibility (Lu et al., 2022; Tsai, 2023). Smog obstructs visibility and negatively impacts tourism.

The second most abundant greenhouse gas, methane, in the United States, released primarily in combination with volatile toxins. Greenhouse gases belong to certain chemical species that increase global temperatures by trapping heat in the atmosphere. The Environmental Protection Agency (EPA) conducted researches, according to which, the industries are the cause of 22% of all greenhouse gases emitted in the atmosphere (Khademi et al., 2022; Severino et al., 2023).

The large amount of acidic species in the atmosphere consequences into the acid rain. The pH value of water can be decreased to 4.2 or 4.4, when it counters with sulfur dioxide (SO_2) and nitrogen oxides (NO_X). Acidic precipitation is then distributed over natural habitats, oceans and urban landscapes, increasing the pH of delicate ecological units. Naturally, fish cannot survive in highly acidic environments that can disrupt the entire food chain. Harsh acidic molecules can strip away important nutrients from forest soils, and aluminum deposits can contaminate neighboring vegetation (Huang et al., 2022; Sharma et al., 2023).

The National Institute of Occupational Safety and Health (NIOSH), the European Occupational Safety and Health Agency (EU-OSHA), and other significant environmental protection authorities have highlighted minimal adverse human impacts as potential hazards of hazardous volatiles. Even at (sub-ppb) values, guidelines have been devised to restrict human exposure to volatile organic chemicals indoors and at work (Epping et al., 2023).

The Impact of Volatile Poisons on Human Health

The volatile poisons when absorbed or inhaled via skin can be the reason of severe health problems in human beings. Aerosols, solvents and exhaust streams employed in unaired areas can produce hazardous pollutants. Humans can experience dizziness, headaches, visual impairment and memory loss, if they exposed to large amounts (Rouf et al., 2022). Compounds in paint stripper, such as methylene chloride, have cancer-causing properties that can cause potentially fatal conditions in the long term. The reaction of volatile compounds with NOx as well as other detrimental molecules to form greatly poisonous particulates. On inhalation, such particulates can cause of bronchitis, emphysema and some other severe respiratory disorders. Smog can cause irritation to our nose, throat and eyes or can intensify the already present heart and respiratory issues like asthma. The people having lung and heart disorders, the elderly and children whose respiratory organs are under-developing are most at danger. Exposed to long-time

Figure 1. Impacts of volatile poisons on environment and human health

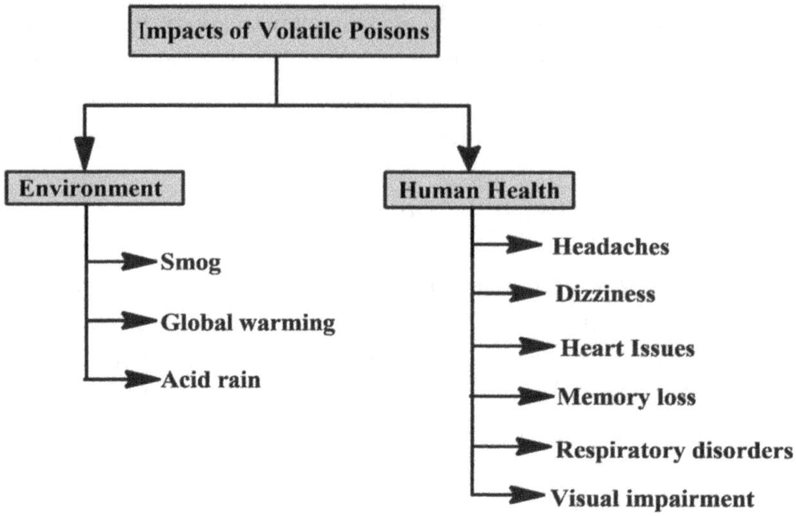

intense smog situation may cause permanent destruction of lung tissue and disturb our immunity (Wang et al., 2022; Ataei et al., 2023).

Volatile substances can have both immediate and long-term consequences on both human health and natural ecosystems. Several volatile substances have been shown to be hazardous or cancer-causing in nature. One well-known example of a carcinogen molecule with significant potential to affect human organs and systems, including the immunological, circulatory, neurological, reproductive, cardiovascular, and respiratory systems, is benzene (Nikhar et al., 2021). Another illustration is Sick Building Syndrome (SBS), a serious health issue in which building occupants endure persistent health issues like coughing, nausea, headaches and eye irritation. Temperature, odour, volatiles, and particulate matter are factors connected to SBS, however no single cause could be found (Mosallaei et al., 2023).

DETECTION OF VOLATILE POISON COMPOUNDS

Therefore, for safety reasons, it is very important to have a device capable of accurately and continuously monitoring the concentration of volatile poisons in gaseous media.

Methods for the Volatile Poison Detection

Any method for the examination of toxic volatile compounds consists of numerous steps:

1. Sampling (for example air traps and collection);
2. Enrichment of the sample;
3. Distribution of volatile compounds from an aerial specimen to the analytical instrument;

4. Detection of every compound.

There are two sampling techniques. Active sampling, which actively draws air through an adsorption device, and passive sampling, which uses the diffusion process of the sample due to a concentration gradient (Khoubnasabjafari et al., 2022). The asorbent is employed to adsorb volatile poisons. Primarily, concentration of the sample provides the sensitivity and selectivity required for analytical instruments. The most preferred method of enriching volatile compounds is the use of solid adsorbents. There are three main types: solid adsorbents, inorganic adsorbents, porous carbon-based materials, and organic polymers.

Volatile poisons can be tested through carbon-based sorbents such as molten carbon filters, graphitic carbon black, activated charcoal and permeable polymers such as phenyl-phenylene oxide polymers and styrene polymers (Gregis et al., 2018). The main methods to separate the sample from the sorbent are solvent extraction and thermal desorption. Solvent extraction is the process of separating compounds according to their different solubility's in organic solvents and water. In thermal desorption, heat is used to increase the volatility of a compound under test and separate it from an adsorbent that is delivered to an analytical device (Chew et al., 2022).

There are several methods used to detect volatile poisons. Proton transfer reaction mass spectrometry (PTR-MS), solid phase micro-extraction (SPME) with accompanying gas chromatography mass spectrometry (SPME-GCMS), nanomaterials sensors, loaded quartz crystal additional variable and polymer-coated contact acoustic wave detectors and ion flow tube mass spectrometry (Sarswat & Free 2014; Nath et al., 2023).

Solid-Phase Micro-Extraction (SPME)

It is a fast, sensitive, simple, and adaptable trial training method which reduces solvent use though assimilating specimen and trial provision stages. It is a broadly employed method for the sample storage and pre-concentration. This storing scheme comprises of a top part, SPME fibers and a sealing part. The fibers are covered with an insulating stage (a solid or liquid for example, Carboxen/PDMS that can separate volatile poisons from sample). For pre-concentration, the collected sample or air must be transported from the selection carrier to a glass vessel, after that SPME fiber is placed in the vial and exposed for a period of time in the gas sample (Figure 2). Stored samples can be examined later by techniques like GC-MS deprived of damage of volatiles. Hence SPME is appropriate for medical uses (Ma et al., 2014). SPME is an excellent alternative to other traditional sample preparation methods due to its fast, simple, sensitive, inexpensive, solvent-free, and easy-to-automate properties. The efficiency of analyte pre-concentration through SPME depends on the volatility and structure, the physicochemical properties of the absorbent fiber, and the analyte/fiber affinity (Piñeiro-García et al., 2020; Aati et al., 2023).

GC or MS-Based Techniques

Mostly such techniques are greatly consistent like GC–MS, and have been broadly employed to detect and analyze volatile poisons. Such systems are usually well-suited with pre-concentration technique, for example SPME, for improved sensitivity. But a lot of of them require a skilled operator and are expensive (Queralto et al., 2014; Sorolla-Rosario et al., 2023).

Figure 2. SPME storage device

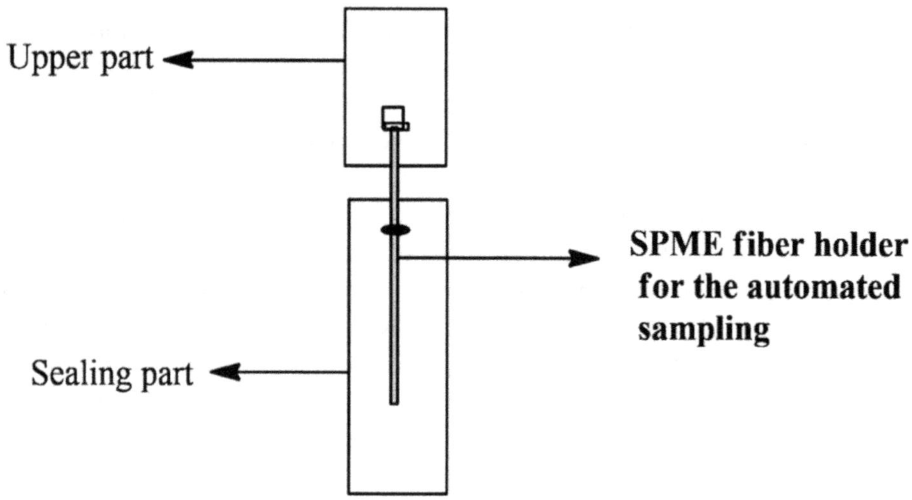

Gas Chromatography-Mass Spectrometry (GC–MS)

GC/MS is the coupling of GC and MS techniques employed to detect various species in test samples. It can be employed to detect unknown volatile toxicity in complex gas combinations and is now renowned as the typical in volatile compound analysis. GC/MS technology integrates the skills and strengths of both analytical methodologies (GC and MS) (Feng et al., 2022). Components are separated from a gas mixture in a gas chromatograph on the basis of the retention time of each component that makes up the chromatogram. Upon entering the mass spectrometer, these components are captured, ionized and identified. Each peak in the chromatogram is therefore resolved into mass spectral component results based on mass-to-charge ratio (Figure 3). This method combines high-resolution separation of components with highly sensitive and selective detection, allowing high-level identification of unknown gases that cannot be recognized by GC or MS alone (Kaikiti et al., 2022). The potential applications of GC/MS techniques to monitor volatile poisons have been explored in several articles.

In the GC–MS techniques, mostly employed the mass spectrometer is the quadrupole mass spectrometer. Additional techniques, like tandem quadrupole MS (MS–MS) and time of flight MS (TOF–MS) have also been employed (Murtada et al., 2022; Yue et al., 2022).

The system has great compassion in ppb array and can produce repeatable outcomes. When the compound is identified, the quantification of volatile compounds is conceivable as well. Though, the system is expensive, slow, and presently immovable, and the samples often require pre-concentration and dehydration. For GC–MS, the real-time measurement is not promising. Consequently, this system is not appropriate for point-of-care usage (Trefz et al., 2013; Lona-Ramirez et al., 2016).

Figure 3. Schematic diagram of GC-MS

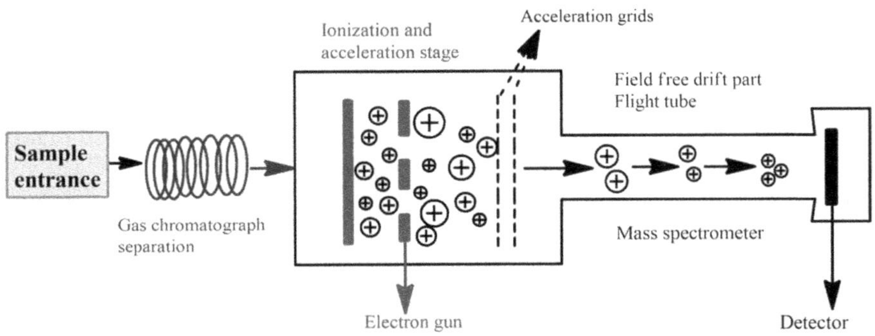

Proton Transfer Reaction-Mass Spectrometry (PTR–MS)

PTR-MS uses the chemical ionization process on the basis of proton-transfer reactions; Hydronium ion (H_3O^+) is employed as reagent ion. Air or sample is continuously directed into the PTR-MS reaction chamber to encounter reagent ions.

Volatile organic poisons comprising unsaturated bonds or a polar functional group have proton attractions greater than water molecule and then reaction will occur with H_3O^+ in a proton transfer reaction where the transfer of proton is occurred between H_3O^+ and the volatile organic poisons (Yu et al., 2020). This reaction can be explained with formula given below:

$$H_3O^+ + V \rightarrow V \cdot H^+ + H_2O$$

Where V is volatile molecule of the sample or analyte.

In this process, reactive ions (H_3O^+) are generated through the hollow cathode from water vapor (Figure 4). The ion basis is associated with the drift tube. A proton trace reaction occurs in the drift tube between H_3O^+ and molecules having greater proton attraction than water. Hence, components of humid air, specifically Water, Nitrogen, Oxygen, and Carbon dioxide will not affect the analysis. The particles will then arrive at the detector (King et al., 2012). PTR–MS has great up to parts per trillion sensitive and great specificity. It is an effective tool to monitor volatile compounds and does not require pre-concentration of the sample. The steps involved in sample pre-concentration are generally time-taking and may affect the sample due to gas adsorption and desorption depending on the nature of the adsorption medium (Righettoni et al., 2013; Zhang et al., 2022).

PTR-MS delivers online, real-time quantification of volatile compounds in indoor air. The device is constructed with a quick response time (approximately 1 second). The device is set up to allow on-line determination of trace elements with concentrations as low as pptv (little parts per volume). This technique makes it happen to quantify volatile compounds that have great proton affinity like acetaldehyde, methanol, acetone, ethanol, xylene, benzene, toluene, Me tert-Bu ether (MTBE) exist in ambient air (Zhao et al., 2020; Sekimoto et al., 2017).

As compare to GC–MS technique, PTR–MS does not require pre-concentration of sample and it is somewhat more sensitive; therefore, its quantitative results are more reliable. Because PTR-MS is time-saving and easy to operate, the sum of themes examined by PTR-MS is informed to be significantly greater as compare to GC-MS, making PTR-MS valuable and attractive for further evaluation, scientific and cancer analysis. As PTR–MS could not detect the volatile compounds exactly, so it could not deliver as much detail as GC–MS. Hence, Bajtarevic et al. (2009) recommended that PTR–MS and SPME–GC–MS balance one another (Bajtarevic et al., 2009).

Merits of PTR-MS technique:

- Real-time quantification
- Real-time measurements
- Easy to handle
- Fast process
- Compact and robust setup.
- Pre-concentration of sample is not required.
- Soft ionization delivers low fragmentation as well as increases the ability of detection.

Demerits of PTR-MS technique:

- All molecules are not capable to detect (Mayhew et al., 2020).

The commercial IONICON PTR-MS has the following considerations:

- Mass series 1–300 amu.
- Resolution <1 amu.
- Response interval 100 ms.
- Recognition confines 300 pptv.
- Linearity range 10 ppmv–300 pptv.

Ion Mobility Spectrometry (IMS)

The ability to identify only VOCs with the help of ion mobility spectrometers makes them special. Other analytical methods cannot do this. IMS is one of the most popular technologies available today because to its simplicity, high sensitivity, analytical versatility, and real-time monitoring capabilities. Moreover, it can be used in other configurations, such as Differential IMS (DIMS) and Field Asymmetric IMS (FAIMS), or in conjunction with other methodologies, such Gas Chromatography or Mass Spectrometry (GC). In this case, the pair combines the greatest features of each technology into a single gadget to produce a new technology that is highly sensitive and selective. Modern devices have the remarkable capacity to detect and measure VOCs in concentrations ranging from parts per billion (ppbv, 109) to parts per trillion (pptv, 1012). IMS is one of the best analyzers because it can characterise VOCs according to their size, with surprising results on the basis of their weight and shape, making it one of the most suitable techniques for air quality assessment and control even though it can only detect and quantify VOCs and is entirely independent of other types of analyzers (Moura et al., 2023; Capitan & Weller, 2021).

Figure 4. Schematic diagram of PTR-MS

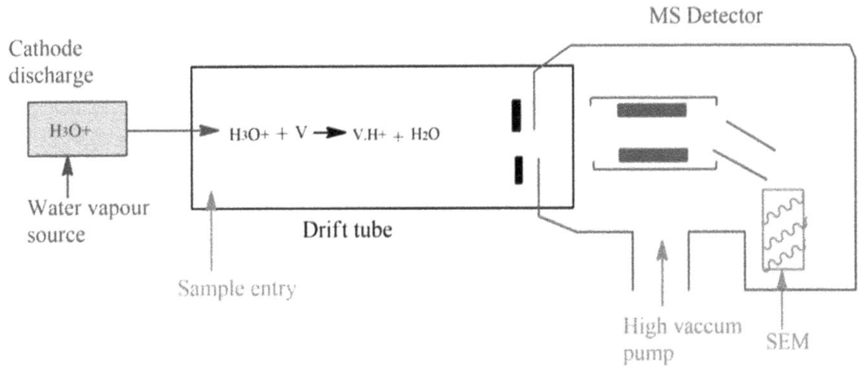

In the IMS process, the ionization of gaseous metabolites takes place at atmospheric pressure, and the resulting molecules are separated in short pulses (approximately 10 to 100 ms) of several centimeters in length in a drift tube. The separation of ions is the consequence of change in their charges and shapes, and speed is affected by both the drift gas and electrical arena. A much smaller amount of current (nA to pA) generated by the Faraday plate constitutes the ion runtime spectrum (Chen et al., 2022). It is particularly beneficial for isomer separation and has a fairly large sensitivity in the threshold of ppm. But with IMS, the identification of compound is not promising and it's not appropriate for actual dimensions as well. To get further detail around the volatile toxins, IMS is mostly combined with GC or MS. The coupling of IMS and gas chromatography columns promises the pre-separation of gaseous metabolites before entering the drift tube (Li et al., 2022; Feng et al., 2022). Unluckily IMS does not recognize unknown species of gas. However, the main benefits of IMS are that it can detect much lower concentrations of volatiles (ranging from ng/L to pg/L and ppmv to pptv) without pre-concentration treatment and reduces the time required for screening. Full sample scan in less than 50 ms, 500 seconds (Wu et al., 2022).

On comparison with GC–MS, IMS systems are cheaper, mobile and sample preparation is not required (Figure 5). The latest developments from IMS make it possible to analyze air or samples quickly, robustly and reliably. In 2015 Brodrick et al. combined a multi-capillary column (MCC) and an ion mobility

Figure 5. Schematic diagram of IMS

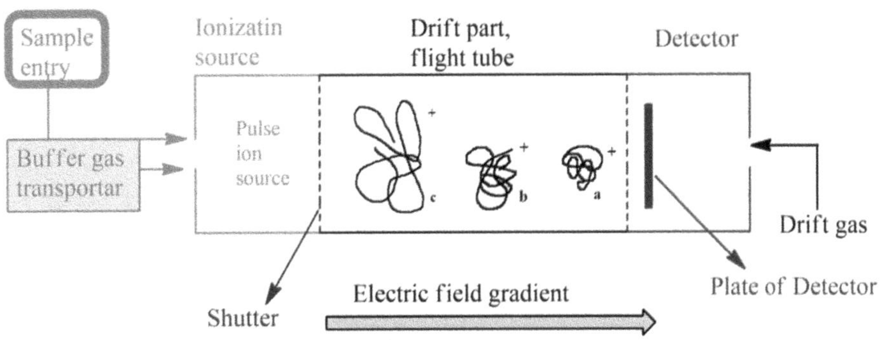

spectrometer, a pre-separation procedure was established and effectively used for sample analysis. This MCC-IMS methodology has been reported accurately, quickly, inexpensively, and can help standardize sample analysis (Brodrick et al., 2015).

Selected Ion Flow Tube Mass Spectrometry (SIFT–MS)

The application of high sensitive SIFT-MS technique can be of advantage in the fast identification of low amounts of several volatile poisons. It is a mass spectrometric quantitative technique which utilizes chemical ionization with carefully selected precursor cations that react with small reaction times with trace volatile gases in air or sample directed into a helium stream in a flow tube (Figure 6). The great grouping of the fast flow tube method, chemical ionization and mass spectrometry permits simultaneous, precise quantitative testing of many trace gases in air or samples up to ppb concentration ranges on an hourly basis (Perkins & Langford, 2022). The facility available for the selection of precursor ions to analyze the specific media explains the flexibility of this technique and reveals its benefit over the other approaches. The comparable PTR-MS system is less versatile because it employs single H_3O^+ precursor ions (Zogka et al., 2022).

SIFT–MS can attain real-time quantification and measurement of trace volatile compounds in moist air. The method is actually established on the chemical ionization of volatiles through the O_2^+ H_3O^+ and NO^+ precursor ions with fast flow tube, quadrupole-mediated detection and quantification, in the complicate mixtures of gases irrespective to content of water vapor in real time (Smith & Spanel, 2022). The moist atmospheric air produced the precursor ions (Figure 5).

By using MS techniques, selective precursor ions are first isolated, after which they are interacted with molecules in the sample or air. A second MS can be used to identify and count precursor and product ions, after which volatile chemical concentrations can be determined.The technology is mobile, fast, and has an incredible level of sensitivity in the ppb range, so it can be used for online testing (Perkins et al., 2021; Roberts et al., 2022).

Figure 6. Schematic diagram of SIFT–MS

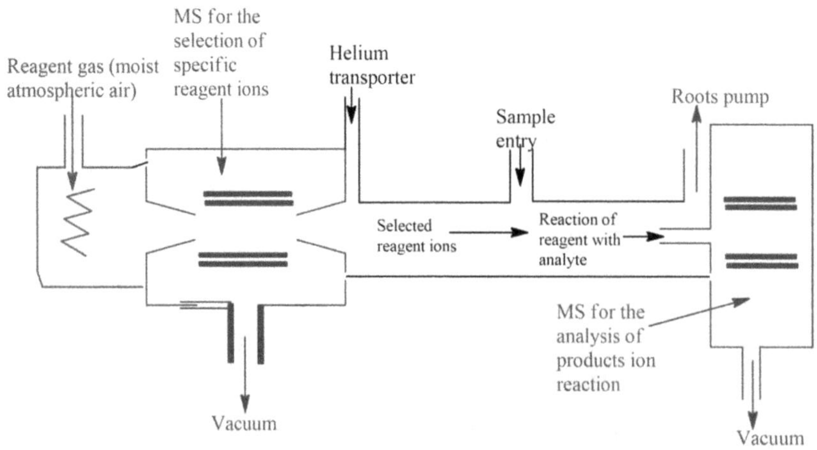

As this technique permits the real-time identification and quantification of volatile compounds in air/sample without any preliminary treatment, Kumar et al. (2013) employed SIFT–MS to examine 17 volatile organic compounds and observed that the concentrations of some specific volatile organic compounds (hexanoic acid, phenol, methylphenol and ethylphenol) were considerably different among persons with esophagogastric cancer and positive control groups. Real-time measurements without sample preparation offer certain advantages to SIFT-MS in clinical situations due to negligible delay and minimal concerns about sample dissociation. However, this technique is expensive and not suitable for chemical detection and broad profiling of volatile compounds. Therefore, other chemical analysis platforms are needed to detect volatile compounds (Kumar et al., 2013).

Gas Chromatography-Flame Ionization Detection (GC–FID)

GC coupled with FID is also a broadly employed system to analyze volatile toxicants. FID is primarily used to identify ions formed during the combustion of compounds. This methodology can identify volatile toxicity with a linear response and is relatively cost-effective and easy to use (Bi et al., 2021). Though, this technique has been informed as a mostly used analytical process to detect volatile poisons in gases. For FID, it is difficult to recognize inorganic compounds and the measurements generally need calibration curves and internal standards. Moreover, such technique often has demerits to detect at a level of ca.10 ppb (Zaric et al., 2014; Wilde et al., 2019; Masite et al., 2022).

Sensor-Based Techniques

Various kinds of sensors have been created to analyze the volatile organic toxins. Volatile poison sensors are generally easy to use, inexpensive, programmable and portable. You can access data in real time with great precision. Therefore, numerous sensor-based volatile poison identification systems have high potential for clinical field use (Nath et al., 2023).

VOC sensors are programmed to recognise minute alterations in a particular gas and transmit the information to a centralised monitoring system. These detectors are used by plant staff to track and stop unwanted emissions spikes. Technicians can calibrate each instrument to detect a particular collection of VOCs or chemicals depending on the type of sensor being utilised (Li et al., 2023).

Custom installation procedures are frequently created for each facility due to the various industrial plants' architecture and the distinct VOCs that each industry emits. Typically, specialised personnel will install sensors in various parts of the facility and link them together with fibre optic cables or wireless networks. Detectors are placed close to leaks, pipe connections, and vent valves to improve results (Hegab et al., 2023).

Nanomaterial-Based Chemoresistive Sensors

Basically the chemoresistive sensors are functioned on the variations of electrical conductivity caused by the change of neighboring air. The identification is primarily determined by the reactions of volatile poisons with groups on the surface (Hung et al., 2021), therefore, the sensing sensitivity determined by the surface-to-volume ratio and the surface area of sensing materials (Steinhauer, 2021). In the last decades, a lot of struggles have been made to enhance the surface-to-volume ratio and to adjust the geometry of the sensing particles, numerous chemoresistive volatile compounds sensors based on the nanomaterials have been manufactured (Zilberman et al., 2011; Amal et al., 2015). Such sensors are generally designed by topping a conductive nanomaterial, like carbon nanotubes and Au/Pt nanoparticles, with organic functional groups (Xu et al., 2013; Kaur Billing, 2021). During the measurement, volatile compounds connect and react designed organic functional groups, altering the connection between conductive inorganic nanomaterials or in some cases causing charge transfer among inorganic nanomaterials and organic functional groups. Such changes will responsible for the alterations in measured conductance (Andreb et al.- 2021). For instance, Peng et al. in 2009, formed sensors array through the designed Au nanoparticles. This array of sensor can analyze sample without its pre-concentration, and differentiate lung cancer patients and healthy subjects. The outcome exposed that Au nanoparticle volatile compound sensors could provide a portable, simple, noninvasive and cost-effective, screening and analysis technique for lung cancer (Peng et al., 2009).

Polymer composite sensors are another type of nanomaterial-based sensor that can be made from a non-conductive polymer layer by adding a conductive material (i.e. carbon black). Specific absorption of volatile compounds can alter the volume of polymer composite and thus change the resistance between electric connections (Queralto et al., 2014). The chemoresistive sensors based on nanomaterial generally show a great sensitivity with fast response, and pre-concentration of the air/sample is not required. However, such sensors are mostly sensitive to temperature or moistness (Hung et al. 2021).

Piezoelectric Sensors

Generally, the piezoelectric sensors operated in reply to applied mechanical stress. The Quartz crystal microbalance (QCM) can be the broadly employed piezoelectric sensors in volatiles examination (Hekiem et al., 2021). In QCM, the quartz surface can be covered with a suitable molecular recognition membrane. So, when the volatile complexes are absorbed on surface, the mass difference changes the basic vibrational frequency of the quartz crystal resonator. Surface acoustic wave (SAW) sensors are another well-recognized kind of sensor. In this sensor, the acoustic waves spread to the elastic substrate's surface and their amplitude diverges exponentially with depth of the substrate. Different specific materials can be used to cover the surface. The adsorption and desorption of the volatiles (in sample) from the layered film can consequence into an alteration in its electrical conductivity (electric field of SAW, linked with acoustic field) and mass of chemical interface. Such changes can affect the phase velocity and amplitude of SAW sensor (Kus et al., 2021).

Speller et al. (2015) developed an idea of employing a QCM based virtual sensor array (VSA) to distinguish a broad variety of volatile compounds. Generally, sensor arrays need numerous sensor materials that show several binding attractions for various volatile toxic compounds. Though, instead of employing chemical attraction, Speller et al. employed the several characteristics of the material, like film thickness and viscoelasticity, as discriminating parameters. A sensor can simulate VSA in this way, and deliver many responses per sample. The sensor is fabricated by leaving a thin layer of ionic liquid on surface of the QCM-D transducer. This is why ionic liquids are greatly tunable, exhibit viscoelastic properties and can arrest the organic fumes reversibly. On the exposure of sensor to several volatile compounds, the variations in frequencies (Δf) were calculated at different harmonics. The technique permitted the volatile compounds to be categorized with about 100% accurateness. The outcomes recommended the VSA scheme and the QCM-D sensor have potential in the identification of volatile compounds (Speller et al., 2015).

The Piezoelectric sensors show great sensitivity in ppt range and they are designed to accurately measure particular volatile organic poisonous species. The sensor's selectivity can be controlled as the resonators can be designed with several layering materials (Kimura et al., 2011; Li et al., 2021). Though, piezoelectric sensors are generally sensitive toward temperature, humidity and vibration, that can influence on resonant frequency of sensor; thus such factors must be accurately controlled for the sample (or breath) testing in order to reduce their influence throughout the samples exposure (Sun et al., 2016).

Nano-Sensors Are Better for the Detection of Volatile Toxins

Researchers have created a variety of volatile organic compounds monitoring techniques, including gas chromatography (GC), mass spectrometry, and high-performance liquid chromatography (HPLC) (MS). Nevertheless, these technologies require pricey hardware, are not portable, and use a lot of power (Bindra et al., 2018). These machines take a lot of time and need operators with extreme specialisation. Today's major issue, the real-time detection of dangerous volatile chemicals, cannot be solved with these technologies. As a result, there is a big market for fast, portable, cheap, and reliable VOC detection equipment (Bezzon et al., 2019). Many VOC sensors have been designed by scientists using various sensing technologies, including resistive, capacitive, optical, and electrochemical (Moseley et al., 2017). Many nanostructured materials have been generated through extensive study that are appropriate for analytical chemistry for VOC detection, where novel methodologies and procedures are being created to satisfy end-user needs for real-time detection. Metal oxide sensors have a straightforward design that is economical, safe, and ecologically benign. The small metal oxide sensor is energy-efficient and quickly produces stable results that are appropriate for green analytical chemistry. Because they are simple to synthesise and provide an adequate response to a variety of volatiles with high recovery and response characteristics, numerous researchers have used various metal oxides (including p-type and n-type) such as SnO_2, WO_3, TiO_2, ZnO, Fe_2O_3, Cr_3O_4, etc. as sensing materials (Hazra et al., 2015). Metal oxides are also better suited for real-time detection of dangerous and poisonous VOCs because to their mechanical stability, environmental protection, and low cost. Because of this, the majority of current sensors are based on semiconducting metal oxides (Xiao et al., 2018; Gakhar et al., 2022).

Mostly, this type of sensors requires complicate and time taking sample pre-treatment. As compare to the conventional analytical technologies, the gas sensors are simple and cost effective devices for the identification and quantification of the volatile poisons present in the environment. The gas sensors are made up by the coupling of active sensing materials with a signal transducer. The sensing material establishment and selection can be executed in simple, less-cost and efficient moveable instrument for the real-time and on-site analysis play a significant part in the designing of high-performance gas sensors. Moreover, the main parameter in the development of the prototype is the measured signal. Thus, photo-luminescent gas sensor methods to identify the organic molecules have showed their superior performance in gas analysis (Songh et al., 2019; Irvani and Verma, 2020). In this perspective, developments in the nanotechnology have permitted to established the unique and upgraded optoelectronic characteristics in nanomaterials.This approach enables the replacement of traditional organic fluorophores and phosphors with other useful luminescent nanomaterials like quantum dots (QDs), which due to their unusual properties have recently been used to develop luminescent bioanalytical methods. This ability has been demonstrated (Ganesan and Nagaraj, 2020; Reji et al., 2023).

Quantum Dots (QDs) as Chemosensor in the Detection of Volatile Poisons

Quantum dots (QDs) are inorganic semiconducting nanoparticles with sizes between 2 and 10 nm that exhibit unusual optical properties because of the quantum confinement effect that occurs when the particles have a small size than excitonic Bohr radius (Zhang et al., 2022). This is generally transformed into great luminescence quantums, slight photoluminescence emission, less photo-bleaching and broad absorption spectra (Reshma and Mohanan, 2019). The mostly common volatile toxins are acetone, toluene, formaldehyde, methanol, ethanol, hydrocarbons, halocarbons and their sources comprise solvent abuse and glue sniffing. The immunoassay approaches are mainly used for the tests of volatile compounds, the Fujiwara test for the trichloro molecules and dichromate test for aldehydes, ketones and ethanol; such compounds are usually estimated through Headspace GC. Chromotropic acid test can be used for the detection of formaldehyde. Moreover, there have been numerous research studies on application of different nano-sensors to detect formaldehyde (Rahman, 2020; Ghadiri et al., 2020; Zhou et al., 2022). QDs have also been employed to detect the volatile organic poisons. A fiber optic probe covered with a layer of CdTe QDs was used to estimate volatiles such as acetone, ethanol and 2-propanol. By means of an electron transfer between the QDs' surface and the volatiles, the PL of the CdTe QD layer was squelched in the presence of organic vapours. In decreasing order, ethanol, acetone, and 2-propanol were found to have a significant impact on the PL intensity of QDs when volatiles were present. Ammonia gas sensing is performed by a gas sensor composed of a continuous array of TiO_2 nanotubes and PbS-QD ion layers with an LOD of 2 ppm (Liua et al., 2016).

Generally, the acetone is obtained from starvation, uncontrolled diabetics, intoxicant brews and 2-propanol toxication. Remarkably, the acetone detection in blood will be beneficial to deal with the coma cases. Mn^{2+}-doped ZnS QDs capped with L-cysteine (CSH) can be effectively used to detect ketones, like acetone having LOD of 0.2 mgL^{-1} (Sotelo-Gonzalez et al., 2012). On the ketones addition, the FL of QDs was considerably quenched. Barroso et al. (2018) established a bio-analytical process to selectively detect methanol in alcoholic beverages via photoelectrochemical (PEC) and FL spectroscopy analysis with L-cysteine (CSH)-capped CdS Quantum dots (Figure 7) (Barroso et al., 2018). The *in situ* production of CdS QDs in existence of CSH, an excellent capping and stabilizing agent. The process is tracked by its quantification with screen-printed carbon electrodes linked with Osmium-PVP polymer that assisted the electric conduction in QDs and electrode. The CdS QDs were employed to produce the photocurrent through the photo-catalyzing the oxidation of thioglycerol (reducing agent). The photocurrent measurement was done as the signal read by PEC. The working principle of the sensor is that analyte-methanol go through the oxidation in the existence of alcohol oxidase (AOx), to make formaldehyde and hydrogen peroxide. As a result, the peroxide oxidizes the L-cysteine (CSH) and convert it into its dimeric structure cystine (CSSC). Removal of the two active –SH groups on the conversion from the monomer (CSH) to cystine (CSSC) having inactive –S-S– group. In consequence, destabilization of QDs' growth occurred which ultimately effects the PEC and the FL intensity (Ganesan & Nagaraaj, 2020).

Because fluorescent quantum dots have superior optical and electrical qualities compared to fluorescent organic dyes, chemical sensors based on them have gained a lot of attention. A CdSe QD polymer-based luminous chemical sensor based on an array of green- or red-emitting CdSe QDs in polycaprolactone as a polymer host matrix was published by Aznar-Gadea et al. (2022). They subjected the two sensors to explosive labels, explosive chemicals, and solvent vapours in order to test the nanocomposites' sensing capacities. Both nanocomposites display extremely quick response speeds of under 30 seconds. For 3-nitrotoluene, 4-nitrotoluene, 2,3-dimethyl-2,3-dinitrobutane, and picric acid, the sensor's detection

Figure 7. FL spectroscopic and PEC analysis of methanol
Source: Barroso et al. (2018)

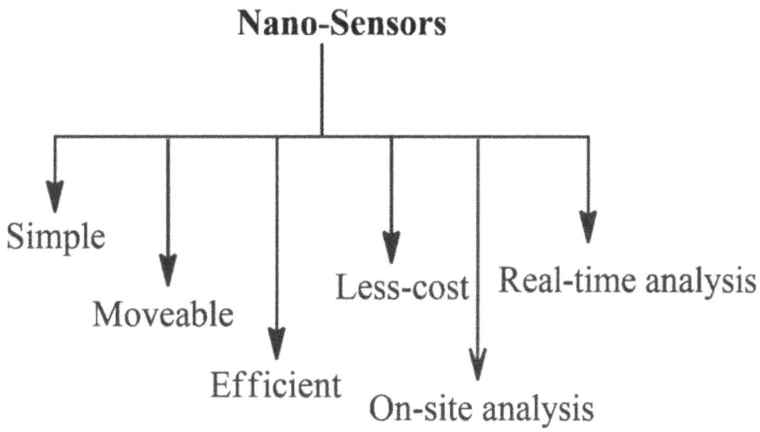

limits were 0.055, 2.7, 0.7, and 916.4 ng, respectively. The sensor array demonstrated sensitivity for identifying chemicals peculiar to picric acid and was a potent tool for distinguishing explosive indicators (3-nitrotoluene, 4-nitrotoluene, and 2,3-dimethyl-2,3-dinitrobutane) (Aznar-Gadea et al., 2022).

BASIC SCHEMES TO DETECT ANALYTE

According to the demands of the optical properties coming from their form, size, emission colour, and position of the absorption band, QDs are used in the manufacturing of sensors. Additionally, a surface change known as functionalization must be made initially to obtain the selectivity of QDs in sensor actions (Wenger et al., 2017). The procedure of attaching and altering chemical molecules that are already exist on the surface of quantum dots is known as functionalization. Salinization, ligand exchange, and the formation of excess coatings or dendrimer structures are examples of physical and chemical methods to accomplish this goal (Liu et al., 2014). Size, shape, and physicochemical characteristics of QDs, such as chemical reactivity and surface charge, are all influenced by the presence of ligands on their surfaces. For use in medical and biological applications, surface modification enables control of colloidal stability and dispersion of QDs in non-polar (mainly synthetic organic solvents) and polar (for example, water) environments. In addition, surface-bound ligands affect whether QDs have the ability to be conjugated with biological molecules (bio-conjugation) or to be used in matrices (Zhou et al., 2017).

QDs are paired with various carriers designed specifically for the analysis to achieve exceptionally great selectivity of the QDs sensors. A sensor that can specifically identify the ingredient used to kill bugs, dicofol, was created by Wells and his colleagues. In order to achieve this, they employed CdS QDs with amino and carboxyl functional groups, glutathione on the surface, and chlorine groups in the dicofol structure. These interactions increased the fluorescence intensity proportionally to the dicofol concentration in the study sample (Nsibande and Forbes, 2016; Walia and Acharya, 2014). Depending

Figure 8. Three examples of the strategy of QDs-based optical sensors: Strategy a) modification of substrate with QDs directed to detection of analyte, strategy b) modification of substrate for detection of analyte-QDs complex, strategy c) using the analyte labeled with appropriate fluorophore

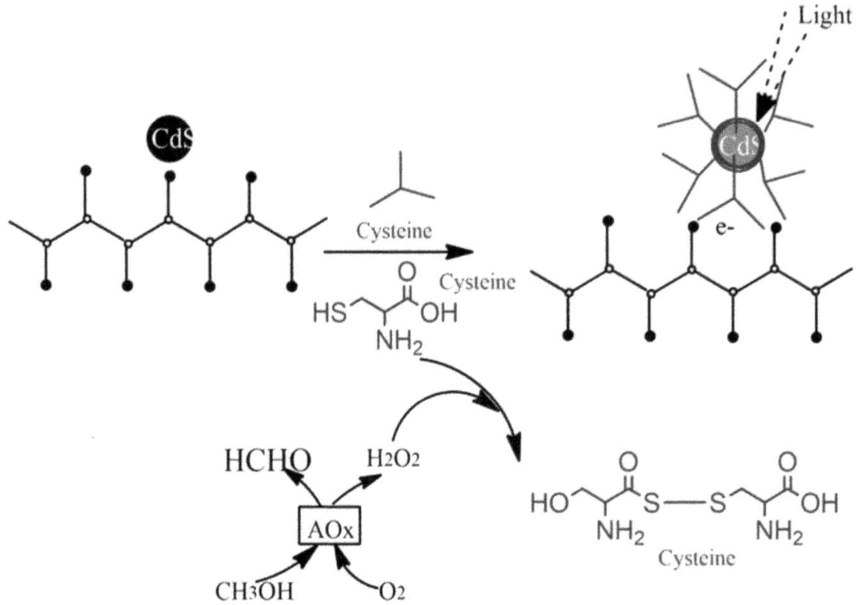

on their sensitivity requirements, the type of analyte detected, cost, or fabrication complexity, quantum dot sensors can be built in a variety of methods. An illustration of a preparation strategy for optical sensors based on CT is shown in Figure 8.

In each situation, the procedure starts with the proper changes to choose the QD surface. Consequently, the goal of QDs is to pinpoint a particular analyte. The availability of substrates that can actively take part in the detecting process is another important quality. The stages III and IV differ between strategies (a) and (b), which are executed in reverse order. Accumulation of QDs and step IV of strategy (a) make up the third and fourth steps, respectively (b). Approaches like sol-gel, layer-by-layer, or electrochemical methods can be used to create this phase (Jiang and Ju, 2007; Wang et al., 2018). Analytic binding occurs in the fourth and third steps of strategy (a) and strategy (b), respectively. This is presumably because the target substrate was already prepared. Jie et al. (2008) suggested utilizing antibodies chosen to an antigen known as human IgG based on CdSe nanocomposites for the combination of analytes with pre-prepared substrates (Jie et al., 2008). CT stimulation for analytic detection is the last stage in any method. As a result, it is possible to evaluate the availability of these materials both quantitatively and qualitatively. Furthermore, it is likely that strategy (c) will integrate the first two approaches employing resonant Förster energy transfer between optical centres (QDs + QDs or dyes). The detection is more difficult and user involvement is increased by the proposed strategy's various numbers of steps (Lesyak et al., 2019; Nath et al., 2023).

PHYSICO-CHEMICAL MECHANISMS TO DETECT ANALYTE

This is one of the most common analytical detection mechanisms are based on QD radiation amplification. In this mechanism, the QD emission intensity is reduced because of the interaction between the QD's surface and the analyte (Figure 9a) (Shang et al., 2009). One more mechanism is based on the increase in the emission of QDs because of the QD surface's stretching when analytes are present (Figure 9b), for example, the introduction of nucleic acids or serum albumin causes the increase in emission of CdS dots layered using mercaptoacetic acid (Wang et al., 2002).

Figure 9. Examples of physico-chemical mechanisms used for analyte optical detection: emission bleaching (a), increase of emission (b), emission localization (c), nanostructures growth's modification (d), emission change (e)

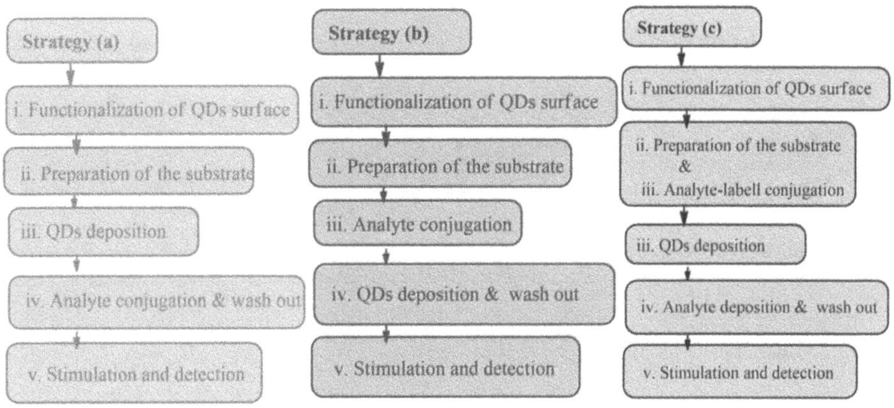

A 3rd mechanism is stimulated aggregation that can be employed to detect analytes (Figure 9c). This mechanism involves dissociation of surface ligands and aggregation of QDs because of the contact between analyte and the QDs surface. Bond formation between activated QDs induced by the analytes can also lead to aggregation (Kavosia et al., 2018).

In addition, there are little-used strategies for detecting analytes based on modification of the growth procedure of nanostructures that introduce analytes during growth. Because of this disorder, the nanostructure could produce various emissions or other detectable characteristics (Figure 9d). There is also a 5th mechanism generally employed to detect analytical on the basis of variations in the optical possessions of QDs. These changes result from the transfer of excitation energy from quantum dots to other optical centers (either quantum dots or dyes). As a result, the donor radiation color changes or the radiation decay time decreases (Figure 8e) (Anikeeva et al., 2008).

Emission Bleaching

The emission bleaching is one of the main mechanisms for detecting the analyte (Figure 10(i)). This strategy can be caused by various physical phenomena, revealed in Fig. 10. In 1st example, the analyte

Figure 10. Examples of physico-chemical mechanisms responsible for quantum dots emission quenching: Energy transfer from QD to analyte (i), charge transfer from analyte to QD (ii), degradation of QD surface (iii)

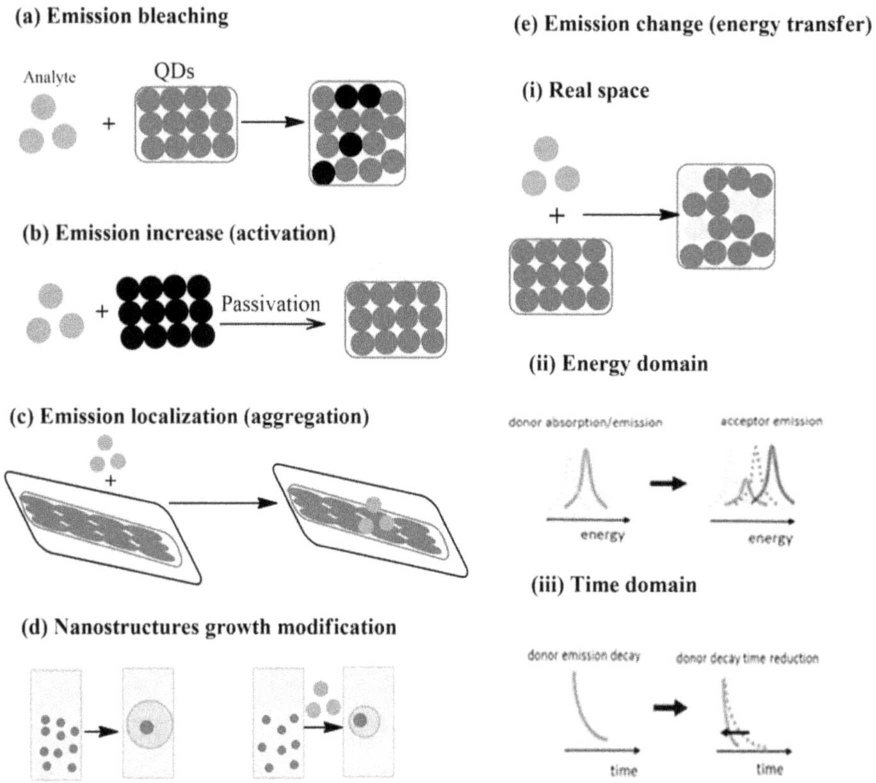

absorption band overlaps the QDs emission band. In this case, the QDs radiation is absorbed by the solvent, which causes the QDs radiation to bleach (Figure 10(ii)). One more scheme (Figure 10(iii)) is based on the transfer of charge from the analyte to the QDs. The Auger process causes the ionization or charging of QDs when excess electrons or holes are present in the excited QDs. In both cases, the dots become non-emissive (black). A 2nd strategy leading to a decrease in QDs emission upon dealings with analytes is analyte decomposition of the QD surface, which leads to a reduction in emission. For instance, thiol-coated ZnS QDs showed significant emission quenching in the presence of peptides (Wang et al., 2002; Biranje et al., 2021).

Emission Change: Nonradiative Förster Resonance Energy Transfer (FRET)

Non-radiative excitation energy transferred from an emitter to a receiver when two optically active cores (emitter and receiver) are near to one another (usually 1 to 10 nm) and the receiver's absorption spectrum matches the emitter's emission spectrum. This phenomenon is known as Förster Resonance Energy Transfer (FRET). For experimentations by the FRET method, a photo stable light emitting body with a quantum efficiency of 0.1 or more and a great luminous intensity (absorption coefficient of 50,000 M-1·cm-1 or more) must be used (Laura, Prieto, 2011). Quantum dots perfectly meet all these conditions. FRET efficiency is inversely proportional to the distance between donor and recipient, defined as:

where: K_{FRET}—FRET efficacy; R—distance between emitter and receiver; R_0—distance where FRET efficacy is equal to 0.5 (Clapp et al., 2006).

The maximum compassion of the FRET signal is observed at distances between 0.5R0 and 1.5R0 between donor and acceptor (Clapp et al., 2006). For many years, scientists have used FRET mechanism to observe the intracellular connections because of its compassion toward molecular reorganizations in 1–10 nm series (which depends on the size of biological macromolecules and on is likely to establish a relationship between them appropriately) (Miyawaki, 2003).

The versatility of the FRET method makes it possible to use it in nano systems as well (Miyawaki, 2003). FRET efficiency is usually calculated by considering one of three parameters of a fluorescent donor: fluorescence intensity, mean fluorescence lifetime, spectral response. Furthermore, FRET has been applied to numerous sensor structures, allowing it to be applied to three analytes recognition approaches. The 1st strategy utilizes analytes as optically active acceptors. In this situation, not only the addition of the analytes but also the `removal of the analyte can be regarded as a change in the ophthalmic gesture. Another strategy uses the analyzer described also in the previous paragraph as an emission reducer (Lasota et al., 2017). The 3rd mechanism is extra complicated and employs a multi-stage power transmission trend. The use of FRET detection between QDs directed to a acceptor-bound linker as a donor, such as one linked to a receptor protein, has been broadly employed to investigate the receptor-ligand interactions and binding to target analytes, Changes in protein structure (Clapp et al., 2004). Therefore, in an assay with multiple receptors, QDs can inter-relate with only one of them, greatly increasing the sensitivity and efficiency of the FRET technique (Patolsky et al., 2003; Goryacheva et al., 2017; Lesiak et al., 2019).

Analyte Stimulated QDs Aggregation

Colloidal solutions of QDs are delicate to excess charges on the surface of the QDs or in the solution, which can lead to aggregation of the QDs. Charge can be induced or introduced by the analyte that ultimately appears in the aggregate of QDs (Coto-García et al., 2011; Freeman et al., 2013; Zhang et al., 2013). In absence of an analyte, fluorescence originates from the entire solution of QDs, whereas emission is localized after analyte-induced aggregation (Xu et al., 2011). Such incentives fall under the category of qualitative tests (Lesiak et al., 2019).

CONCLUSION AND FUTURE PERSPECTIVES

Some solids or liquids emit volatile toxins as gases. These volatile organic poisons are made up of a variety of compounds, some of which have the potential to cause cancer, that can have both immediate and long-term negative impacts on the environment and human health. The analysis of volatile toxins has great potential because samples may be easily, safely, and often obtained. Nanomaterials are crucial in the detection of poisonous and volatile compounds. One of the most useful nanomaterials, quantum dots exhibit exceptional optical characteristics. It has a wide range of uses, including as active ingredients in nanostructure-biomolecular complexes and new fluorescence probes. A number of photoconversion quantum dot application situations have been successfully investigated, allowing for high selectivity, sensitivity, and a range of detection in test samples. The use of quantum dots in sensors to detect the presence of ions, molecules, and pH changes has been made possible by the development of analytical procedures to identify diverse chemical or biological components. The fundamentals of volatile toxicant analysis and the methods used for sample collection, pre-concentration, and detection are outlined in this chapter. GC-MS is currently regarded as the gold standard among detection techniques, and sensor technologies based on diverse nanomaterials have been created. The ideal instrument for volatile toxicity testing and cancer diagnosis should be economical, quick, portable, reusable, simple to use, and adaptable to different forms of disease in order to be used for environmental management, clinical point-of-care use, and population-wide screening reasons. It must be highly sensitive and particular, suited for a range of temperature and humidity conditions. Future advancements will include novel approaches or combinations of cutting-edge technology for the sampling, detection, and analysis of volatile toxins, as well as the standardisation and validation of these techniques for application in actual clinical settings.

REFERENCES

Aati, H. Y., Attia, H., Babtin, R., Al-Qahtani, N., & Wanner, J. (2023). Headspace Solid Phase Micro-Extraction of Volatile Constituents Produced from Saudi Ruta chalepensis and Molecular Docking Study of Potential Antioxidant Activity. *Molecules (Basel, Switzerland)*, *28*(4), 1891. doi:10.3390/molecules28041891 PMID:36838880

Almaie, S., Vatanpour, V., Rasoulifard, M. H., & Koyuncu, I. (2022). Volatile organic compounds (VOCs) removal by photocatalysts: A review. *Chemosphere*, *306*, 135655. doi:10.1016/j.chemosphere.2022.135655 PMID:35817187

Amal, H., Shi, D. Y., Ionescu, R., Zhang, W., Hua, Q. L., Pan, Y. Y., & Haick, H. (2015). Assessment of ovarian cancer conditions from exhaled breath. *International Journal of Cancer*, *136*(6), E614–E622. doi:10.1002/ijc.29166 PMID:25159530

Andre, R. S., Ngo, Q. P., Fugikawa-Santos, L., Correa, D. S., & Swager, T. M. (2021). Wireless tags with hybrid nanomaterials for volatile amine detection. *ACS Sensors*, *6*(6), 2457–2464. doi:10.1021/acssensors.1c00812 PMID:34110807

Anikeeva, P. O., Madigan, C. F., Halpert, J. E., Bawendi, M. G., & Bulović, V. J. P. R. B. (2008). Electronic and excitonic processes in light-emitting devices based on organic materials and colloidal quantum dots. *Physical Review B: Condensed Matter and Materials Physics*, *78*(8), 085434. doi:10.1103/PhysRevB.78.085434

Arduini, F., Cinti, S., Scognamiglio, V., & Moscone, D. (2020). Nanomaterial-based sensors. In *Handbook of nanomaterials in analytical chemistry* (pp. 329–359). Elsevier. doi:10.1016/B978-0-12-816699-4.00013-X

Ataei, Y., Sun, Y., Liu, W. S., Ellie, A., Dong, H., & Ahmad, U. M. (2023). Health Effects of Exposure to Indoor Semi-Volatile Organic Compounds in Chinese Building Environment: A Systematic Review. *International Journal of Environmental Research and Public Health*, *20*(1), 678. doi:10.3390/ijerph20010678 PMID:36613006

Aznar-Gadea, E., Rodriguez-Canto, P. J., Sánchez, S. A., Martinez-Pastor, J. P., & Abargues, R. (2022). Luminescent CdSe Quantum Dot Arrays for Rapid Sensing of Explosive Taggants. *ACS Applied Nano Materials*, *5*(5), 6717–6725. doi:10.1021/acsanm.2c00743

Bajtarevic, A., Ager, C., Pienz, M., Klieber, M., Schwarz, K., Ligor, M., & Amann, A. (2009). Noninvasive detection of lung cancer by analysis of exhaled breath. *BMC Cancer*, *9*(1), 1–16. doi:10.1186/1471-2407-9-348 PMID:19788722

Barroso, J., Diez-Buitrago, B., Saa, L., Möller, M., Briz, N., & Pavlov, V. (2018). Specific bioanalytical optical and photoelectrochemical assays for detection of methanol in alcoholic beverages. *Biosensors & Bioelectronics*, *101*, 116–122. doi:10.1016/j.bios.2017.10.022 PMID:29055193

Bezzon, V. D., Montanheiro, T. L., de Menezes, B. R., Ribas, R. G., Righetti, V. A., Rodrigues, K. F., & Thim, G. P. (2019). Carbon nanostructure-based sensors: A brief review on recent advances. *Advances in Materials Science and Engineering*, *2019*, 2019. doi:10.1155/2019/4293073

Bi, C., Krechmer, J. E., Frazier, G. O., Xu, W., Lambe, A. T., Claflin, M. S., Lerner, B. M., Jayne, J. T., Worsnop, D. R., Canagaratna, M. R., & Isaacman-VanWertz, G. (2021). Coupling a gas chromatograph simultaneously to a flame ionization detector and chemical ionization mass spectrometer for isomer-resolved measurements of particle-phase organic compounds. *Atmospheric Measurement Techniques*, *14*(5), 3895–3907. doi:10.5194/amt-14-3895-2021

Bindra, P., Gangopadhyay, S., & Hazra, A. (2018). Au/TiO 2 Nanotubes/Ti-based solid-state vapor sensor: Efficient sensing in resistive and capacitive modes. *IEEE Transactions on Electron Devices*, *65*(5), 1918–1924. doi:10.1109/TED.2018.2809677

Biranje, A., Azmi, N., Tiwari, A., & Chaskar, A. (2021). Quantum dots based fluorescent probe for the selective detection of heavy metal ions. *Journal of Fluorescence*, *31*(5), 1241–1250. doi:10.100710895-021-02755-8 PMID:34181146

Brodrick, E., Davies, A., Neill, P., Hanna, L., & Williams, E. M. (2015). Breath analysis: Translation into clinical practice. *Journal of Breath Research*, *9*(2), 027109. doi:10.1088/1752-7155/9/2/027109 PMID:25971863

Capitain, C., & Weller, P. (2021). Non-targeted screening approaches for profiling of volatile organic compounds based on gas chromatography-ion mobility spectroscopy (GC-IMS) and machine learning. *Molecules (Basel, Switzerland)*, *26*(18), 5457. doi:10.3390/molecules26185457 PMID:34576928

Chen, Y. P., Cai, D., Li, W., Blank, I., & Liu, Y. (2022). Application of gas chromatography-ion mobility spectrometry (GC-IMS) and ultrafast gas chromatography electronic-nose (uf-GC E-nose) to distinguish four Chinese freshwater fishes at both raw and cooked status. *Journal of Food Biochemistry*, *46*(6), e13840. doi:10.1111/jfbc.13840 PMID:34189733

Chew, B. S., Contreras, R. P., McCartney, M. M., Borras, E., Kenyon, N. J., & Davis, C. E. (2022). A low cost, easy-to-assemble, open-source modular mobile sampler design for thermal desorption analysis of breath and environmental VOCs. *Journal of Breath Research*, *16*(3), 036005. doi:10.1088/1752-7163/ac6c9f PMID:35508102

Clapp, A. R., Medintz, I. L., & Mattoussi, H. (2006). Förster resonance energy transfer investigations using quantum-dot fluorophores. *ChemPhysChem, 7*(1), 47–57. doi:10.1002/cphc.200500217 PMID:16370019

Clapp, A. R., Medintz, I. L., Mauro, J. M., Fisher, B. R., Bawendi, M. G., & Mattoussi, H. (2004). Fluorescence resonance energy transfer between quantum dot donors and dye-labeled protein acceptors. *Journal of the American Chemical Society, 126*(1), 301–310. doi:10.1021/ja037088b PMID:14709096

Coto-García, A. M., Sotelo-González, E., Fernández-Argüelles, M. T., Pereiro, R., Costa-Fernández, J. M., & Sanz-Medel, A. (2011). Nanoparticles as fluorescent labels for optical imaging and sensing in genomics and proteomics. *Analytical and Bioanalytical Chemistry, 399*(1), 29–42. doi:10.100700216-010-4330-3 PMID:21052647

Cumeras, R., & Correig, X. (Eds.). (2018). *Volatile organic compound analysis in biomedical diagnosis applications*. CRC Press. doi:10.1201/9780429433580

Epping, R., & Koch, M. (2023). On-Site Detection of Volatile Organic Compounds (VOCs). *Molecules (Basel, Switzerland), 28*(4), 1598. doi:10.3390/molecules28041598 PMID:36838585

Feng, X., Wang, H., Wang, Z., Huang, P., & Kan, J. (2022). Discrimination and characterization of the volatile organic compounds in eight kinds of huajiao with geographical indication of China using electronic nose, HS-GC-IMS and HS-SPME-GC–MS. *Food Chemistry, 375*, 131671. doi:10.1016/j.foodchem.2021.131671 PMID:34865919

Freeman, R., Girsh, J., & Willner, I. (2013). Nucleic acid/quantum dots (QDs) hybrid systems for optical and photoelectrochemical sensing. *ACS Applied Materials & Interfaces, 5*(8), 2815–2834. doi:10.1021/am303189h PMID:23425022

Gakhar, T., Basu, S., & Hazra, A. (2022). Carbon-metal oxide nanocomposites for selective detection of toxic and hazardous volatile organic compounds (VOC)-A review. *Green Analytical Chemistry*.

Ganesan, M., & Nagaraaj, P. (2020). Quantum dots as nanosensors for detection of toxics: A literature review. *Analytical Methods, 12*(35), 4254–4275. doi:10.1039/D0AY01293A PMID:32940270

Ghadiri, M., Ghashghaee, M., & Ghambarian, M. (2020). Defective phosphorene for highly efficient formaldehyde detection: Periodic density functional calculations. *Physics Letters. [Part A], 384*(31), 126792. doi:10.1016/j.physleta.2020.126792

Goryacheva, O. A., Beloglazova, N. V., Vostrikova, A. M., Pozharov, M. V., Sobolev, A. M., & Goryacheva, I. Y. (2017). Lanthanide-to-quantum dot Förster resonance energy transfer (FRET): Application for immunoassay. *Talanta, 164*, 377–385. doi:10.1016/j.talanta.2016.11.054 PMID:28107944

Gregis, G., Sanchez, J. B., Bezverkhyy, I., Guy, W., Berger, F., Fierro, V., & Celzard, A. (2018). Detection and quantification of lung cancer biomarkers by a micro-analytical device using a single metal oxide-based gas sensor. *Sensors and Actuators. B, Chemical, 255*, 391–400. doi:10.1016/j.snb.2017.08.056

Han, S., Tan, Y., Gao, Y., Li, X., Ho, S. S. H., Wang, M., & Lee, S. C. (2023). Volatile organic compounds at a roadside site in Hong Kong: Characteristics, chemical reactivity, and health risk assessment. *The Science of the Total Environment, 866*, 161370. doi:10.1016/j.scitotenv.2022.161370 PMID:36621478

Hazra, A., Bhowmik, B., Dutta, K., Chattopadhyay, P. P., & Bhattacharyya, P. (2015). Stoichiometry, length, and wall thickness optimization of TiO2 nanotube array for efficient alcohol sensing. *ACS Applied Materials & Interfaces*, *7*(18), 9336–9348. doi:10.1021/acsami.5b01785 PMID:25918822

Hegab, H., Khanna, N., Monib, N., & Salem, A. (2023). Design for sustainable additive manufacturing: A review. *Sustainable. Materials Technology.*

Hekiem, N. L. L., Ralib, A. A. M., Ahmad, F. B., Nordin, A. N., Ab Rahim, R., & Za'bah, N. F. (2021). Advanced vapour sensing materials: Existing and latent to acoustic wave sensors for VOCs detection as the potential exhaled breath biomarkers for lung cancer. *Sensors and Actuators. A, Physical*, *329*, 112792. doi:10.1016/j.sna.2021.112792

Huang, Y., Li, X., Chen, X., Wang, W., Wang, Y., Liu, Z., & Tang, G. (2022). Low-molecular-weight carbonyl volatile organic compounds on the North China Plain. *Atmospheric Environment*, *275*, 119000. doi:10.1016/j.atmosenv.2022.119000

Hung, C. M., Van Duy, L., Le, D. T. T., Nguyen, H., Van Duy, N., & Hoa, N. D. (2021). ZnO coral-like nanoplates decorated with Pd nanoparticles for enhanced VOC gas sensing. *Journal of Science. Advanced Materials and Devices*, *6*(3), 453–461. doi:10.1016/j.jsamd.2021.05.005

Iravani, S., & Varma, R. S. (2020). Green synthesis, biomedical and biotechnological applications of carbon and graphene quantum dots. A review. *Environmental Chemistry Letters*, *18*(3), 703–727. doi:10.100710311-020-00984-0 PMID:32206050

Jiang, H., & Ju, H. (2007). Electrochemiluminescence sensors for scavengers of hydroxyl radical based on its annihilation in CdSe quantum dots film/peroxide system. *Analytical Chemistry*, *79*(17), 6690–6696. doi:10.1021/ac071061j PMID:17676926

Jie, G., Zhang, J., Wang, D., Cheng, C., Chen, H. Y., & Zhu, J. J. (2008). Electrochemiluminescence immunosensor based on CdSe nanocomposites. *Analytical Chemistry*, *80*(11), 4033–4039. doi:10.1021/ac800052g PMID:18435547

Kaikiti, C., Stylianou, M., & Agapiou, A. (2022). TD-GC/MS analysis of indoor air pollutants (VOCs, PM) in hair salons. *Chemosphere*, *294*, 133691. doi:10.1016/j.chemosphere.2022.133691 PMID:35065178

Kaur Billing, B. (2021). Carbon nanotubes and its potential application in sensing. *ChemistrySelect*, *6*(36), 9571–9590. doi:10.1002lct.202102636

Kavosi, B., Navaee, A., & Salimi, A. (2018). Amplified fluorescence resonance energy transfer sensing of prostate specific antigen based on aggregation of CdTe QDs/antibody and aptamer decorated of AuNPs-PAMAM dendrimer. *Journal of Luminescence*, *204*, 368–374. doi:10.1016/j.jlumin.2018.08.012

Khademi, F., Samaei, M. R., Shahsavani, A., Azizi, K., Mohammadpour, A., Derakhshan, Z., & Bilal, M. (2022). Investigation of the Presence Volatile Organic Compounds (BTEX) in the Ambient Air and Biogases Produced by a Shiraz Landfill in Southern Iran. *Sustainability (Basel)*, *14*(2), 1040. doi:10.3390u14021040

Khoubnasabjafari, M., Mogaddam, M. R. A., Rahimpour, E., Soleymani, J., Saei, A. A., & Jouyban, A. (2022). Breathomics: Review of sample collection and analysis, data modeling and clinical applications. *Critical Reviews in Analytical Chemistry*, *52*(7), 1461–1487. doi:10.1080/10408347.2021.1889961 PMID:33691552

Kimura, M., Liu, Y., Sakai, R., Sato, S., Hirai, T., Fukawa, T., & Mihara, T. (2011). Detection of volatile organic compounds by analyses of polymer-coated quartz crystal microbalance sensor arrays. *Sensors and Materials*, *23*(7), 359–368.

King, J., Kupferthaler, A., Frauscher, B., Hackner, H., Unterkofler, K., Teschl, G., & Högl, B. (2012). Breath acetone and isoprene concentration profiles in normal human sleep. *Physiological Measurement*, *33*, 413–428. doi:10.1088/0967-3334/33/3/413 PMID:22370046

Kumar, S., Huang, J., Abbassi-Ghadi, N., Španěl, P., Smith, D., & Hanna, G. B. (2013). Selected ion flow tube mass spectrometry analysis of exhaled breath for volatile organic compound profiling of esophago-gastric cancer. *Analytical Chemistry*, *85*(12), 6121–6128. doi:10.1021/ac4010309 PMID:23659180

Kus, F., Altinkok, C., Zayim, E., Erdemir, S., Tasaltin, C., & Gurol, I. (2021). Surface acoustic wave (SAW) sensor for volatile organic compounds (VOCs) detection with calix [4] arene functionalized Gold nanorods (AuNRs) and silver nanocubes (AgNCs). *Sensors and Actuators. B, Chemical*, *330*, 129402. doi:10.1016/j.snb.2020.129402

Lasota, S., Baster, Z., Witko, T., Zimoląg, E., Sroka, J., Rajfur, Z., & Madeja, Z. (2017). Zastosowanie biosensorów typu FRET w badaniach mikroskopowych procesu migracji komórkowej. *Postepy Biochemii*, *63*(1), 16–33. PMID:28409572

Lesiak, A., Drzozga, K., Cabaj, J., Bański, M., Malecha, K., & Podhorodecki, A. (2019). Optical sensors based on II-VI quantum dots. *Nanomaterials (Basel, Switzerland)*, *9*(2), 192. doi:10.3390/nano9020192 PMID:30717393

Li, C., Wan, H., Wu, X., Yin, J., Zhu, L., Chen, H., & Li, Z. (2022). Discrimination and Characterization of the Volatile Organic Compounds in Schizonepetae Spica from Six Regions of China Using HS-GC-IMS and HS-SPME-GC-MS. *Molecules (Basel, Switzerland)*, *27*(14), 4393. doi:10.3390/molecules27144393 PMID:35889268

Li, D., Xie, Z., Qu, M., Zhang, Q., Fu, Y., & Xie, J. (2021). Virtual sensor array based on butterworth–van dyke equivalent model of QCM for selective detection of volatile organic compounds. *ACS Applied Materials & Interfaces*, *13*(39), 47043–47051. doi:10.1021/acsami.1c13046 PMID:34546706

Li, Z., Hou, S., Zhang, H., Song, Q., Wang, S., & Guo, H. (2023). *Recent advances in fluorescent and colorimetric sensing for volatile organic amines and biogenic amines in food*. Advanced Agrochem. doi:10.1016/j.aac.2023.02.001

Liu, Y., Wang, L., Wang, H., Xiong, M., Yang, T., & Zakharova, G. S. (2016). Highly sensitive and selective ammonia gas sensors based on PbS quantum dots/TiO_2 nanotube arrays at room temperature. *Sensors and Actuators. B, Chemical*, *236*, 529–536. doi:10.1016/j.snb.2016.06.037

Lona-Ramirez, F. J., Gonzalez-Alatorre, G., Rico-Ramírez, V., Perez-Perez, M. C. I., & Castrejón-González, E. O. (2016). Gas chromatography/mass spectrometry for the determination of nitrosamines in red wine. *Food Chemistry, 196*, 1131–1136. doi:10.1016/j.foodchem.2015.09.090 PMID:26593598

Loura, L. M., & Prieto, M. (2011). FRET in membrane biophysics: An overview. *Frontiers in Physiology, 2*, 82. doi:10.3389/fphys.2011.00082 PMID:22110442

Lu, Y., Pang, X., Lyu, Y., Li, J., Xing, B., Chen, J., & Wu, H. (2022). Characteristics and sources analysis of ambient volatile organic compounds in a typical industrial park: Implications for ozone formation in 2022 Asian Games. *The Science of the Total Environment, 848*, 157746. doi:10.1016/j.scitotenv.2022.157746 PMID:35926610

Ma, H., Li, X., Chen, J., Wang, H., Cheng, T., Chen, K., & Xu, S. (2014). Analysis of human breath samples of lung cancer patients and healthy controls with solid-phase microextraction (SPME) and flow-modulated comprehensive two-dimensional gas chromatography (GC× GC). *Analytical Methods, 6*(17), 6841–6849. doi:10.1039/C4AY01220H

Masite, N. S., Ncube, S., Mtunzi, F. M., Madikizela, L. M., & Pakade, V. E. (2022). Determination of furanic compounds in Mopane worms, corn, and peanuts using headspace solid-phase microextraction with gas chromatography-flame ionisation detector. *Food Chemistry, 369*, 130944. doi:10.1016/j.foodchem.2021.130944 PMID:34464834

Mayhew, C. A., Herbig, J., & Beauchamp, J. D. (2020). Proton transfer reaction–mass spectrometry. In *Breathborne Biomarkers and the Human Volatilome* (pp. 155–170). Elsevier. doi:10.1016/B978-0-12-819967-1.00010-4

Miyawaki, A. (2003). Visualization of the spatial and temporal dynamics of intracellular signaling. *Developmental Cell, 4*(3), 295–305. doi:10.1016/S1534-5807(03)00060-1 PMID:12636912

Miyawaki, A. (2008). Green fluorescent protein glows gold. *Cell, 135*(6), 987–990. doi:10.1016/j.cell.2008.11.025 PMID:19070562

Molaei, M. J. (2020). Principles, mechanisms, and application of carbon quantum dots in sensors: A review. *Analytical Methods, 12*(10), 1266–1287. doi:10.1039/C9AY02696G

Montero-Montoya, R., López-Vargas, R., & Arellano-Aguilar, O. (2018). Volatile organic compounds in air: Sources, distribution, exposure and associated illnesses in children. *Annals of Global Health, 84*(2), 225–238. doi:10.29024/aogh.910 PMID:30873816

Mosallaei, S., Hashemi, H., Hoseini, M., Dehghani, M., & Naz, A. (2023). Polycyclic Aromatic hydrocarbons (PAHs) in household dust: The association between PAHs, cancer risk and sick building syndrome. *Building and Environment, 229*, 109966. doi:10.1016/j.buildenv.2022.109966

Moseley, P. T. (2017). Progress in the development of semiconducting metal oxide gas sensors: A review. *Measurement Science & Technology, 28*(8), 082001. doi:10.1088/1361-6501/aa7443

Moura, P. C., Vassilenko, V., & Ribeiro, P. A. (2023). Ion Mobility Spectrometry Towards Environmental Volatile Organic Compounds Identification and Quantification: a Comparative Overview over Infrared Spectroscopy. *Emission Control Science and Technology*, 1-22.

Munteanu, I. G., & Apetrei, C. (2023). Classification and Antioxidant Activity Evaluation of Edible Oils by Using Nanomaterial-Based Electrochemical Sensors. *International Journal of Molecular Sciences*, *24*(3), 3010. doi:10.3390/ijms24033010 PMID:36769346

Murtada, K., Bowman, D., Edwards, M., & Pawliszyn, J. (2022). Thin-film microextraction combined with comprehensive two-dimensional gas chromatography time-of-flight mass spectrometry screening for presence of multiclass organic pollutants in drinking water samples. *Talanta*, *242*, 123301. doi:10.1016/j.talanta.2022.123301 PMID:35167962

Nath, N., Kumar, A., Chakroborty, S., Soren, S., Barik, A., Pal, K., & de Souza, F. G. Jr. (2023). Carbon Nanostructure Embedded Novel Sensor Implementation for Detection of Aromatic Volatile Organic Compounds: An Organized Review. *ACS Omega*, *8*(5), 4436–4452. doi:10.1021/acsomega.2c05953 PMID:36777592

Nikhar, S., Sahu, P., Rarotra, S., & Kumar, P. (2021). Biological metal organic framework (Bio-MOF) for detection of voltaic organic compounds (VOC's). *Inorganic Chemistry Communications*, *130*, 108711. doi:10.1016/j.inoche.2021.108711

Nsibande, S. A., & Forbes, P. B. C. (2016). Fluorescence detection of pesticides using quantum dot materials–a review. *Analytica Chimica Acta*, *945*, 9–22. doi:10.1016/j.aca.2016.10.002 PMID:27968720

Patolsky, F., Gill, R., Weizmann, Y., Mokari, T., Banin, U., & Willner, I. (2003). Lighting-up the dynamics of telomerization and DNA replication by CdSe– ZnS quantum dots. *Journal of the American Chemical Society*, *125*(46), 13918–13919. doi:10.1021/ja035848c PMID:14611202

Peng, G., Tisch, U., Adams, O., Hakim, M., Shehada, N., Broza, Y. Y., & Haick, H. (2009). Diagnosing lung cancer in exhaled breath using gold nanoparticles. *Nature Nanotechnology*, *4*(10), 669–673. doi:10.1038/nnano.2009.235 PMID:19809459

Perkins, M. J., & Langford, V. S. (2021). Application of routine analysis procedures to a direct mass spectrometry technique: Selected ion flow tube mass spectrometry (SIFT-MS). *Reviews in Separation Sciences*, *3*(1), e21003–e21003. doi:10.17145/rss.21.003

Perkins, M. J., & Langford, V. S. (2022). Multiple headspace extraction-selected ion flow tube mass spectrometry (MHE-SIFT-MS). Part 1: A protocol for method development and transfer to routine analysis. *Reviews in Separation Sciences.*

Piñeiro-García, A., González-Alatorre, G., Vega-Díaz, S. M., Pérez-Pérez, M. C. I., Tristan, F., & Patiño-Herrera, R. (2020). Reduced graphene oxide coating with high performance for the solid phase micro-extraction of furfural in espresso coffee. *Journal of Food Measurement and Characterization*, *14*(1), 314–321. doi:10.100711694-019-00293-3

Queralto, N., Berliner, A. N., Goldsmith, B., Martino, R., Rhodes, P., & Lim, S. H. (2014). Detecting cancer by breath volatile organic compound analysis: A review of array-based sensors. *Journal of Breath Research*, *8*(2), 027112. doi:10.1088/1752-7155/8/2/027112 PMID:24862241

Rahman, M. M. (2020). Efficient formaldehyde sensor development based on Cu-codoped ZnO nanomaterial by an electrochemical approach. *Sensors and Actuators. B, Chemical*, *305*, 127541. doi:10.1016/j.snb.2019.127541

Reji, R. P., Balaji, S. K. C., Sivalingam, Y., Kawazoe, Y., & Velappa Jayaraman, S. (2023). First-Principles Density Functional Theory Calculations on the Potential of Sc_2CO_2 MXene Nanosheets as a Dual-Mode Sensor for Detection of Volatile Organic Compounds in Exhaled Human Breath. *ACS Applied Nano Materials*, *6*(7), 5345–5356. doi:10.1021/acsanm.2c05474

Reshma, V. G., & Mohanan, P. V. (2019). Quantum dots: Applications and safety consequences. *Journal of Luminescence*, *205*, 287–298. doi:10.1016/j.jlumin.2018.09.015

Righettoni, M., Schmid, A., Amann, A., & Pratsinis, S. E. (2013). Correlations between blood glucose and breath components from portable gas sensors and PTR-TOF-MS. *Journal of Breath Research*, *7*(3), 037110. doi:10.1088/1752-7155/7/3/037110 PMID:23959908

Roberts, I. J., Carpenter, L. J., Shaw, M. D., & Langford, V. S. (2022). Selected Ion Flow Tube–Mass Spectrometry (SIFT-MS) study of the reactions of H3O+, NO+ and O2+ with a range of oxygenated volatile organic carbons (OVOCs). *International Journal of Mass Spectrometry*, *479*, 116892. doi:10.1016/j.ijms.2022.116892

Rouf, Z., Dar, I. Y., Javaid, M., Dar, M. Y., & Jehangir, A. (2022). Volatile Organic Compounds Emission from Building Sector and Its Adverse Effects on Human Health. In *Ecological and Health Effects of Building Materials* (pp. 67–86). Springer. doi:10.1007/978-3-030-76073-1_5

Sarswat, P., & Free, M. L. (2014). Quantum Dot Based Sensing Platform for Selective Detection of Volatile Organic Compounds and Biomarkers. *ECS Transactions*, *61*(19), 11–16. doi:10.1149/06119.0011ecst

Sekimoto, K., Li, S. M., Yuan, B., Koss, A., Coggon, M., Warneke, C., & de Gouw, J. (2017). Calculation of the sensitivity of proton-transfer-reaction mass spectrometry (PTR-MS) for organic trace gases using molecular properties. *International Journal of Mass Spectrometry*, *421*, 71–94. doi:10.1016/j.ijms.2017.04.006

Severino, M. I., Al Mohtar, A., Vieira Soares, C., Freitas, C., Sadovnik, N., Nandi, S., Mouchaham, G., Pimenta, V., Nouar, F., Daturi, M., Maurin, G., Pinto, M. L., & Serre, C. (2023). MOFs with Open Metal (III) Sites for the Environmental Capture of Polar Volatile Organic Compounds. *Angewandte Chemie International Edition*, *62*(6), e202211583. doi:10.1002/anie.202211583 PMID:36468308

Shang, L., Zhang, L., & Dong, S. (2009). Turn-on fluorescent cyanide sensor based on copper ion-modified CdTe quantum dots. *Analyst (London)*, *134*(1), 107–113. doi:10.1039/B812458B PMID:19082182

Sharma, M., Chauhan, P., Nehra, S., Sharma, R., & Kumar, D. (2023). Volatile Organic Compounds Removal by Nanofiltration from Groundwater. In *Nanofiltration Membrane for Water Purification* (pp. 119–139). Springer Nature Singapore. doi:10.1007/978-981-19-5315-6_7

Smith, D., & Španěl, P. (2022). Ternary association reactions of H3O+, NO+ and O2+• with N2, O2, CO2 and H2O; implications for selected ion flow tube mass spectrometry analyses of air and breath. *Rapid Communications in Mass Spectrometry*, *36*(6), e9241. doi:10.1002/rcm.9241 PMID:34904315

Song, H., Su, Y., Zhang, L., & Lv, Y. (2019). Quantum dots-based chemiluminescence probes: An overview. *Luminescence*, *34*(6), 530–543. doi:10.1002/bio.3633 PMID:31025479

Sorolla-Rosario, D., Llorca-Porcel, J., Pérez-Martínez, M., Lozano-Castelló, D., & Bueno-López, A. (2023). Microplastics' analysis in water: Easy handling of samples by a new Thermal Extraction Desorption-Gas Chromato graphy-Mass Spectrometry (TED-GC/MS) methodology. *Talanta*, *253*, 123829. doi:10.1016/j.talanta.2022.123829 PMID:36087410

Sotelo-Gonzalez, E., Fernandez-Argüelles, M. T., Costa-Fernandez, J. M., & Sanz-Medel, A. (2012). Mn-doped ZnS quantum dots for the determination of acetone by phosphorescence attenuation. *Analytica Chimica Acta*, *712*, 120–126. doi:10.1016/j.aca.2011.11.023 PMID:22177074

Speller, N. C., Siraj, N., Regmi, B. P., Marzoughi, H., Neal, C., & Warner, I. M. (2015). Rational design of QCM-D virtual sensor arrays based on film thickness, viscoelasticity, and harmonics for vapor discrimination. *Analytical Chemistry*, *87*(10), 5156–5166. doi:10.1021/ac5046824 PMID:25913394

Steinhauer, S. (2021). Gas sensors based on copper oxide nanomaterials: A review. *Chemosensors (Basel, Switzerland)*, *9*(3), 51. doi:10.3390/chemosensors9030051

Sun, X., Shao, K., & Wang, T. (2016). Detection of volatile organic compounds (VOCs) from exhaled breath as noninvasive methods for cancer diagnosis. *Analytical and Bioanalytical Chemistry*, *408*(11), 2759–2780. doi:10.100700216-015-9200-6 PMID:26677028

Timbrell, J., & Barile, F. A. (2023). *Introduction to toxicology*. CRC Press. doi:10.1201/9781003188575

Trefz, P., Rösner, L., Hein, D., Schubert, J. K., & Miekisch, W. (2013). Evaluation of needle trap microextraction and automatic alveolar sampling for point-of-care breath analysis. *Analytical and Bioanalytical Chemistry*, *405*(10), 3105–3115. doi:10.100700216-013-6781-9 PMID:23388692

Tsai, W. T. (2023). A Survey on Toxic Volatile Organic Compounds (VOCs): Toxicological Profiles, Health Exposure Risks, and Regulatory Strategies for Mitigating Emissions from Stationary Sources in Taiwan. *Atmosphere (Basel)*, *14*(2), 242. doi:10.3390/atmos14020242

Walia, S., & Acharya, A. (2014). Fluorescent cadmium sulfide nanoparticles for selective and sensitive detection of toxic pesticides in aqueous medium. *Journal of Nanoparticle Research*, *16*(12), 1–10. doi:10.100711051-014-2778-3

Wang, L. Y., Kan, X. W., Zhang, M. C., Zhu, C. Q., & Wang, L. (2002). Fluorescence for the determination of protein with functionalized nano-ZnS. *Analyst (London)*, *127*(11), 1531–1534. doi:10.1039/b206587h PMID:12475047

Wang, N., Ernle, L., Bekö, G., Wargocki, P., & Williams, J. (2022). Emission rates of volatile organic compounds from humans. *Environmental Science & Technology*, *56*(8), 4838–4848. doi:10.1021/acs.est.1c08764 PMID:35389619

Wang, W., Kang, S., Zhou, W., & Vikesland, P. J. (2023). Environmental routes of virus transmission and the application of nanomaterial-based sensors for virus detection. *Environmental Science. Nano*, *10*(2), 393–423. doi:10.1039/D2EN00600F

Wang, Z., Zhang, Y., Zhang, B., & Lu, X. (2018). Mn2+ doped ZnS QDs modified fluorescence sensor based on molecularly imprinted polymer/sol-gel chemistry for detection of Serotonin. *Talanta*, *190*, 1–8. doi:10.1016/j.talanta.2018.07.065 PMID:30172484

Wenger, W. N., Bates, F. S., & Aydil, E. S. (2017). Functionalization of cadmium selenide quantum dots with poly (ethylene glycol): Ligand exchange, surface coverage, and dispersion stability. *Langmuir*, *33*(33), 8239–8245. doi:10.1021/acs.langmuir.7b01924 PMID:28768415

Wilde, M. J., Cordell, R. L., Salman, D., Zhao, B., Ibrahim, W., Bryant, L., & Monks, P. S. (2019). Breath analysis by two-dimensional gas chromatography with dual flame ionisation and mass spectrometric detection–method optimisation and integration within a large-scale clinical study. *Journal of Chromatography. A*, *1594*, 160–172. doi:10.1016/j.chroma.2019.02.001 PMID:30755317

Wu, X., Li, W., Li, C., Yin, J., Wu, T., Zhang, M., & Li, Z. (2022). Discrimination and characterization of the volatile organic compounds of Acori tatarinowii rhizoma based on headspace-gas chromatography-ion mobility spectrometry and headspace solid phase microextraction-gas chromatography-mass spectrometry. *Arabian Journal of Chemistry*, *15*(11), 104231. doi:10.1016/j.arabjc.2022.104231

Xiao, Z., Kong, L. B., Ruan, S., Li, X., Yu, S., Li, X., Jiang, Y., Yao, Z., Ye, S., Wang, C., Zhang, T., Zhou, K., & Li, S. (2018). Recent development in nanocarbon materials for gas sensor applications. *Sensors and Actuators. B, Chemical*, *274*, 235–267. doi:10.1016/j.snb.2018.07.040

Xing, L. I. U., & Yang, L. U. O. (2014). Surface modifications technology of quantum dots based biosensors and their medical applications. *Chinese Journal of Analytical Chemistry*, *42*(7), 1061–1069. doi:10.1016/S1872-2040(14)60753-2

Xu, X., Liu, X., Nie, Z., Pan, Y., Guo, M., & Yao, S. (2011). Label-free fluorescent detection of protein kinase activity based on the aggregation behavior of unmodified quantum dots. *Analytical Chemistry*, *83*(1), 52–59. doi:10.1021/ac102786c PMID:21128608

Xu, Z. Q., Broza, Y. Y., Ionsecu, R., Tisch, U., Ding, L., Liu, H., & Haick, H. (2013). A nanomaterial-based breath test for distinguishing gastric cancer from benign gastric conditions. *British Journal of Cancer*, *108*(4), 941–950. doi:10.1038/bjc.2013.44 PMID:23462808

Yang, L., Li, Y., Sun, Y., Wang, W., & Shao, Z. (2022). Perovskite oxides in catalytic combustion of volatile organic compounds: Recent advances and future prospects. *Energy & Environmental Materials*, *5*(3), 751–776. doi:10.1002/eem2.12256

Yang, X., Gao, L., Zhao, S., Pan, G., Fan, G., Xia, Z., & Jin, X. (2023). Volatile Organic Compounds in the North China Plain: Characteristics, Sources, and Effects on Ozone Formation. *Atmosphere (Basel)*, *14*(2), 318. doi:10.3390/atmos14020318

Yu, Z., Liu, C., Niu, H., Wu, M., Gao, W., Zhou, Z., & Li, X. (2020). Real time analysis of trace volatile organic compounds in ambient air: A comparison between membrane inlet single photon ionization mass spectrometry and proton transfer reaction mass spectrometry. *Analytical Methods*, *12*(35), 4343–4350. doi:10.1039/D0AY01102A PMID:32844845

Yue, K. X., Li, H. L., Pi, Z. F., Song, F. R., Wang, Y. P., & Liu, Z. Y. (2022). Multiresidue screening of pesticides in Panax Ginseng CA Meyer by ultra-high-performance liquid chromatography with quadrupole time-of-flight mass spectrometry. *Journal of Separation Science*, *45*(10), 1702–1710. doi:10.1002/jssc.202100933 PMID:35263500

Zarić, B. L., Petrović, S., Bjekić, M., Rajić, I., Popović, A. R., & Đorđević, D. S. (2014). Analysis of human exhaled breath in a population of young volunteers. *Archives of Biological Sciences*, *66*(4), 1529–1538. doi:10.2298/ABS1404529Z

Zhang, H., Zhang, L., Liang, R. P., Huang, J., & Qiu, J. D. (2013). Simultaneous determination of Concanavalin A and Peanut agglutinin by dual-color quantum dots. *Analytical Chemistry*, *85*(22), 10969–10976. doi:10.1021/ac402496e PMID:24128387

Zhang, L., Yang, H., Tang, Y., Xiang, W., Wang, C., Xu, T., & Zhang, J. (2022). High-performance CdSe/CdS@ ZnO quantum dots enabled by ZnO sol as surface ligands: A novel strategy for improved optical properties and stability. *Chemical Engineering Journal*, *428*, 131159. doi:10.1016/j.cej.2021.131159

Zhang, Q., Bao, X., Liang, Q., Sun, Q., Xu, W., Zou, X., & Chu, Y. (2022). Evaluation of a New DC-Ion Funnel Drift Tube for Use in Proton Transfer Reaction Mass Spectrometry. *Analytical Chemistry*, *94*(20), 7174–7180. doi:10.1021/acs.analchem.1c05086 PMID:35536750

Zhao, Z., Dai, J., Wang, T., Niu, G., He, F., & Duan, Y. (2020). Development of microwave plasma proton transfer reaction mass spectrometry (MWP-PTR-MS) for on-line monitoring of volatile organic compounds: Design, characterization and performance evaluation. *Talanta*, *208*, 120468. doi:10.1016/j.talanta.2019.120468 PMID:31816683

Zhou, J., Liu, Y., Tang, J., & Tang, W. (2017). Surface ligands engineering of semiconductor quantum dots for chemosensory and biological applications. *Materials Today*, *20*(7), 360–376. doi:10.1016/j.mattod.2017.02.006

Zhou, X., Hu, Y., Cao, Y., Liu, Y., & Qian, T. (2022). A novel signal-on fluorometric sensor based on metal ion-mediated carbon dots for formaldehyde determination and lysosome-targeted bioimaging. *New Journal of Chemistry*, *46*(36), 17540–17547. doi:10.1039/D2NJ03599E

Zilberman, Y., Ionescu, R., Feng, X. L., Mullen, K., & Haick, H. (2011). Nanoarray of polycyclic aromatic hydrocarbons and carbon nanotubes for accurate and predictive detection in real-world environmental humidity. *ACS Nano*, *5*(8), 6743–6753. doi:10.1021/nn202314k PMID:21774511

Zogka, A. G., Romanias, M. N., & Thevenet, F. (2022). Formaldehyde and glyoxal measurement deploying a selected ion flow tube mass spectrometer (SIFT-MS). *Atmospheric Measurement Techniques*, *15*(7), 2001–2019. doi:10.5194/amt-15-2001-2022

Chapter 14
Current Challenges and Future Perspective of Nanotechnology in Forensic Investigation

Sareen Akhtar
University of Health Sciences, Pakistan

Danish Rao
Xi'an Jiaotong University, China

Salah Ud Din
University of Health Sciences, Pakistan

ABSTRACT

Nanotechnology includes different kinds of nano-materials including the nano-composites, nano-hybrid, and nanostructured materials. There are immense and broader views based on pragmatism and cross-fertilization of nanotechnology. Because of the development in nanotechnology, the progression in technical sector is also increasing. Similarly, the role of nanotechnology is increasing in the field of forensic sciences. There are numerous branches of forensics covered by nanotechnology including the trend of nanotechnology enhancing already existing technologies in forensics. In the upcoming era, advanced nanotechnology in instruments for illicit drug screening and fingerprint detection is increasing because nano-based techniques hold an immense future in fingerprinting and security features. Therefore, this chapter offers an application of nanoparticles in numerous areas of forensics.

INTRODUCTION

In every sector, nanotechnology is making a significant impact. Being able to control and characterise the matter at the level of individual and small atoms makes it a commonly used method (Kesarwani, 2020). It is generally understood to be the investigation, design, production, synthesis, manipulation, and application of useful materials, systems, and devices by controlling matter at the nanoscale. It has

DOI: 10.4018/978-1-6684-8325-1.ch014

been utilised in a variety of academic fields, such as electronic engineering, physical sciences, medicinal sciences, and many others (Chen, 2011). The term 'Nano' implies dwarf or little, that specifically relates to a specific size that will be one billionth which is nearly to the one nanometer nm as per calculation. As compared to the other values, it is almost 50,000 times thinner than the thickness of a virus or hair of humanoid, or around 3-5 atoms broad (100 nm). As a result, nanotechnology focuses on newly developed materials with sizes between 1 and 100 nm (Pitkethly, 2016).

Vandan Prasada (2016) presented the limited concept of innovative nano-technique through the lecture presented by him and it was named as "There's Plenty of Room at the Bottom." He proposed things by using different tools that could be operated surely and accurately at the level of atomics employ single atoms and close molecules. Current known nanomaterials include nanoparticles, quantum dots (semiconductor nanomaterials),4 carbon nanotubes, self-assembled peptide nanotubes, and numerous others. As per the definated presented by the United States NNI: "Nanotechnology is the basic knowledge as well as regulation of matters at specific dimensions that is almost 1 to 100 nanometers in size, where exclusive phenomena permit innovative applications." They can be utilised in consumer goods, environmental engineering, communication, medicine, and catalysis (Chen, 2011).

Since nanotechnology will have a significant impact on practically all businesses and all facets of civilization in its advanced form, it is frequently referred to as a universal purpose technology.3 As a result, there are many opportunities for this technology to be developed in forensic investigation (Hazarika, 2008).

Nanotechnology is the term used to designate the study area of science, engineering, and technology carried out at a scale that varies from 1 to 100 nanometers. Due to its small size, vast surface area, and improved reactivity, nanotechnology has the potential to be used in electronics, diagnostics, biosensing, imaging, optical devices, and drug delivery (Prasad, 2020). It is now considered a technology with a universal purpose due to its widespread applicability in practically all fields. As a result, there are several applications regarding this universal technique specifically in the branch of forensic research (Paikrao Hariprasad, 2021).

The newest and fastest-growing area of novelty which deals in the use of nanotechnology is called nano-forensics. With the development of nanosensors, huge, heavy instruments have been replaced by considerably smaller chip-based platforms. This significantly helps with criminal investigations, also identifies unspecified evidence in a quicker examination which is considered as the best way in the process of investigation in more accurate and sensitive way to solve crimes (Pandya, 2018), (Chauhan, 2017).

A completely new field of forensic research, nano-forensic is a beautiful fusion of forensic science with nanotechnology. As mentioned previously, nanoforensics aids in the identification and examination of evidence at the nanoscale level; yet, due to the instrument's detection limit, it is not able to evaluate the crucial evidence. Nano analysis is revolutionising the investigative process by improving the techniques' accuracy, speed, and sensitivity. Applications for fingerprint analysis, explosive detection, drug screening, dangerous substance analysis, and DNA analysis are all prominent uses of - and other technologies (Lodha, 2016).

FORENSIC INVESTIGATION AND NANOTECHNOLOGY

Identification, appraisal, crime scene investigation, and establishing links between specific pieces of evidence and offenders are the core concerns of forensic science. A brand-new, highly developed field of forensic science called nano-forensics examines the activity of terrorist by looking at the existence of specific items including, explosive gases, organic matters, and remains in the process of criminal activity (Sirivdiya, 2016). An essential benefit of adopting nanotechnology in forensic science is that it makes concealed evidence visible, that might be help the forensic scientists conclude their investigations.

The use of nanotechnology in forensic science, in addition to the use of sophisticated analysis techniques, has been crucial in locking up the accused without leaving any traces of the investigation. Nanotechnology plays a significant role in a number of processes, including the development of latent fingerprints, the identification of illegal drugs, the measurement of alcohol in drunk drivers, the detection of explosives, nerve gas, saliva, inorganic pigments in hit-and-run accidents, and many more (Bokhari, 2018).

In two different ways, nanotechnology benefits forensic sciences. The ability to identify and analyse samples at the nanoscale allows for the analysis and use of crucial evidence that was previously impossible to gather and analyse because of the manufacturing of instruments was not specific and limited. In addition, namomaterials have brand-new qualities that can help with evidence collecting and detection that wasn't before possible. Tiny amounts of gunshot residue, heavy metals, explosives, DNA on fingerprints or palmprints, and soon are a few instances.

According to the Foreign law enforcement administration's there is need to think about the funding to promote the practical on the forensic uses of nanotechnology (Allard, 2016). According to the National Institute of Justice (NIJ) of the United States Department of Justice (USDOJ) has dual discrete areas of project that integrate nanotechnique, including the Research of DNA and Development ($1.0 million) and Chemical and Biological Defense ($1.0 million), according to the National Nanotechnology Initiative's fiscal year 2006 budget request. Fundamental research and the demonstration of chip-based or microdevice technologies for DNA analysis in forensic applications make up the DNA Research and Development programme.

The Chemical and Biological Defense initiative is focused on creating a wearable, affordable device to warn users when they may be exposed to unanticipated chemical and biological threats so they can take appropriate precautions (Chen, 2011).

Figure 1. Nanotechnology in forensics
Source: Alok Pandya (2018)

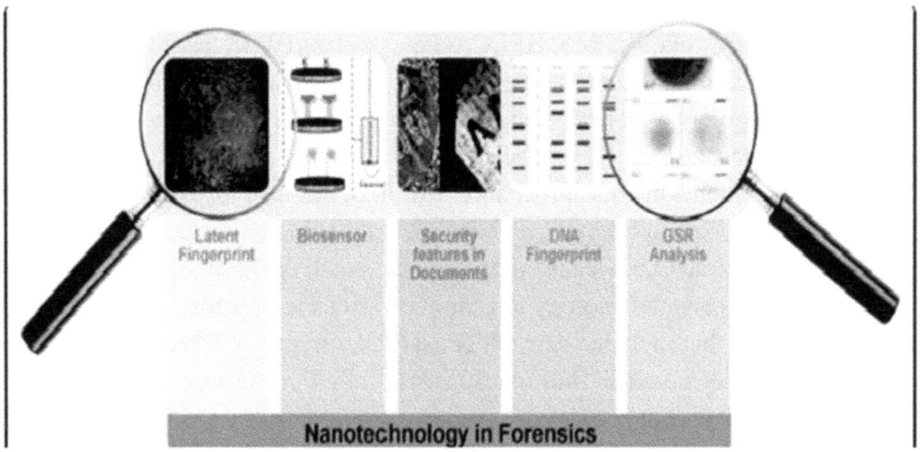

Table 1. Nanotechnology and forensic science

Nanotechnology		Forensic Science
Nano sensors		Forensic Toxicology
Nano Particles		Screening of drug-facilitated crime
Nanotechnology of DNA	Forensic Nanotechnology	Forensic DNA Examination
Nanolithography		Forensic Fingerprint Visualization
Scanning probe microscopy		Forensic Explosive and Blast Residues Analysis
Nanorobotics		Gunshot residue
Nano Beams		Question Document analysis
Nano Plates		Time Since Death TSD

In the identification of crimes using nanotechnology, nano-analysis is frequently used. These analysis methods include Raman microscopy, Atomic Force Microscopy, Dynamic Light Scattering, Transmission Electron Microscopy, Scanning Electron Microscopy, and Atomic Force Microscopy (Arita, 2004). These methods help forensic investigators in two ways: first, by enabling the analysis of nanoscaled samples, and second, by utilising the unique properties of nanomaterial to locate and gather evidence that would not have been possible with earlier methods. One of the cutting-edge techniques that makes it easier for forensic scientists to offer conclusive proof is the extraction of DNA through fingerprints, prints of palm, explosive, gun residues, and heavy metals (Ashraf, 2016).

APPLICATION OF FORENSIC NANOTECHNOLOGY

To paint a whole picture of how nanomaterials might be used in forensic research, however, application and improvement insinuations should also be observed. Through this way, any probable for danger or ambiguity will be completely investigated (Shukla 2013). Following are some branches of Forensics in which future perspective of Nano technology is clear;

Forensic Toxicological Analysis

By the use of nano-technology like Tof-MS, HPLC, XPS as well as FT-IR has an extensive variety of both authorized medications including the loratadine and paracetamol also the prohibited substances includes ecstasy and cocaine are regularly distinguished and acknowledged. Nano-grams of pharmaceuticals and formulations of drug can now be found in a variety of media that includes the hairs, urine, as well as blood thanks to modern advancements in these procedures. Modern forensic investigators now have access to a wealth of data about drugs, paraphernalia, and excipients that includes the morphology, composition of components and stiffness surface because to the advancement of these techniques (Irvin, 2003).

Forensic toxicology uses nanotechnology to examine various toxic substances from a variety of significant forensic evidences that includes hair, blood, saliva, vitreous humour, and even the remains of skeletal and samples evidence obtained through fingerprinting. For the sake of increasing the detection limit some of the nanoparticles are recurrently employed including the gold titanium oxide and silver (Cleaveland, 2001). Also the vitreous humour of the body, which contains Lidocaine hydrochloride can be detected. The low-cost, active, stable, and time-limited nanosensor produced with these novel methodologies might be employed as an instant spot test and a significant replacement for on-the-field testing methodologies regarding forensic toxicological drug screening. The usefulness of the nano sensors for toxicological investigation was demonstrated using forensic nanotechnology on genuine specimens (Basu, 2004).

Screening of Drug-Facilitated Crime

The demand for illicit narcotics is always increasing. Cannabis, amphetamine-type compounds, cocaine, and heroin continue to be the most frequent illicit drugs, but relatively, new psychoactive agents have also been increasing in the market (Binnewies, 2006). Date rape is the specific tern that is used to refer the drug-facilitated crime (DFC). These crimes include burglary, money exaction, and physical damage to individuals of all kind of ages whereas under the stimulus of psychotropic constituents. The purpose of this crime is to disturb an individual who is consuming illicit substances to stabilize their behavior, awareness, or capability to make conclusions.

This kind of shortages possibly are the main lead part to have benefit of the incapacitated individual, significantly or mistakenly. Subsequently illicit constituents are not the part of the sample that is being questioned and also brought to the forensic science research workplace for the sake of investigation and detection. The predictable procedures of illicit substance testing are linked through numerous complications that includes sensitivity, limitations of equipment, and most importantly cost-effectiveness. In comparison to that, application of nanotechnology in the pharmacological business is critical for drug examination in order to evade these difficulties.

A "smart" approach was recently created for the quick recognition and codeine (COD) quantification sulphate using smartphone cameras and citrate-stabilized gold nanoparticles (AuNPs) as a probe. Camera of smartphone is being used by this technique for the detection of ultrasensitive colorimetric fluctuations transported on through the existence or nonappearance of codeine sulphate in the specific probe. This procedure makes the discovery of illegal substances rapid, easy, and transportable (Budowle, 2005). Additionally, a comparable research team published a highly accurate, cost-efficient, and selective approach for the detection of minute quantities of clonazepam in the occurrence of melamine based on AuNPs.

Forensic DNA Examination

One of the most problematic thing faced by our modern civilization is and increasing rate of dreadful crimes. The civil and criminal judicial systems stand to gain much from DNA analysis. In forensic investigations, it establishes a person's identify. DNA analysis is now influenced by nanoparticle-based techniques because of its affordability, simplicity of automation, and ease of usage. Because of their improved sensitivity and high DNA yield, magnetic nanoparticles are specifically utilised to extract DNA.

Nucleic acid was extracted from four distinct sources, including bacteria, yeast, human blood, and viruses, by (Biteen, 2016) using Fe3O4 nanoparticles. The outcomes demonstrated that magnetic nanoparticles might be used to harvest DNA. The latest and cutting-edge tools used for forensic DNA analysis are microfluidic devices. These devices have the advantages of quicker inspection times, lower contamination risks, and direct application to crime scenes. Microfluidic chip technology is another crucial advancement that has already shown promise in medical settings, including point-of-care applications.

DNA Typing in Forensic

A finest-quality PCR extraction is vital for forensic DNA typing. Copper, silica and magnetic nanoparticles are employed to extract good quality, PCR-ready DNA samples from skeletal remains and suggestive bodily fluids that are purchased for forensic investigation. Application of magnetite nanoparticles containing carboxylic compounds as adsorbents for PCR amplification helps to isolate DNA (Dance, 2009). The polymerase chain reaction (PCR) is a popular technique for making duplicates of particular DNA fragments. By using PCR, a single DNA molecule can be multiplied into billions of molecules using Polymerase chain reaction. Consequently, PCR is a useful technology in forensic investigation related to DNA that can be used to identify an individual or a group of individuals.

Gold nanoparticles have been found to significantly increase the polymerase chain reaction's (PCR) effectiveness. When a specific quantity of PCR reagent is added to the reaction observations says the time is condensed and the rate of heating and cooling of thermal cycle raised. This entire situation takes place when the quantity of reagents is 0.7 nm of 13 nm (Budowle, 2003).

Forensic Fingerprint Visualization

Fingerprints utilization for the purpose of crime examination is not new it has been utilised as distinguishing evidence since ancient times. Latter it has been used on Babylonian mud drugs as a business dealing. At crime scene, generally came across to plastic, latent and patent fingerprints. The majority of fingerprints found at crime scenes are latent, meaning they cannot be seen with the naked eye and must be processed further (Pitkethly 2010). There are lots of colored material through which latent fingerprints can be produced. For instance, aluminium fragments on a shady or dusky background or carbon black on a white or bright back surface.

The weakness of these kind of constituents is their capability to attribute to the background of the fingerprint in accumulation to help in the process of latent fingerprint lifting. One of the main issue while capturing a clear copy of the fingerprint since they attribute both the print as well as its background (Arita, 2004). Recently, it has become clear that using nanoparticles can create the nano fingerprints, the newest generation of fingerprinting methods. The latent fingerprints are currently being developed using a variety of methods, including cyanoacrylate, silver nitrate, ninhydrin, and iodine fuming.

However, in cases where the prints are even older, the environment may contaminate the latent fingerprints. In such cases, the sensitivity of the fingerprints to these procedures may be diminished. For better detection of fingerprints, nanoscientists have been utilising more complex approaches to leverage the effectiveness of nanomaterials like Gold, Cadmium sulphide and zinc sulphide, for the development of fingerprints that cannot be visualized with naked eye i.e latent fingerprints (Ashraf, 2016).

Since silver nanoparticles were introduced in 1970, latent fingerprints have been created using nanoparticles. The small size of nanoparticles allows them to bond with the minute ridges of a fingerprint, greatly strengthening the ridge features on fingerprints and sweat pores. Silver nanoparticles can be employed as physical developers to see latent fingerprints on paper, according to Morris. As a result of the reaction, organic fingerprint components were converted into silver nanoparticles, which developed the fingerprint as a dark grey or black sliver picture on the porous surface.

Due to their inertness, excellent selectivity, and sensitivity, gold nanoparticles are useful for latent fingerprint analysis. The benefit is that latent prints made by gold nanoparticles can be preserved for a longer period of time. Due to these characteristics, gold nanoparticles are utilised in single metal depo-

Figure 2. Fingerprint visualization using functionalized nanoparticles
Source: Chen (2011)

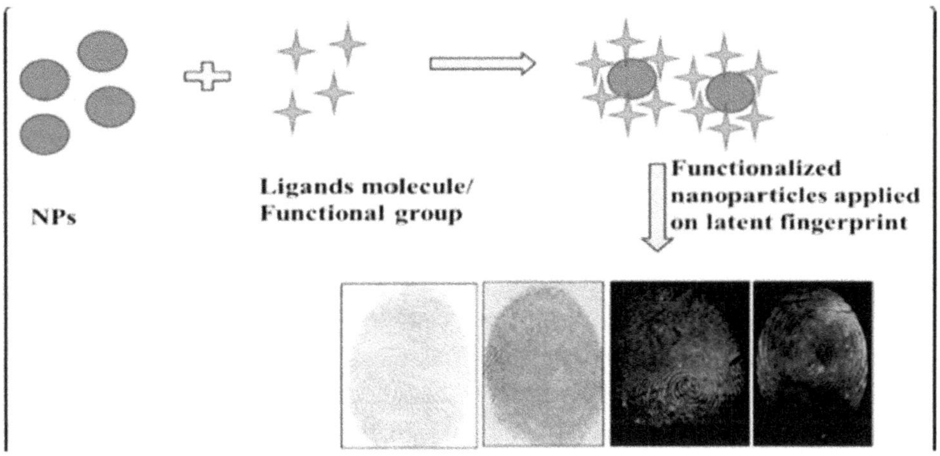

sition and multi-metal deposition to enhance the visibility of latent fingerprints. Due to their small size and exceptional fluorescence intensity, quantum dots and fluorescent materials have attracted a lot of attention. Quantum dots can visualise, according to a study by Dr. Roland Menzel. Bloody fingerprints can also be created using quantum dots. Bloody fingerprints are more likely to smear and be contaminated, which can ruin the ridge details (Blattner, 1997).

Quantum dots, which have the fluorescence property needed to analyse bloody prints, were able to defeat it. According to the study, adding a little amount of ZnOSiO2 nanoparticles to powder improves the ability to see latent print features on the third-level ridge. For the visibility of prints against multicoloured or patterned backgrounds, carbon nano powder has been developing (Bossi, 2004). Nowadays, several categories of have been employed having the properties of nanoparticles. These are employed to different surfaces for the purpose of visualization of the latent fingerprints. A unique ZnO-SiO2 nanopowder was recently created by researchers, and it was employed in combination with the small particle reagent (SPR) approach to create latent fingerprints.

In comparison to other traditional powder processes, this technique produced highly resolved features of finger ridge features. Additionally, it has been noted that titanium dioxide (TiO2-NPs) powder can leave fingerprints on bloody prints as well as the adhesive side of black tape. Because of this, it has been reported that the application of nanoparticles such as Gold, Zinc Sulphide, Cadmium sulphide, Iron oxide, Titanium dioxide, molybdenum disulfide, europium oxide, and noble metals nanocrystals has significantly improved latent fingerprinting methods on a variety of porous and non-porous surfaces by improving the element of sensitivity (Lane, 2007).

Forensic Explosive and Blast Residues Analysis

Global terrorist activity has expanded over the past two decades, bringing attention to the necessity for innovative, cutting-edge technologies to counteract such actions and safeguard the public. Over 4400 of the 12000 terrorist assaults carried out worldwide used explosives. These attacks resulted in over

20,000 fatalities and over 30,000 injuries globally. Since 2005, the prevalence of suicide bombings has increased by 60% (Jain, 2007).

For the law enforcing agencies, the aptness to discern armaments used for mass destruction is the core challenge due to numerous variables like wide range of materials could be used as detonators, the excessive number of implementations, lack of cheap sensors giving amplified sensitivity and selectivity concurrently. Therefore, having improved selectivity and sensitivity as well as the capacity to make it economical associated with the fabrication and deployment of sensors is crucial for success in the fight against terrorism using explosives (Engler, 2002).

Explosives like bombs, grenades, and IEDs are frequently used to commit acts of mass destruction. Different sensing tools have been created during the past year to find explosive traces. Nanomaterials are frequently employed to create low-cost sensors because to their distinctive electrical and optical capabilities. Biologically based sensors are typically used to detect explosives. The TNT can be detected with excellent sensitivity using immunosensing techniques having limit of detection 0.09g/ml. Using the AntiTNT antibody, constructed a capillary immunosensor for the precise detection of TNT (Farrow, 2013).

Nanomaterials derived from carbon like graphene, carbon nanoparticles, and carbon nanotubes have unique electrical properties that make them suitable for use as sensors. They are also chemically inert, have low cytotoxicity, and are highly biocompatible. In order to identify nitroaromatic explosives, Chen et al. employed graphene oxide, which is graphene's oxidized derivative. The graphene-based sensor was built, and it was used to test substances including trinitrobenzene (TNB), dinitrobenzene, and dinitrotoluene (DNT) (Jain, 2005).

When applied to explosives, polymer particles and nanoparticles, as well as different nanosensors such as electronic noses, nanowires or nanotubes, and lasing plasmon nanocavities, create an alteration in one of their measurably changing possessions through the fact of high specificity recognition. Curcumin used as nanocurcumin has been described as a highly efficient method of detecting trace Trinitrotoluene (TNT). Additionally, SERS's non-destructive nature and significant Raman enhancement factors have made it a prime candidate for explosive detection.

Gunshot Residue Analysis

Applying nanotechnology to the analysis of gunshot residue is possible. Following the discharge of a handgun, a shooter's hands frequently have some microscopic gunshot remnants on them. It can be on his clothing or any of the shooter's possessions. The GSR analysis employs high-resolution SEM photography to identify residue particles and X-ray spectroscopy to ascertain the components included in each particle (Hedberg, 2006).

Examination of Ion Beam

For the examination of forensic evidences one of the main instrumental technique is named as Ion beam analysis. This technique more specifically is designed to locate the forensic evidences that can't be seen through naked eye. Explosives, guns, dirt, ink, and fingerprint remnants can all be located, categorised, and examined using this technique. This method can assist to develop a link among the one or more suspect, the crime scene, and the illegitimate conduct (Hahn, 2005). This technique assists in elemental chemical characterization by correlating the chemical compositions of components discovered during the investigation to materials provides as reference. Ion beam analysis is an extremely delicate methodology that can quantitatively test most elements, in contrast to old-style procedures.

The electron beams analysis is a state-of-the-art technique that analyses samples taken from crime scenes using ion beams form milli electron Volt (meV) The ion beam analysis method is used for both element detection and elemental depth profiling in the sub monolayer. The extent of profiling spans from a few nanometers to ten nanometers (Fletcher, 2006). In case of ion beam examination, charged particles from the beam's impact on the sample's surface interacts with the subatomic particles of the atoms like the electrons and nuclei. The following sequence describes how particle emission happens. Both quantitative and qualitative representation of the sample's individual composition can be obtained from these particle or radiation emissions.

Given technique exploits the use of spectrometric methods for elemental analysis of different substances by using spectral peaks and to calculate the optical densities (OD) values of the elements to determine how much of each element is present. This technique can also be used to determine the thickness, composition, and placement of the pieces. Numerous ion beam methodology combined with particulate induced X-ray fluorescence and particulate gamma-ray emission are utilised to evaluate the samples qualitatively and quantitatively. Depending on the nature of crime scene and as per the samples obtained, investigators might select a range of methodologies for analysing samples in this manner (Fornaciari, 2017).

Trackers Through Nanotechnology

Numbers and equipment are used to combat crime. Trackers, which take the shape of a concealed pattern on the item, are used as a preventative technique to deter theft. Apart from preventative measure, sensors can also be helpful to track down those objects that are taken or stolen. Since they can be used to stop prisoners since avoidance, another purpose in which trackers can also be is the security. In this scenario, nano trackers are implanted into the inmates' bodies to aid in the capture of them if they escape. It is now possible to monitor inmates after their release thanks to nano trackers. Nano trackers make it simpler to find them if they carry out another criminal act (Geho, 2006).

Barcodes also assist in deterring crime. Nanotechnology can be used to determine whether an object is real. Several businesses could employ nanotechnology to protect themselves from inferior or illegal products. Products that are duplicated or counterfeited result in huge financial losses and damage the reputation of the individual businesses and their countries. If a customer purchases fake or duplicate items, the authenticity of the products and the firm's reputation will be compromised (Hosseini, 2010). The dependence of the products and the company's reputation are directly associated. Customers and businesses could be harmed as a result of companies manufacturing phoney goods and passing them off as genuine.

The use of watermarks might be beneficial to avoid product duplication by allowing consumers to distinguish between genuine and fake goods (Gilbride, 2006). Through nanotechnology, several instruments are being established that might be able to detect between genuine and counterfeit goods. This is one method that nanotechnology helps to stop counterfeiting. Nanotechnology has produced nanoparticles and nanocrystals with such a wide range of colours to detect counterfeit goods. These nanosheets and nanocrystals detect variety of light sources in the field of forensic to decode the codes. Thus, this method has successfully eliminated fraud-related crime (Eysenbach, 2006).

Moreover, there are lots of crimes including fraud, hit-and-run as well as robbery that can be examined through the support of nano technique. Nano technique is also important for investigating petroleum hydrocarbon traces in cases where petroleum related products are implicated, for instance in arson investigations. Furthermore, situations involving cyber (that includes mobile and internet) as well as sting procedures are relevant for nanotechnology (verification of different types of tapes also the identification of orators). In these situations, the detectives examine the cassettes that are suspected by looking for a coating of nanoparticles on their surface that serves as an authentication mark (Godoy, 2003).

Time Since Death (TSD)

An essential step in a medico-legal investigation is determining the time since death (TSD). To determine the cause and time of death is the main task and its proper evaluation can be done with the help of nanomaterials. The older method can be used to calculate the TSD utilising a variety of parameters. These criteria include the variations in the eye, rigour mortis, algor mortis, postnatal epistasis, fluctuations in degradation, abdominal region and intestinal fillings, pee bladder components, and anecdotal evidence.

With the help of these variables only estimation of death time can be find out. The alterations in the body fluid that are observable includes the pericardiac fluid, blood, fluid of synovial, fluif of spinal region, aqueous humour as well as vitreous humour. Nanomaterials are beneficial to find out body fluids changes to define TSD, instantly or lately after death (Budowle, 2005). Vitreous humour (VH) is the only bodily fluid that doesn't change even after a long stretch of time of death. It is established that most of the biological fluctuations more specifically the level of amino acid comes under the biochemical alterations. So, it is finalized through literature that biochemical examination is the important factor to examine the post mortem changes or TSD (Harrison, 2014).

The amino acid cysteine can be easily determined using a smart, rapid, subtle, inexpensive, and lab-on-chip technique that was lately established (Budowle, 2005). This technique is the central part for the estimation of TSD approximately to 96 hours until cystine intensity since VH upsurges knowingly and shows a direct association among growth of TSD. This methodology can be used to generate a microfluidic scheme through keen approach-based finding in the nearby future that will be able to assess the cysteine content and associates through the estimation of TSD (Ilic, 2004). Currently, Williams along with the coworkers have established a flow cytometry-based approach to assess TSD.

In entire procedure, rate of DNA degradation of spleen as well as tissues of brain were tested up to 96 h and brain tissues are acceptable in comparison to spleen. Additionally, a potential use of flow cytometry for TSD estimate was carefully explained in another work. In this work, erythrocytes in polluted VH could be found using a flow cytometry-based approach. This technique aided in the screening of samples whose biochemical outcomes would not be reliable (Malik, 2022).

Flow cytometry is the most widely used approach nowadays for estimating TSD because to its increased sensitivity when compared to older methods (Engler, 2002). According to this flow cytometry-based approach would speed up the practise of authorized medicine, beneficial to prevent erroneous interpretations, and decrease the amount of court cases still waiting (Hahn, 2005). In upcoming future, fluorescent nanoparticles will be able to detect the concentration of amino acid from VH, and subsequent quantification will be evaluated by the usage of flow cytometry, which could be a method to quantify TSD exactly.

Figure 3. Smartphone enabled platform for the time since death determination from vitreous humor cystine
Source: Chen (2011)

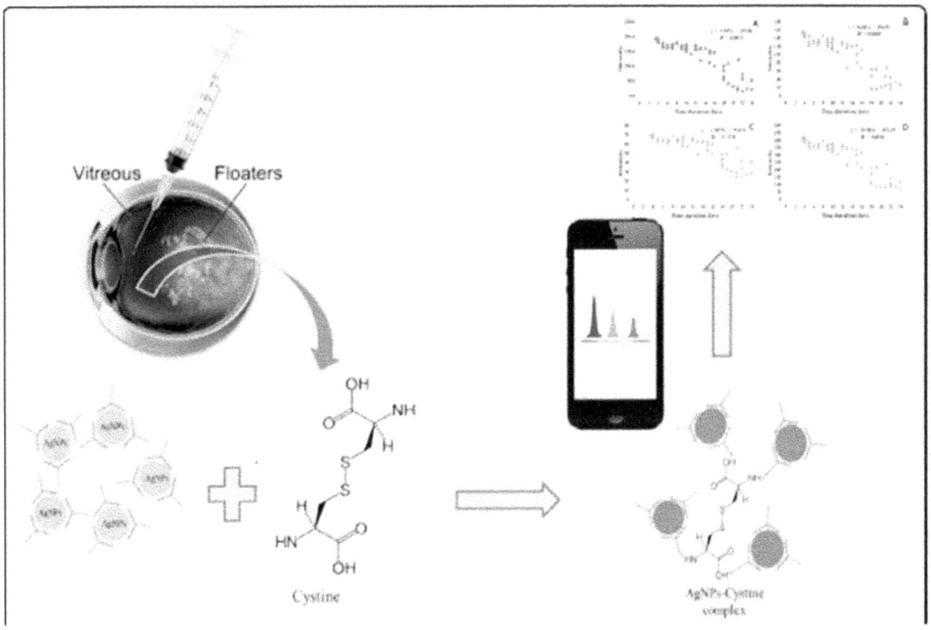

Question Document Analysis

With the development of methods and tools that can manipulate molecules on the nanoscale, technologies for studying forensic constituents have correspondingly advanced at the level of nanoscale (in the order of nm). One such cutting-edge tool that can accomplish this is AFM. This is a particular kind of microscopic technique which can see crossing of ink through examining the page to ascertain the order of the ink blows (Kostoff, 2007). Additionally, the subject offers 3-D exterior morphology that is helpful aids in determining critical details like the order of ballpoint pen strokes or ribbon dye.

In addition, metal, carbon, and silicon-based nanostructured materials have been created with the of help mass spectroscopy in surface-assisted laser desorption/ionization mass spectrometry (SALDI-MS). Solvent-free gold nanoparticle have been developed by the researchers and these are helpful in LDI-MS approach to identify inks of picture also the visual dyes on banknotes and disputed document that need to be analyzed. They then analysed the order of several strokes of ink to trace fabrication or modifications (Partouche, 2017).

Security

Nanoparticles in the field of forensic science as well as cybercrime can modernise the way an enquiry is being carried out through the reduction of the time, cost, and degree of ability required, as well as by boosting the accurateness and approachability of nano gathering and investigative devices as presented in (Chen, 2012). Applied research in nanotechn is still in its early stages, many implementation strategies and drifts are still being expected yet not been demonstrated to work. Numerous challenges, including

Figure 4. Ink crossing scan area; red and blue color of ink
Source: Chen (2011)

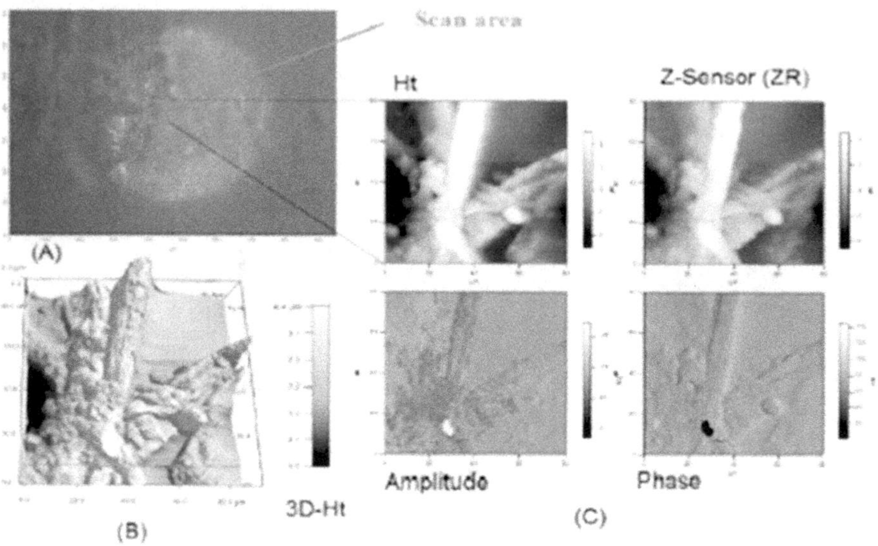

the absence of uniform terminology, the possibility for "nanoparticles" to disrupt the power balance in the globe," and health and safety concerns, are posed by the usage of nanotechnology in daily life (Sastry, 2011).

Conflicts in politics, morality, and law could be researched by linking the security domain more deeply. The greatest societal problem in India is the overwhelming ignorance about the security sector. Conducting a growing number of educational programmes at all levels, educating individuals about the safety segment and giving them admittance to their data in total safety, may close this gap. The necessity for particular regulations to avoid illegal contact or vanishing is unquestionably the result, though.

Additionally, the transfer of authority from the government and non-official stakeholders who continue to possess the expertise and knowledge to construct the technology used may raise the issue regarding how to manage and control and make sure the quality assurance of technology's incorporation of legal standards, that could be more problematic through the ambiguity of nanotechnology. The emergence of nanotechnology has also raised some of the most commonly mentioned moral and lawful matters. This is due to the development of tiny, covert technologies that will enable the gathering and dissemination of information about individuals or companies to rivals (Bajpai, 2018), (Pavlov, 2014).

However, because to nanodevices' modest size and eventual widespread accessibility, both of the government as well as private organizations will be able to obtain sensitive personal data without violating anyone's privacy (Shukla, 2019). Accountability is another vital subject under consideration since nanotechnologies may influence and control human behaviour. It could be challenging to determine the motivation behind a behaviour when it is externally managed. Furthermore, when there is a dealing through a lot of non-essential info and options, individual freedom can be questioned.

Additional research into the gadgets' practicality and development would interfere with already-running systems, such as police and government management (Katrina 2008). How nanomaterials and neuroscience are integrated was explained by Katrina. She emphasised that nanomaterials may have an effect on sensible psychology by supplying new information about the alleged individual's psychological states (modification, suppression, or integrating), which is relevant to the case process of investigation.

Not to mention, while technical progress has always included some dangers, benefits have typically surpassed hazards. This is also achievable thanks to nanotechnologies, which present a wealth of ground-breaking security protection opportunities (Kapoor, 2021). The theoretical and actual prospects for research will both be enhanced by all of these factors. Extensive regulation that takes into account all difficulties that have been recognised and strict oversight of design advancement can lessen hazards from a legal and ethical perspective without imposing superfluous rules (Singh, 2021).

CURRENT CHALLENGES AND FUTURE ASPECT OF NANOTECHNOLOGY IN FORENSIC INVESTIGATION

With numerous applications to conventional techniques in the domains of medicine, electronics, engineering, and other disciplines, nanotechnology has managed to show its superiority. In forensic science, improved nanosensors, nanomanipulators, and nanoimaging tools for visualisation have been created as a result of nanotechnology research, opening up new perspectives for real-time crime investigation. With great precision, sensitivity, and a reduction in the amount of time needed, nanotechnology frequently helps in boosting and increasing the effectiveness of already applied and current forensic techniques.

As a result, present study examines the critical uses of traditional nanotechnology techniques for verifying documents, determining the age of bloodstains, and determining the length of time since a death, as well as how they might be used to enhance PCR effectiveness, DNA analysis, and identification of explosive. It also makes an effort to showcase the usage of cutting-edge nanotechnology tools for drug testing. Fingerprinting and security measures have a bright future thanks to nano-based technology. As a result, present work also provides information about how nanoparticles can be used to find latent fingerprints.

Nanotechnology, which involves manipulating matter at the level of atom and molecule level for the creation of innovative constituents with the help of unusually diverse and distinctive properties, is a rapidly growing field of study through enormous possibility in a variety of industries, including production, physical science, and health care. It offers to revolutionise the use of micro-fluidic classifications in forensic research for DNA extraction media, post-polymerase chain reaction (PCR) measurement, and fingerprint imaging utilising nanopowder. Generally, forensic nanotechnology will reduce analysis times and convert cumbersome instruments into compact chip-based platforms, making investigations more pertinent, timely, and sensitive.

The advances in DNA reporting in bioengineering from the past 50 years has had a parallel influence on current expertise as has nanotechnology. Just to make sure and categorize, personalize, and assess suggestion, forensic science has the capability to apply information and methodologies through natural science. Formerly, through the help of evidence, crime scenes can be reconstructed, investigations can also be engaged, and criminals will be accused. These purposes have been attended through the use of nanotechnology. Nano-analysis is one of the nanotechnologies that is most frequently used in forensic science, including tools like the atomic force microscope as (AFM), TEM as, transmission electron microscope, Raman microspectroscopy, (Micro-Raman) as well as scanning electron microscope (SEM). The ideas of nanotechnology, composition, and well-known forensic applications are being covered through this chapter. Additionally, future prospects and the toxicities of nanomaterials examination is also covered.

REFERENCES

Allard, M. W. (2016). Commentary: The Future of Whole-Genome Sequencing for Public Health and the Clinic. *Journal of Clinical Microbiology, 54*(8), 1946–1948. doi:10.1128/JCM.01082-16 PMID:27307454

Alok Pandya, R. K. (2018). New perspective of nanotechnology: Role in preventive forensic. *Egyptian Journal of Forensic Sciences*, 2–6.

Arita, I., Nakane, M., Kojima, K., Yoshihara, N., Nakano, T., & El-Gohary, A. (2004). Role of a sentinel surveillance system in the context of global surveillance of infectious diseases. *The Lancet. Infectious Diseases, 4*(3), 171–177. doi:10.1016/S1473-3099(04)00942-9 PMID:14998504

. Ashraf, S., Pelaz, B., Pino, P. D., Carril, M., Escudero, A., Parak, W. J., ... Carrillo-Carrion, C. (2016). Gold-based nanomaterials for applications in nanomedicine. *Light-Responsive Nanostructured Systems for Applications in Nanomedicine*, 169-202.

Bajpai, V. K., Kamle, M., Shukla, S., Mahato, D. K., Chandra, P., Hwang, S. K., & Han, Y. K. (2018). Prospects of using nanotechnology for food preservation, safety, and security. *Yao Wu Shi Pin Fen Xi, 26*(4), 1201–1214. PMID:30249319

Basu, M., Seggerson, S., Henshaw, J., Jiang, J., del A Cordona, R., Lefave, C., Boyle, P. J., Miller, A., Pugia, M., & Basu, S. (2004). Nano-biosensor development for bacterial detection during human kidney infection: Use of glycoconjugate-specific antibody-bound gold NanoWire arrays (GNWA). *Glycoconjugate Journal, 21*(8), 487–496. doi:10.100710719-004-5539-1 PMID:15750790

Binnewies, T. T., Motro, Y., Hallin, P. F., Lund, O., Dunn, D., La, T., & Ussery, D. W. (2006). Ten years of bacterial genome sequencing: Comparative-genomics-based discoveries. *Functional & Integrative Genomics, 6*(3), 165–185. doi:10.100710142-006-0027-2 PMID:16773396

Biteen, J. S., Blainey, P. C., Cardon, Z. G., Chun, M., Church, G. M., Dorrestein, P. C., & Young, T. D. (2016). Tools for the microbiome: Nano and beyond. *ACS Nano, 10*(1), 6–37. doi:10.1021/acsnano.5b07826 PMID:26695070

Blattner, F. R., Plunkett III, G., Bloch, C. A., Perna, N. T., Burland, V., Riley, M., & Shao, Y. (1997). The complete genome sequence of Escherichia coli K-12. *Science, 277*(5331), 1453-1462.

Bokhari, H. (2018). Exploitation of microbial forensics and nanotechnology for the monitoring of emerging pathogens. *Critical Reviews in Microbiology, 44*(4), 504–521. doi:10.1080/1040841X.2018.1444013 PMID:29513060

Bossi, P., Tegnell, A., Baka, A., Van Loock, F., Hendriks, J., Werner, A., & Gouvras, G. (2004). Bichat guidelines for the clinical management of glanders and melioidosis and bioterrorism-related glanders and melioidosis. *Eurosurveillance, 9*(12), E17–E18. doi:10.2807/esm.09.12.00507-en PMID:15677841

Budowle, B., Johnson, M. D., Fraser, C. M., Leighton, T. J., Murch, R. S., & Chakraborty, R. (2005). Genetic analysis and attribution of microbial forensics evidence. *Critical Reviews in Microbiology*, *31*(4), 233–254. doi:10.1080/10408410500304082 PMID:16417203

Budowle, B., Schutzer, S. E., Einseln, A., Kelley, L. C., Walsh, A. C., Smith, J. A., Marrone, B. L., Robertson, J., & Campos, J. (2003). Building microbial forensics as a response to bioterrorism. *Science*, *301*(5641), 1852–1853. doi:10.1126cience.1090083 PMID:14512607

Chauhan, V., Singh, V., & Tiwari, A. (2017). Applications of nanotechnology in forensic investigation. *Int. J. Life. Sci. Scienti. Res*, *3*(3), 1047–1051.

Chen, S. Z., Tsai, T. L., & Chen, Y. F. (2012). Forensic application of atomic force microscopy—Questioned Document. *Journal of the Chinese Chemical Society (Taipei)*, *59*(3), 283–288. doi:10.1002/jccs.201100739

Chen, Y. (2011). Forensic Applications of Nanotechnology. *Journal of the Chinese Chemical Society (Taipei)*, *58*(6), 828–835. doi:10.1002/jccs.201190129

Cleaveland, S., Laurenson, M. K., & Taylor, L. H. (2001). Diseases of humans and their domestic mammals: Pathogen characteristics, host range and the risk of emergence. *Philosophical Transactions of the Royal Society of London. Series B, Biological Sciences*, *356*(1411), 991–999. doi:10.1098/rstb.2001.0889 PMID:11516377

Dance, D. A. B. (2009). Melioidosis and glanders as possible biological weapons. In *Bioterrorism and infectious agents: a new dilemma for the 21st century* (pp. 99–145). Springer. doi:10.1007/978-1-4419-1266-4_4

Engler, K. H., Efstratiou, A., Norn, D., Kozlov, R. S., Selga, I., Glushkevich, T. G., & George, R. C. (2002). Immunochromatographic strip test for rapid detection of diphtheria toxin: Description and multicenter evaluation in areas of low and high prevalence of diphtheria. *Journal of Clinical Microbiology*, *40*(1), 80–83. doi:10.1128/JCM.40.1.80-83.2002 PMID:11773096

Eysenbach, G. (2006). Infodemiology: Tracking flu-related searches on the web for syndromic surveillance. *AMIA ... Annual Symposium Proceedings - AMIA Symposium. AMIA Symposium*, *2006*, 244. PMID:17238340

Farrow, B., Hong, S. A., Romero, E. C., Lai, B., Coppock, M. B., Deyle, K. M., & Heath, J. R. (2013). A chemically synthesized capture agent enables the selective, sensitive, and robust electrochemical detection of anthrax protective antigen. *ACS Nano*, *7*(10), 9452–9460. doi:10.1021/nn404296k PMID:24063758

Fletcher, J., Bender, C., Budowle, B., Cobb, W. T., Gold, S. E., Ishimaru, C. A., Luster, D., Melcher, U., Murch, R., Scherm, H., Seem, R. C., Sherwood, J. L., Sobral, B. W., & Tolin, S. A. (2006). Plant pathogen forensics: Capabilities, needs, and recommendations. *Microbiology and Molecular Biology Reviews*, *70*(2), 450–471. doi:10.1128/MMBR.00022-05 PMID:16760310

Fornaciari, A. (2017). Environmental microbial forensics and archaeology of past pandemics. *Microbiology Spectrum*, *5*(1), EMF-0011–EMF-2016. doi:10.1128/microbiolspec.EMF-0011-2016 PMID:28233511

Geho, D. H., Jones, C. D., Petricoin, E. F., & Liotta, L. A. (2006). Nanoparticles: Potential biomarker harvesters. *Current Opinion in Chemical Biology*, *10*(1), 56–61. doi:10.1016/j.cbpa.2006.01.003 PMID:16413816

Gilbride, K. A., Lee, D. Y., & Beaudette, L. A. (2006). Molecular techniques in wastewater: Understanding microbial communities, detecting pathogens, and real-time process control. *Journal of Microbiological Methods*, *66*(1), 1–20. doi:10.1016/j.mimet.2006.02.016 PMID:16635533

Godoy, D., Randle, G., Simpson, A. J., Aanensen, D. M., Pitt, T. L., Kinoshita, R., & Spratt, B. G. (2003). Multilocus sequence typing and evolutionary relationships among the causative agents of melioidosis and glanders, Burkholderia pseudomallei and Burkholderia mallei. *Journal of Clinical Microbiology*, *41*(5), 2068–2079. doi:10.1128/JCM.41.5.2068-2079.2003 PMID:12734250

Hahn, M. A., Tabb, J. S., & Krauss, T. D. (2005). Detection of single bacterial pathogens with semiconductor quantum dots. *Analytical Chemistry*, *77*(15), 4861–4869. doi:10.1021/ac050641i PMID:16053299

Harrison, J. W., Dung, T. T. N., Siddiqui, F., Korbrisate, S., Bukhari, H., Tra, M. P. V., Hoang, N. V. M., Carrique-Mas, J., Bryant, J., Campbell, J. I., Studholme, D. J., Wren, B. W., Baker, S., Titball, R. W., & Champion, O. L. (2014). Identification of possible virulence marker from Campylobacter jejuni isolates. *Emerging Infectious Diseases*, *20*(6), 1026–1029. doi:10.3201/eid2006.130635 PMID:24856088

Hazarika, P. J., Jickells, S. M., Wolff, K., & Russell, D. A. (2008). Imaging of latent fingerprints through the detection of drugs and metabolites. Angew. *Angewandte Chemie International Edition*, *47*(52), 10167–10170. doi:10.1002/anie.200804348

Hedberg, C. W., & Besser, J. M. (2006). Commentary: Cluster evaluation, PulseNet, and public health practice. *Foodborne Pathogens and Disease*, *3*(1), 32–35. doi:10.1089/fpd.2006.3.32 PMID:16602977

Hosseini, S. J. F., Dehyouri, S., & Mirdamadi, S. M. (2010). The perception of agricultural researchers about the role of nanotechnology in achieving food security. *African Journal of Biotechnology*, *9*(37), 6152–6157.

Ilic, B., Yang, Y., & Craighead, H. G. (2004). Virus detection using nanoelectromechanical devices. *Applied Physics Letters*, *85*(13), 2604–2606. doi:10.1063/1.1794378

Irvin, C. B., Nouhan, P. P., & Rice, K. (2003). Syndromic analysis of computerized emergency department patients' chief complaints: An opportunity for bioterrorism and influenza surveillance. *Annals of Emergency Medicine*, *41*(4), 447–452. doi:10.1067/mem.2003.104 PMID:12658241

Jain, K. K. (2005). Nanotechnology in clinical laboratory diagnostics. *Clinica Chimica Acta*, *358*(1-2), 37–54. doi:10.1016/j.cccn.2005.03.014 PMID:15890325

Jain, K. K. (2007). Applications of nanobiotechnology in clinical diagnostics. *Clinical Chemistry*, *53*(11), 2002–2009. doi:10.1373/clinchem.2007.090795 PMID:17890442

Kapoor, N., Sulke, P., Shukla, R. K., Kakad, R., Pardeshi, P., & Badiye, A. (2021). Forensic analytical approaches to the dating of documents: An overview. *Microchemical Journal*, *170*, 106722. doi:10.1016/j.microc.2021.106722

. Kesarwani, S., Parihar, K., Sankhla, M. S., & Kumar, R. (2020). Nano-forensic: new perspective and extensive applications in solving crimes. *Latent in Applied Nanobioscience, 10*(1), 1792-1798.

Kostoff, R. N., Koytcheff, R. G., & Lau, C. G. (2007). Technical structure of the global nanoscience and nanotechnology literature. *Journal of Nanoparticle Research, 9*(5), 701–724. doi:10.100711051-007-9224-8

Lane, C. R., Hobden, E., Walker, L., Barton, V. C., Inman, A. J., Hughes, K. J. D., & Barker, I. (2007). Evaluation of a rapid diagnostic field test kit for identification of Phytophthora species, including P. ramorum and P. kernoviae at the point of inspection. *Plant Pathology, 56*(5), 828–835. doi:10.1111/j.1365-3059.2007.01615.x

Lodha, A. S., Pandya, A., & Shukla, R. K. (2016). Nanotechnology: An applied and robust approach for forensic investigation. *Forensic Res Criminol Int J, 2*(1), 00044.

Malik, A. A., & Hassan, S. M. (2022). The ZnO Nanoparticles: Synthesis, Characterization and Applications in Forensics. *International Journal for Electronic Crime Investigation, 6*(1), 8–8. doi:10.54692/ijeci.2022.060195

Paikrao Hariprasad, D. A. (2021). *Nano-Forensics a Comprehensive Review.* Crimson Publishers.

Pandya, A., & Shukla, R. K. (2018). New perspective of nanotechnology: Role in preventive forensic. *Egyptian Journal of Forensic Sciences, 8*(1), 1–11. doi:10.118641935-018-0088-0

Partouche, C. F., & Bois, R. S. (2017, August). Questioned Documents. In *17th Interpol International Forensic Science Managers Symposium, Lyon 8 th-10 th October 2013* (*Vol. 526*, p. 854). Academic Press.

Pavlov, A. J., & Batova, V. N. (2014). Ensuring economic security of the innovative development of nanotechnology in the Russian Federation and foreign countries. *Life Science Journal, 11*(6), 322.

Pitkethly, M. (2009). Nanotechnology and forensics. *Materials Today, 12*(6), 6. doi:10.1016/S1369-7021(09)70167-1

Prasad, V. L., Lukose, S., Agarwal, P., & Prasad, L. (2020). Role of Nanomaterials for Forensic Investigation and Latent Fingerprinting—A Review. *Journal of Forensic Sciences, 65*(1), 26–36. doi:10.1111/1556-4029.14172 PMID:31454084

Sastry, R. K., Rashmi, H. B., & Rao, N. H. (2011). Nanotechnology for enhancing food security in India. *Food Policy, 36*(3), 391–400. doi:10.1016/j.foodpol.2010.10.012

. Shukla, V., Chaturvedi, A., & Srivastava, N. (2019). Nanotechnology and cryptographic protocols: issues and possible solutions. *Nanomaterials and Energy, 8*(1), 78-83.

Singh, S., & Samal, N. (2021). Nanotechnology: A Powerful Tool in Forensic Science for Solving Criminal Cases. *Arab Journal of Forensic Sciences & Forensic Medicine, 3*(2), 273–296. doi:10.26735/YDDJ4516

Vandan Prasada, S. L. (2016). Emerging Forensic Applications of Nanotechnology. *Int J Eng Allied Sci,* 1-8.

Chapter 15
International Activities in Nano-Forensics

Ahtisham Shuja Abbasi
University of Healh Sciences, Pakistan

Rahat Abdul Rehman
University of Health Sciences, Pakistan

Amna Anwar
University of Health Sciences, Pakistan

Umme Ammarah Mehak
University of Health Sciences, Pakistan

ABSTRACT

Nanotechnology has proven its supremacy with enormous applications to outdated methods in medical, electronic, engineering, and other fields. In forensic science, nanotechnology research has delivered a new viewpoint for real-time crime investigation and established progressive nano sensors, nano-manipulators, and nano-imaging tools for visualization. Often, nanotechnology helps in augmenting and improving the productivity of already existing and applied forensic techniques with high precision, sensitivity, and reducing the time required. So, this chapter reviews international activities in nano forensic in different branches of forensic science (e.g., for inspecting questioned documents, age of bloodstains, and time since death, along with its application in improving PCR efficiency, DNA analysis, and explosive detection). Therefore, in this chapter, an effort has been made to highlight the new outlook or knowledge of nanotechnology with its applications in several divisions of forensic science at an international level.

SUMMARY

Nanotechnology has consistently demonstrated its unparalleled quality through massive applications to obsolete techniques in clinical, electronic, design, and other fields. In the field of forensic science, nanotechnology research has given another viewpoint to continuous misconduct examination and estab-

DOI: 10.4018/978-1-6684-8325-1.ch015

lished reasonable Nano sensors, Nano-controllers, and Nano-imaging devices for perception. Frequently, nanotechnology aids in the expansion and enhancement of previously existing and practical methods with high accuracy, awareness, and reduced time requirements. In that method, this section examines Global practices in Nano technology in various areas of forensic sciences, such as reviewing addressed reports, age of bloodstains, and post mortem interval, as well its application further developing PCR productivity, DNA examination, and sensitive identification. Furthermore makes an effort to highlight the use of cutting-edge nanotechnology tools for illegal medication screening. Nano-based practices have a bright future in fingerprinting and security highlights. As a result, this chapter also provides information on the applications of nanoparticles in recognising idle fingerprints. The development of Nano sensors replaces large cumbersome instruments with much smaller chip-based stages. This greatly aids false examination as well as recognising unknown proof in a more limited examination technique for examination with greater precision and responsiveness for resolving violations. As a result, this chapter is an effort to present a new perspective or information on nanotechnology and its applications in a few divisions of Forensic sciences on a global scale.

WHAT ARE NANOTECHNOLOGY AND NANO FORENSICS?

Introduction

Nanotechnology is a branch of science, with important applications in a variety of other fields. Nanotechnology also has a important impact on forensic science. Nano-forensics is the use of nanotechnology-based methods to solve forensic science cases. (Kumar & Sharma, 2021). This is known as Nano-forensics, and it is the most recent advancement in the field of forensic science. Nanotechnology aids technology that deals with nanometer-sized matters. The term "nanoparticle" is typically applied to inorganic materials opposed to single molecules. Ultrafine particles are the same as nanoparticles and range between 1 and 100 nm in size, whereas fine particles range between 100 and 2,500 nm, and coarse particles range between 2,500 and 10,000 nm. (Brayner et al., 2012). There are many ways to make nanoparticles, including chemical, physical, and biological processes. While the chemical technique of synthesis only requires a little time to produce a high number of nanoparticles, this method requires capping agents to maintain the size of the nanoparticles. Nanoparticle production and stabilisation chemicals are toxic and produce byproducts that are bad for the environment. The demand for green nanotechnology is driven by the need for environmentally safe, non-toxic synthetic processes for the manufacture of nanoparticles. (Bar et al., 2009).

Nanoparticles put forward fundamentally advantageous platform, validating unique properties with potentially wide-ranging application (Murray, Kagan, & Bawendi, 2000). Many research teams have using biological systems for the synthesis of nanoparticles due to the numerous advantages they offer over non-biological systems. Due to a number of reasons, comprising their similar size to biomolecules like proteins and poly nucleic acids, nanoparticles have unique capabilities and are useful (Akhlaghi,Peng, Yao, & Tam, 2013).

Figure 1. To advance the science of criminal detection and analysis, nano-forensic integrates nanotechnology into the forensic sector
Source: Prasad, Lukose, Agarwal, and Prasad (2020)

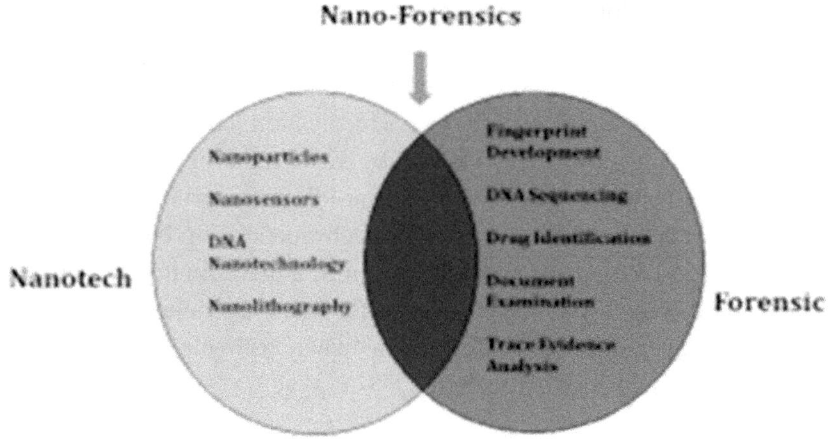

History of Nanotechnology

Famous physicist Richard Feynman initially proposed the ideas that gave rise to nanotechnology in 1959 in his lecture "There's Plenty of Space at the Bottom," in which he declared the potential of synthesis through direct atom manipulation. Norio Taniguchi coined the phrase "nanotechnology" for the first time in 1974, albeit it was not widely used at the time.

K. Eric Drexler coined the term "nanotechnology" in his 1986 book Engines of Creation: The Coming Era of Nanotechnology, which put forth the concept of a Nano scale "assembler" capable of creating copies of itself and other objects of arbitrary complexity with atomic control. Drexler was inspired by Feynman's ideas. Thus, Drexler's theoretical and public work, which created and popularized a conceptual framework for nanotechnology, and highly visible experimental advancements, which increased public awareness of the potential for atomic control of matter, directed to the arise of nanotechnology as a field in the 1980s. Since its popularity surged in the 1980s, the majority of nanotechnology research has focused on exploring various methods for creating mechanical devices out of a light number of atoms. (Lee et al., 2009).

Application of Nanotechnology in Forensic Science

Today, the use of Nanoforensic science has effectively altered the investigation processes by making them quicker, more accurate, more effective, more sensitive, and simpler to use, which emphasises unquestionable importance of this technologies. The use of reactive ('smart') materials, microchip technology, nanomanipulators, and visualisation tools in forensic science has received little attention from research teams (Pandya & Shukla, 2018). Regarding its application in many forensic scientific fields, such as forensic DNA typing, forensic finger print visualisation, and others, nanotechnology has a variety of roles to play. Nanotechnology's use in the many forensic science fields (Lodha, Pandya, & Shukla, 2016).

Figure 2.

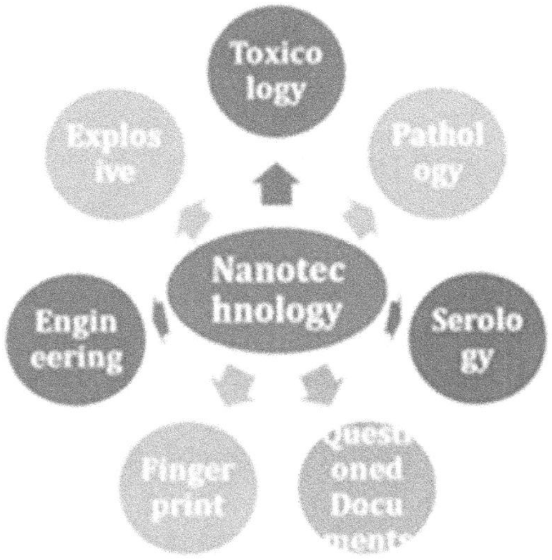

Electrochemical Sensors and Biosensors in Forensic Science

In order to create electrochemical sensors and biosensors, a diversity of nanoparticle arrangements, containing oxide, metal, and semiconductor nanoparticles, have been produced. These nanoparticles play a variety of roles in various sensing systems in various forensic scene settings. Biomolecules can be immobilised, electrochemical reactions can be catalysed, electron transport between proteins and electrode surfaces is improved, biomolecules can be labelled, and nanoparticles can still operate as reactants. Nanoparticles are incredibly appropriate for building new and upgraded sensing devices, particularly electrochemical sensors and biosensors, due to their distinctive chemical and physical features (Liu *et al.*, 2003). Forensic DNA typing most typically employs the use of gold nanoparticles to immobilise proteins. Gold electrodes that had been modified with a cyst amine monolayer were first coated with gold nanoparticles, and the horseradish peroxidase was subsequently successfully immobilised on these

Figure 3. Application of nano sensors in different areas of forensic sciences

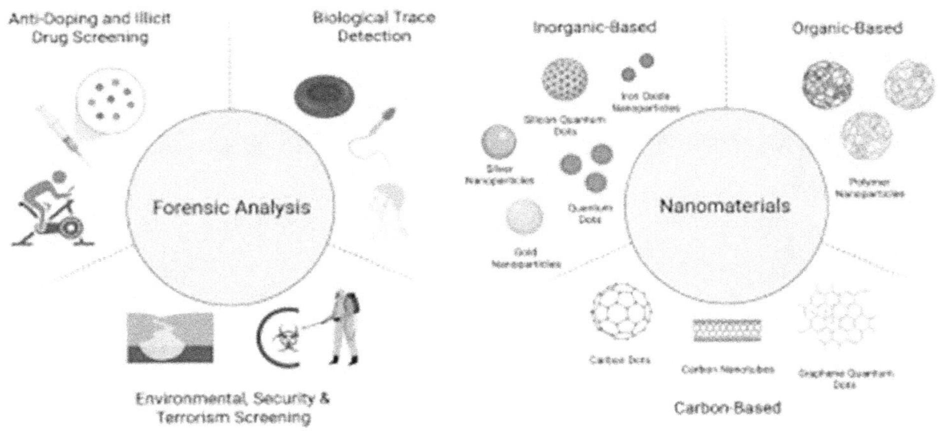

nanoparticles. DNA is a different class of biomolecule that can also be immobilised using nanoparticles to make electrochemical DNA sensors. The DNA strands are frequently modified with precise functional groups that can cooperate effectively with the nanoparticles in order to immobilise DNA onto the surfaces of nanoparticles.

Nano-Forensic in Genetics and DNA Typing

One of the most significant and distinctive pieces of evidence that could pinpoint a person's presence at the crime scene is DNA (genetic material) (Smith, 2022). To increase PCR efficiency, first used gold nanoparticles with a size of 0.7 nm of 13 m Au. Excellent heat transmission properties and the ability to boost electrochemiluminescence are both found in AuNPs (ECL). Hence, the response time during the thermal cycle's heating or cooling is reduced, increasing speed. When those nanoparticles are utilised, the sensitivity of conventional PCR is improved by around 5–10 times, and by as much as 10,000 times in real-time PCR (Rohatgi & Dominica, 2022).

In forensic genetics individual typing is very affected steps to identified person involved in criminal activity through successful isolation of nucleic acid from biological material and well performing PCR. There are various techniques are present for the nucleic acid isolation for example commercially available kit but these techniques are time consuming and are not e enough to controlled contamination successfully, in recently year many significant progress and development in different technique of DNA isolation is required currently. Nanotechnology is a latest research for DNA isolation, in this technology different type of nano particles are used which are coated with various polymer in order to isolate biomolecules like protein, DNA and RNA from different biological material like blood, fluid, semen, saliva and hair the Nano technique are cost effective, comprehensive, robust and not affected by DNA degradation and low quantity during purification (Yonar, Karatas, Erkan, & Yukseloglu, 2022). The most modern and cutting-edge tools for forensic DNA examination are microfluidic devices, which have

Figure 4. Different steps of DNA profiling

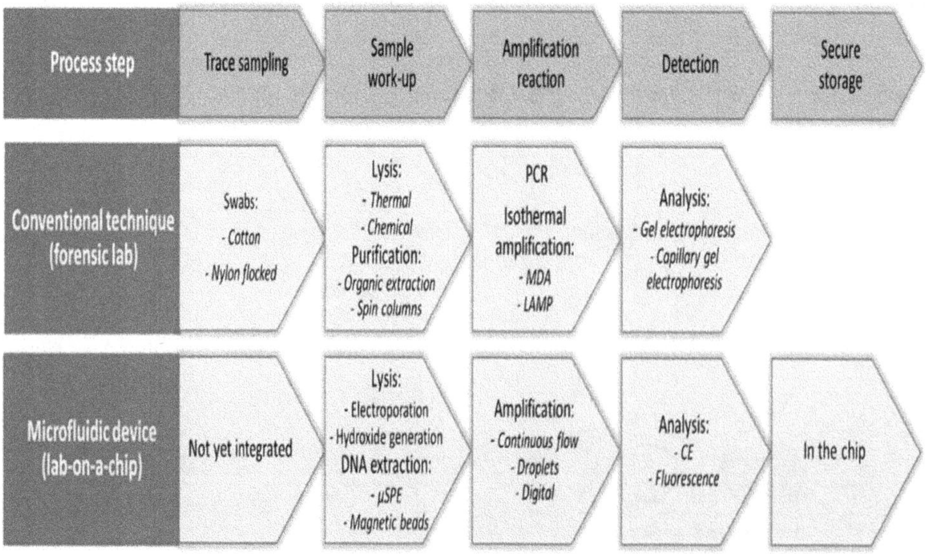

the advantages of quicker analysis times, lower contamination risks, and direct application at the site of the crime. Furthermore helpful and efficient in medical applications (Bruijns, Van Asten, Tiggelaar, & Gardeniers, 2016). To extract PCR with outstanding quality, copper and silica-based magnetic nanoparticles are utilised. As adsorbents for PCR amplification, carboxylic-containing magnetite nanoparticles are employed (Tang et al., 2020).

Nano-Forensic in Fingerprinting and Visualization

Finger markings are the remnants of the finger's ruptured ridges. These fingerprints make it possible to recognize a person at crime scenes, the three types of fingerprints—plastic, patent, and latent—are frequently observed... Most often, latent fingerprint can't be seen with the naked eye or that need additional processing are discovered at crime scenes. In most cases, latent fingermarks are created utilising fingerprint powders like carbon black on white or light backgrounds and aluminium flake on black or dark backgrounds. Nevertheless, one drawback of these materials is their lack of ability to attach to the backdrop of the latent fingerprint. Because they stick to both the print and the backdrop, it is difficult to get a clear image of the fingerprint, which makes identification challenging. In this circumstance, nanoparticles that have been manufactured or designed are being employed to remedy this problem. (Pandya & Shukla, 2018; Srividya, 2016). In the past few years, a number of initiatives to provide quick, accurate, and affordable methods for the proof of identity of latent fingerprints have developed. The most used techniques to date are powder dusting, fluorescent color staining, ninhydrin splashing, vacuum metal deposition, cyanoacrylate/iodine fuming, and small particle reagent approach because of its simplicity, effectiveness, and usability.

Even though the conventional approaches are effective in the majority of cases, there is an crucial requirement for simple and effective ways to produce the latent fingerprints with better contrast, sensitivity, and selectivity. These difficulties include attempting to distinguish latent fingerprints on multicoloured backgrounds with great contrast and minimal background interference.

Due to their highe surface area, improved stability, Brightness, nanoparticle materials have sparked a lot of interest. Even at very low concentrations of the analytes, fluorescence detection typically has a high sensitivity. Quantum dots, metal colloids, and fluorescent up conversion nanoparticles, for instance, have been used to successfully visualize fingerprints However, conventional fluorescent materials are typically monochromatic, making it challenging to apply them to objects with intricate multicolor backgrounds and difficult to tell fingerprint lines from multicolor backgrounds .

An all-inclusive nanomaterial with high responsiveness, low foundation impedance, and tunable radiance hue is needed against a multicoloured background in order to identify fingerprints on items with complex multicoloured backgrounds. To identify latent fingerprints, new hybrid polystyrene nanoparticles (PSNPs) were created. All hybrid compounds have exhibit brilliant, steady fluorescence, which qualifies them for sensitive fingerprint visualisation (X. Wang et al., 2022).

Numerous additional types of nanomaterials are used, such as metal nanoparticles, metallic oxide nanoparticles, semiconductor quantum specks, carbon dabs, polymer spots, fluorescent silica nanoparticles, fluorescent mesoporous silica nanoparticles, fluorescent silica nanoparticles, formed polyelectrolyte dots uncommon earth fluorescence nanoparticles, and aggregation induced emission luminogens molecule incorporated nanomaterials, Some of the nanomaterials used to identify latent fingerprints did not produce accurate fingerprint images (Prabakaran & Pillay, 2021). However, there have also been reports of the employment of specialist tools like different kinds of microscopes or Raman spectrometers. While some

Figure 5.

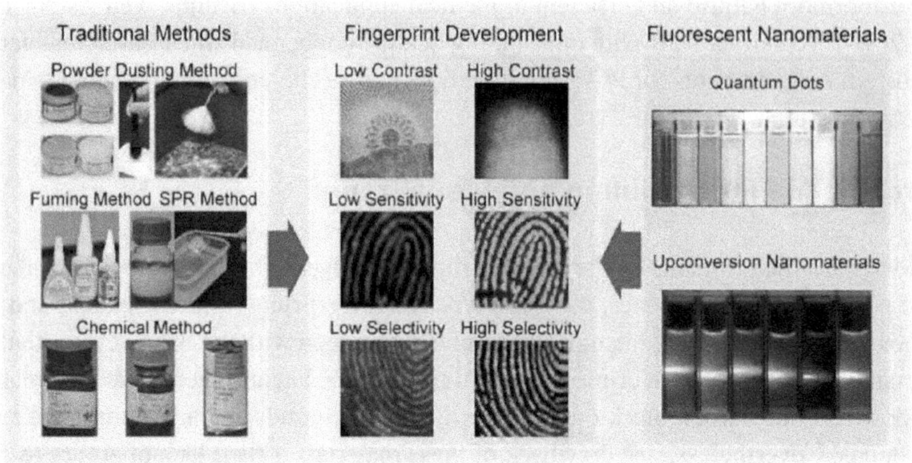

Table 1. Explain different nanoparticle application for fingerprint development

Nano-Particles	Methods
Poly(p-phenylenevinylene) (PPV) Nanoparticles	**Immersing**
Silica Nanoparticle	**Dusting and Spray**
Silica Nanoparticle	**Dusting method**
Aluminum oxide nanoparticles	**Dusting method**
Organosilicon fluorescent nanoparticles	**Spray Method**
Oligomer/silica hybrid nanoparticles	**Spray**

of the suggested approaches seem unworkable for actual use, others have great potential. But generally speaking, it can be claimed that most NPs are not yet ready to be used in operational forensic facilities for the purpose of fingermark detection (Kanodarwala, Moret, Spindler, Lennard, & Roux, 2019).

Nano-Forensic in Explosives

Explosive based terrorism has been greatly increased in recent times due to their ease of availability and use. There are the many sensitive techniques for their detection but at present most of these techniques are only applicable inside of the laboratory setting. In this scenario there is the need of on spot, selective, sensitive and robust detection technique that can also work on the real crime scenes samples. Different conventional techniques are already present and in use to check the explosives involved in bomb blasts and other criminal activities. Nanotechnology gained scientists attraction to provide the solution with all these mentioned aspects and provide effective methods of criminal investigations that can aid the forensic community at large.

The application of modern nano-sensors based instruments can produce robust results, on spot testing of evidence on the sight of crime rather than proper collection, packaging and transportation of the

Figure 6.

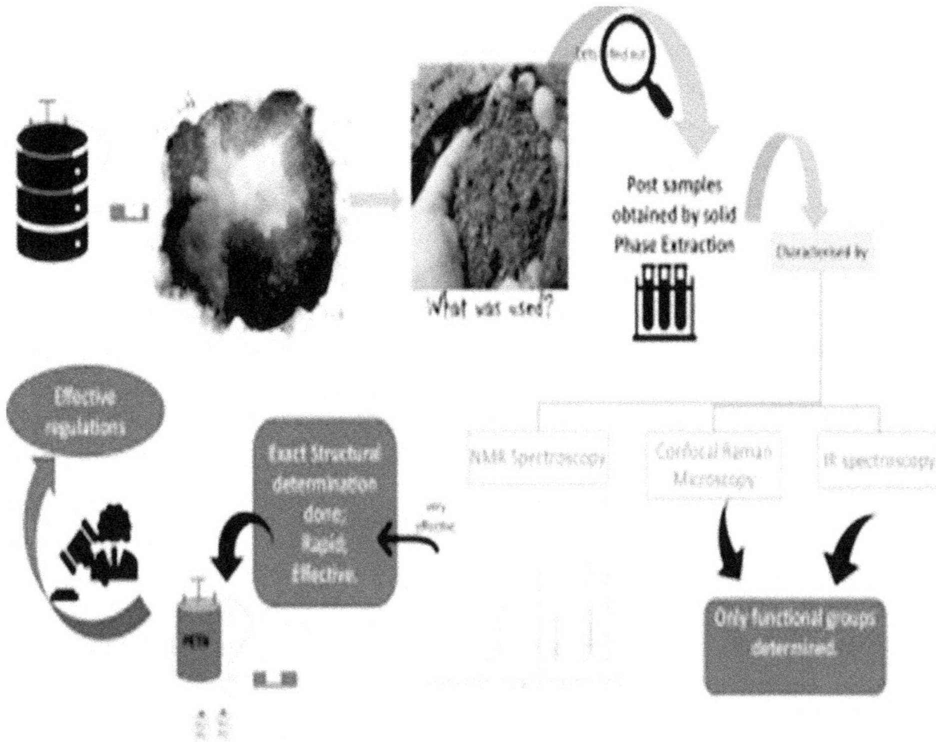

sample to laboratory. Importantly the results can be produced in a small period of time which will increase sensitivity and accuracy of the investigation process and results in the quick resolving of crimes. Nanoparticles synthesized from inorganic transition metal compounds such as molybdenum disulphide (MoS_2) have band gaps that can be sized from both directions in the form of single layers, allowing applications such as electroluminescent devices, transistors and photo detectors. These nanoparticles based sensors can be develop in to photoluminescence based devices that displayed a high sensitivity toward nitroaromatic compounds (having nitro group as functional group attached with aromatic ring) which are highly explosive. These sensors can target molecules like 2, 4, 6- Tri-NitroToluene (TNT), 2, 4, 6 Tri-Nitro-Phenol (TNP) and other phenols. The uniqueness of molybdenum based nanoparticle sensors will results in quick detection of explosive materials that will give information from the critical and fragile evidences collected from crime scene.

Through different scientific techniques and methods, we can analyze the evidence collected from the crime scene and connect different things and solve the crime (Pandya and Shukla, 2018). Meanwhile, the need for explosive materials either for military operations or for terrorist activities has been increased. Certain crimes scenes involve particularly the use of explosive material in the context of the crime scene(Prasad et al., 2020) (Prasad et al., 2019). One of the chemical groups of explosive nature is the nitro group. When the nitro group is attached as a functional group with another chemical compound that compound is changed into an explosive material. Similarly, when the nitro group is attached with aromatic compounds then nitroaromatic compounds are synthesized (To, Ben-Jaber, & Parkin, 2020).

In the last decades, nanotechnology has emerged and there is an increasing demand for inorganic nanoparticles in forensic sciences. In forensic science, the development of Inorganic Nanosensors, Nano-manipulators and Nano-imaging devices have opened up new ways for visualization that provides effective and quick ways for the crime scene investigation and later processing (Bhatt et al., 2020). Nanotechnology is not only improving the efficiency of already existing techniques but has also aided in providing the new forensic techniques with an increased amount of sensitivity, specificity and less processing time (Bhatt et al., 2020).

These nano-based sensors comprise a sensitive technique that is inexpensive and can result in the detection of a vast variety of materials. Hence, nanosensors have high specificity and high sensitivity and a low limit of detection (LOD) with relatively low processing cost (Rawtani et al., 2019, To et al., 2020). The use of modern nano-sensors-based devices such as electronic noses, nano-tubes, nano-curcumin-based probes and nanowires have specific features that can result in their adherence with materials and cause their rapid and specific detection. Modern research on nano-particles has also reported the use of nano curcumin as nanoparticles of curcumin for a very inexpensive method of detecting toxic materials like TNP, TNT and phenol (To et al., 2020).

Nano-Forensic in Question Document

Question document examination starts with classifying the kind of device used to fabricate the document tracked by physical and chemical exploration of inks in order to discern, if some transcribed entries might have been added or altered, association of two or more inks, linear sequence of two or more ink contributions that interconnect each other and the relative ages/backdating of inks(Jones, Cody, & Mc-Clelland, 2006). Nano-forensics is the modern revolution and has a far-reaching influence in the field of forensic science. Incognito evidence which are worrisome to identify by conventional techniques could be easily related with nano-sensors and nano-devices (Prasad et al., 2020).

Figure 7.

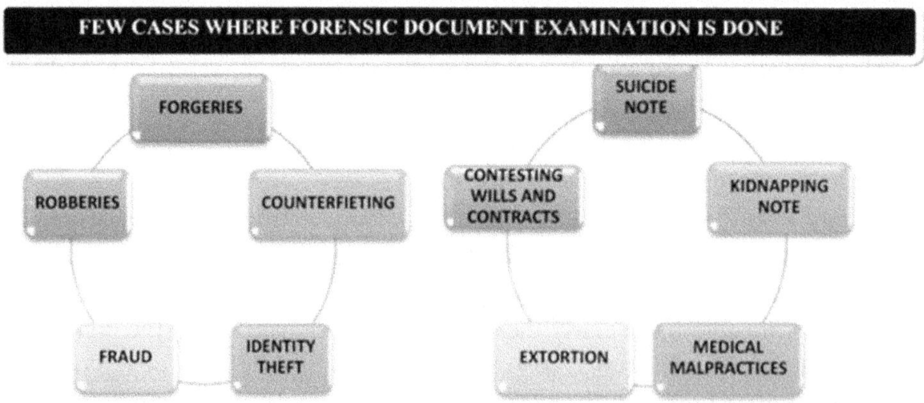

Technology has advanced thanks to questioned document (QD) analysis, enabling nanoscale analysis of the subject matter (nm) (Shukla, 2016). Investigating the microscopic components of the ink might provide a wealth of information in cases of fake documents using ink. AFM is one such cutting-edge technique that can accomplish this. Page scanning to determine the arrangement of ink strokes, this type of microscope can visualise ink crossing. Also, it withdraws 3-D surface morphology, which aids in determining critical details regarding the order of ballpoint pen strokes or ribbon dye. (Huynh, Phelps, Golden, & Verbeck, 2016; Jones et al., 2006). Furthermore, silicon, carbon, and metal nanostructured components developed to help mass spectroscopy in surface-assisted laser desorption/ionization mass spectrometry (SALDI-MS). Researchers developed a solvent-free gold nanoparticle assisted LDI-MS approach to identify visible dyes, image inks on interrogated documents, such as banknotes, and have classified distinct ink strokes to look for forgeries or alterations (Han et al., 2015).

Nano-Forensic in Automobile Paint

In criminal investigations, painted objects are frequently seen. By examining the paint found at the crime scene, a link between individuals, places, and matters can be established. Examining paint typically entails comparing the paint from the crime scene with a sample of paint that has been taken from a known source and providing crucial clues for the investigation (Wu et al., 2022). In rare circumstances, paint examinations may be helpful. Sample of paint cases of hit-and-run, burglary, and assault (Caddy, 2001). The layered structure of the automotive paint coating may be seen on its cross-section. The type of coating determines the figure, color, viscosity, and chemical makeup of the different layers. The vehicle's body has the most intricate construction, and a tool's surface, like a crowbar, has the simplest. Initial microscopic analyses enable one to see the strat of paint coating particles to examine the chemical arrangement of each layer, the technique of infrared micro-spectrometry and X-ray micro spectrometer is first used to categories the kinds of polymers that make up the coating and identify the principal colours and fillers. This allows for the identification of the automobile varnish type (acrylic, nitrocellulose, alkyd, etc.) (Zięba-Palus, 2017). There are various technique used for examining paint in forensic level but Raman spectroscopy is a highly helpful instrument for forensically examining several colored evi-

Figure 8. Various nanocomposite coating types

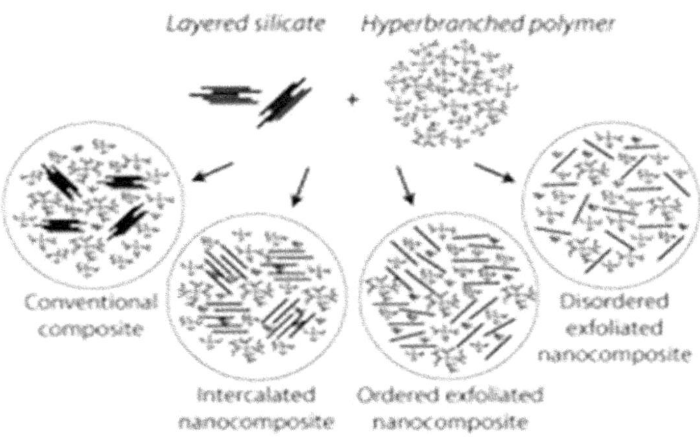

dence, covering its application to pinpoint the pigments and dyes in paint and inks. Forensic scientists often employ a battery of various tests to define paint and inks because both substances are composed of complex heterogeneous matrices. Each test provides data that will be assessed and combined to create an overall compositional profile (Buzzini, Massonnet, & Monard Sermier, 2006; Buzzini & Suzuki, 2016).

Various coatings, various pigments utilized in paint manufacture, and a particular attention is focused on the application of coatings using nanomaterials and fillers. To extend the life of the tire and provide the black color, common nanoparticles like carbon black are employed in automobile tire. There are various nanoparticle are present in automobile vehicles (Mathiazhagan & Joseph, 2011).

Nano-Forensic in Serology

Analyses of body fluids are an important aspect of forensic investigation; therefore, rapid and sensitive detection is required for their analyses as well as characterization. This Chapter presents an outline of the applications of different nanomaterials with respect to the biological material detection. If we talk about forensic biological samples, we especially need to handle them with care, because such samples from crime scenes are always in traces. This chapter briefly discusses the use of functional nanoparticles for the detection of biological materials. Genetic material obtained from crime scenes is one of the most important pieces of evidence in forensic investigations and the chapter focuses on the application of functional nanomaterials for their detection.

Classification of Biological Evidence

There are two major categories of biological evidence found at a crime scene, which are further divided based on their areas of origin, 1) Body Fluids 2) Non-Body Fluids. The biological evidence that is most commonly found and analyzed from crime scenes includes blood, semen, saliva, vaginal secretion, and urine. DNA can be isolated from all biological materials. The type of evidence that can be found and recovered from a crime scene depends on the nature of the crime. In Table 2, several possible locations from which different types of biological fluids can be recovered during crime scene investigation are listed, as well as the possible amount of DNA that can be isolated from the respective biological fluid. However, the amount of DNA extracted may vary depending on the condition of the crime scene, collection, preservation, and the applied analytical techniques.

Nano-Material-Based Biodetection

With the use of nanomaterials of different types, such as Nano chips, Nano probes, and nano-bio sensors, the use of nanotechnology for forensic investigation ranges from the detection of illicit drugs, toxins, and poisons to the detection of fingerprints, analysis of forensic documents, and examination of DNA and body fluids (Amitkumar, 2016). For DNA-based applications of nanotechnology, analytical techniques used in forensic sample analyses, such as polymerase chain reaction (PCR), are optimized by incorporating various nanoparticles to increase the sensitivity of the technique. This resulted in nanomaterial-assisted PCR (nano-PCR). Different nanoparticles, such as AuNPs, are used in nano-PCR to increase the sensitivity, specificity, and efficiency of analytical reactions (D. Pan, 2012). For fingerprint analysis, various nanoparticles, such as AgNPs, AuNPs, Al2O3-NPs, ZnO-NPs, and fluorescent carbon-based nanoparticles have been exploited for developing and detecting fingerprints (Vandana, 2020). Quantum

Table 2.

Biological Fluids	Possible Location of Recovery	Amount of DNA
Blood	i. Pool of blood next to body of crime ii. Weapon of Offense Fingertips Wall Splatter and Trail in Floor iii. Gloves iv. Bullet Casings and Projectile	Liquid Blood: 20000–40000ng/ml Blood Stain: 250–500ng/cm^2
Semen/vaginal fluid	i. Used Condom ii. Clothing iii. Floor and Bed of crime scene	Liquid semen: 150000–300000ng/ml Post-coital vaginal swab: 10–3000ng/swab
Saliva	i. Food ii. Rim of Glass, Bottle, Cans iii. Toothpick iv. Fingernails v. End of Pen vi. Corner of Currency	Liquid saliva: 1000–10000ng/ml Oral swab; 100–1500ng/ Swab
Sweat	i. Clothing ii. Footwear iii. Place of Seating (Chair, backrest, handle) iv. Personal Belongings (Wallet, Watches, Jewelry, Phone Cover)	-

dots of carbon, graphene, and silicon have been utilized for screening and sensing illicit drugs to aid forensic investigations (Ganesan, 2020). For example, graphene-based quantum dots have been reported to detect cyanide ions (Hou, 2013), and Mn-doped ZnS quantum dots have been reported to successfully detect cocaine in saliva (María Pilar, 2018). On the other hand, luminescent quantum dots have been utilized to detect counterfeit inks in forensic examination of documents (Kesarwani, 2020) whereas; quantum dot-based nanohybrid have been developed to detect trinitrotoluene (TNT) (Peng Wu, 2015).

Carbon Nanodots for Biodetection

The carbon dots (CDs) are nanomaterials with a size of less than 10 nm (Bae, 2010). Carbon nanodots (CNDs) are a new class of carbon dots that have high luminescence and water solubility. These CNDs have been utilized in electrochemical sensors because of their DNA sensing ability (Verhagen, 2020; Mulpur, 2015; Loo, 2016).

Carbon Black Nanomaterials for Biodetection

Carbon based nanomaterials have been studied for quite a long time because of their highly customizable properties and cost-effective material. A new type of carbon nanomaterial called Carbon Black has recently piqued the interest of researchers, which is different from carbon nanotube, nanopore and nanorod because of its exploitative functionalization (F. Arduini, 2020). To increase sensitivity and specificity for the analysis of DNA sequences, atomic force microscopy techniques have been optimized by utilizing carbon nanotubes. Carbon black nanomaterials are used in biosensors for detection of various biological materials including DNA, enzymes and antibodies (Aydin, 2018).

Hybrid Nanocomposites Biodetection

Biosensors developed through hybrid nanocomposites have a vast potential to detect biological materials in forensic investigation (Gao, 2017). The hybrid sensing probes of hybrid nanocomposites detect biological material through thermal, optical, or electrical signals (Stewart, 2015; Takahashi, 2018).

DNA Detection

Conventional techniques of DNA detection and analysis are destructive and in forensic investigation, evidence in the form of DNA is most often in minute quantities. Nanomaterials are useful for such trace amount of DNA found at crime scene as they provide rapid, selective and efficient results (Li Fan, 2013). For this purpose, several platforms have already been developed where AuNPs are being exploited for the sensing of DNA. SiNP are reported to have been used for the same purpose owing to their unique geometric properties and higher stablilty (Y. Wang, 2007). Dendrimers, nanowires and quantum dots have been utilized for the detection of DNA (Rawtani, 2019). For the optimization of analytical techniques employed during analysis of DNA, mainly colorimetric technique, AuNPs and AgNPs have been used to functionalize Halloysite Nanotubes (HNT) for detection of DNA (Vikrant, 2019).

Optical Biosensor for Trace Biological Detection

Body fluids frequently found in trace amounts at a crime scene include blood, semen and saliva. The on-site tests performed in routine for detection of these trace biological evidences are the presumptive tests which are highly destructive in nature and thus hinders the forensic investigation. Majority of nanoparticles incorporated biosensors that are used to detect biological evidence found in literature function on electrochemical properties. Functional nanoparticle based biosensors are required that function on optical transduction mechanism (Hayley, 2022). Currently, very limited literature data is present that targets the optical properties of biosensors and nanoparticles for detection of biological materials.

Biosensor for Blood Detection

CdSe/ZnS Quantum Dots were used as nanomaterial which are conjugated with an anti-glycophorin an antibody that targets the glycophorin a marker present in human blood. It was observed that treatment with quantum dots in the biosensor did not have a degrading effect on DNA and full DNA profiles were obtained after detection.

Biosensor for Semen Detection

For the detection of human semen found at crime scenes, various nanomaterials have been utilized in biosensors that target different markers present in semen. Zirconium metal-organic frameworks and Gold nanoparticles have been used in colorimetric biosensor for semen detection which gave positive results with 90% accuracy. Another marker present in semen is the prostate specific antigen (PSA) which has been targeted using polyamine nanoparticles functionalized with magnetic iron oxide cores. The principal of PSA detection is based on the mechanism designed by Frascione et al. for PSA biosensor which is illustrated in Figure 9.

Figure 9. Acid phosphatase present in semen has been targeted using CuInS2 quantum dots based on fluorescent "turn off-on" biosensor application. Presence of acid phosphatase does not allow the aggregation of quantum dots of CuInS2 on the Biosensor and the quantum dots give off fluorescence which indicates a positive detection.

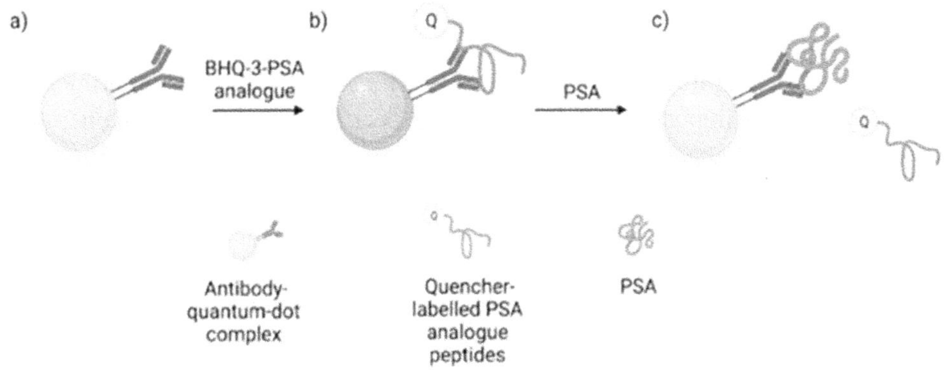

Biosensor for Saliva Detection

Instead of targeting a specific marker present in saliva, a study targeted the specific bacteria unique to saliva fluid. The optical biosensor developed using bacteria as a specific target for saliva detection used silicon carbide nanoparticles which were stabilized using bovine serum albumin and functionalized using antibacterial peptide GH12. The targeted bacteria unique to saliva fluid for its detection is Streptococcus salivarius. A schematic diagram of basic principal of how the biosensor works is given in Figure 10.

Figure 10.

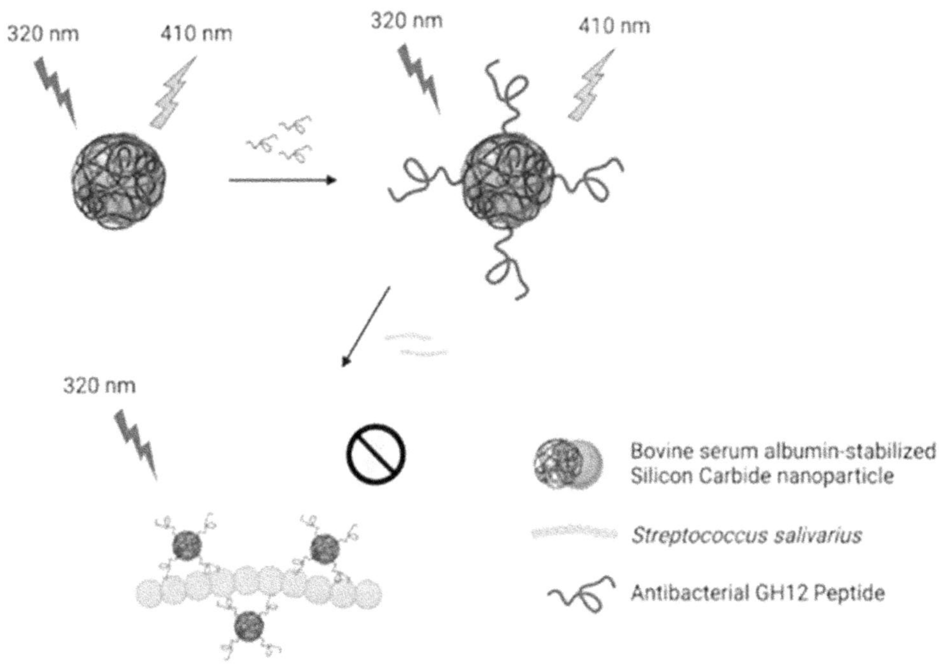

Nano Forensic in Forensic Medicine

Nanotechnology have broad application in forensic medicine to determine postmortem interval and cause of death. Postmortem interval (PMI) is the time that has passed after a human has died. It is used to calculate the range of time of death and constricted the scope of suspects in forensic analysis (Li et al., 2016). Exact determination of PMI remains a task in forensic medicine worldwide. There is no specific selective method to find PMI through validate scientific parameters. In homicidal crimes to find accurate time since death is very important its help to find suspect and time of crime scene. Accurate time span after death help law enforcement agency to resolve many questions (Q. Wang, Lin, Xu, Huang, & Wang, 2018).

Nano biosensor detection of different constituents is also an established practice for determination of time since death. A novel colorimetric detection based smartphone technique was found to be time effective method of estimating postmortem interval from vitreous cystine level (Ansari, Lodha, & Menon, 2016). A silver nanoparticle based technique for measuring vitreous cystine level along with other molecules within the chemical composition of the vitreous was found a promising method of determining postmortem interval. The tryptophan a biomolecule was also detected and estimated for determination of postmortem interval alongside cystine in the vitreous humor through the synthesized nanoparticles(Singh, Sharma, Pathak, & Dutta, 2022). The most widely acceptable and worldly prevalent technique is development and utilization of some nanohybrid sensing platform helping in determining PMI. The technology involves production of bimetallic Nano sensor particles to consume a certain concentration of vitreous potassium in order to predict PMI. Since vitreous potassium is subject to constant rise with increasing time, the fluorescent nanoparticles sensors can sense the concentration of the potassium and can answer the question of PMI. These are instant, evidence based, universal and scientific mean of determining the PM

Table 3. Explain different method to detect biomolecule relative to nanoparticles

Sr No	Methods	Detection ug/ml
1	Flow Injection Spectrofluorometric	0.024
2	Colorimetric	0.2
3	RP-HPLC	3
4	Capillary electrophoresis	24
5	Spectrophotometric	2.4
6	Gold nanoparticles	0.240
7	Silver nanoparticles	0.007

Nano Forensic in Forensic Toxicology

In the most recent advancement, the use of nanotechnology in forensic toxicology has been successful in exposing several dangerous substances and quantifying them from a variety of forensic evidence, including urine, vitreous humour, blood, hair, saliva, and even fingerprints and skeletal remains. Progressive nanotechnologies like HPLC, XPS, and Tof-MS are used to identify a variety of substances, including

Figure 11.

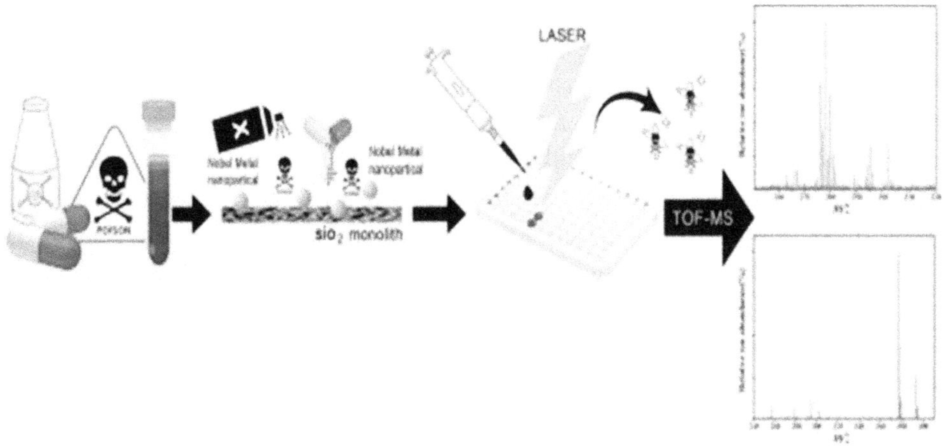

legal medications e.g paracetamol and illegal substances like cocaine, heroin, and other barbiturates. Also, capillary electrophoresis, nano-gold (Au-NPs), nano-silver, and nano-titanium dioxide (TiO2) particles together with SEM, TEM, and FTIR are used in the development (Chen et al., 2015).

Beginning in the late 1990s, functionalized magnetic specks containing antibodies were employed to detect numerous substances and metabolites, including morphine (heroin metabolite), benzoylecgonine (cocaine metabolite), and other illicit drugs (Hazarika, Jickells, Wolff, & Russell, 2008). The ability to distinguish between illicit drugs in saliva utilising a micro fluid system and silver nanoparticles has also been demonstrated. (Andreou, Hoonejani, Barmi, Moskovits, & Meinhart, 2013). Scientists have also originate a quick method for drug recognition in human urine samples by using Dynamic-surface enhanced Raman spectroscopy (D-SERS) and gold nanorods (GNRs) along with the classification algorithm known as support vector machine (SVM)(Dong, Weng, Yang, & Liu, 2015) . A recent study found that employing methods like micro-extraction by packed sorbent (MEPS) and direct analysis at real-time (DART) Tof-MS enhanced the speed of drug detection from urine samples by less than a minute and extraction by less than two minutes. Recently, a "smart" system was created that uses a "smartphone camera" as an analytical tool and citrate-stabilized Au-NPs as a probe to swiftly detect and quantify the amount of codeine sulphate. The smartphone camera was capable of differentiating especially sensitive colour changes, or ultrasensitive colorimetric changes in the nanoprobe, depending on whether codeine sulphate was present or not. This system of nanotechnology provides quick, portable, and on-the-spot drug detection tools. (Anand Lodha, Pandya, Sutariya, & Menon, 2014).

Problem and Solution in Nano-Forensic

It is yet unclear law enforcement organisations and personnel are prepared to trained and use this nanotechnology-based equipment for serious scientific research or if it will remain the domain of science fiction. Knowing that nanotechnology has the ability to benefit society and not only contribute to crime prevention but also crime solving is reassuring. The different fields of forensic science, such as virtual autopsy, fingerprint recognition, crime scene investigation, questioned documents, ballistics, and toxicology, may soon benefit from the inventive and preventive capabilities of nanotechnology. Products manufactured with nanotechnology are supposedly more valuable because developing countries have fewer access to funding to produce them. Nanotechnology is genuinely precious and developing it can bring you a lot of benefits as well as being sufficiently delicate to make.

Being very small these particles can cause environment pollution and also hazardous to human body. Nanoparticles should be plant base. Nano particles should be ecofriendly.

In order to be more environmentally friendly, single-pot processes that do not require extra surfactants, capping agents, or templates have typically been utilised in NP synthesis. Researchers will benefit from the biosynthesis of NPs using waste materials not only to create safer nanomaterials but also to advance knowledge of the health and safety implications of NPs. Because the biomaterial-based methods do not require the use of hazardous chemicals, useful products can be easily manufactured even at acceptable scale. The use of several types of algae in the synthesis of NPs has fostered the development of straightforward, environmentally friendly, affordable, and time-efficient procedures, therefore reducing the need for chemicals and solvents. Algal membranes contain polysaccharides, proteins, and lipids that serve as capping agents, limiting the use of commercial surfactants that are difficult to remove after the synthesis of NPs and are not biodegradable. However, the use of algae is restricted since not all species may be used for synthesis because some of them contain harmful chemicals and because the method of synthesis has not yet been fully uncovered.

The use of waste materials in the synthesis of NPs has been made possible as a result of this limitation. The waste materials, particularly fruit waste, are readily available and don't need to be prepared beforehand. The synthesis process is quite straightforward, requiring little time and effort and predictable mechanisms. This creates a chance for the usage of biodegradable materials, notably when making metal oxide nanoparticles (NPs). Diverse capping agents, including biomolecules and polysaccharides, which can function as chelating/reducing and capping agents for the production of NPs, are the subject of considerable research.

The resultant particles are thereby shielded from further reactions and aggregation, increasing their stability and lifetime. In areas where NPs have significant uses, we think there are substantial potential for establishing industrial scale manufacturing. Overall, the green synthesis of nanoparticles and nanomaterials using algae and waste materials is a new and interesting area of nanotechnology that could have a big impact on future developments in nanoscience. Also, with the availability of computational approaches, the field of green synthesis and the production of nanomaterials will be expanded and may one day be employed as "Nano medicines" in the field of medicine.

Future Aspects of Nano Forensic

In various areas of forensic science, health sciences, and automotive engineering, the development of nanotechnology benefits from the integration of these scientific fields. By combining the requirements of current applications of biomedical and forensic research with the most cutting-edge chemical and physical technologies. The majority of scientists and analysts are now able to pursue effective strategic goals and possess solid nanotechnology skills due to the growing deman (Chauhan, Singh, & Tiwari, 2017).

Table 4. Some of future prospective in nano forensic

1	**Design and creation of novel contrast agents and nanostructured materials for multimodal imaging**
2	**study of thin films for energy transformation and medicinal applications**
3	**Molecularly imprinted polymer (MIPS) preparation for analytical applications as nanoparticles and Nano surfaces**
4	**study of surface phenomena in the use of nanostructures for contemporary molecular separation technologies**
5	**Characterization and physicochemical analysis of the functional materials research into the vibrational characteristics and structure of useful materials**
6	**Cellular nano-engineering**

CONCLUSION

It is established that the numerous applications of nanoparticles, nano-based techniques, and devices in the field of forensic science have produced quick, more precise, sensitive, selective, and effective techniques for investigating crime scenes and examining forensic evidence, which ultimately contribute to solving crimes. With more sophisticated and sensitive methods of investigation, nanotechnology is therefore not only expected to play a significant part in the future of forensic science but also in many other sectors (Kesarwani, Parihar, Sankhla, & Kumar, 2020).

REFERENCES

Akhlaghi, S. P., Peng, B., Yao, Z., & Tam, K. C. (2013). Sustainable nanomaterials derived from polysaccharides and amphiphilic compounds. *Soft Matter*, *9*(33), 7905–7918. doi:10.1039/c3sm50358e

Andreou, C., Hoonejani, M. R., Barmi, M. R., Moskovits, M., & Meinhart, C. D. (2013). Rapid detection of drugs of abuse in saliva using surface enhanced Raman spectroscopy and microfluidics. *ACS Nano*, *7*(8), 7157–7164. doi:10.1021/nn402563f PMID:23859441

Ansari, N., Lodha, A., & Menon, S. K. (2016). Smart platform for the time since death determination from vitreous humor cystine. *Biosensors & Bioelectronics*, *86*, 115–121. doi:10.1016/j.bios.2016.06.042 PMID:27343833

Aydın, M., Aydın, E. B., & Sezgintürk, M. K. (2018). A Highly Selective Electrochemical Immunosensor Based on Conductive Carbon Black and Star PGMA Polymer Composite Material for IL-8 Biomarker Detection in Human Serum and Saliva. *Biosensors & Bioelectronics*, *117*, 720–728. doi:10.1016/j.bios.2018.07.010 PMID:30014946

Bar, H., Bhui, D. K., Sahoo, G. P., Sarkar, P., De, S. P., & Misra, A. (2009). Green synthesis of silver nanoparticles using latex of Jatropha curcas. *Colloids and Surfaces. A, Physicochemical and Engineering Aspects*, *339*(1-3), 134–139. doi:10.1016/j.colsurfa.2009.02.008

Brayner, R., Coradin, T., Beaunier, P., Grenèche, J.-M., Djediat, C., Yéprémian, C., Couté, A., & Fiévet, F. (2012). Intracellular biosynthesis of superparamagnetic 2-lines ferri-hydrite nanoparticles using Euglena gracilis microalgae. *Colloids and Surfaces. B, Biointerfaces*, *93*, 20–23. doi:10.1016/j.colsurfb.2011.10.014 PMID:22225944

Bruijns, B., Van Asten, A., Tiggelaar, R., & Gardeniers, H. (2016). Microfluidic devices for forensic DNA analysis: A review. *Biosensors (Basel)*, *6*(3), 41. doi:10.3390/bios6030041 PMID:27527231

Buzzini, P., Massonnet, G., & Monard Sermier, F. (2006). The micro Raman analysis of paint evidence in criminalistics: case studies. *Journal of Raman Spectroscopy, 37*(9), 922-931.

Buzzini, P., & Suzuki, E. (2016). Forensic applications of Raman spectroscopy for the in situ analyses of pigments and dyes in ink and paint evidence. *Journal of Raman Spectroscopy : JRS*, *47*(1), 16–27. doi:10.1002/jrs.4818

Caddy, B. (2001). *Forensic examination of glass and paint: Analysis and interpretation*. CRC Press. doi:10.1201/9780203483589

Chauhan, V., Singh, V., & Tiwari, A. (2017). Applications of nanotechnology in forensic investigation. *Int. J. Life. Sci. Scienti. Res*, *3*(3), 1047–1051.

Chen, X., Tang, Y., Wang, S., Song, Y., Tang, F., & Wu, X. (2015). Field-amplified sample injection in capillary electrophoresis with amperometric detection for the ultratrace analysis of diastereomeric ephedrine alkaloids. *Electrophoresis*, *36*(16), 1953–1961. doi:10.1002/elps.201500024 PMID:25873262

Dong, R., Weng, S., Yang, L., & Liu, J. (2015). Detection and direct readout of drugs in human urine using dynamic surface-enhanced Raman spectroscopy and support vector machines. *Analytical Chemistry*, *87*(5), 2937–2944. doi:10.1021/acs.analchem.5b00137 PMID:25634247

Ganesan, M., & Nagaraaj, P. (2020). Quantum Dots as Nanosensors for Detection of Toxics: A Literature Review. *Analytical Methods*, *12*(35), 4254–4275. doi:10.1039/D0AY01293A PMID:32940270

Han, Z., Liu, H., Wang, B., Weng, S., Yang, L., & Liu, J. (2015). Three-dimensional surface-enhanced Raman scattering hotspots in spherical colloidal superstructure for identification and detection of drugs in human urine. *Analytical Chemistry*, *87*(9), 4821–4828. doi:10.1021/acs.analchem.5b00176 PMID:25853724

Hazarika, P., Jickells, S. M., Wolff, K., & Russell, D. A. (2008). Imaging of latent fingerprints through the detection of drugs and metabolites. *Angewandte Chemie International Edition*, *47*(52), 10167–10170. doi:10.1002/anie.200804348 PMID:19039814

Hou, X., Zeng, F., Du, F., & Wu, S. (2013). Carbon-Dot-Based Fluorescent Turn-on Sensor for Selectively Detecting Sulfide Anions in Totally Aqueous Media and Imaging inside Live Cells. *Nanotechnology*, *24*(33), 335502. doi:10.1088/0957-4484/24/33/335502 PMID:23892368

Huynh, V., Phelps, M. S., Golden, T. D., & Verbeck, G. F. (2016). Direct analyte-probed nanoextraction (DAPNe) coupled to matrix-assisted laser desorption ionization (MALDI) for examination of the ink chemistry on documents. *Forensic Chemistry (Amsterdam, Netherlands)*, *2*, 86–92. doi:10.1016/j.forc.2016.10.007

Jones, R. W., Cody, R. B., & McClelland, J. F. (2006). Differentiating writing inks using direct analysis in real time mass spectrometry. *Journal of Forensic Sciences*, *51*(4), 915–918. doi:10.1111/j.1556-4029.2006.00162.x PMID:16882239

Kanodarwala, F. K., Moret, S., Spindler, X., Lennard, C., & Roux, C. (2019). Nanoparticles used for fingermark detection—A comprehensive review. *WIREs Forensic Science, 1*(5), e1341. doi:10.1002/wfs2.1341

Kesarwani, S., Parihar, K., Sankhla, M. S., & Kumar, R. (2020). Nano-forensic: new perspective and extensive applications in solving crimes. *Latent in Applied Nanobioscience, 10*(1), 1792-1798.

Kumar, N., & Sharma, A. (2021). *Nano-Forensics: The New Perspective in Precision Forensic Science. In Nanotechnology Applications in Health and Environmental Sciences*. Springer.

Lee, J. Y., Hong, B. H., Kim, W. Y., Min, S. K., Kim, Y., Jouravlev, M. V., ... Kaufman, L. J. (2009). Near-field focusing and magnification through self-assembled nanoscale spherical lenses. *Nature, 460*(7254), 498–501. doi:10.1038/nature08173

Li, C., Wang, Q., Zhang, Y., Lin, H., Zhang, J., Huang, P., & Wang, Z. (2016). Research progress in the estimation of the postmortem interval by Chinese forensic scholars. *Forensic Sciences Research, 1*(1), 3–13. doi:10.1080/20961790.2016.1229377 PMID:30483604

Lodha, A., Pandya, A., & Shukla, R. (2016). Nanotechnology: An applied and robust approach for forensic investigation. *Forensic Res Criminol Int J, 2*(1), 00044.

Lodha, A., Pandya, A., Sutariya, P. G., & Menon, S. K. (2014). A smart and rapid colorimetric method for the detection of codeine sulphate, using unmodified gold nanoprobe. *RSC Advances, 4*(92), 50443–50448. doi:10.1039/C4RA06269H

Mathiazhagan, A., & Joseph, R. (2011). Nanotechnology-a New prospective in organic coating-review. *International Journal of Chemical Engineering and Applications, 2*(4), 225–237. doi:10.7763/IJCEA.2011.V2.108

Mulpur, P., Yadavilli, S., Mulpur, P., Kondiparthi, N., Sengupta, B., Rao, A. M., Podila, R., & Kamisetti, V. (2015). Flexible Ag–C 60 Nano-Biosensors Based on Surface Plasmon Coupled Emission for Clinical and Forensic Applications. *Physical Chemistry Chemical Physics, 17*(38), 25049–25054. doi:10.1039/C5CP04268B PMID:26345678

Murray, C. B., Kagan, C. R., & Bawendi, M. G. (2000). Synthesis and characterization of monodisperse nanocrystals and close-packed nanocrystal assemblies. *Annual Review of Materials Science, 30*(1), 545–610. doi:10.1146/annurev.matsci.30.1.545

Pandya, A., & Shukla, R. K. (2018). New perspective of nanotechnology: Role in preventive forensic. *Egyptian Journal of Forensic Sciences, 8*(1), 1–11. doi:10.118641935-018-0088-0

Prabakaran, E., & Pillay, K. (2021). Nanomaterials for latent fingerprint detection: A review. *Journal of Materials Research and Technology, 12*, 1856–1885. doi:10.1016/j.jmrt.2021.03.110

Prasad, V., Lukose, S., Agarwal, P., & Prasad, L. (2020). Role of nanomaterials for forensic investigation and latent fingerprinting—A review. *Journal of Forensic Sciences, 65*(1), 26–36. doi:10.1111/1556-4029.14172 PMID:31454084

Rao, Pandey, & Tharmavaram. (2020). Biological Evidence and Their Handling. *Technology in Forensic Science*.

Rawtani, D., Tharmavaram, M., Pandey, G., & Hussain, C. M. (2019). Functionalized Nanomaterial for Forensic Sample Analysis. *Trends in Analytical Chemistry, 120*, 115661. doi:10.1016/j.trac.2019.115661

Rohatgi, R., & Dominica, a. F. (2022). Application of Nanotechnology in Forensic Science. *Journal of Forensic Research, 13*(8).

Shukla, R. (2016). Nanotechnology: An Applied and Robust Approach for Forensic Investigation. *Foresic Res. Criminol. Int. J.*, *2*(1), 35–37. doi:10.15406/frcij.2016.02.00044

Singh, G., Sharma, D., Pathak, D., & Dutta, S. (2022). Comparative Study on Post Mortem Analysis of Sodium and Potassium Levels of Vitreous Humour and Synovial Fluid in Determining Time Since Death. *Indian Journal of Forensic Medicine Toxicology*, *16*(3), 393–398.

Smith, J. (2022). *Forensic DNA Investigation. In Handbook of DNA Profiling*. Springer.

Srividya, B. (2016). Nanotechnology in forensics and its application in forensic investigation. *Res. Rev. J. Pharm. Nanotechnol*, *4*(2), 1–7.

Tang, C., He, Z., Liu, H., Xu, Y., Huang, H., Yang, G., ... Deng, Y. (2020). Application of magnetic nanoparticles in nucleic acid detection. *Journal of Nanobiotechnology*, *18*(1), 1–19. doi:10.118612951-020-00613-6 PMID:32316985

To, K. C., Ben-Jaber, S., & Parkin, I. P. (2020). Recent developments in the field of explosive trace detection. *ACS Nano*, *14*(9), 10804–10833. doi:10.1021/acsnano.0c01579 PMID:32790331

Wang, Q., Lin, H., Xu, J., Huang, P., & Wang, Z. (2018). Current Research and Prospects on Postmortem Interval Estimation. *Fa Yi Xue Za Zhi*, *34*(5), 459–467. PMID:30468046

Wang, X., Liao, T., Wang, H., Hao, H., Yang, Q., Zhou, H., Ma, Y., Zhi, M., Wang, J., & Fan, R. (2022). Novel Organic-Inorganic Hybrid Polystyrene Nanoparticles with Trichromatic Luminescence for the Detection of Latent Fingerprints. *International Journal of Analytical Chemistry*, *2022*, 2022. doi:10.1155/2022/2230360 PMID:35295922

Wu, D., Wu, Q., Lu, Y., Wang, C., Yv, S., Wang, L., ... Gao, S. (2022). A novel approach for forensic identification of automotive paints using optical coherence tomography and multivariate statistical methods. *Journal of Forensic Sciences*, *67*(6), 2253–2266. doi:10.1111/1556-4029.15114 PMID:35913098

Yonar, F. C., Karatas, O., Erkan, I., & Yukseloglu, E. H. (2022). Nanoscience and forensic genetics. *Understanding Crime Through Forensic Sciences*, 54.

Zięba-Palus, J. (2017). Forensic examinations of micro-traces. *Forensic Science and Criminology*, *2*(2), 1–3. doi:10.15761/FSC.1000113

Compilation of References

. Ashraf, S., Pelaz, B., Pino, P. D., Carril, M., Escudero, A., Parak, W. J., ... Carrillo-Carrion, C. (2016). Gold-based nanomaterials for applications in nanomedicine. *Light-Responsive Nanostructured Systems for Applications in Nanomedicine*, 169-202.

. Kesarwani, S., Parihar, K., Sankhla, M. S., & Kumar, R. (2020). Nano-forensic: new perspective and extensive applications in solving crimes. *Latent in Applied Nanobioscience, 10*(1), 1792-1798.

. Shukla, V., Chaturvedi, A., & Srivastava, N. (2019). Nanotechnology and cryptographic protocols: issues and possible solutions. *Nanomaterials and Energy, 8*(1), 78-83.

Aati, H. Y., Attia, H., Babtin, R., Al-Qahtani, N., & Wanner, J. (2023). Headspace Solid Phase Micro-Extraction of Volatile Constituents Produced from Saudi Ruta chalepensis and Molecular Docking Study of Potential Antioxidant Activity. *Molecules (Basel, Switzerland), 28*(4), 1891. doi:10.3390/molecules28041891 PMID:36838880

Abdulrahman, A., & Ghanem, A. (2018). Recent advances in chromatographic purification of plasmid DNA for gene therapy and DNA vaccines: A review. *Analytica Chimica Acta, 1025*, 41–57. doi:10.1016/j.aca.2018.04.001 PMID:29801607

Abeer, S. (2012). Future medicine: Nanomedicine. *Jimsa, 25*(3), 187–192.

Adabi, M., Naghibzadeh, M., Adabi, M., Zarrinfard, M. A., Esnaashari, S. S., Seifalian, A. M., Faridi-Majidi, R., Tanimowo Aiyelabegan, H., & Ghanbari, H. (2017). Biocompatibility and nanostructured materials: Applications in nanomedicine. *Artificial Cells, Nanomedicine, and Biotechnology, 45*(4), 833–842. doi:10.1080/21691401.2016.1178134 PMID:27247194

Adegoke, O., Mckenzie, C., & Daeid, N. N. (2019). Multi-shaped cationic gold nanoparticle-L-cysteine-ZnSeS quantum dots hybrid nanozyme as an intrinsic peroxidase mimic for the rapid colorimetric detection of cocaine. *Sensors and Actuators. B, Chemical, 287*, 416–427. doi:10.1016/j.snb.2019.02.074

Adegoke, O., Zolotovskaya, S., Abdolvand, A., & Daeid, N. N. (2022). Fabrication of a near-infrared fluorescence-emitting SiO2-AuZnFeSeS quantum dots-molecularly imprinted polymer nanocomposite for the ultrasensitive fluorescence detection of levamisole. *Colloids and Surfaces. A, Physicochemical and Engineering Aspects, 646*, 129013. doi:10.1016/j.colsurfa.2022.129013

Adhikari, S., Ampadu, E. K., Kim, M., Noh, D., Oh, E., & Lee, D. (2021). Detection of explosives by SERS platform using metal nanogap substrates. *Sensors (Basel), 21*(16), 5567. doi:10.339021165567 PMID:34451009

Adhikary, R. R., & Banerjee, R. (2021). Development of smart core-shell nanoparticle-based sensors for the point-of-care detection of alpha amylase in diagnostics and forensics. *Biosensors & Bioelectronics, 184*, 113244. doi:10.1016/j.bios.2021.113244 PMID:33934052

Adserias-Garriga, J., Hernández, M., Quijada, N. M., Lázaro, D. R., Steadman, D., & Garcia-Gil, J. (2017). Daily thanatomicrobiome changes in soil as an approach of postmortem interval estimation: An ecological perspective. *Forensic Science International, 278*, 388–395. doi:10.1016/j.forsciint.2017.07.017 PMID:28818754

Aggarwal, P., & Chitkara, M. (2022). Detection of Metabolites in Latent Fingerprints Through Green Nanoparticles for Both Forensic Investigations and Medical Purposes. *ECS Transactions, 107*(1), 10593–10602. doi:10.1149/10701.10593ecst

Ahangaran, F., & Navarchian, A. H. (2020). Recent advances in chemical surface modification of metal oxide nanoparticles with silane coupling agents: A review. *Advances in Colloid and Interface Science, 286*, 102298. doi:10.1016/j.cis.2020.102298 PMID:33171357

Ahire, S. A., Bachhav, A. A., Pawar, T. B., Jagdale, B. S., Patil, A. V., & Koli, P. B. (2022). The augmentation of nanotechnology era: A concise review on fundamental concepts of nanotechnology and applications in material science and technology. *Results in Chemistry, 4*, 100633. doi:10.1016/j.rechem.2022.100633

Ahmed, Z. E. (2022). Forensic documents examination. *Asian Journal of Forensic Sciences, 1*(1).

Ai, J., Biazar, E., Jafarpour, M., Montazeri, M., Majdi, A., Aminifard, S., ... Rad, H. G. (2011). Nanotoxicology and nanoparticle safety in biomedical designs. *International Journal of Nanomedicine, 6*, 1117. PMID:21698080

Akhlaghi, S. P., Peng, B., Yao, Z., & Tam, K. C. (2013). Sustainable nanomaterials derived from polysaccharides and amphiphilic compounds. *Soft Matter, 9*(33), 7905–7918. doi:10.1039/c3sm50358e

Akhoundian, M., Alizadeh, T., Ganjali, M. R., & Norouzi, P. (2019). Ultra-trace detection of methamphetamine in biological samples using FFT-square wave voltammetry and nano-sized imprinted polymer/MWCNTs-modified electrode. *Talanta, 200*, 115–123. doi:10.1016/j.talanta.2019.02.027 PMID:31036164

Akpan, E., Shen, X., Wetzel, B., & Friedrich, K. (2019). *Design and synthesis of polymer nanocomposites. In Polymer composites with functionalized nanoparticles*. Elsevier.

Al Thaher, Y., Satoof, A., Kamal, A., Almani, D., Shaban, D., Kassab, G., & Ajaleh, S. A. (2021). *Instrumental analytical techniques for physicochemical characterization of bio-nanomaterials. In Handbook on Nanobiomaterials for Therapeutics and Diagnostic Applications*. Elsevier.

Al-Ameri, M. A. A., Ciylan, B., & Mahmood, B. (2022). Spectral Data Analysis for Forgery Detection in Official Documents: A Network-Based Approach. *Electronics (Basel), 11*(23), 23. Advance online publication. doi:10.3390/electronics11234036

Albanese, A., Tang, P. S., & Chan, W. C. (2012). The effect of nanoparticle size, shape, and surface chemistry on biological systems. *Annual Review of Biomedical Engineering, 14*(1), 1–16. doi:10.1146/annurev-bioeng-071811-150124 PMID:22524388

Aleksandrova, E. (2002). Structural properties of polyimides and the mechanism of carrier photogeneration in polyimide-based molecular complexes. *Optics and Spectroscopy, 93*(1), 118–125. doi:10.1134/1.1496734

Ali, A., Chiang, Y. W., & Santos, R. M. (2022). X-ray diffraction techniques for mineral characterization: A review for engineers of the fundamentals, applications, and research directions. *Minerals (Basel), 12*(2), 205. doi:10.3390/min12020205

Allard, M. W. (2016). Commentary: The Future of Whole-Genome Sequencing for Public Health and the Clinic. *Journal of Clinical Microbiology, 54*(8), 1946–1948. doi:10.1128/JCM.01082-16 PMID:27307454

Allen, G. (1997). *Protein: A Comprehensive Treatise*. Academic Press.

Almaie, S., Vatanpour, V., Rasoulifard, M. H., & Koyuncu, I. (2022). Volatile organic compounds (VOCs) removal by photocatalysts: A review. *Chemosphere*, *306*, 135655. doi:10.1016/j.chemosphere.2022.135655 PMID:35817187

Almeida, L., Felzenszwalb, I., Marques, M., & Cruz, C. (2020). Nanotechnology activities: Environmental protection regulatory issues data. *Heliyon*, *6*(10), e05303. doi:10.1016/j.heliyon.2020.e05303 PMID:33102878

Amal, H., Shi, D. Y., Ionescu, R., Zhang, W., Hua, Q. L., Pan, Y. Y., & Haick, H. (2015). Assessment of ovarian cancer conditions from exhaled breath. *International Journal of Cancer*, *136*(6), E614–E622. doi:10.1002/ijc.29166 PMID:25159530

Amendola, V., & Meneghetti, M. (2009). Laser ablation synthesis in solution and size manipulation of noble metal nanoparticles. *Physical Chemistry Chemical Physics*, *11*(20), 3805–3821. doi:10.1039/b900654k PMID:19440607

Amreen, K., & Goel, S. (2022). Microfluidic Devices with Integrated Conductive Polymeric Electrodes for Biosensing Applications. In Conducting Polymers (pp. 273-286). CRC Press. doi:10.1201/9781003205418-19

Anderson, T., Petranker, R., Rosenbaum, D., Weissman, C. R., Dinh-Williams, L.-A., Hui, K., Hapke, E., & Farb, N. A. S. (2019). Microdosing psychedelics: Personality, mental health, and creativity differences in microdosers. *Psychopharmacology*, *236*(2), 731–740. doi:10.100700213-018-5106-2 PMID:30604183

Andreou, C., Hoonejani, M. R., Barmi, M. R., Moskovits, M., & Meinhart, C. D. (2013). Rapid detection of drugs of abuse in saliva using surface enhanced Raman spectroscopy and microfluidics. *ACS Nano*, *7*(8), 7157–7164. doi:10.1021/nn402563f PMID:23859441

Andre, R. S., Ngo, Q. P., Fugikawa-Santos, L., Correa, D. S., & Swager, T. M. (2021). Wireless tags with hybrid nanomaterials for volatile amine detection. *ACS Sensors*, *6*(6), 2457–2464. doi:10.1021/acssensors.1c00812 PMID:34110807

Anikeeva, P. O., Madigan, C. F., Halpert, J. E., Bawendi, M. G., & Bulović, V. J. P. R. B. (2008). Electronic and excitonic processes in light-emitting devices based on organic materials and colloidal quantum dots. *Physical Review B: Condensed Matter and Materials Physics*, *78*(8), 085434. doi:10.1103/PhysRevB.78.085434

Ansari, S., & Masoum, S. (2021). *A hybrid imprinted polymer based on magnetic graphene oxide and carbon dots for ultrasonic assisted dispersive solid-phase microextraction of oxycodone*. Academic Press.

Ansari, N., Lodha, A., & Menon, S. K. (2016). Smart platform for the time since death determination from vitreous humor cystine. *Biosensors & Bioelectronics*, *86*, 115–121. doi:10.1016/j.bios.2016.06.042 PMID:27343833

Anslinger, K., Bayer, B., Danilov, S. M., & Metzger, R. (2008). Application of sperm-specific antibodies for the separation of sperm from cell mixtures. *Forensic Science International. Genetics Supplement Series*, *1*(1), 394–395. doi:10.1016/j.fsigss.2007.10.140

Anuradha & Bhatia, T. (2022). Novel nanomaterials in forensic investigations: A review. *Materials Today: Proceedings*, *50*, 1071–1079. https://doi.org/https://doi.org/10.1016/j.matpr.2021.07.466

Aparna, R. S., Anjali Devi, J. S., Anjana, R. R., Nebu, J., & George, S. (2019). Zn(II) ion modulated red emitting copper nanocluster probe for the fluorescence turn on sensing of RDX. *Sensors and Actuators. B, Chemical*, *291*, 298–305. doi:10.1016/j.snb.2019.04.051

Arabali, V., Malekmohammadi, S., & Karimi, F. (2020). *Surface amplification of pencil graphite electrode using CuO nanoparticle/polypyrrole nanocomposite; a powerful electrochemical strategy for determination of tramadol*. Academic Press.

Arduini, F., Cinti, S., Mazzaracchio, V., Scognamiglio, V., Amine, A., & Moscone, D. (2020). Carbon black as an outstanding and affordable nanomaterial for electrochemical (bio) sensor design. *Biosensors & Bioelectronics, 156*, 112033. doi:10.1016/j.bios.2020.112033 PMID:32174547

Arduini, F., Cinti, S., Scognamiglio, V., & Moscone, D. (2020). Nanomaterial-based sensors. In *Handbook of nanomaterials in analytical chemistry* (pp. 329–359). Elsevier. doi:10.1016/B978-0-12-816699-4.00013-X

Area, M. C., & Cheradame, H. (2011). Paper aging and degradation: Recent findings and research methods. *BioResources, 6*, 5307–5337.

Arif, T., Nisa, N., Amin, S. S., Shoib, S., & Mushtaq, R. (2015). Therapeutic and Diagnostic Applications of Nanotechnology in Dermatology and Cosmetics. *Journal of Nanomedicine & Biotherapeutic Discovery, 5*, 134.

Arita, I., Nakane, M., Kojima, K., Yoshihara, N., Nakano, T., & El-Gohary, A. (2004). Role of a sentinel surveillance system in the context of global surveillance of infectious diseases. *The Lancet. Infectious Diseases, 4*(3), 171–177. doi:10.1016/S1473-3099(04)00942-9 PMID:14998504

Arnaldi, S., & Muratorio, A. (2013). *Nanotechnology, uncertainty and regulation. A guest editorial* (Vol. 7). Springer.

Arshad, A., Farrukh, M. A., Ali, S., Khaleeq-ur-Rahman, M., & Tahir, M. A. (2015). Development of latent fingermarks on various surfaces using ZnO-SiO2 nanopowder. *Journal of Forensic Sciences, 60*(5), 1182–1187. doi:10.1111/1556-4029.12890 PMID:26348223

Artioli, G. (2010). *Scientific Methods and Cultural Heritage: An introduction to the application of materials science to archaeometry and conservation science*. OUP Oxford. doi:10.1093/acprof:oso/9780199548262.001.0001

Ataei, Y., Sun, Y., Liu, W. S., Ellie, A., Dong, H., & Ahmad, U. M. (2023). Health Effects of Exposure to Indoor Semi-Volatile Organic Compounds in Chinese Building Environment: A Systematic Review. *International Journal of Environmental Research and Public Health, 20*(1), 678. doi:10.3390/ijerph20010678 PMID:36613006

Atta, N. F., Galal, A., El-Ads, E. H., & Hassan, S. H. (2019). Cobalt oxide nanoparticles/graphene/ionic liquid crystal modified carbon paste electrochemical sensor for ultra-sensitive determination of a narcotic drug. *Advanced Pharmaceutical Bulletin, 9*(1), 110–121. doi:10.15171/apb.2019.014 PMID:31011565

Aydın, M., Aydın, E. B., & Sezgintürk, M. K. (2018). A highly selective electrochemical immunosensor based on conductive carbon black and star PGMA polymer composite material for IL-8 biomarker detection in human serum and saliva. *Biosensors & Bioelectronics, 117*, 720–728. doi:10.1016/j.bios.2018.07.010 PMID:30014946

Azimi, S., & Docoslis, A. (2022). Recent Advances in the Use of Surface-Enhanced Raman Scattering for Illicit Drug Detection. *Sensors (Basel), 22*(10), 3877. doi:10.339022103877 PMID:35632286

Azman, A. M. (2020). Fast, easy, reproducible method for planting fingerprints for ninhydrin, iodine development. *Journal of Chemical Education, 97*(2), 571–573. doi:10.1021/acs.jchemed.9b00357

Aznar-Gadea, E., Rodriguez-Canto, P. J., Sánchez, S. A., Martinez-Pastor, J. P., & Abargues, R. (2022). Luminescent CdSe Quantum Dot Arrays for Rapid Sensing of Explosive Taggants. *ACS Applied Nano Materials, 5*(5), 6717–6725. doi:10.1021/acsanm.2c00743

Bae, G., Kim, M., Jang, M., Kwon, Y. M., Yim, S., Song, W., Myung, S., Lee, S. S., Hasan, T., & An, K. S. (2022). Photon-Pen Writing for Metal Oxide-Based Versatile Nanosensors. *Advanced Functional Materials, 32*(38), 2204821. doi:10.1002/adfm.202204821

Bae, S., Kim, H., Lee, Y., Xu, X., Park, J.-S., Zheng, Y., Balakrishnan, J., Lei, T., Ri Kim, H., Song, Y., Kim, Y.-J., Kim, K. S., Özyilmaz, B., Ahn, J.-H., Hong, B. H., & Iijima, S. (2010). Roll-to-roll production of 30-inch graphene films for transparent electrodes. *Nature Nanotechnology*, *5*(8), 574–578. doi:10.1038/nnano.2010.132 PMID:20562870

Bagban, M. A., Mansuri, R., & Jain, N. (2018). A novel nanoparticle-based method to isolate DNA from dried saliva and semen samples. *Journal of Advanced Medical Sciences and Applied Technologies*, *4*(1), 13–20.

Bahadır, E. B., & Sezgintürk, M. K. (2015). Electrochemical biosensors for hormone analyses. *Biosensors & Bioelectronics*, *68*, 62–71. doi:10.1016/j.bios.2014.12.054 PMID:25558874

Baig, N., Kammakakam, I., & Falath, W. (2021). Nanomaterials: A review of synthesis methods, properties, recent progress, and challenges. *Materials Advances*, *2*(6), 1821–1871. doi:10.1039/D0MA00807A

Bajorowicz, B., Kobylanski, M. P., Golabiewska, A., Nadolna, J., Zaleska-Medynska, A., & Malankowska, A. (2018). Quantum dot-decorated semiconductor micro- and nanoparticles: A review of their synthesis, characterization and application in photocatalysis. *Advances in Colloid and Interface Science*, *256*, 352–372. doi:10.1016/j.cis.2018.02.003 PMID:29544654

Bajpai, V. K., Kamle, M., Shukla, S., Mahato, D. K., Chandra, P., Hwang, S. K., & Han, Y. K. (2018). Prospects of using nanotechnology for food preservation, safety, and security. *Yao Wu Shi Pin Fen Xi*, *26*(4), 1201–1214. PMID:30249319

Bajtarevic, A., Ager, C., Pienz, M., Klieber, M., Schwarz, K., Ligor, M., & Amann, A. (2009). Noninvasive detection of lung cancer by analysis of exhaled breath. *BMC Cancer*, *9*(1), 1–16. doi:10.1186/1471-2407-9-348 PMID:19788722

Baker, S. N., & Baker, G. A. (2010). Luminescent carbon nanodots: emergent nanolights. Angewandte Chemie-Int, 49(38), 6726–6744.

Balbus, J. M., Florini, K., Denison, R. A., & Walsh, S. A. (2006). Getting it right the first time: Developing nanotechnology while protecting workers, public health, and the environment. *Annals of the New York Academy of Sciences*, *1076*(1), 331–342. doi:10.1196/annals.1371.027 PMID:17119213

Balbus, J. M., Florini, K., Denison, R. A., & Walsh, S. A. (2007). Protecting workers and the environment: An environmental NGO's perspective on nanotechnology. *Journal of Nanoparticle Research*, *9*(1), 11–22. doi:10.100711051-006-9173-7

Balbus, J., Denison, R., Florini, K., & Walsh, S. (2005). Getting nanotechnology right the first time. *Issues in Science and Technology*, *21*(4), 65–71.

Barabadi, H., Najafi, M., Samadian, H., Azarnezhad, A., Vahidi, H., Mahjoub, M. A., Koohiyan, M., & Ahmadi, A. (2019). A systematic review of the genotoxicity and antigenotoxicity of biologically synthesized metallic nanomaterials: Are green nanoparticles safe enough for clinical marketing? *Medicina*, *55*(8), 439. doi:10.3390/medicina55080439 PMID:31387257

Bar, H., Bhui, D. K., Sahoo, G. P., Sarkar, P., De, S. P., & Misra, A. (2009). Green synthesis of silver nanoparticles using latex of Jatropha curcas. *Colloids and Surfaces. A, Physicochemical and Engineering Aspects*, *339*(1-3), 134–139. doi:10.1016/j.colsurfa.2009.02.008

Barhoum, A., García-Betancourt, M. L., Jeevanandam, J., Hussien, E. A., Mekkawy, S. A., Mostafa, M., ... Bechelany, M. (2022). Review on natural, incidental, bioinspired, and engineered nanomaterials: History, definitions, classifications, synthesis, properties, market, toxicities, risks, and regulations. *Nanomaterials (Basel, Switzerland)*, *12*(2), 177. doi:10.3390/nano12020177 PMID:35055196

Barroso, J., Diez-Buitrago, B., Saa, L., Möller, M., Briz, N., & Pavlov, V. (2018). Specific bioanalytical optical and photoelectrochemical assays for detection of methanol in alcoholic beverages. *Biosensors & Bioelectronics*, *101*, 116–122. doi:10.1016/j.bios.2017.10.022 PMID:29055193

Barua, R., Datta, S., Chowdhury, A. R., & Datta, P. (2018). Advances in MEMS and Micro-Scale technologies for application in controlled Drug-Dosing systems: MEMS-Based drug delivery systems. In *Design and Development of Affordable Healthcare Technologies* (pp. 165–179). IGI Global. doi:10.4018/978-1-5225-4969-7.ch007

Basu, M., Seggerson, S., Henshaw, J., Jiang, J., del A Cordona, R., Lefave, C., Boyle, P. J., Miller, A., Pugia, M., & Basu, S. (2004). Nano-biosensor development for bacterial detection during human kidney infection: Use of glycoconjugate-specific antibody-bound gold NanoWire arrays (GNWA). *Glycoconjugate Journal*, *21*(8), 487–496. doi:10.100710719-004-5539-1 PMID:15750790

Bawa, R. (2007). *Patents and nanomedicine*. Academic Press.

Bawa, R. (2008). *Nanoparticle-based therapeutics in humans: A survey*. Academic Press.

Ben-Jaber, S., Peveler, W. J., Quesada-Cabrera, R., Sol, C. W., Papakonstantinou, I., & Parkin, I. P. (2017). Sensitive and specific detection of explosives in solution and vapour by surface-enhanced Raman spectroscopy on silver nanocubes. *Nanoscale*, *9*(42), 16459–16466. doi:10.1039/C7NR05057G PMID:29063930

Benya, R., Quintana, J., & Brundage, B. (1989). Adverse reactions to indocyanine green: A case report and a review of the literature. *Catheterization and Cardiovascular Diagnosis*, *17*(4), 231–233. doi:10.1002/ccd.1810170410 PMID:2670244

Betzig, E., Patterson, G. H., Sougrat, R., Lindwasser, O. W., Olenych, S., Bonifacino, J. S., Davidson, M. W., Lippincott-Schwartz, J., & Hess, H. F. (2006). Imaging intracellular fluorescent proteins at nanometer resolution. *Science*, *313*(5793), 1642–1645. doi:10.1126cience.1127344 PMID:16902090

Bezzon, V. D., Montanheiro, T. L., de Menezes, B. R., Ribas, R. G., Righetti, V. A., Rodrigues, K. F., & Thim, G. P. (2019). Carbon nanostructure-based sensors: A brief review on recent advances. *Advances in Materials Science and Engineering*, *2019*, 2019. doi:10.1155/2019/4293073

Bhardwaj, N., & Kundu, S. (2012). *Chondrogenic differentiation of rat MSCs on porous scaffolds of silk fibroin/chitosan blends*. Academic Press.

Bhatia, T. (2022). Novel nanomaterials in forensic investigations: A review. *Materials Today: Proceedings*, *50*, 1071–1079. doi:10.1016/j.matpr.2021.07.466

Bhushan, B. (2017). Introduction to nanotechnology. Springer handbook of nanotechnology, 1-19.

Bi, C., Krechmer, J. E., Frazier, G. O., Xu, W., Lambe, A. T., Claflin, M. S., Lerner, B. M., Jayne, J. T., Worsnop, D. R., Canagaratna, M. R., & Isaacman-VanWertz, G. (2021). Coupling a gas chromatograph simultaneously to a flame ionization detector and chemical ionization mass spectrometer for isomer-resolved measurements of particle-phase organic compounds. *Atmospheric Measurement Techniques*, *14*(5), 3895–3907. doi:10.5194/amt-14-3895-2021

Bindra, P., Gangopadhyay, S., & Hazra, A. (2018). Au/TiO 2 Nanotubes/Ti-based solid-state vapor sensor: Efficient sensing in resistive and capacitive modes. *IEEE Transactions on Electron Devices*, *65*(5), 1918–1924. doi:10.1109/TED.2018.2809677

Binnewies, T. T., Motro, Y., Hallin, P. F., Lund, O., Dunn, D., La, T., & Ussery, D. W. (2006). Ten years of bacterial genome sequencing: Comparative-genomics-based discoveries. *Functional & Integrative Genomics*, *6*(3), 165–185. doi:10.100710142-006-0027-2 PMID:16773396

Biranje, A., Azmi, N., Tiwari, A., & Chaskar, A. (2021). Quantum dots based fluorescent probe for the selective detection of heavy metal ions. *Journal of Fluorescence*, *31*(5), 1241–1250. doi:10.100710895-021-02755-8 PMID:34181146

Biteen, J. S., Blainey, P. C., Cardon, Z. G., Chun, M., Church, G. M., Dorrestein, P. C., & Young, T. D. (2016). Tools for the microbiome: Nano and beyond. *ACS Nano*, *10*(1), 6–37. doi:10.1021/acsnano.5b07826 PMID:26695070

Blattner, F. R., Plunkett III, G., Bloch, C. A., Perna, N. T., Burland, V., Riley, M., & Shao, Y. (1997). The complete genome sequence of Escherichia coli K-12. *Science, 277*(5331), 1453-1462.

Bodmeier, R., & McGinity, J. W. (1987). Polylactic acid microspheres containing quinidine base and quinidine sulphate prepared by the solvent evaporation technique. II. Some process parameters influencing the preparation and properties of microspheres. *Journal of Microencapsulation*, *4*(4), 289–297. doi:10.3109/02652048709021821 PMID:3504510

Bogner, A., Jouneau, P. H., Thollet, G., Basset, D., & Gauthier, C. (2007). A history of scanning electron microscopy developments: Towards "wet-STEM" imaging. *Micron (Oxford, England)*, *38*(4), 390–401. doi:10.1016/j.micron.2006.06.008 PMID:16990007

Bojko, K., Roux, C., & Reedy, B. J. (2008). An examination of the sequence of intersecting lines using attenuated total reflectance–Fourier transform infrared spectral imaging. *Journal of Forensic Sciences*, *53*(6), 1458–1467. PMID:18624894

Bokhari, H. (2018). Exploitation of microbial forensics and nanotechnology for the monitoring of emerging pathogens. *Critical Reviews in Microbiology*, *44*(4), 504–521. doi:10.1080/1040841X.2018.1444013 PMID:29513060

Bossi, P., Tegnell, A., Baka, A., Van Loock, F., Hendriks, J., Werner, A., & Gouvras, G. (2004). Bichat guidelines for the clinical management of glanders and melioidosis and bioterrorism-related glanders and melioidosis. *Eurosurveillance*, *9*(12), E17–E18. doi:10.2807/esm.09.12.00507-en PMID:15677841

Boumba, V. A., & Vougiouklakis, T. (2015). Impact of blood collection tubes on erroneous 1-propanol detection and on forensic ethanol analysis. *Journal of Forensic Toxicology and Pharmacology*, *4*(1). Advance online publication. doi:10.4172/2325-9841.1000134

Bowman, D. M., & Hodge, G. A. (2006a). Nanotechnology: Mapping the wild regulatory frontier. *Futures*, *38*(9), 1060–1073. doi:10.1016/j.futures.2006.02.017

Bowman, D. M., & Hodge, G. A. (2006b). A small matter of regulation: An international review of nanotechnology regulation. *Colum. Sci. & Tech. L. Rev.*, *8*, 1.

Brayner, R., Coradin, T., Beaunier, P., Grenèche, J.-M., Djediat, C., Yéprémian, C., Couté, A., & Fiévet, F. (2012). Intracellular biosynthesis of superparamagnetic 2-lines ferri-hydrite nanoparticles using Euglena gracilis microalgae. *Colloids and Surfaces. B, Biointerfaces*, *93*, 20–23. doi:10.1016/j.colsurfb.2011.10.014 PMID:22225944

Braz, A., Lopez-Lopez, M., & Garcia-Ruiz, C. (2013). Raman spectroscopy for forensic analysis of inks in questioned documents. *Forensic Science International*, *232*(1–3), 206–212. doi:10.1016/j.forsciint.2013.07.017 PMID:24053882

Brewer, J. D., Hagan, K. A., & Egan, J. M. (2005). Forensic analysis of blue ballpoint pen inks using capillary electrophoresis. *Research and Technology, 7*(3).

Brodrick, E., Davies, A., Neill, P., Hanna, L., & Williams, E. M. (2015). Breath analysis: Translation into clinical practice. *Journal of Breath Research*, *9*(2), 027109. doi:10.1088/1752-7155/9/2/027109 PMID:25971863

Bruijns, B., Van Asten, A., Tiggelaar, R., & Gardeniers, H. (2016). Microfluidic devices for forensic DNA analysis: A review. *Biosensors (Basel)*, *6*(3), 41. doi:10.3390/bios6030041 PMID:27527231

Brunelle, R. L., & Crawford, K. R. (2003). *Advances in the Forensic Analysis and Dating of Writing Ink.* Charles C Thomas Publishers.

Budowle, B., Johnson, M. D., Fraser, C. M., Leighton, T. J., Murch, R. S., & Chakraborty, R. (2005). Genetic analysis and attribution of microbial forensics evidence. *Critical Reviews in Microbiology, 31*(4), 233–254. doi:10.1080/10408410500304082 PMID:16417203

Budowle, B., Schutzer, S. E., Einseln, A., Kelley, L. C., Walsh, A. C., Smith, J. A., Marrone, B. L., Robertson, J., & Campos, J. (2003). Building microbial forensics as a response to bioterrorism. *Science, 301*(5641), 1852–1853. doi:10.1126cience.1090083 PMID:14512607

Buffler, A., & Tickner, J. (2010). *Detecting contraband using neutrons: challenges and future directions.* Academic Press.

Bügler, J. H., Buchner, H., & Dallmayer, A. (2005). Characterization of ballpoint pen inks by thermal desorption and gas chromatography-mass spectrometry. *Journal of Forensic Sciences, 50*(5), 2004486–6. doi:10.1520/JFS2004486 PMID:16225233

Bunaciu, A. A., Udriştioiu, E., & Aboul-Enein, H. Y. (2015). X-ray diffraction: Instrumentation and applications. *Critical Reviews in Analytical Chemistry, 45*(4), 289–299. doi:10.1080/10408347.2014.949616 PMID:25831472

Bunschoten, A., Buckle, T., Kuil, J., Luker, G. D., Luker, K. E., Nieweg, O. E., & van Leeuwen, F. W. B. (2012). Targeted non-covalent self-assembled nanoparticles based on human serum albumin. *Biomaterials, 33*(3), 867–875. doi:10.1016/j.biomaterials.2011.10.005 PMID:22024362

Buszewski, B., & Szultka, M. (2012). Past, Present, and Future of Solid Phase Extraction: A Review. *Critical Reviews in Analytical Chemistry, 42*(3), 198–213. doi:10.1080/07373937.2011.645413

Butt, B. Z., & Naseer, I. (2020). Nanofertilizers. *Nanoagronomy*, 125-152.

Buzea, C., Pacheco, I. I., & Robbie, K. (2007). *Nanomaterials and nanoparticles: Sources and toxicity.* Academic Press.

Buzzini, P., Massonnet, G., & Monard Sermier, F. (2006). The micro Raman analysis of paint evidence in criminalistics: case studies. *Journal of Raman Spectroscopy, 37*(9), 922-931.

Buzzini, P., & Suzuki, E. (2016). Forensic applications of Raman spectroscopy for the in situ analyses of pigments and dyes in ink and paint evidence. *Journal of Raman Spectroscopy : JRS, 47*(1), 16–27. doi:10.1002/jrs.4818

Caddy, B. (2001). *Forensic examination of glass and paint: Analysis and interpretation.* CRC Press. doi:10.1201/9780203483589

Calcerrada, M., & García-Ruiz, C. (2015). Analysis of questioned documents: A review. *Analytica Chimica Acta, 853*, 143–166. doi:10.1016/j.aca.2014.10.057 PMID:25467455

Cantu, A. A. (1995). A sketch of analytical methods for document dating. Part I. The static approach: Determining age independent analytical profiles. *International Journal of Forensic Document Examiners, 1*(1), 40–51.

Cantú, A. A. (2012). A study of the evaporation of a solvent from a solution—Application to writing ink aging. *Forensic Science International, 219*(1–3), 119–128. doi:10.1016/j.forsciint.2011.12.008 PMID:22261554

Cao, G., & Wang, Y. (2004). Nanostructures and nanomaterials: Synthesis. *Properties and Applications, 2*, 94.

Capitain, C., & Weller, P. (2021). Non-targeted screening approaches for profiling of volatile organic compounds based on gas chromatography-ion mobility spectroscopy (GC-IMS) and machine learning. *Molecules (Basel, Switzerland), 26*(18), 5457. doi:10.3390/molecules26185457 PMID:34576928

Carbonaro, C. M., Corpino, R., Salis, M., Mocci, F., Thakkar, S. V., Olla, C., & Ricci, P. C. (2019). On the Emission Properties of Carbon Dots: Reviewing Data and Discussing Models. *C, 5*(4), 4. Advance online publication. doi:10.3390/c5040060

Carter, D. O., Tomberlin, J. K., Benbow, M. E., & Metcalf, J. L. (2017). Forensic microbiology. *Forensic Microbiol, 36*(7), 1–391.

Carvalho, P. M., Felício, M. R., Santos, N. C., Gonçalves, S., & Domingues, M. M. (2018). Application of light scattering techniques to nanoparticle characterization and development. *Frontiers in Chemistry, 6*, 237. doi:10.3389/fchem.2018.00237 PMID:29988578

Casciano, D. A., & Sahu, S. C. (2009). *Nanotoxicity: from in vivo and in vitro models to health risks*. John Wiley & Sons.

Cedervall, T., Lynch, I., Foy, M., Berggård, T., Donnelly, S. C., Cagney, G., Linse, S., & Dawson, K. A. (2007). Detailed identification of plasma proteins adsorbed on copolymer nanoparticles. *Angewandte Chemie International Edition in English, 46*(30), 5754–5756. doi:10.1002/anie.200700465 PMID:17591736

Chansuvarn, W., Tuntulani, T., & Imyim, A. (2015). Colorimetric detection of mercury(II) based on gold nanoparticles, fluorescent gold nanoclusters and other gold-based nanomaterials. *Trends in Analytical Chemistry, 65*, 83–96. doi:10.1016/j.trac.2014.10.013

Chantada–Vázquez, M. P., de–Becerra–Sánchez, C., Fernández–del–Río, A., Sánchez–González, J., Bermejo, A. M., Bermejo–Barrera, P., & Moreda–Piñeiro, A. (2018). Development and application of molecularly imprinted polymer – Mn-doped ZnS quantum dot fluorescent optosensing for cocaine screening in oral fluid and serum. *Talanta, 181*, 232–238. doi:10.1016/j.talanta.2018.01.017 PMID:29426506

Chan, W. C. (2009). *Bio-applications of nanoparticles* (Vol. 620). Springer Science & Business Media.

Chan, W., Wong, G. F., Hung, C., Wong, Y., Fung, K., Lee, W., Dao, K.-L., Leung, C., Lo, K., Lee, W., & Cheung, B. K. (2020). Interpol review of toxicology 2016–2019. *Forensic Science International. Synergy, 2*, 563–607. doi:10.1016/j.fsisyn.2020.01.018 PMID:33385147

Charles, P. P., & Frank, J. O. (2003). *Introduction to nanotechnology*. Academic Press.

Charles, P. T., Wadhwa, V., Kouyate, A., Mesa-Donado, K. J., Adams, A. A., Deschamps, J. R., & Kusterbeck, A. W. (2018). A High Aspect Ratio Bifurcated 128-Microchannel Microfluidic Device for Environmental Monitoring of Explosives. *Sensors (Basel), 18*(5), 1568. doi:10.339018051568 PMID:29762499

Chaudhary, L. S., Ghatmale, P. R., & Chavan, S. (2016). Review on: Application of nanotechnology in computer science. *International Journal of Scientific Research, 5*, 1542–1545.

Chauhan, A., & Chauhan, P. (2014). Powder XRD technique and its applications in science and technology. *Journal of Analytical & Bioanalytical Techniques, 5*(5), 1–5. doi:10.4172/2155-9872.1000212

Chauhan, V., Singh, V., & Tiwari, A. (2017). Applications of nanotechnology in forensic investigation. *Int. J. Life. Sci. Scienti. Res, 3*(3), 1047–1051.

Chávez, D., Garcia, C. R., Oliva, J., & Diaz-Torres, L. A. (2021). A review of phosphorescent and fluorescent phosphors for fingerprint detection. *Ceramics International, 47*(1), 10–41. doi:10.1016/j.ceramint.2020.08.259

Chen, A., & Chatterjee, S. (2013). Nanomaterials based electrochemical sensors for biomedical applications. *Chemical Society Reviews, 42*(12), 5425–5438. doi:10.1039/c3cs35518g PMID:23508125

Chen, C. H., Wang, C. C., Ko, P. Y., & Chen, Y. L. (2020). Nanomaterial-based adsorbents and optical sensors for illicit drug analysis. *Yao Wu Shi Pin Fen Xi, 28*(4), 654. doi:10.38212/2224-6614.1137 PMID:35696141

Chen, C., Song, M., Du, Y., Yu, Y., Li, C., Han, Y., Yan, F., Shi, Z., & Feng, S. (2021). Tumor-Associated-Macrophage-Membrane-Coated Nanoparticles for Improved Photodynamic Immunotherapy. *Nano Letters, 21*(13), 5522–5531. doi:10.1021/acs.nanolett.1c00818 PMID:34133181

Cheng, C. J., Tietjen, G. T., Saucier-Sawyer, J. K., & Saltzman, W. M. (2015). A holistic approach to targeting disease with polymeric nanoparticles. *Nature Reviews. Drug Discovery, 14*(4), 239–247. doi:10.1038/nrd4503 PMID:25598505

Cheng, L. T. J. (2013).. . *Nano Today, 8*, 290–312. doi:10.1016/j.nantod.2013.04.007 PMID:23997809

Chen, H.-S., Meng, H.-H., & Cheng, K.-C. (2002). A survey of methods used for the identification and characterization of inks. *Forensic Science Journal, 1*(1), 1–14.

Chen, L., McCrate, J. M., Lee, J. C., & Li, H. (2011). The role of surface charge on the uptake and biocompatibility of hydroxyapatite nanoparticles with osteoblast cells. *Nanotechnology, 22*(10), 105708. doi:10.1088/0957-4484/22/10/105708 PMID:21289408

Chen, R., & Chen, C. (2012). *Nanotoxicity. In The nanobiotechnology handbook.* CRC Press.

Chen, S. Z., Tsai, T. L., & Chen, Y. F. (2012). Forensic application of atomic force microscopy—Questioned Document. *Journal of the Chinese Chemical Society (Taipei), 59*(3), 283–288. doi:10.1002/jccs.201100739

Chen, T. W., Sheng, Z. H., Wang, K., Wang, F. B., & Xia, X. H. (2011). Determination of explosives using electrochemically reduced graphene. *Chemistry, an Asian Journal, 6*(5), 1210–1216. doi:10.1002/asia.201000836 PMID:21387564

Chen, X., Tang, Y., Wang, S., Song, Y., Tang, F., & Wu, X. (2015). Field-amplified sample injection in capillary electrophoresis with amperometric detection for the ultratrace analysis of diastereomeric ephedrine alkaloids. *Electrophoresis, 36*(16), 1953–1961. doi:10.1002/elps.201500024 PMID:25873262

Chen, X., Wu, M., Liu, X., Wang, D., Liu, F., Chen, Y., Yi, F., Huang, W., & Wang, S. (2019). Tuning the Doping Ratio and Phase Transition Temperature of VO2 Thin Film by Dual-Target Co-Sputtering. *Nanomaterials (Basel, Switzerland), 9*(6), 834. doi:10.3390/nano9060834 PMID:31159352

Chen, Y. P., Cai, D., Li, W., Blank, I., & Liu, Y. (2022). Application of gas chromatography-ion mobility spectrometry (GC-IMS) and ultrafast gas chromatography electronic-nose (uf-GC E-nose) to distinguish four Chinese freshwater fishes at both raw and cooked status. *Journal of Food Biochemistry, 46*(6), e13840. doi:10.1111/jfbc.13840 PMID:34189733

Chen, Y.-F. (2011). Forensic Applications of Nanotechnology. *Journal of the Chinese Chemical Society (Taipei), 58*(6), 828–835. doi:10.1002/jccs.201190129

Cheong, K. H., Yi, D. K., Lee, J. G., Park, J. M., Kim, M. J., Edel, J. B., & Ko, C. (2008). Gold nanoparticles for one step DNA extraction and real-time PCR of pathogens in a single chamber. *Lab on a Chip, 8*(5), 810–813. doi:10.1039/b717382b PMID:18432353

Cheon, J., Chan, W., & Zuhorn, I. (2019). The Future of Nanotechnology: Cross-disciplined Progress to Improve Health and Medicine. *Accounts of Chemical Research, 52*(9), 2405–2405. doi:10.1021/acs.accounts.9b00423 PMID:31525881

Chew, B. S., Contreras, R. P., McCartney, M. M., Borras, E., Kenyon, N. J., & Davis, C. E. (2022). A low cost, easy-to-assemble, open-source modular mobile sampler design for thermal desorption analysis of breath and environmental VOCs. *Journal of Breath Research, 16*(3), 036005. doi:10.1088/1752-7163/ac6c9f PMID:35508102

Chiang, C.-H., Lee, H.-H., Chen, B.-H., Lin, Y.-C., Chao, Y.-Y., & Huang, Y.-L. (2019). Using ambient mass spectrometry and LC–MS/MS for the rapid detection and identification of multiple illicit street drugs. *Journal of Food and Drug Analysis, 27*(2), 439-450.

Chirayil, C. J., Abraham, J., Mishra, R. K., George, S. C., & Thomas, S. (2017). *Instrumental techniques for the characterization of nanoparticles. In Thermal and Rheological Measurement Techniques for Nanomaterials Characterization.* Elsevier.

Cho, E. C., Glaus, C., Chen, J. Y., Welch, M. J., & Xia, Y. N. (2010). Inorganic nanoparticle-based contrast agents for molecular imaging. *Trends in Molecular Medicine, 16*(12), 561–573. doi:10.1016/j.molmed.2010.09.004 PMID:21074494

Christian, F., Adityawarman, D., & Indarto, A. (2013). Application of nanotechnologies in the energy sector: A brief and short review. *Frontiers in Energy, 7*(1), 6–18. doi:10.100711708-012-0219-5

Christoffersen, S. (2015). The importance of microbiological testing for establishing cause of death in 42 forensic autopsies. *Forensic Science International, 250*, 27–32. doi:10.1016/j.forsciint.2015.02.020 PMID:25769131

Chun, L. P., Miguel, M. J., Junkins, E. N., Forbes, S. L., & Carter, D. O. (2015). An initial investigation into the ecology of culturable aerobic postmortem bacteria. *Science & Justice, 55*(6), 394–401. doi:10.1016/j.scijus.2015.07.003 PMID:26654073

Clapp, A. R., Medintz, I. L., & Mattoussi, H. (2006). Förster resonance energy transfer investigations using quantum-dot fluorophores. *ChemPhysChem, 7*(1), 47–57. doi:10.1002/cphc.200500217 PMID:16370019

Clapp, A. R., Medintz, I. L., Mauro, J. M., Fisher, B. R., Bawendi, M. G., & Mattoussi, H. (2004). Fluorescence resonance energy transfer between quantum dot donors and dye-labeled protein acceptors. *Journal of the American Chemical Society, 126*(1), 301–310. doi:10.1021/ja037088b PMID:14709096

Cleaveland, S., Laurenson, M. K., & Taylor, L. H. (2001). Diseases of humans and their domestic mammals: Pathogen characteristics, host range and the risk of emergence. *Philosophical Transactions of the Royal Society of London. Series B, Biological Sciences, 356*(1411), 991–999. doi:10.1098/rstb.2001.0889 PMID:11516377

Clunan, A., Rodine-Hardy, K., Hsueh, R., Kosal, M. E., & McManus, I. (2014). Nanotechnology in a globalized world: strategic assessments of an emerging technology. Academic Press.

Cohen Hyams, T., Mam, K., & Killingsworth, M. C. (2020). Scanning electron microscopy as a new tool for diagnostic pathology and cell biology. *Micron (Oxford, England), 130*, 102797. doi:10.1016/j.micron.2019.102797 PMID:31862481

Coles, D., & Frewer, L. J. (2013). Nanotechnology applied to European food production–A review of ethical and regulatory issues. *Trends in Food Science & Technology, 34*(1), 32–43. doi:10.1016/j.tifs.2013.08.006

Committee for Medicinal Products for Human Use. (2013). *Reflection paper on surface coatings: General issues for consideration regarding parenteral administration of coated nanomedicine products.* Author.

Conlan, A. P., Moldovan, G., Bruas, L., Monroy, E., & Cooper, D. (2021). Electron beam induced current microscopy of silicon p–n junctions in a scanning transmission electron microscope. *Journal of Applied Physics, 129*(13), 135701. doi:10.1063/5.0040243

Cook, E., Labiento, G., & Chauhan, B. P. (2021). Fundamental Methods for the Phase Transfer of Nanoparticles. *Molecules (Basel, Switzerland), 26*(20), 6170. doi:10.3390/molecules26206170 PMID:34684750

Cornelis, S., Tytgat, O., Fauvart, M., Gansemans, Y., Vander Plaetsen, A. S., Wiederkehr, R. S., & Stakenborg, T. (2019). Silicon µPCR chip for forensic STR profiling with hybeacon probe melting curves. *Scientific Reports, 9*(1), 1–9. doi:10.103841598-019-43946-5 PMID:31089203

Costa-Fernández, J. M., Menéndez-Miranda, M., Bouzas-Ramos, D., Encinar, J. R., & Sanz-Medel, A. (2016). Mass spectrometry for the characterization and quantification of engineered inorganic nanoparticles. *Trends in Analytical Chemistry*, *84*, 139–148. doi:10.1016/j.trac.2016.06.001

Costanzo, H., Gooch, J., & Frascione, N. (2023). Nanomaterials for optical biosensors in forensic analysis. *Talanta*, *253*, 123945. doi:10.1016/j.talanta.2022.123945 PMID:36191514

Coto-García, A. M., Sotelo-González, E., Fernández-Argüelles, M. T., Pereiro, R., Costa-Fernández, J. M., & Sanz-Medel, A. (2011). Nanoparticles as fluorescent labels for optical imaging and sensing in genomics and proteomics. *Analytical and Bioanalytical Chemistry*, *399*(1), 29–42. doi:10.100700216-010-4330-3 PMID:21052647

Cova, L., Bigini, P., Diana, V., Sitia, L., Ferrari, R., Pesce, R. M., Khalaf, R., Bossolasco, P., Ubezio, P., Lupi, M., Tortarolo, M., Colombo, L., Giardino, D., Silani, V., Morbidelli, M., Salmona, M., & Moscatelli, D. (2013). Biocompatible fluorescent nanoparticles for *in vivo* stem cell tracking. *Nanotechnology*, *24*(24), 245603. doi:10.1088/0957-4484/24/24/245603 PMID:23690139

Crepeault, H., Socias, M. E., Tobias, S., Lysyshyn, M., Custance, A., Shapiro, A., & Ti, L. (2023). Examining fentanyl and its analogues in the unregulated drug supply of British Columbia, Canada using drug checking technologies. *Drug and Alcohol Review*, *42*(3), 538–543. doi:10.1111/dar.13580 PMID:36423900

Cumeras, R., & Correig, X. (Eds.). (2018). *Volatile organic compound analysis in biomedical diagnosis applications.* CRC Press. doi:10.1201/9780429433580

Cusato, J., Allegra, S., Nicolò, A. D., Calcagno, A., & D'Avolio, A. (2018). *Precision medicine for HIV: where are we?* Academic Press.

da Silva, J., & Goncalves, H. M. R. (2011). Analytical and bioanalytical applications of carbon dots. *Trends in Analytical Chemistry*, *30*(8), 1327–1336. doi:10.1016/j.trac.2011.04.009

Dabbousi, B., Rodriguez-Viejo, J., Mikulec, F. V., Heine, J., Mattoussi, H., Ober, R., Jensen, K., & Bawendi, M. (1997). (CdSe) ZnS core-shell quantum dots: Synthesis and characterization of a size series of highly luminescent nanocrystallites. *The Journal of Physical Chemistry B*, *101*(46), 9463–9475. doi:10.1021/jp971091y

Daeid, N. N., Hackman, L., Adegoke, O., & Nic Daeid, N. (2021). Colorimetric optical nanosensors for trace explosive detection using metal nanoparticles: Advances, pitfalls, and future perspective. *Emerging Topics in Life Sciences*, *5*(3), 367–379. doi:10.1042/ETLS20200281 PMID:33960382

Dagar, M., Yadav, S., Sai, V., Satija, J., & Bhatia, H. (2022). Emerging trends in point-of-care sensors for illicit drugs analysis. *Talanta*, *238*, 123048. doi:10.1016/j.talanta.2021.123048 PMID:34801905

Dahiya, K., Sharma, H., Biswas, L., & Verma, A. K. (2022). *Nanotechnology in Forensic Science: Extensive Applications and New Perspective*. Academic Press.

Dance, D. A. B. (2009). Melioidosis and glanders as possible biological weapons. In *Bioterrorism and infectious agents: a new dilemma for the 21st century* (pp. 99–145). Springer. doi:10.1007/978-1-4419-1266-4_4

Darwish, M. S., Mostafa, M. H., & Al-Harbi, L. M. (2022). Polymeric nanocomposites for environmental and industrial applications. *International Journal of Molecular Sciences*, *23*(3), 1023. doi:10.3390/ijms23031023 PMID:35162946

Das, S., Sen, B., & Debnath, N. (2015). *Recent trends in nanomaterials applications in environmental monitoring and remediation.* Academic Press.

Davies, J. C. (2008). Nanotechnology oversight. Washington, DC: Project on Emerging Nanotechnologies.

Dazzi, A., & Prater, C. B. (2017). AFM-IR: Technology and applications in nanoscale infrared spectroscopy and chemical imaging. *Chemical Reviews*, *117*(7), 5146–5173. doi:10.1021/acs.chemrev.6b00448 PMID:27958707

Dazzi, C., & Pedrabissi, L. (2009). Graphology and personality: An empirical study on validity of handwriting analysis. *Psychological Reports*, *105*(3), 1255–1268. doi:10.2466/PR0.105.F.1255-1268 PMID:20229925

de Beijer, R. P., de Graaf, C., van Weert, A., van Leeuwen, T. G., Aalders, M. C. G., & van Dam, A. (2018). Identification and detection of protein markers to differentiate between forensically relevant body fluids. *Forensic Science International*, *290*, 196–206. doi:10.1016/j.forsciint.2018.07.013 PMID:30071450

De Jong, W. H., Hagens, W. I., Krystek, P., Burger, M. C., Sips, A. J., & Geertsma, R. E. (2008). Particle size-dependent organ distribution of gold nanoparticles after intravenous administration. *Biomaterials*, *29*(12), 1912–1919. doi:10.1016/j.biomaterials.2007.12.037 PMID:18242692

De Villiers, M. M., Aramwit, P., & Kwon, G. S. (2008). *Nanotechnology in drug delivery*. Springer Science & Business Media.

de Vlieger, J. S., Crommelin, D. J., Tyner, K., Drummond, D. C., Jiang, W., McNeil, S. E., & Shah, V. P. (2019). *Report of the AAPS guidance forum on the FDA draft guidance for industry: "Drug products, including biological products, that contain nanomaterials"*. Springer.

Decuzzi, P., Pasqualini, R., Arap, W., & Ferrari, M. (2009). Intravascular delivery of particulate systems: Does geometry really matter? *Pharmaceutical Research*, *26*(1), 235–243. doi:10.100711095-008-9697-x PMID:18712584

Delgado-Ramos, G. C. (2014). Nanotechnology in Mexico: Global trends and national implications for policy and regulatory issues. *Technology in Society*, *37*, 4–15. doi:10.1016/j.techsoc.2013.09.005

Dhanyamraju, P., Sharma, A., Dovat, S., Amin, S., & Dhimant, D. (2020). Nanotherapeutics: A mini review. *Modern Approach Drug Design*, *2*(5), 1–3.

Dhawan, A., & Sharma, V. (2010). Toxicity assessment of nanomaterials: Methods and challenges. *Analytical and Bioanalytical Chemistry*, *398*(2), 589–605. doi:10.100700216-010-3996-x PMID:20652549

Díez-Pascual, A. M. (2021). Development of graphene-based polymeric nanocomposites: A brief overview. *Polymers*, *13*(17), 2978. doi:10.3390/polym13172978 PMID:34503017

Díez-Pascual, A. M., Cruz, D. L., & Redondo, A. L. (2022). Advanced Carbon-Based Polymeric Nanocomposites for Forensic Analysis. *Polymers*, *14*(17), 17. Advance online publication. doi:10.3390/polym14173598 PMID:36080673

Ding, D., Goh, C. C., Feng, G. X., Zhao, Z. J., Liu, J., Liu, R. R., Tomczak, N., Geng, J. L., Tang, B. Z., Ng, L. G., & Liu, B. (2013). Ultrabright Organic Dots with Aggregation-Induced Emission Characteristics for Real-Time Two-Photon Intravital Vasculature Imaging. *Advanced Materials*, *25*(42), 6083–6088. doi:10.1002/adma.201301938 PMID:24038281

Ding, D., Liu, J., Feng, G. X., Li, K., Hu, Y., & Liu, B. (2013). Bright Far-Red/Near-Infrared Conjugated Polymer Nanoparticles for In Vivo Bioimaging. *Small*, *9*(18), 3093–3102. doi:10.1002mll.201300171 PMID:23625815

Dingman, J. (2008). Guest commentary: nanotechnology: its impact on food safety. *Journal of Environmental Health*, *70*(6), 47–50. PMID:18236938

Ding, Y., Li, X., Guo, Y., Duan, W., Ling, J., Zha, L., Yan, J., Zou, Y., & Cai, J. (2017). Estimation of postmortem interval by vitreous potassium evaluation with a novel fluorescence aptasensor. *Scientific Reports*, *7*(1), 1–9. doi:10.103841598-017-02027-1 PMID:28500353

Dirwono, W., Park, J. S., Agustin-Camacho, M. R., Kim, J., Park, H. M., Lee, Y., & Lee, K. B. (2010). Application of micro-attenuated total reflectance FTIR spectroscopy in the forensic study of questioned documents involving red seal inks. *Forensic Science International, 199*(1–3), 6–8. doi:10.1016/j.forsciint.2010.02.009 PMID:20413232

Djozan, D., Baheri, T., Karimian, G., & Shahidi, M. (2008). Forensic discrimination of blue ballpoint pen inks based on thin layer chromatography and image analysis. *Forensic Science International, 179*(2–3), 199–205. doi:10.1016/j.forsciint.2008.05.013 PMID:18639403

Donaldson, A. E., & Lamont, I. L. (2014). Estimation of post-mortem interval using biochemical markers. *The Australian Journal of Forensic Sciences, 46*(1), 8–26. doi:10.1080/00450618.2013.784356

Dong, R., Weng, S., Yang, L., & Liu, J. (2015). Detection and direct readout of drugs in human urine using dynamic surface-enhanced Raman spectroscopy and support vector machines. *Analytical Chemistry, 87*(5), 2937–2944. doi:10.1021/acs.analchem.5b00137 PMID:25634247

Drake, J., Charles, C., Bourgeois, J. W., Daniel, E. S., & Kwend, E. M. (2020). Exploring the impact of the opioid epidemic in Black and Hispanic communities in the United States. *Drug Science, Policy and Law, 6*, 2050324520940428. doi:10.1177/2050324520940428

Dreher, K. L. (2004). Health and environmental impact of nanotechnology: Toxicological assessment of manufactured nanoparticles. *Toxicological Sciences, 77*(1), 3–5. doi:10.1093/toxsci/kfh041 PMID:14756123

Dube, S., Tharmavaram, M., Pandey, G., Rawtani, D., & Mustansar Hussain, C. (2020). Sensors for the Detection of Biological Fluids. *Technology in Forensic Science: Sampling, Analysis, Data and Regulations*, 239–258.

Duran, A., Perez-Rodriguez, J. L., Espejo, T., Franquelo, M. L., Castaing, J., & Walter, P. (2009). Characterization of illuminated manuscripts by laboratory-made portable XRD and micro-XRD systems. *Analytical and Bioanalytical Chemistry, 395*(7), 1997–2004. doi:10.100700216-009-2992-5 PMID:19685044

Eaton, P., & West, P. (2010). *Atomic force microscopy*. Oxford university press. doi:10.1093/acprof:oso/9780199570454.001.0001

Efimov, A. E., Tonevitsky, A. G., Dittrich, M., & Matsko, N. B. (2007). Atomic force microscope (AFM) combined with the ultramicrotome: A novel device for the serial section tomography and AFM/TEM complementary structural analysis of biological and polymer samples. *Journal of Microscopy, 226*(3), 207–216. doi:10.1111/j.1365-2818.2007.01773.x PMID:17535260

Efros, A. L., & Efros, A. L. (1982). Interband absorption of light in a semiconductor sphere. *Soviet Phys Semicond USSR, 16*(7), 772–775.

Egerton, R. F. (2005). *Physical principles of electron microscopy* (Vol. 56). Springer. doi:10.1007/b136495

Ekimov, A., & Onushchenko, A. (1982). Quantum size effect in the opticalspectra of semiconductor micro-crystals. *Soviet Phys Semicond USSR, 16*(7), 775–778.

El Kirat, K., Burton, I., Dupres, V., & Dufrene, Y. (2005). Sample preparation procedures for biological atomic force microscopy. *Journal of Microscopy, 218*(3), 199–207. doi:10.1111/j.1365-2818.2005.01480.x PMID:15958012

El-Ansary, A., & Faddah, L. M. (2010). Nanoparticles as biochemical sensors. *Nanotechnology, Science and Applications, 3*, 65–76. doi:10.2147/NSA.S8199 PMID:24198472

Elcock, D. (2007). *Potential impacts of nanotechnology on energy transmission applications and needs*. Academic Press.

El-Hussein, A. (2016). Study DNA damage after photodynamic therapy using silver nanoparticles with A549 cell line. *Journal of Nanomedicine & Nanotechnology*, *7*(1), 1000346.

Elijah, V. T., & Odiyo, J. (2019). Perception of environmental spillovers across scale in climate change adaptation planning: The case of small-scale farmers' irrigation strategies. Academic Press.

El-Sayed, A., & Kamel, M. (2020). *Advances in nanomedical applications: diagnostic, therapeutic, immunization, and vaccine production*. Academic Press.

El-Shabouri, M. H. (2002). Positively charged nanoparticles for improving the oral bioavailability of cyclosporin-A. *International Journal of Pharmaceutics*, *249*(1-2), 101–108. doi:10.1016/S0378-5173(02)00461-1 PMID:12433438

Ema, M., Gamo, M., & Honda, K. (2016). *A review of toxicity studies of single-walled carbon nanotubes in laboratory animals*. Academic Press.

Engler, K. H., Efstratiou, A., Norn, D., Kozlov, R. S., Selga, I., Glushkevich, T. G., & George, R. C. (2002). Immunochromatographic strip test for rapid detection of diphtheria toxin: Description and multicenter evaluation in areas of low and high prevalence of diphtheria. *Journal of Clinical Microbiology*, *40*(1), 80–83. doi:10.1128/JCM.40.1.80-83.2002 PMID:11773096

Epping, R., & Koch, M. (2023). On-Site Detection of Volatile Organic Compounds (VOCs). *Molecules (Basel, Switzerland)*, *28*(4), 1598. doi:10.3390/molecules28041598 PMID:36838585

Epp, J. (2016). 4 - X-ray diffraction (XRD) techniques for materials characterization. In G. Hübschen, I. Altpeter, R. Tschuncky, & H.-G. Herrmann (Eds.), *Materials Characterization Using Nondestructive Evaluation (NDE) Methods* (pp. 81–124). Woodhead Publishing. doi:10.1016/B978-0-08-100040-3.00004-3

Erdoğan, A. (2022). Analysis and Chemical Imaging of Blue Inks for the Investigation of Document Forgery by XPS. *Microchemical Journal*, *183*, 108062. doi:10.1016/j.microc.2022.108062

Eshed, E., Abd El Majid, S., Bamberger, M., & Osovski, S. (2020). TEM and high resolution TEM investigation of phase formation in high entropy alloy AlCrFe2Ni2. *Frontiers in Materials*, *7*, 284. doi:10.3389/fmats.2020.00284

Estracanholli, É. S., Kurachi, C., Vicente, J. R., de Menezes, P. F. C., & Bagnato, V. S. (2009). Determination of post-mortem interval using in situ tissue optical fluorescence. *Optics Express*, *17*(10), 8185–8192. doi:10.1364/OE.17.008185 PMID:19434150

Eysenbach, G. (2006). Infodemiology: Tracking flu-related searches on the web for syndromic surveillance. *AMIA ... Annual Symposium Proceedings - AMIA Symposium. AMIA Symposium*, *2006*, 244. PMID:17238340

Ezcurra, M., Góngora, J. M., Maguregui, I., & Alonso, R. (2010). Analytical methods for dating modern writing instrument inks on paper. *Forensic Science International*, *197*(1–3), 1–20. doi:10.1016/j.forsciint.2009.11.013 PMID:20061099

Fadel, T. R., Farrell, D. F., Friedersdorf, L. E., Griep, M. H., Hoover, M. D., Meador, M. A., & Meyyappan, M. (2016). Toward the responsible development and commercialization of sensor nanotechnologies. *ACS Sensors*, *1*(3), 207–216. doi:10.1021/acssensors.5b00279 PMID:28261665

Fakhrullin, R., Nigamatzyanova, L., & Fakhrullina, G. (2021). Dark-field/hyperspectral microscopy for detecting nanoscale particles in environmental nanotoxicology research. *The Science of the Total Environment*, *772*, 145478. doi:10.1016/j.scitotenv.2021.145478 PMID:33571774

Fang, L., Jenk, T. M., Singer, T., Hou, S., & Schwikowski, M. (2021). Radiocarbon dating of alpine ice cores with the dissolved organic carbon (DOC) fraction. *The Cryosphere*, *15*(3), 1537–1550. doi:10.5194/tc-15-1537-2021

Fan, M., Andrade, G. F., & Brolo, A. G. (2020). A review on recent advances in the applications of surface-enhanced Raman scattering in analytical chemistry. *Analytica Chimica Acta*, *1097*, 1–29. doi:10.1016/j.aca.2019.11.049 PMID:31910948

Farahani, M., Khademabbasi, K., & Mollatayefeh, N. (2018). A selective morphine nanosensor derived from functionalized CdS quantum dots. *Materials Letters*, *228*, 68–71. doi:10.1016/j.matlet.2018.05.120

Faria, M., Björnmalm, M., Thurecht, K. J., Kent, S. J., Parton, R. G., Kavallaris, M., . . . Boyd, B. J. (2018). Minimum information reporting in bio–nano experimental literature. *Nature Nanotechnology*, *13*(9), 777–785.

Farokhzad, O. C., & Langer, R. (2009). Impact of nanotechnology on drug delivery. *ACS Nano*, *3*(1), 16–20. doi:10.1021/nn900002m PMID:19206243

Farrow, B., Hong, S. A., Romero, E. C., Lai, B., Coppock, M. B., Deyle, K. M., & Heath, J. R. (2013). A chemically synthesized capture agent enables the selective, sensitive, and robust electrochemical detection of anthrax protective antigen. *ACS Nano*, *7*(10), 9452–9460. doi:10.1021/nn404296k PMID:24063758

Fathi-Achachelouei, M., Knopf-Marques, H., Ribeiro da Silva, C. E., Barthès, J., Bat, E., Tezcaner, A., & Vrana, N. E. (2019). Use of Nanoparticles in Tissue Engineering and Regenerative Medicine. *Frontiers in Bioengineering and Biotechnology*, *7*, 113. doi:10.3389/fbioe.2019.00113 PMID:31179276

FDA. (2014). *Guidance for Industry: Considering Whether an FDA-Regulated Product Involves the Application of Nanotechnology*. FDA.

Feng, J., Lepetre-Mouelhi, S., Gautier, A., Mura, S., Cailleau, C., Coudore, F., Hamon, M., & Couvreur, P. (2019). A new painkiller nanomedicine to bypass the blood-brain barrier and the use of morphine. *Science Advances*, *5*(2), eaau5148. doi:10.1126ciadv.aau5148 PMID:30788432

Feng, X. T., Ashley, J., Zhou, T. C., & Sun, Y. (2018). Fluorometric determination of doxycycline based on the use of carbon quantum dots incorporated into a molecularly imprinted polymer. *Mikrochimica Acta*, *185*(11), 9. doi:10.100700604-018-2999-8 PMID:30293175

Feng, X., Wang, H., Wang, Z., Huang, P., & Kan, J. (2022). Discrimination and characterization of the volatile organic compounds in eight kinds of huajiao with geographical indication of China using electronic nose, HS-GC-IMS and HS-SPME-GC–MS. *Food Chemistry*, *375*, 131671. doi:10.1016/j.foodchem.2021.131671 PMID:34865919

Fikiet, M. A., Khandasammy, S. R., Mistek, E., Ahmed, Y., Halámková, L., Bueno, J., & Lednev, I. K. (2018). Surface enhanced Raman spectroscopy: A review of recent applications in forensic science. *Spectrochimica Acta. Part A: Molecular and Biomolecular Spectroscopy*, *197*, 255–260. doi:10.1016/j.saa.2018.02.046 PMID:29496406

Finnie, K. S., Bartlett, J. R., Barbé, C. J. A., & Kong, L. (2007). Formation of silica nanoparticles in microemulsions. *Langmuir*, *23*(6), 3017–3024. doi:10.1021/la0624283 PMID:17300209

Fischer, H. C., & Chan, W. C. (2007). Nanotoxicity: The growing need for in vivo study. *Current Opinion in Biotechnology*, *18*(6), 565–571. doi:10.1016/j.copbio.2007.11.008 PMID:18160274

Fleck, R. A., & Humbel, B. M. (2019). Biological Field Emission. *Scanning Electron Microscopy*, 2.

Fletcher, J., Bender, C., Budowle, B., Cobb, W. T., Gold, S. E., Ishimaru, C. A., Luster, D., Melcher, U., Murch, R., Scherm, H., Seem, R. C., Sherwood, J. L., Sobral, B. W., & Tolin, S. A. (2006). Plant pathogen forensics: Capabilities, needs, and recommendations. *Microbiology and Molecular Biology Reviews*, *70*(2), 450–471. doi:10.1128/MMBR.00022-05 PMID:16760310

Florea, A., Cowen, T., Piletsky, S., & De Wael, K. (2019). *Electrochemical sensing of cocaine in real samples based on electrodeposited biomimetic affinity ligands*. Academic Press.

Florea, A., Cowen, T., Piletsky, S., & De Wael, K. (2019). Electrochemical sensing of cocaine in real samples based on electrodeposited biomimetic affinity ligands. *Analyst (London), 144*(15), 4639–4646. doi:10.1039/C9AN00618D PMID:31250860

Foldvari, M., & Bagonluri, M. (2008). Carbon nanotubes as functional excipients for nanomedicines: II. Drug delivery and biocompatibility issues. *Nanomedicine (London), 4*(3), 183–200. doi:10.1016/j.nano.2008.04.003 PMID:18550450

Fong, K. E., & Yung, L.-Y. L. (2013). Localized surface plasmon resonance: A unique property of plasmonic nanoparticles for nucleic acid detection. *Nanoscale, 5*(24), 12043–12071. doi:10.1039/c3nr02257a PMID:24166199

Fornaciari, A. (2017). Environmental microbial forensics and archaeology of past pandemics. *Microbiology Spectrum, 5*(1), EMF-0011–EMF-2016. doi:10.1128/microbiolspec.EMF-0011-2016 PMID:28233511

Frascione, N., Pinto, V., & Daniel, B. (2012). Development of a biosensor for human blood: New routes to body fluid identification. *Analytical and Bioanalytical Chemistry, 404*(1), 23–28. doi:10.100700216-012-6111-7 PMID:22644154

Freeman, R., Girsh, J., & Willner, I. (2013). Nucleic acid/quantum dots (QDs) hybrid systems for optical and photoelectrochemical sensing. *ACS Applied Materials & Interfaces, 5*(8), 2815–2834. doi:10.1021/am303189h PMID:23425022

Furton, K. G., & Myers, L. (2001). *The scientific foundation and efficacy of the use of canines as chemical detectors for explosives*. Academic Press.

Gakhar, T., Basu, S., & Hazra, A. (2022). Carbon-metal oxide nanocomposites for selective detection of toxic and hazardous volatile organic compounds (VOC)-A review. *Green Analytical Chemistry*.

Ganash, A., Alshammari, S., & Ganash, E. (2023). Development of a Novel Electrochemical Sensor Based on Gold Nanoparticle-Modified Carbon-Paste Electrode for the Detection of Congo Red Dye. *Molecules (Basel, Switzerland), 28*(1), 1. Advance online publication. doi:10.3390/molecules28010019 PMID:36615215

Gandhi, S., Banga, I., Maurya, P. K., & Eremin, S. A. (2018). A gold nanoparticle-single-chain fragment variable antibody as an immunoprobe for rapid detection of morphine by dipstick. *RSC Advances, 8*(3), 1511–1518. doi:10.1039/C7RA12810J PMID:35540925

Ganesan, M., & Nagaraaj, P. (2020). Quantum dots as nanosensors for detection of toxics: A literature review. *Analytical Methods, 12*(35), 4254–4275. doi:10.1039/D0AY01293A PMID:32940270

Ganesh, E. N. (2016). Application of nanotechnology in forensic science. *Environment, 1*(1).

Ganguly, J., Saha, S., Bera, A., & Ghosh, M. (2017). Exploring electro-optic effect and third-order nonlinear optical susceptibility of impurity doped quantum dots: Interplay between hydrostatic pressure, temperature and noise. *Optics Communications, 387*, 166–173. doi:10.1016/j.optcom.2016.11.040

Ganson, K. T., Murray, S. B., & Nagata, J. M. (2021). Associations between eating disorders and illicit drug use among college students. *International Journal of Eating Disorders, 54*(7), 1127–1134. doi:10.1002/eat.23493 PMID:33638571

Gao, D., Li, F., Song, J., Xu, X., Zhang, Q., & Niu, L. (2009). One step to detect the latent fingermarks with gold nanoparticles. *Talanta, 80*(2), 479–483. doi:10.1016/j.talanta.2009.07.007 PMID:19836507

Gao, L., Xiang, W., Deng, Z., Shi, K., Wang, H., & Shi, H. (2020). Cocaine detection using aptamer and molybdenum disulfide-gold nanoparticle-based sensors. *Nanomedicine (London), 15*(04), 325–335. doi:10.2217/nnm-2019-0046 PMID:31976806

Gao, W., Wang, C., Muzyka, K., Kitte, S. A., Li, J., Zhang, W., & Xu, G. (2017). Artemisinin-luminol chemiluminescence for forensic bloodstain detection using a smart phone as a detector. *Analytical Chemistry*, *89*(11), 6160–6165. doi:10.1021/acs.analchem.7b01000 PMID:28492080

Gardner, J. W. (2004). *Review of conventional electronic noses and their possible application to the detection of explosives. In Electronic Noses & Sensors for the Detection of Explosives*. Springer. doi:10.1007/1-4020-2319-7

Gavrilă, A. M., Stoica, E. B., Iordache, T. V., & Sârbu, A. (2022). Modern and Dedicated Methods for Producing Molecularly Imprinted Polymer Layers in Sensing Applications. *Applied Sciences (Basel, Switzerland)*, *12*(6), 3080. doi:10.3390/app12063080

Gawande, M. B., Monga, Y., Zboril, R., & Sharma, R. K. (2015). Silica-decorated magnetic nanocomposites for catalytic applications. *Coordination Chemistry Reviews*, *288*, 118–143. doi:10.1016/j.ccr.2015.01.001

Gebre, S. H., & Sendeku, M. G. (2019). New frontiers in the biosynthesis of metal oxide nanoparticles and their environmental applications: An overview. *SN Applied Sciences*, *1*(8), 1–28. doi:10.100742452-019-0931-4

Geho, D. H., Jones, C. D., Petricoin, E. F., & Liotta, L. A. (2006). Nanoparticles: Potential biomarker harvesters. *Current Opinion in Chemical Biology*, *10*(1), 56–61. doi:10.1016/j.cbpa.2006.01.003 PMID:16413816

Ge, L., Li, S. P., & Lisak, G. (2020). Advanced sensing technologies of phenolic compounds for pharmaceutical and biomedical analysis. *Journal of Pharmaceutical and Biomedical Analysis*, *179*, 112913. doi:10.1016/j.jpba.2019.112913 PMID:31679844

Gelderman, H., Boer, L., Naujocks, T., IJzermans, A., & Duijst, W. (2018). The development of a post-mortem interval estimation for human remains found on land in the Netherlands. *International Journal of Legal Medicine*, *132*(3), 863–873. doi:10.100700414-017-1700-9 PMID:29110084

Geminiani, L., Campione, F. P., Corti, C., Luraschi, M., Motella, S., Recchia, S., & Rampazzi, L. (2022). Differentiating between Natural and Modified Cellulosic Fibres Using ATR-FTIR Spectroscopy. *Heritage*, *5*(4), 4. Advance online publication. doi:10.3390/heritage5040213

Gerion, D., Pinaud, F., Williams, S. C., Parak, W. J., Zanchet, D., Weiss, S., & Alivisatos, A. P. (2001). Synthesis and properties of biocompatible water-soluble silica-coated CdSe/ZnS semiconductor quantum dots. *The Journal of Physical Chemistry B*, *105*(37), 8861–8871. doi:10.1021/jp0105488

Gessner, A., Lieske, A., Paulke, B., & Müller, R. (2002). Influence of surface charge density on protein adsorption on polymeric nanoparticles: Analysis by two-dimensional electrophoresis. *European Journal of Pharmaceutics and Biopharmaceutics*, *54*(2), 165–170. doi:10.1016/S0939-6411(02)00081-4 PMID:12191688

Ghadiri, M., Ghashghaee, M., & Ghambarian, M. (2020). Defective phosphorene for highly efficient formaldehyde detection: Periodic density functional calculations. *Physics Letters. [Part A]*, *384*(31), 126792. doi:10.1016/j.physleta.2020.126792

Giannoukos, S., Brkic, B., Taylor, S., Marshall, A., & Verbeck, G. F. (2016). Chemical sniffing instrumentation for security applications. *Chemical Reviews*, *116*(14), 8146–8172. doi:10.1021/acs.chemrev.6b00065 PMID:27388215

Gilbride, K. A., Lee, D. Y., & Beaudette, L. A. (2006). Molecular techniques in wastewater: Understanding microbial communities, detecting pathogens, and real-time process control. *Journal of Microbiological Methods*, *66*(1), 1–20. doi:10.1016/j.mimet.2006.02.016 PMID:16635533

Giunipero, L. C., Hooker, R. E., Joseph-Matthews, S., Yoon, T. E., & Brudvig, S. (2008). A decade of SCM literature: Past, present and future implications. *The Journal of Supply Chain Management*, *44*(4), 66–86. doi:10.1111/j.1745-493X.2008.00073.x

Godoy, D., Randle, G., Simpson, A. J., Aanensen, D. M., Pitt, T. L., Kinoshita, R., & Spratt, B. G. (2003). Multilocus sequence typing and evolutionary relationships among the causative agents of melioidosis and glanders, Burkholderia pseudomallei and Burkholderia mallei. *Journal of Clinical Microbiology*, *41*(5), 2068–2079. doi:10.1128/JCM.41.5.2068-2079.2003 PMID:12734250

Gokus, T., Nair, R. R., Bonetti, A., Boehmler, M., Lombardo, A., Novoselov, K. S., Geim, A. K., Ferrari, A. C., & Hartschuh, A. (2009). Making Graphene Luminescent by Oxygen Plasma Treatment. *ACS Nano*, *3*(12), 3963–3968. doi:10.1021/nn9012753 PMID:19925014

Göl, E., & Çok, I. (2019). New psychoactive substances in Turkey: Narcotics cases assessed by the Council of Forensic Medicine between 2016 and 2017 in Ankara, Turkey. *Forensic Science International*, *294*, 113–123. doi:10.1016/j.forsciint.2018.11.003 PMID:30502695

Gooch, J., Goh, H., Daniel, B., Abbate, V., & Frascione, N. (2016). Monitoring criminal activity through invisible fluorescent “peptide coding” taggants. *Analytical Chemistry*, *88*(8), 4456–4460. doi:10.1021/acs.analchem.6b00263 PMID:27010696

Gopi, S., Amalraj, A., Haponiuk, J. T., & Thomas, S. (2016). Introduction of nanotechnology in herbal drugs and nutraceutical: A review. *Journal of Nanomedicine & Biotherapeutic Discovery*, *6*(2), 1–8. doi:10.4172/2155-983X.1000143

Gorshkova, K. O., Rossinskaya, E. R., Kirillova, N. P., Fogel, A. A., Kochemirovskaia, S. V., & Kochemirovsky, V. A. (2020). Investigation of the new possibility of mathematical processing of Raman spectra for dating documents. *Science & Justice*, *60*(5), 451–465. doi:10.1016/j.scijus.2020.06.007 PMID:32873385

Gorup, L. F., Amorin, L. H., Camargo, E. R., Sequinel, T., Cincotto, F. H., Biasotto, G., & La Porta, F. A. (2020). *Methods for design and fabrication of nanosensors: the case of ZnO-based nanosensor. In Nanosensors for smart cities*. Elsevier.

Goryacheva, O. A., Beloglazova, N. V., Vostrikova, A. M., Pozharov, M. V., Sobolev, A. M., & Goryacheva, I. Y. (2017). Lanthanide-to-quantum dot Förster resonance energy transfer (FRET): Application for immunoassay. *Talanta*, *164*, 377–385. doi:10.1016/j.talanta.2016.11.054 PMID:28107944

Graziano, S., Anzillotti, L., Mannocchi, G., Pichini, S., & Busardò, F. P. (2019). Screening methods for rapid determination of new psychoactive substances (NPS) in conventional and non-conventional biological matrices. *Journal of Pharmaceutical and Biomedical Analysis*, *163*, 170–179. doi:10.1016/j.jpba.2018.10.011 PMID:30316062

Greenham, N. C., Peng, X., & Alivisatos, A. P. (1996). Charge separation and transport in conjugated-polymer/semiconductor-nanocrystal composites studied by photoluminescence quenching and photoconductivity. *Physical Review. B*, *54*(24), 17628–17637. doi:10.1103/PhysRevB.54.17628 PMID:9985889

Gregis, G., Sanchez, J. B., Bezverkhyy, I., Guy, W., Berger, F., Fierro, V., & Celzard, A. (2018). Detection and quantification of lung cancer biomarkers by a micro-analytical device using a single metal oxide-based gas sensor. *Sensors and Actuators. B, Chemical*, *255*, 391–400. doi:10.1016/j.snb.2017.08.056

Grim, D. M., Siegel, J., & Allison, J. (2001). *Evaluation of desorption/ionization mass spectrometric*. Academic Press.

Grimshaw, J., Eccles, M., & Tetroe, J. (2004). Implementing clinical guidelines: Current evidence and future implications. *The Journal of Continuing Education in the Health Professions*, *24*(S1), S31–S37. doi:10.1002/chp.1340240506 PMID:15712775

Gui, L., Chen, B., Zhou, Z., Liang, Y., He, M., & Hu, B. (2022). Phytic acid functionalized magnetic adsorbents for facile enrichment of trace rare earth elements in environmental water, digested atmospheric particulates and the extracts followed by inductively coupled plasma mass spectrometry detection. *Talanta*, *244*, 244123426. doi:10.1016/j.talanta.2022.123426 PMID:35381498

Guo, W. H., Li, J. J., Wang, Y. A., & Peng, X. G. (2003). Luminescent CdSe/CdS core/shell nanocrystals in dendron boxes: Superior chemical, photochemical and thermal stability. *Journal of the American Chemical Society, 125*(13), 3901–3909. doi:10.1021/ja028469c PMID:12656625

Hachenberger, Y. U., Rosenkranz, D., Kromer, C., Krause, B. C., Dreiack, N., Kriegel, F. L., ... Bierkandt, F. S. (2023). Nanomaterial Characterization in Complex Media—Guidance and Application. *Nanomaterials (Basel, Switzerland), 13*(5), 922. doi:10.3390/nano13050922 PMID:36903800

Haddad, Y., Dostalova, S., Kudr, J., Zitka, O., Heger, Z., & Adam, V. (2017). DNA-magnetic particle binding analysis by dynamic and electrophoretic light scattering. *Journal of Visualized Experiments*, 129. PMID:29155773

Hafner, A., Lovrić, J., Lakoš, G. P., & Pepić, I. (2014). Nanotherapeutics in the EU: An overview on current state and future directions. *International Journal of Nanomedicine, 9*, 1005–1023. doi:10.2147/IJN.S55359 PMID:24600222

Hagarová, I. (2020). Magnetic Solid Phase Extraction as a Promising Technique for Fast Separation of Metallic Nanoparticles and Their Ionic Species: A Review of Recent Advances. *Journal of Analytical Methods in Chemistry, 8847565*, 1–12. Advance online publication. doi:10.1155/2020/8847565 PMID:32963882

Hahn, M. A., Tabb, J. S., & Krauss, T. D. (2005). Detection of single bacterial pathogens with semiconductor quantum dots. *Analytical Chemistry, 77*(15), 4861–4869. doi:10.1021/ac050641i PMID:16053299

Hakonen, A., Wu, K., Schmidt, M. S., Andersson, P. O., Boisen, A., & Rindzevicius, T. (2018). Detecting forensic substances using commercially available SERS substrates and handheld Raman spectrometers. *Talanta, 189*, 649–652. doi:10.1016/j.talanta.2018.07.009 PMID:30086972

Han, J., Zhuo, Y., Chai, Y.-Q., Xiang, Y., & Yuan, R. (2015). *New type of redox nanoprobe: C60-based nanomaterial and its application in electrochemical immunoassay for doping detection.* Academic Press.

Han, S., Tan, Y., Gao, Y., Li, X., Ho, S. S. H., Wang, M., & Lee, S. C. (2023). Volatile organic compounds at a roadside site in Hong Kong: Characteristics, chemical reactivity, and health risk assessment. *The Science of the Total Environment, 866*, 161370. doi:10.1016/j.scitotenv.2022.161370 PMID:36621478

Han, Z., Liu, H., Wang, B., Weng, S., Yang, L., & Liu, J. (2015). Three-dimensional surface-enhanced Raman scattering hotspots in spherical colloidal superstructure for identification and detection of drugs in human urine. *Analytical Chemistry, 87*(9), 4821–4828. doi:10.1021/acs.analchem.5b00176 PMID:25853724

Harrison, J. W., Dung, T. T. N., Siddiqui, F., Korbrisate, S., Bukhari, H., Tra, M. P. V., Hoang, N. V. M., Carrique-Mas, J., Bryant, J., Campbell, J. I., Studholme, D. J., Wren, B. W., Baker, S., Titball, R. W., & Champion, O. L. (2014). Identification of possible virulence marker from Campylobacter jejuni isolates. *Emerging Infectious Diseases, 20*(6), 1026–1029. doi:10.3201/eid2006.130635 PMID:24856088

Hassellöv, M., Readman, J. W., Ranville, J. F., & Tiede, K. (2008). Nanoparticle analysis and characterization methodologies in environmental risk assessment of engineered nanoparticles. *Ecotoxicology (London, England), 17*(5), 344–361. doi:10.100710646-008-0225-x PMID:18483764

Hauther, K. A., Cobaugh, K. L., Jantz, L. M., Sparer, T. E., & DeBruyn, J. M. (2015). Estimating time since death from postmortem human gut microbial communities. *Journal of Forensic Sciences, 60*(5), 1234–1240. doi:10.1111/1556-4029.12828 PMID:26096156

Hazarika, P. J., Jickells, S. M., Wolff, K., & Russell, D. A. (2008). Imaging of latent fingerprints through the detection of drugs and metabolites. Angew. *Angewandte Chemie International Edition, 47*(52), 10167–10170. doi:10.1002/anie.200804348

Hazra, A., Bhowmik, B., Dutta, K., Chattopadhyay, P. P., & Bhattacharyya, P. (2015). Stoichiometry, length, and wall thickness optimization of TiO2 nanotube array for efficient alcohol sensing. *ACS Applied Materials & Interfaces*, *7*(18), 9336–9348. doi:10.1021/acsami.5b01785 PMID:25918822

He, C., Zhu, W., Xu, Y., Zhong, Y., Zhou, J., & Qian, X. (2010). Ratiometric and reusable fluorescent nanoparticles for Zn2+ and H 2 PO 4− detection in aqueous solution and living cells. *Journal of Materials Chemistry*, *20*(47), 10755–10764. doi:10.1039/c0jm01925a

Hedberg, C. W., & Besser, J. M. (2006). Commentary: Cluster evaluation, PulseNet, and public health practice. *Foodborne Pathogens and Disease*, *3*(1), 32–35. doi:10.1089/fpd.2006.3.32 PMID:16602977

Hegab, H., Khanna, N., Monib, N., & Salem, A. (2023). Design for sustainable additive manufacturing: A review. *Sustainable. Materials Technology.*

Heidinger, S., Spranger, F., Dostál, J., Zhang, C., & Klaus, C. (2022). Material Treatment in the Pulsation Reactor—From Flame Spray Pyrolysis to Industrial Scale. *Sustainability (Basel)*, *14*(6), 3232. doi:10.3390u14063232

He, J., Guo, J., Fu, X., & Cai, J. (2019). Potential use of high-throughput sequencing of bacterial communities for postmortem submersion interval estimation. *Brazilian Journal of Microbiology*, *50*(4), 999–1010. doi:10.100742770-019-00119-w PMID:31364013

He, K., Chen, N., Wang, C., Wei, L., & Chen, J. (2018). Method for determining crystal grain size by x-ray diffraction. *Crystal Research and Technology*, *53*(2), 1700157. doi:10.1002/crat.201700157

Hekiem, N. L. L., Ralib, A. A. M., Ahmad, F. B., Nordin, A. N., Ab Rahim, R., & Za'bah, N. F. (2021). Advanced vapour sensing materials: Existing and latent to acoustic wave sensors for VOCs detection as the potential exhaled breath biomarkers for lung cancer. *Sensors and Actuators. A, Physical*, *329*, 112792. doi:10.1016/j.sna.2021.112792

Henningfield, J. E., Coe, M. A., Griffiths, R. R., Belouin, S. J., Berger, A., Coker, A. R., ... Nichols, C. D. (2022). Psychedelic drug abuse potential assessment research for new drug applications and Controlled Substances Act scheduling. *Neuropharmacology*, *218*, 109220. doi:10.1016/j.neuropharm.2022.109220 PMID:35987353

Hettler, S., Wagner, J., Dries, M., Oster, M., Wacker, C., Schröder, R. R., & Gerthsen, D. (2015). On the role of inelastic scattering in phase-plate transmission electron microscopy. *Ultramicroscopy*, *155*, 27–41. doi:10.1016/j.ultramic.2015.04.001 PMID:25879156

He, X., Ma, Y., Li, M., Zhang, P., Li, Y., & Zhang, Z. (2013). Quantifying and imaging engineered nanomaterials in vivo: Challenges and techniques. *Small*, *9*(9-10), 1482–1491. doi:10.1002mll.201201502 PMID:23027545

He, Z., Zhang, Z., & Bi, S. (2020). Nanoparticles for organic electronics applications. *Materials Research Express*, *7*(1), 1–13. doi:10.1088/2053-1591/ab636f

Hines, M. A., & Guyot-Sionnest, P. (1996). Synthesis and characterization of strongly luminescing ZnS-capped CdSe nanocrystals. *Journal of Physical Chemistry*, *100*(2), 468–471. doi:10.1021/jp9530562

Hobson, D. W. (2009). Commercialization of nanotechnology. *Wiley Interdisciplinary Reviews. Nanomedicine and Nanobiotechnology*, *1*(2), 189–202. doi:10.1002/wnan.28 PMID:20049790

Holder, C. F., & Schaak, R. E. (2019). *Tutorial on powder X-ray diffraction for characterizing nanoscale materials* (Vol. 13). ACS Publications.

Hondebrink, L., Nugteren-van Lonkhuyzen, J. J., Hunault, C. C., van den Berg, J., Van Der Gouwe, D., & van Riel, A. J. (2020). New psychoactive substances (NPS) in the Netherlands: Occurrence in forensic drug samples, consumer drug samples and poisons center exposures between 2013 and 2017. *Addiction (Abingdon, England)*, *115*(4), 716–725. doi:10.1111/add.14868 PMID:31656050

Hong, Y., Zhou, X., Xu, B., Huang, Y., He, W., Wang, S., Wang, C., Zhou, G., Chen, Y., & Gong, T. (2020). Optoplasmonic hybrid materials for trace detection of methamphetamine in biological fluids through SERS. *ACS Applied Materials & Interfaces*, *12*(21), 24192–24200. doi:10.1021/acsami.0c00853 PMID:32351116

Horváth, T., & Szatai, J. Z. (2020). A History of Detection of Explosive Devices 1. *Land Forces Academy Review*, *25*(3), 189–200. doi:10.2478/raft-2020-0023

Hosseini, S. J. F., Dehyouri, S., & Mirdamadi, S. M. (2010). The perception of agricultural researchers about the role of nanotechnology in achieving food security. *African Journal of Biotechnology*, *9*(37), 6152–6157.

Hou, X., Zeng, F., Du, F., & Wu, S. (2013). Carbon-dot-based fluorescent turn-on sensor for selectively detecting sulfide anions in totally aqueous media and imaging inside live cells. *Nanotechnology*, *24*(33), 335502. doi:10.1088/0957-4484/24/33/335502 PMID:23892368

Howard, A. L., Kennedy, T. M., Mitchell, J. T., Sibley, M. H., Hinshaw, S. P., Arnold, L. E., Roy, A., Stehli, A., Swanson, J. M., & Molina, B. S. (2020). Early substance use in the pathway from childhood attention-deficit/hyperactivity disorder (ADHD) to young adult substance use: Evidence of statistical mediation and substance specificity. *Psychology of Addictive Behaviors*, *34*(2), 281–292. doi:10.1037/adb0000542 PMID:31886682

Hsiao, I. L., & Huang, Y.-J. (2011). Effects of various physicochemical characteristics on the toxicities of ZnO and TiO2 nanoparticles toward human lung epithelial cells. *The Science of the Total Environment*, *409*(7), 1219–1228. doi:10.1016/j.scitotenv.2010.12.033 PMID:21255821

Huang, L., Peng, Z., Guo, Y., & Porter, A. L. (2010). Identifying the emerging roles of nanoparticles in biosensors. *Journal of Business Chemistry*, *7*(1).

Huang, S., Chen, J., Zhang, T., Dai, X., Wang, X., Zhou, J., Kong, W., Liu, Q., Qu, J., & Shao, Y. (2022). Recent Advances in Surface Plasmon Resonance Microscopy. *Chemosensors (Basel, Switzerland)*, *10*(12), 509. doi:10.3390/chemosensors10120509

Huang, Y., Li, X., Chen, X., Wang, W., Wang, Y., Liu, Z., & Tang, G. (2022). Low-molecular-weight carbonyl volatile organic compounds on the North China Plain. *Atmospheric Environment*, *275*, 119000. doi:10.1016/j.atmosenv.2022.119000

Hui, Y. W., Mahat, N. A., Ismail, D., & Ibrahim, R. K. R. (2019). Laser-induced breakdown spectroscopy (LIBS) for printing ink analysis coupled with principle component analysis (PCA). AIP Conf. Proc. https://doi.org/ doi:10.1063

Hulla, J., Sahu, S., & Hayes, A. (2015). *Nanotechnology: History and future*. Academic Press.

Hulla, J., Sahu, S., & Hayes, A. (2015). Nanotechnology: History and future. *Human and Experimental Toxicology*, *34*(12), 1318–1321. doi:10.1177/0960327115603588 PMID:26614822

Hullmann, A. (2006). *The economic development of nanotechnology-An indicators based analysis.* EU Report.

Hung, C. M., Van Duy, L., Le, D. T. T., Nguyen, H., Van Duy, N., & Hoa, N. D. (2021). ZnO coral-like nanoplates decorated with Pd nanoparticles for enhanced VOC gas sensing. *Journal of Science. Advanced Materials and Devices*, *6*(3), 453–461. doi:10.1016/j.jsamd.2021.05.005

Hussain, S., Hess, K., Gearhart, J., Geiss, K., & Schlager, J. (2005). *In vitro toxicity of nanoparticles in BRL 3A rat liver cells*. Academic Press.

Hussain, S. M., Warheit, D. B., Ng, S. P., Comfort, K. K., Grabinski, C. M., & Braydich-Stolle, L. K. (2015). At the Crossroads of Nanotoxicology in vitro: Past Achievements and Current Challenges. *Toxicological Sciences*, *147*(1), 5–16. doi:10.1093/toxsci/kfv106 PMID:26310852

Huynh, V., Phelps, M. S., Golden, T. D., & Verbeck, G. F. (2016). Direct analyte-probed nanoextraction (DAPNe) coupled to matrix-assisted laser desorption ionization (MALDI) for examination of the ink chemistry on documents. *Forensic Chemistry (Amsterdam, Netherlands)*, *2*, 86–92. doi:10.1016/j.forc.2016.10.007

ICH. (1997). Guideline: Preclinical Safety Evaluation of Biotechnology-Derived Pharmaceuticals. Academic Press.

Ijaz, I., Gilani, E., Nazir, A., & Bukhari, A. (2020). Detail review on chemical, physical and green synthesis, classification, characterizations and applications of nanoparticles. *Green Chemistry Letters and Reviews*, *13*(3), 223–245. doi:10.1080/17518253.2020.1802517

Ilic, B., Yang, Y., & Craighead, H. G. (2004). Virus detection using nanoelectromechanical devices. *Applied Physics Letters*, *85*(13), 2604–2606. doi:10.1063/1.1794378

Iravani, S., & Varma, R. S. (2020). Green synthesis, biomedical and biotechnological applications of carbon and graphene quantum dots. A review. *Environmental Chemistry Letters*, *18*(3), 703–727. doi:10.100710311-020-00984-0 PMID:32206050

Irvin, C. B., Nouhan, P. P., & Rice, K. (2003). Syndromic analysis of computerized emergency department patients' chief complaints: An opportunity for bioterrorism and influenza surveillance. *Annals of Emergency Medicine*, *41*(4), 447–452. doi:10.1067/mem.2003.104 PMID:12658241

Ishikawa-Ankerhold, H. C., Ankerhold, R., & Drummen, G. P. (2012). Advanced fluorescence microscopy techniques—Frap, Flip, Flap, Fret and flim. *Molecules (Basel, Switzerland)*, *17*(4), 4047–4132. doi:10.3390/molecules17044047 PMID:22469598

Ivanova, I., Spiridonov, I., & Lasheva, V. (2022). Spectroscopic technique for studying the characteristics of paper. Academic Press.

Jadhav, S. S., & Jadhav, S. V. (2018). *Application of nanotechnology in modern computers.* Paper presented at the 2018 International Conference on Circuits and Systems in Digital Enterprise Technology (ICCSDET). 10.1109/ICCSDET.2018.8821086

Jain, K. K. (2005). Nanotechnology in clinical laboratory diagnostics. *Clinica Chimica Acta*, *358*(1-2), 37–54. doi:10.1016/j.cccn.2005.03.014 PMID:15890325

Jain, K. K. (2007). Applications of nanobiotechnology in clinical diagnostics. *Clinical Chemistry*, *53*(11), 2002–2009. doi:10.1373/clinchem.2007.090795 PMID:17890442

Jain, N., Jain, R., Thakur, N., Gupta, B. P., Jain, D. K., Banveer, J., & Jain, S. (2010). Nanotechnology: A safe and effective drug delivery system. *Asian Journal of Pharmaceutical and Clinical Research*, *3*(3), 159–165.

Jakubowska, J., Maciejewska, A., Bielawski, K. P., & Pawłowski, R. (2014). mRNA heptaplex protocol for distinguishing between menstrual and peripheral blood. *Forensic Science International. Genetics*, *13*, 53–60. doi:10.1016/j.fsigen.2014.07.006 PMID:25082137

Jamkhande, P. G., Ghule, N. W., Bamer, A. H., & Kalaskar, M. G. (2019). Metal nanoparticles synthesis: An overview on methods of preparation, advantages and disadvantages, and applications. *Journal of Drug Delivery Science and Technology*, *53*, 101174. doi:10.1016/j.jddst.2019.101174

Jan, N., Majeed, N., Ahmad, M., Lone, W. A., & John, R. (2022). Nano-pollution: Why it should worry us. *Chemosphere*, *302*, 134746. doi:10.1016/j.chemosphere.2022.134746 PMID:35489464

Javaid, M., Haleem, A., Singh, R. P., Rab, S., & Suman, R. (2021). Exploring the potential of nanosensors: A brief overview. *Sensors International*, *2*, 100130. doi:10.1016/j.sintl.2021.100130

Jeevanandam, J., Barhoum, A., Chan, Y. S., Dufresne, A., & Danquah, M. K. (2018). Review on nanoparticles and nanostructured materials: History, sources, toxicity and regulations. *Beilstein Journal of Nanotechnology*, *9*(1), 1050–1074. doi:10.3762/bjnano.9.98 PMID:29719757

Jeevanandam, J., Chan, Y. S., & Danquah, M. K. (2016). Biosynthesis of metal and metal oxide nanoparticles. *ChemBioEng Reviews*, *3*(2), 55–67. doi:10.1002/cben.201500018

Jeon, S., Oberreit, D. R., Van Schooneveld, G., & Hogan, C. J. (2016). Nanomaterial size distribution analysis via liquid nebulization coupled with ion mobility spectrometry (LN-IMS). *Analyst (London)*, *141*(4), 1363–1375. doi:10.1039/C5AN02150B PMID:26750519

Jeyaraj, M., Gurunathan, S., Qasim, M., Kang, M.-H., & Kim, J.-H. (2019). A comprehensive review on the synthesis, characterization, and biomedical application of platinum nanoparticles. *Nanomaterials (Basel, Switzerland)*, *9*(12), 1719. doi:10.3390/nano9121719 PMID:31810256

Jiang, H., & Ju, H. (2007). Electrochemiluminescence sensors for scavengers of hydroxyl radical based on its annihilation in CdSe quantum dots film/peroxide system. *Analytical Chemistry*, *79*(17), 6690–6696. doi:10.1021/ac071061j PMID:17676926

Jiang, X., Röcker, C., Hafner, M., Brandholt, S., Dörlich, R. M., & Nienhaus, G. U. (2010). Endo-and exocytosis of zwitterionic quantum dot nanoparticles by live HeLa cells. *ACS Nano*, *4*(11), 6787–6797. doi:10.1021/nn101277w PMID:21028844

Jiang, Z., Li, L., Huang, H., He, W., & Ming, W. (2022). Progress in Laser Ablation and Biological Synthesis Processes:"Top-Down" and "Bottom-Up" Approaches for the Green Synthesis of Au/Ag Nanoparticles. *International Journal of Molecular Sciences*, *23*(23), 14658. doi:10.3390/ijms232314658 PMID:36498986

Jia, Z., Li, J., Gao, L., Yang, D., & Kanaev, A. (2023). Dynamic Light Scattering: A Powerful Tool for In Situ Nanoparticle Sizing. *Colloids and Interfaces*, *7*(1), 15. doi:10.3390/colloids7010015

Jie, G., Zhang, J., Wang, D., Cheng, C., Chen, H. Y., & Zhu, J. J. (2008). Electrochemiluminescence immunosensor based on CdSe nanocomposites. *Analytical Chemistry*, *80*(11), 4033–4039. doi:10.1021/ac800052g PMID:18435547

Jones, D., Caballero, S., & Davidov-Pardo, G. (2019). Bioavailability of nanotechnology-based bioactives and nutraceuticals. *Advances in Food and Nutrition Research*, *88*, 235–273. doi:10.1016/bs.afnr.2019.02.014 PMID:31151725

Jones, R. W., Cody, R. B., & Mcclelland, J. F. (2006). Differentiating writing inks using direct analysis in real time mass spectrometry. *Journal of Forensic Sciences*, *51*(4), 915–918. doi:10.1111/j.1556-4029.2006.00162.x PMID:16882239

Joshi, J., Homburg, S. V., & Ehrmann, A. (2022). Atomic Force Microscopy (AFM) on Biopolymers and Hydrogels for Biotechnological Applications—Possibilities and Limits. *Polymers*, *14*(6), 1267. doi:10.3390/polym14061267 PMID:35335597

Joudeh, N., & Linke, D. (2022). Nanoparticle classification, physicochemical properties, characterization, and applications: A comprehensive review for biologists. *Journal of Nanobiotechnology*, *20*(1), 262. doi:10.118612951-022-01477-8 PMID:35672712

Joy, J., Krishnamoorthy, A., Tanna, A., Kamathe, V., Nagar, R., & Srinivasan, S. (2022). Recent Developments on the Synthesis of Nanocomposite Materials via Ball Milling Approach for Energy Storage Applications. *Applied Sciences (Basel, Switzerland)*, *12*(18), 9312. doi:10.3390/app12189312

Jung, H.-S., Cho, K.-J., Ryu, S.-J., Takagi, Y., Roche, P. A., & Neuman, K. C. (2020). Biocompatible fluorescent nanodiamonds as multifunctional optical probes for latent fingerprint detection. *ACS Applied Materials & Interfaces*, *12*(5), 6641–6650. doi:10.1021/acsami.9b19245 PMID:31939655

Kaastra, J., & Bleeker, J. (2016). Optimal binning of X-ray spectra and response matrix design. *Astronomy & Astrophysics*, *587*, A151. doi:10.1051/0004-6361/201527395

Kahru, A., & Dubourguier, H.-C. (2010). *From ecotoxicology to nanoecotoxicology*. Academic Press.

Kaikiti, C., Stylianou, M., & Agapiou, A. (2022). TD-GC/MS analysis of indoor air pollutants (VOCs, PM) in hair salons. *Chemosphere*, *294*, 133691. doi:10.1016/j.chemosphere.2022.133691 PMID:35065178

Kango, S., Kalia, S., Celli, A., Njuguna, J., Habibi, Y., & Kumar, R. (2013). Surface modification of inorganic nanoparticles for development of organic–inorganic nanocomposites—A review. *Progress in Polymer Science*, *38*(8), 1232–1261. doi:10.1016/j.progpolymsci.2013.02.003

Kanodarwala, F. K., Moret, S., Spindler, X., Lennard, C., & Roux, C. (2019). Nanoparticles used for fingermark detection—A comprehensive review. *WIREs Forensic Science*, *1*(5), e1341. doi:10.1002/wfs2.1341

Kapoor, N., Sulke, P., Shukla, R. K., Kakad, R., Pardeshi, P., & Badiye, A. (2021). Forensic analytical approaches to the dating of documents: An overview. *Microchemical Journal*, *170*, 170106722. doi:10.1016/j.microc.2021.106722

Kargozar, S., & Mozafari, M. (2018). Nanotechnology and Nanomedicine: Start small, think big. *Materials Today: Proceedings*, *5*(7), 15492–15500. doi:10.1016/j.matpr.2018.04.155

Karimadom, B. R., & Kornweitz, H. (2021). Mechanism of producing metallic nanoparticles, with an emphasis on silver and gold nanoparticles, using bottom-up methods. *Molecules (Basel, Switzerland)*, *26*(10), 2968. doi:10.3390/molecules26102968 PMID:34067624

Karimi, P. (2016). Effects of structure and partially localization of the pi electron clouds of single-walled carbon nanotubes on the cation-pi Interactions. Iran J Chem Chem Eng Int, 35(3), 35–43.

Kaspar, A. A., & Reichert, J. M. (2013). Future directions for peptide therapeutics development. *Drug Discovery Today*, *18*(17), 807–817. doi:10.1016/j.drudis.2013.05.011 PMID:23726889

Kaur Billing, B. (2021). Carbon nanotubes and its potential application in sensing. *ChemistrySelect*, *6*(36), 9571–9590. doi:10.1002lct.202102636

Kaur, R., Saini, K., & Sood, N. C. (2013). Sequencing the intersections of printed strokes with writing instrument strokes using DocuCentre expert (PIA 6000-EDF). *Science & Justice*, *53*(2), 206–211. doi:10.1016/j.scijus.2012.08.005 PMID:23601730

Kaushik, M., Mahendru, S., Chaudhary, S., & Kukreti, S. (2017). DNA fingerprints: Advances in their forensic analysis using nanotechnology. *Journal of Forensic Biomechanics*, *8*(1), 1–4. doi:10.4172/2090-2697.1000131

Kavosi, B., Navaee, A., & Salimi, A. (2018). Amplified fluorescence resonance energy transfer sensing of prostate specific antigen based on aggregation of CdTe QDs/antibody and aptamer decorated of AuNPs-PAMAM dendrimer. *Journal of Luminescence*, *204*, 368–374. doi:10.1016/j.jlumin.2018.08.012

Kawasaki, H., Kosaka, Y., Myoujin, Y., Narushima, T., Yonezawa, T., & Arakawa, R. (2011). Microwave-assisted polyol synthesis of copper nanocrystals without using additional protective agents. *Chemical Communications (Cambridge), 47*(27), 7740–7742. doi:10.1039/c1cc12346g PMID:21647527

Kay, L., & Shapira, P. (2009). Developing nanotechnology in Latin America. *Journal of Nanoparticle Research, 11*(2), 259–278. doi:10.100711051-008-9503-z PMID:21170134

Ke, W., Crist, R. M., Clogston, J. D., Stern, S. T., Dobrovolskaia, M. A., Grodzinski, P., & Jensen, M. (2022). *Trends and Patterns in Cancer Nanotechnology Research: A Survey of NCI's caNanoLab and Nanotechnology Characterization Laboratory*. Academic Press.

Keck, C. M., Jansch, M., & Müller, R. H. (2012). Protein Adsorption Patterns and Analysis on IV Nanoemulsions-The Key Factor Determining the Organ Distribution. *Pharmaceutics, 5*(1), 36–68. doi:10.3390/pharmaceutics5010036 PMID:24300396

Keller, K. H. (2007). Nanotechnology and society. *Journal of Nanoparticle Research, 9*(1), 5–10. doi:10.100711051-006-9193-3

Kesarwani, S., Parihar, K., Sankhla, M. S., & Kumar, R. (2020). Nano-forensic: new perspective and extensive applications in solving crimes. *Latent in Applied Nanobioscience, 10*(1), 1792-1798.

Kesarwani, S., Parihar, K., Sankhla, M. S., & Kumar, R. (2020). Nano-forensic: New perspective and extensive applications in solving crimes. *Latent in Applied Nanobioscience, 10*(1), 1792–1798. doi:10.33263/LIANBS101.17921798

Kethineni, S., & Cao, Y. (2020). The rise in popularity of cryptocurrency and associated criminal activity. *International Criminal Justice Review, 30*(3), 325–344. doi:10.1177/1057567719827051

Khademi, F., Samaei, M. R., Shahsavani, A., Azizi, K., Mohammadpour, A., Derakhshan, Z., & Bilal, M. (2022). Investigation of the Presence Volatile Organic Compounds (BTEX) in the Ambient Air and Biogases Produced by a Shiraz Landfill in Southern Iran. *Sustainability (Basel), 14*(2), 1040. doi:10.3390u14021040

Khan, S. H. (2020). Green nanotechnology for the environment and sustainable development. *Green materials for wastewater treatment*, 13-46.

Khan, N. A., Khan, S. U., Ahmed, S., Farooqi, I. H., Dhingra, A., Hussain, A., & Changani, F. (2019). Applications of nanotechnology in water and wastewater treatment: A review. *Asian Journal of Water, Environment and Pollution, 16*(4), 81–86. doi:10.3233/AJW190051

Khan, S., Choi, Y., Ahn, H. Y., Han, J. H., Ju, B. K., Chung, J., & Cho, S. H. (2020). Control of particle size in flame spray pyrolysis of Tb-doped Y(2)O(3)for bioimaging. *Materials (Basel), 13*(13), 2987. doi:10.3390/ma13132987 PMID:32635558

Kher, A., Mulholland, M., Green, E., & Reedy, B. (2006). Forensic classification of ballpoint pen inks using high performance liquid chromatography and infrared spectroscopy with principal components analysis and linear discriminant analysis. *Vibrational Spectroscopy, 40*(2), 270–277. doi:10.1016/j.vibspec.2005.11.002

Khofar, P. N. A., Karim, U. K. A., Elias, E., Safian, M. F., & Halim, M. I. A. (n.d.). Trends of Forensic Analysis of Pen Ink Using Attenuated Total Reflectance Fourier Transform Infrared (ATR- FTIR) Spectroscopy. *Indonesian Journal of Chemistry*.

Khorablou, Z., Shahdost-Fard, F., Razmi, H., Yola, M. L., & Karimi-Maleh, H. (2021). Recent advances in developing optical and electrochemical sensors for analysis of methamphetamine: A review. *Chemosphere, 278*, 130393. doi:10.1016/j.chemosphere.2021.130393 PMID:33823350

Khoubnasabjafari, M., Mogaddam, M. R. A., Rahimpour, E., Soleymani, J., Saei, A. A., & Jouyban, A. (2022). Breathomics: Review of sample collection and analysis, data modeling and clinical applications. *Critical Reviews in Analytical Chemistry, 52*(7), 1461–1487. doi:10.1080/10408347.2021.1889961 PMID:33691552

Kievit, F. M., & Zhang, M. Q. (2011). Cancer Nanotheranostics: Improving Imaging and Therapy by Targeted Delivery Across Biological Barriers. *Advanced Materials, 23*(36), H217–H247. doi:10.1002/adma.201102313 PMID:21842473

Kilin, V. N., Anton, H., Anton, N., Steed, E., Vermot, J., Vandamme, T. E., Mely, Y., & Klymchenko, A. S. (2014). Counterion-enhanced cyanine dye loading into lipid nano-droplets for single-particle tracking in zebrafish. *Biomaterials, 35*(18), 4950–4957. doi:10.1016/j.biomaterials.2014.02.053 PMID:24661553

Kim, B. J., Han, C., Moon, H., Kwon, J., Jang, I.-S., Lim, S.-K., Park, K.-W., Choi, J.-S., & An, H. J. (2018). Monitoring of post-mortem changes of saliva N-glycosylation by nano LC/MS. *Analytical and Bioanalytical Chemistry, 410*(1), 45–56. doi:10.100700216-017-0702-2 PMID:29164281

Kim, B. Y., Rutka, J. T., & Chan, W. C. (2010). Nanomedicine. *The New England Journal of Medicine, 363*(25), 2434–2443. doi:10.1056/NEJMra0912273 PMID:21158659

Kim, E. M., & Jeong, H. J. (2017). Current Status and Future Direction of Nanomedicine: Focus on Advanced Biological and Medical Applications. *Nuclear Medicine and Molecular Imaging, 51*(2), 106–117. doi:10.100713139-016-0435-8 PMID:28559935

Kim, J.-H., Bak, J. G., & Kim, C.-K. (2020). Electrical Resistivity of Ni–Fe Wires Coated with Sn Using low-Pressure Chemical Vapor Deposition. *Coatings, 10*(4), 317. doi:10.3390/coatings10040317

Kimura, M., Liu, Y., Sakai, R., Sato, S., Hirai, T., Fukawa, T., & Mihara, T. (2011). Detection of volatile organic compounds by analyses of polymer-coated quartz crystal microbalance sensor arrays. *Sensors and Materials, 23*(7), 359–368.

King, J., Kupferthaler, A., Frauscher, B., Hackner, H., Unterkofler, K., Teschl, G., & Högl, B. (2012). Breath acetone and isoprene concentration profiles in normal human sleep. *Physiological Measurement, 33*, 413–428. doi:10.1088/0967-3334/33/3/413 PMID:22370046

Klaessig, F., Marrapese, M., & Abe, S. (2011). Current perspectives in nanotechnology terminology and nomenclature. *Nanotechnology Standards*, 21-52.

Klimov, V. I. (2017). *Nanocrystal quantum dots*. CRC Press. doi:10.1201/9781420079272

Koczkur, K. M., Mourdikoudis, S., Polavarapu, L., & Skrabalak, S. E. (2015). Polyvinylpyrrolidone (PVP) in nanoparticle synthesis. *Dalton Transactions (Cambridge, England), 44*(41), 17883–17905. doi:10.1039/C5DT02964C PMID:26434727

Kong, W., Zhao, X., Zhu, Q., Gao, L., & Cui, H. (2017). Highly chemiluminescent magnetic beads for label-free sensing of 2, 4, 6-trinitrotoluene. *Analytical Chemistry, 89*(13), 7145–7151. doi:10.1021/acs.analchem.7b01111 PMID:28551993

Kong, X., Chong, X., Squire, K., & Wang, A. X. (2018). Microfluidic diatomite analytical devices for illicit drug sensing with ppb-Level sensitivity. *Sensors and Actuators. B, Chemical, 259*, 587–595. doi:10.1016/j.snb.2017.12.038 PMID:29755211

Kong, X., Squire, K., Chong, X., & Wang, A. X. (2017). Ultra-sensitive lab-on-a-chip detection of Sudan I in food using plasmonics-enhanced diatomaceous thin film. *Food Control, 79*, 258–265. doi:10.1016/j.foodcont.2017.04.007 PMID:29056826

Koppenhaver, K. M. (2007). *Forensic document examination*. Humana Press Incorporated.

Kopra, E. I., Ferris, J. A., Rucker, J. J., McClure, B., Young, A. H., Copeland, C. S., & Winstock, A. R. (2022). Adverse experiences resulting in emergency medical treatment seeking following the use of lysergic acid diethylamide (LSD). *Journal of Psychopharmacology (Oxford, England)*, *36*(8). doi:10.1177/02698811221099650 PMID:35672900

Kostoff, R. N., Koytcheff, R. G., & Lau, C. G. (2007). Technical structure of the global nanoscience and nanotechnology literature. *Journal of Nanoparticle Research*, *9*(5), 701–724. doi:10.100711051-007-9224-8

Kothamasu, P., Kanumur, H., Ravur, N., Maddu, C., Parasuramrajam, R., & Thangavel, S. (2012). Nanocapsules: The weapons for novel drug delivery systems. *BioImpacts*, *2*(2), 71–81. doi:10.5681/bi.2012.011 PMID:23678444

Kriznar, A., Ager, F., Macías, L. R., Ortega Feliu, I., Tubío, B. G., & Respaldiza, M. Á. (2022). Pigments and Inks Applied in Juan Vespucci's Portolan Chart (1520). *Colorants*, *1*(4), 4. Advance online publication. doi:10.3390/colorants1040026

Kudr, J., Haddad, Y., Richtera, L., Heger, Z., Cernak, M., Adam, V., & Zitka, O. (2017). Magnetic Nanoparticles: From Design and Synthesis to Real World Applications. In Nanomaterials (Vol. 7, Issue 9). doi:10.3390/nano7090243

Kuhn, S., Hakanson, U., Rogobete, L., & Sandoghdar, V. (2006). Enhancement of single-molecule fluorescence using a gold nanoparticle as an optical nanoantenna. *Physical Review Letters*, *97*(1), 4. doi:10.1103/PhysRevLett.97.017402 PMID:16907406

Kumar, A., Gupta, K., Dixit, S., Mishra, K., & Srivastava, S. (2019). A review on positive and negative impacts of nanotechnology in agriculture. *International Journal of Environmental Science and Technology*, *16*(4), 2175–2184. doi:10.100713762-018-2119-7

Kumar, D., & Elias, A. J. (2019). The Explosive Chemistry of Nitrogen: A Fascinating Journey From 9th Century to the Present. *Resonance*, *24*(11), 1253–1271. doi:10.100712045-019-0893-2

Kumar, N., & Kumbhat, S. (2016). *Essentials in nanoscience and nanotechnology*. John Wiley & Sons. doi:10.1002/9781119096122

Kumar, N., & Sharma, A. (2021). *Nano-Forensics: The New Perspective in Precision Forensic Science. In Nanotechnology Applications in Health and Environmental Sciences*. Springer.

Kumar, R., & Sharma, V. (2017). A novel combined approach of diffuse reflectance UV-Vis-NIR spectroscopy and multivariate analysis for non-destructive examination of blue ballpoint pen inks in forensic application. *Spectrochimica Acta. Part A: Molecular and Biomolecular Spectroscopy*, *175*, 17567–17575. doi:10.1016/j.saa.2016.12.008 PMID:28024250

Kumar, S., Huang, J., Abbassi-Ghadi, N., Španěl, P., Smith, D., & Hanna, G. B. (2013). Selected ion flow tube mass spectrometry analysis of exhaled breath for volatile organic compound profiling of esophago-gastric cancer. *Analytical Chemistry*, *85*(12), 6121–6128. doi:10.1021/ac4010309 PMID:23659180

Kus, F., Altinkok, C., Zayim, E., Erdemir, S., Tasaltin, C., & Gurol, I. (2021). Surface acoustic wave (SAW) sensor for volatile organic compounds (VOCs) detection with calix [4] arene functionalized Gold nanorods (AuNRs) and silver nanocubes (AgNCs). *Sensors and Actuators. B, Chemical*, *330*, 129402. doi:10.1016/j.snb.2020.129402

Kyrkou, S. G., Vrettos, E. I., Gorpas, D., Crook, T., Syed, N., & Tzakos, A. G. (2022). Design principles governing the development of theranostic anticancer agents and their nanoformulations with photoacoustic properties. *Pharmaceutics*, *14*(2), 362. doi:10.3390/pharmaceutics14020362 PMID:35214094

Lad, A. N., Pandya, A., & Agrawal, Y. K. (2016). Overview of nano-enabled screening of drug-facilitated crime: A promising tool in forensic investigation. *Trends in Analytical Chemistry*, *80*, 458–470. doi:10.1016/j.trac.2015.07.016

Lane, C. R., Hobden, E., Walker, L., Barton, V. C., Inman, A. J., Hughes, K. J. D., & Barker, I. (2007). Evaluation of a rapid diagnostic field test kit for identification of Phytophthora species, including P. ramorum and P. kernoviae at the point of inspection. *Plant Pathology*, *56*(5), 828–835. doi:10.1111/j.1365-3059.2007.01615.x

Langford, J. B., & Lurie, I. S. (2022). Use of micro, capillary, and nano liquid chromatography for forensic analysis. *Journal of Separation Science*, *45*(1), 38–50. doi:10.1002/jssc.202100631 PMID:34626162

Lantam, A., Limbut, W., Thiagchanya, A., & Phonchai, A. (2020). A portable optical colorimetric sensor for the determination of promethazine in lean cocktail and pharmaceutical doses. *Microchemical Journal*, *159*, 105519. doi:10.1016/j.microc.2020.105519

LaPorte, G. M. (2015). Chemical analysis for the scientific examination of questioned documents. In Fundam. Appl (pp. 318–353). John Wiley & Sons, Ltd. doi:10.1002/9781118897768.ch8

Lasota, S., Baster, Z., Witko, T., Zimoląg, E., Sroka, J., Rajfur, Z., & Madeja, Z. (2017). Zastosowanie biosensorów typu FRET w badaniach mikroskopowych procesu migracji komórkowej. *Postepy Biochemii*, *63*(1), 16–33. PMID:28409572

Laurila, T., Sainio, S., & Caro, M. (2017). *Hybrid carbon based nanomaterials for electrochemical detection of biomolecules*. Academic Press.

Leal, T. A., Ferreira, C., Quintas, A., & Bernardo, A. (2019). Dating inks on paper through chromatographic analysis of volatile compounds: A mini-review. *Ann. Med*, *51*(sup1), 180. . doi:10.1080/07853890.2018.1562750

Lea, T., Amada, N., & Jungaberl, E. H. (2020). Psychedelic microdosing: A subreddit analysis. *Journal of Psychoactive Drugs*, *52*(2), 101–112. doi:10.1080/02791072.2019.1683260 PMID:31648596

Lee, H., Lee, B. P., & Messersmith, P. (2007). *A reversible wet/dry adhesive inspired by mussels and geckos*. Academic Press.

Lee, H.-J., Lee, S. J., Uthaman, S., Thomas, R. G., Hyun, H., Jeong, Y. Y., . . . Park, I.-K. (2015). *Biomedical applications of magnetically functionalized organic/inorganic hybrid nanofibers*. Academic Press.

Lee, A. J., Wang, X. Y., Carlson, L. J., Smyder, J. A., Loesch, B., Tu, X. M., Zheng, M., & Krauss, T. D. (2011). Bright fluorescence from individual single-walled carbon nanotubes. *Nano Letters*, *11*(4), 1636–1640. doi:10.1021/nl200077t PMID:21417364

Lee, H., Choi, J. I., Park, J., Jang, S. S., & Lee, S. W. (2020). Role of anions on electrochemical exfoliation of graphite into graphene in aqueous acids. *Carbon*, *167*, 816–825. doi:10.1016/j.carbon.2020.06.044

Lee, J. E., Lee, N., Kim, T., Kim, J., & Hyeon, T. (2011). Multifunctional mesoporous silica nanocomposite nanoparticles for theranostic applications. *Accounts of Chemical Research*, *44*(10), 893–902. doi:10.1021/ar2000259 PMID:21848274

Lee, J. Y., Hong, B. H., Kim, W. Y., Min, S. K., Kim, Y., Jouravlev, M. V., ... Kaufman, L. J. (2009). Near-field focusing and magnification through self-assembled nanoscale spherical lenses. *Nature*, *460*(7254), 498–501. doi:10.1038/nature08173

Lehnert, A. L., & Kearfott, K. (2010). *The detection of explosive materials: review of considerations and methods*. Academic Press.

Lesiak, A., Drzozga, K., Cabaj, J., Bański, M., Malecha, K., & Podhorodecki, A. (2019). Optical sensors based on II-VI quantum dots. *Nanomaterials (Basel, Switzerland)*, *9*(2), 192. doi:10.3390/nano9020192 PMID:30717393

Liao, L., Xing, Y., Xiong, X., Gan, L., Hu, L., Zhao, F., Tong, Y., & Deng, S. (2020). An electrochemical biosensor for hypoxanthine detection in vitreous humor: A potential tool for estimating the post-mortem interval in forensic cases. *Microchemical Journal*, *155*, 104760. doi:10.1016/j.microc.2020.104760

Li, C., Wang, Q., Zhang, Y., Lin, H., Zhang, J., Huang, P., & Wang, Z. (2016). Research progress in the estimation of the postmortem interval by Chinese forensic scholars. *Forensic Sciences Research*, *1*(1), 3–13. doi:10.1080/20961790.2016.1229377 PMID:30483604

Li, C., Wan, H., Wu, X., Yin, J., Zhu, L., Chen, H., & Li, Z. (2022). Discrimination and Characterization of the Volatile Organic Compounds in Schizonepetae Spica from Six Regions of China Using HS-GC-IMS and HS-SPME-GC-MS. *Molecules (Basel, Switzerland)*, *27*(14), 4393. doi:10.3390/molecules27144393 PMID:35889268

Li, D., Xie, Z., Qu, M., Zhang, Q., Fu, Y., & Xie, J. (2021). Virtual sensor array based on butterworth–van dyke equivalent model of QCM for selective detection of volatile organic compounds. *ACS Applied Materials & Interfaces*, *13*(39), 47043–47051. doi:10.1021/acsami.1c13046 PMID:34546706

Li, F., Pei, H., Wang, L., Lu, J., Gao, J., Jiang, B., Zhao, X., & Fan, C. (2013). Nanomaterial-based fluorescent DNA analysis: A comparative study of the quenching effects of graphene oxide, carbon nanotubes, and gold nanoparticles. *Advanced Functional Materials*, *23*(33), 4140–4148. doi:10.1002/adfm.201203816

Li, J. L., Tang, B., Yuan, B., Sun, L., & Wang, X. G. (2013). A review of optical imaging and therapy using nanosized graphene and graphene oxide. *Biomaterials*, *34*(37), 9519–9534. doi:10.1016/j.biomaterials.2013.08.066 PMID:24034502

Li, J., & Sun, D. (2022). Molecularly imprinted ratiometric fluorescence nanosensors. *Langmuir*, *38*(44), 13305–13312. doi:10.1021/acs.langmuir.2c01925 PMID:36278924

Li, J., Zhu, J.-J., & Xu, K. (2014). Fluorescent metal nanoclusters: From synthesis to applications. *Trends in Analytical Chemistry*, *58*, 90–98. doi:10.1016/j.trac.2014.02.011

Lin, P. C., Lin, S., Wang, P. C., & Sridhar, R. (2014). Techniques for physicochemical characterization of nanomaterials. *Biotechnology Advances*, *32*(4), 711–726. doi:10.1016/j.biotechadv.2013.11.006 PMID:24252561

Li, R. (2008). *Forensic biology: identification and DNA analysis of biological evidence*. CRC Press. doi:10.4324/9781420043440

Li, S. Q., Pan, R. F., Mehdi, Y. A., Xiao, D. L., & He, H. (2015). One-step spontaneous synthesis of fluorescent carbon nanoparticles with thermosensitivity from polyethylene glycol. *New Journal of Chemistry*, *39*(9), 7033–7039. doi:10.1039/C5NJ01129A

Liu, H. P., Ye, T., & Mao, C. D. (2007). Fluorescent carbon nanoparticles derived from candle soot. Angewandte Chemie-Int Ed, 46(34), 6473–6475.

Liu, S., Xu, L., Zhang, T., Ren, G., & Yang, Z. (2010). *Oxidative stress and apoptosis induced by nanosized titanium dioxide in PC12 cells*. Academic Press.

Liu, Y., Sowerby, B., & Tickner, J. (2008). *Comparison of neutron and high-energy X-ray dual-beam radiography for air cargo inspection*. Academic Press.

Liu, C. M., He, H. Y., Xu, L., & Hua, Z. D. (2021). New qualitative analysis strategy for illicit drugs using Raman spectroscopy and characteristic peaks method. *Drug Testing and Analysis*, *13*(3), 720–728. doi:10.1002/dta.2963 PMID:33142047

Liu, C., Wu, S., Yan, Y., Dong, Y., Shen, X., & Huang, C. (2019). Application of magnetic particles in forensic science. *Trends in Analytical Chemistry*, *121*, 115674. doi:10.1016/j.trac.2019.115674

Liu, J. N., Bu, W. B., & Shi, J. L. (2015). Silica coated upconversion nanoparticles: A versatile platform for the development of efficient theranostics. *Accounts of Chemical Research*, *48*(7), 1797–1805. doi:10.1021/acs.accounts.5b00078 PMID:26057000

Liu, Q., & Guo, B. D. (2013). Article. *Nano Letters*, *13*, 2436–2441. doi:10.1021/nl400368v PMID:23675758

Liu, R., Li, Z., Huang, Z., Li, K., & Lv, Y. (2019). Biosensors for explosives: State of art and future trends. *Trends in Analytical Chemistry*, *118*, 123–137. doi:10.1016/j.trac.2019.05.034

Liu, Y., Chen, C.-L., Zhang, Y., Sonkusale, S. R., Wang, M. L., & Dokmeci, M. R. (2012). SWNT based nanosensors for wireless detection of explosives and chemical warfare agents. *IEEE Sensors Journal*, *13*(1), 202–210. doi:10.1109/JSEN.2012.2211007

Liu, Y., Wang, L., Wang, H., Xiong, M., Yang, T., & Zakharova, G. S. (2016). Highly sensitive and selective ammonia gas sensors based on PbS quantum dots/TiO_2 nanotube arrays at room temperature. *Sensors and Actuators. B, Chemical*, *236*, 529–536. doi:10.1016/j.snb.2016.06.037

Liu, Z. Z., Xu, Q. F., Zhang, C. T. F., Sun, Q. Y., Wang, C. B., Dong, M. D., Wang, Z. G., Ohmori, H., Kosinova, M., Goto, T., Tu, R., & Zhang, S. (2020). Laser-induced growth of large-area epitaxial graphene with low sheet resistance on 4H-SiC(0001). *Applied Surface Science*, *514*, 145938. doi:10.1016/j.apsusc.2020.145938

Liu, Z., Lin, Z., Liu, L., & Su, X. (2015). A convenient and label-free fluorescence "turn off–on" nanosensor with high sensitivity and selectivity for acid phosphatase. *Analytica Chimica Acta*, *876*, 83–90. doi:10.1016/j.aca.2015.03.027 PMID:25998462

Li, W., Chang, Y., Cheng, Z., Ling, J., Han, L., Li, X., & Ding, Y. (2018). Vitreous humor: A review of biochemical constituents in postmortem interval estimation. *Journal of Forensic Science and Medicine*, *4*(2), 85. doi:10.4103/jfsm.jfsm_13_18

Li, X., Ding, Y., Ling, J., Yao, W., Zha, L., Li, N., Chang, Y., Wang, Y., & Cai, J. (2019). Bacteria-targeting BSA-stabilized SiC nanoparticles as a fluorescent nanoprobe for forensic identification of saliva. *Mikrochimica Acta*, *186*(12), 1–10. doi:10.100700604-019-3890-y PMID:31707552

Li, X., Li, J., Ling, J., Wang, C., Ding, Y., Chang, Y., Li, N., Wang, Y., & Cai, J. (2020). A smartphone-based bacteria sensor for rapid and portable identification of forensic saliva sample. *Sensors and Actuators. B, Chemical*, *320*, 128303. doi:10.1016/j.snb.2020.128303

Li, X., Wang, H., Shimizu, Y., Pyatenko, A., Kawaguchi, K., & Koshizaki, N. (2011). Preparation of carbon quantum dots with tunable photoluminescence by rapid laser passivation in ordinary organic solvents. *Chemical Communications (Cambridge)*, *47*(3), 932–934. doi:10.1039/C0CC03552A PMID:21079826

Li, X., Zhang, X. N., Li, X. D., & Chang, J. (2016). Multimodality imaging in nanomedicine and nanotheranostics. *Cancer Biology & Medicine*, *13*(3), 339–348. doi:10.20892/j.issn.2095-3941.2016.0055 PMID:27807501

Li, Y., Hu, Y., Zhao, Y., Shi, G. Q., Deng, L. E., Hou, Y. B., & Qu, L. T. (2011). An electrochemical avenue to green-luminescent graphene quantum dots as potential electron-acceptors for photovoltaics. *Advanced Materials*, *23*(6), 776–780. doi:10.1002/adma.201003819 PMID:21287641

Li, Z. H., Wu, L. Y., Hu, P. R., Han, S. H., Zhang, T., Fan, H. L., Jin, W., Jin, Q. H., & Mu, Y. (2012). Soft nanomaterial-based targeting polymersomes for near-infrared fluorescence multispectral in vivo imaging. *Nanoscale*, *4*(22), 7097–7105. doi:10.1039/c2nr32047a PMID:23069779

Li, Z., Hou, S., Zhang, H., Song, Q., Wang, S., & Guo, H. (2023). *Recent advances in fluorescent and colorimetric sensing for volatile organic amines and biogenic amines in food.* Advanced Agrochem. doi:10.1016/j.aac.2023.02.001

Lloyd-Hughes, H., Shiatis, A. E., Pabari, A., Mosahebi, A., & Seifalian, A. (2015). Current and future nanotechnology applications in the management of melanoma: A review. *Journal of Nanomedicine & Nanotechnology*, *6*(6), 1. doi:10.4172/2157-7439.1000334

Lloyd-Parry, O., Downing, C., Aleisaei, E., Jones, C., & Coward, K. (2018). Nanomedicine applications in women's health: State of the art. *International Journal of Nanomedicine*, *13*, 1963–1983. doi:10.2147/IJN.S97572 PMID:29636611

Lo Giudice, M. C., Herda, L. M., Polo, E., & Dawson, K. A. (2016). In situ characterization of nanoparticle biomolecular interactions in complex biological media by flow cytometry. *Nature Communications*, *7*(1), 1–10. doi:10.1038/ncomms13475 PMID:27845346

Lodha, A. (2018). Forensic nanotechnology in forensic genetics. *Peer Rev J Forensic Genet Sci*, *1*(2), 19–22.

Lodha, A. S., Pandya, A., & Shukla, R. K. (2016). Nanotechnology: An applied and robust approach for forensic investigation. *Forensic Res Criminol Int J*, *2*(1), 44.

Lodha, A., Pandya, A., Sutariya, P. G., & Menon, S. K. (2013). Melamine modified gold nanoprobe for "on-spot" colorimetric recognition of clonazepam from biological specimens. *Analyst (London)*, *138*(18), 5411–5416. doi:10.1039/c3an00184a PMID:23807945

Lodha, A., Pandya, A., Sutariya, P. G., & Menon, S. K. (2014). A smart and rapid colorimetric method for the detection of codeine sulphate, using unmodified gold nanoprobe. *RSC Advances*, *4*(92), 50443–50448. doi:10.1039/C4RA06269H

Lombardo, D., Kiselev, M. A., & Caccamo, M. T. (2019). Smart nanoparticles for drug delivery application: Development of versatile nanocarrier platforms in biotechnology and nanomedicine. *Journal of Nanomaterials*. doi:10.1155/2019/3702518

Lona-Ramirez, F. J., Gonzalez-Alatorre, G., Rico-Ramírez, V., Perez-Perez, M. C. I., & Castrejón-González, E. O. (2016). Gas chromatography/mass spectrometry for the determination of nitrosamines in red wine. *Food Chemistry*, *196*, 1131–1136. doi:10.1016/j.foodchem.2015.09.090 PMID:26593598

Loo, A. H., Sofer, Z., Bouša, D., Ulbrich, P., Bonanni, A., & Pumera, M. (2016). Carboxylic carbon quantum dots as a fluorescent sensing platform for DNA detection. *ACS Applied Materials & Interfaces*, *8*(3), 1951–1957. doi:10.1021/acsami.5b10160 PMID:26762211

Loura, L. M., & Prieto, M. (2011). FRET in membrane biophysics: An overview. *Frontiers in Physiology*, *2*, 82. doi:10.3389/fphys.2011.00082 PMID:22110442

Louro, H. (2018). Relevance of Physicochemical Characterization of Nanomaterials for Understanding Nano-cellular Interactions. *Advances in Experimental Medicine and Biology*, *1048*, 123–142. doi:10.1007/978-3-319-72041-8_8 PMID:29453536

Lu, Y., Pang, X., Lyu, Y., Li, J., Xing, B., Chen, J., & Wu, H. (2022). Characteristics and sources analysis of ambient volatile organic compounds in a typical industrial park: Implications for ozone formation in 2022 Asian Games. *The Science of the Total Environment*, *848*, 157746. doi:10.1016/j.scitotenv.2022.157746 PMID:35926610

Lu, Z., & Yin, Y. (2012). Colloidal nanoparticle clusters: Functional materials by design. *Chemical Society Reviews*, *41*(21), 6874–6887. doi:10.1039/c2cs35197h PMID:22868949

Ma, C., Li, C., Wang, F., Ma, N., Li, X., Li, Z., Deng, Y., Wang, Z., Xi, Z., Tang, Y., & He, N. (2013). Magnetic nanoparticles-based extraction and verification of nucleic acids from different sources. *Journal of Biomedical Nanotechnology*, *9*(4), 703–709. doi:10.1166/jbn.2013.1566 PMID:23621032

Macutkevic, J., Seliuta, D., Valusis, G., Adomavicius, R., Krotkus, A., Kuzhir, P., Paddubskaya, A., Maksimenko, S., Kuznetsov, V., Mazov, I., & Simonova, I. (2012). Multi-walled carbon nanotubes/PMMA composites for THz applications. *Diamond and Related Materials*, *25*, 13–18. doi:10.1016/j.diamond.2012.02.002

Maffre, P., Nienhaus, K., Amin, F., Parak, W. J., & Nienhaus, G. U. (2011). Characterization of protein adsorption onto FePt nanoparticles using dual-focus fluorescence correlation spectroscopy. *Beilstein Journal of Nanotechnology*, *2*, 374–383. doi:10.3762/bjnano.2.43 PMID:22003445

Ma, H., Li, X., Chen, J., Wang, H., Cheng, T., Chen, K., & Xu, S. (2014). Analysis of human breath samples of lung cancer patients and healthy controls with solid-phase microextraction (SPME) and flow-modulated comprehensive two-dimensional gas chromatography (GC× GC). *Analytical Methods*, *6*(17), 6841–6849. doi:10.1039/C4AY01220H

Ma, H., & Shen, Z. G. (2020). Exfoliation of graphene nanosheets in aqueous media. *Ceramics International*, *46*(14), 21873–21887. doi:10.1016/j.ceramint.2020.05.314

Mahbub, T., & Hoque, M. E. (2020). *Nanofabrication Smart Nanosensor Applications*. Elsevier.

Majdinasab, M., Mitsubayashi, K., & Marty, J. L. (2019). Optical and electrochemical sensors and biosensors for the detection of quinolones. *Trends in Biotechnology*, *37*(8), 898–915. doi:10.1016/j.tibtech.2019.01.004 PMID:30777309

Majeed, M. I., Nawaz, H., & Arshad, F. N. (2020). Detecting the Presence of Illicit Drugs Using Biosensors. *Nanobiosensors: From Design to Applications*, 223-253.

Majerič, P., & Rudolf, R. (2020). Advances in ultrasonic spray pyrolysis processing of noble metal nanoparticles. *Materials (Basel)*, *13*(16), 3485. doi:10.3390/ma13163485 PMID:32784637

Ma, L., He, S., Huang, J., Cao, L., Yang, F., & Li, L. (2009). Maximizing specificity and yield of PCR by the quantum dot itself rather than property of the quantum dot surface. *Biochimie*, *91*(8), 969–973. doi:10.1016/j.biochi.2009.04.020 PMID:19442702

Malefane, M. (2020). Applications of nanotechnology towards detection and treatment of HIV/AIDs: A review article. *Research & Development in Material Science*, *12*(5), 1315–1321. doi:10.31031/RDMS.2020.12.000796

Malenica, M., Vukomanović, M., Kurtjak, M., Masciotti, V., Dal Zilio, S., Greco, S., ... Jelovica Badovinac, I. (2021). Perspectives of microscopy methods for morphology characterisation of extracellular vesicles from human biofluids. *Biomedicines*, *9*(6), 603. doi:10.3390/biomedicines9060603 PMID:34073297

Malik, A. A., & Hassan, S. M. (2022). The ZnO Nanoparticles: Synthesis, Characterization and Applications in Forensics. *International Journal for Electronic Crime Investigation*, *6*(1), 8–8. doi:10.54692/ijeci.2022.060195

Malik, S., Muhammad, K., & Waheed, Y. (2023). Nanotechnology: A Revolution in Modern Industry. *Molecules (Basel, Switzerland)*, *28*(2), 661. doi:10.3390/molecules28020661 PMID:36677717

Malinovskaya, Y., Melnikov, P., Baklaushev, V., Gabashvili, A., Osipova, N., Mantrov, S., Ermolenko, Y., Maksimenko, O., Gorshkova, M., Balabanyan, V., Kreuter, J., & Gelperina, S. (2017). Delivery of doxorubicin-loaded PLGA nanoparticles into U87 human glioblastoma cells. *International Journal of Pharmaceutics*, *524*(1-2), 77–90. doi:10.1016/j.ijpharm.2017.03.049 PMID:28359811

Manoharan, M. (2008). Research on the frontiers of materials science: The impact of nanotechnology on new material development. *Technology in Society*, *30*(3-4), 401–404. doi:10.1016/j.techsoc.2008.04.016

Mansoori, G. A., & Soelaiman, T. F. (2005). *Nanotechnology—An introduction for the standards community*. ASTM International.

Mao, J., Kang, Y., Yu, D., & Zhou, J. (2021). Surface-enhanced Raman spectroscopy integrated with aligner mediated cleavage strategy for ultrasensitive and selective detection of methamphetamine. *Analytica Chimica Acta*, *1146*, 124–130. doi:10.1016/j.aca.2020.12.028 PMID:33461707

Mao, K., Zhang, H., Pan, Y., Zhang, K., Cao, H., Li, X., & Yang, Z. (2020). Nanomaterial-based aptamer sensors for analysis of illicit drugs and evaluation of drugs consumption for wastewater-based epidemiology. *Trends in Analytical Chemistry*, *130*, 115975. doi:10.1016/j.trac.2020.115975 PMID:32834242

Markovic, D. S., Zivkovic, D., Cvetkovic, D., & Popovic, R. (2012). Impact of nanotechnology advances in ICT on sustainability and energy efficiency. *Renewable & Sustainable Energy Reviews*, *16*(5), 2966–2972. doi:10.1016/j.rser.2012.02.018

Marques, L., Félix, L., Cruz, G., Coelho, V., Caetano, J., Vale, A., Cruz, C., Alves, L., & Vaz, P. (2023). Neutron and Gamma-Ray Detection System Coupled to a Multirotor for Screening of Shipping Container Cargo. *Sensors (Basel)*, *23*(1), 329. doi:10.339023010329 PMID:36616926

Marshall, B., Bland, M. K., Hulla, R., & Gatchel, R. J. (2019). Considerations in addressing the opioid epidemic and chronic pain within the USA. *Pain Management*, *9*(2), 131–138. doi:10.2217/pmt-2018-0070 PMID:30806566

Masite, N. S., Ncube, S., Mtunzi, F. M., Madikizela, L. M., & Pakade, V. E. (2022). Determination of furanic compounds in Mopane worms, corn, and peanuts using headspace solid-phase microextraction with gas chromatography-flame ionisation detector. *Food Chemistry*, *369*, 130944. doi:10.1016/j.foodchem.2021.130944 PMID:34464834

Massaro, S., & Lorenzoni, G. (2021). *Nanomedicine: A socio-technical system*. Academic Press.

Mast, M. P., Modh, H., Champanhac, C., Wang, J. W., Storm, G., Krämer, J., Mailänder, V., Pastorin, G., & Wacker, M. G. (2021). Nanomedicine at the crossroads - A quick guide for IVIVC. *Advanced Drug Delivery Reviews*, *179*, 113829. doi:10.1016/j.addr.2021.113829 PMID:34174332

Mathiazhagan, A., & Joseph, R. (2011). Nanotechnology-a New prospective in organic coating-review. *International Journal of Chemical Engineering and Applications*, *2*(4), 225–237. doi:10.7763/IJCEA.2011.V2.108

Mathur, S., Choudhary, B. S. K., & Vyas, C. J. M. (2016). Effect of disguise on fundamental frequency of voice. *Journal of Forensics Research*, *7*(327), 2. doi:10.4172/2157-7145.1000327

Matsuzawa, S., Itoh, Y., Kimura, H., Kobayashi, R., Nakagawa, T., & Ohno, S. (1994). Rapid detection of human seminal plasma proteins by membrane aspiration test (MAT). *Forensic Science International*, *64*(2), 119–124. doi:10.1016/0379-0738(94)90221-6 PMID:8175082

Mayhew, C. A., Herbig, J., & Beauchamp, J. D. (2020). Proton transfer reaction–mass spectrometry. In *Breathborne Biomarkers and the Human Volatilome* (pp. 155–170). Elsevier. doi:10.1016/B978-0-12-819967-1.00010-4

Mazzei, F., Antiochia, R., Botrè, F., Favero, G., & Tortolini, C. (2014). *Affinity-based biosensors in sport medicine and doping control analysis*. Academic Press.

McCord, B. (2006). Nanotechnology and its potential in forensic DNA analysis. *Profiles in DNA*, *9*(2), 7–9.

McGee, E. (2009). *Nanomedicine: Ethical concerns beyond diagnostics, drugs, and techniques*. Academic Press.

Medina, C., Santos-Martinez, M., Radomski, A., Corrigan, O., & Radomski, M. (2007). Nanoparticles: Pharmacological and toxicological significance. *British Journal of Pharmacology*, *150*(5), 552–558. doi:10.1038j.bjp.0707130 PMID:17245366

Meixner, A. J., Fleischer, M., Kern, D. P., Sheremet, E., & McMillan, N. (2022). *Applications*. Walter de Gruyter GmbH & Co KG. doi:10.1515/9783110442908

Mejia de Grubb, M. C., Salemi, J. L., Gonzalez, S. J., Chima, C. C., Kowalchuk, A. A., & Zoorob, R. J. (2020). Opioid, cocaine, and amphetamine use disorders are associated with higher30-day inpatient readmission rates in the United States. *Substance Abuse*, *41*(3), 365–374. doi:10.1080/08897077.2019.1635964 PMID:31295052

Menaa, F. (2014). Global financial model for responsible research and development of the fast growing nanotechnology business. *Journal of Business & Financial Affairs*, *3*(1), e139. doi:10.4172/2167-0234.1000e139

Meng, Z., Stolz, R. M., Mendecki, L., & Mirica, K. A. (2019). Electrically-transduced chemical sensors based on two-dimensional nanomaterials. *Chemical Reviews*, *119*(1), 478–598. doi:10.1021/acs.chemrev.8b00311 PMID:30604969

Menzel, E. R. (2001). Fingerprint detection with photoluminescent. *Advances in fingerprint technology*, 456.

Menzyk, A., Zadora, G., & Sajewicz, M. (2015). Physicochemical analysis of ink – dating and establishing the sequence of intersecting lines of ink entries. *Z Zagadnien Nauk Sadowych*, *104*, 279–302.

Metcalf, J. L., Wegener Parfrey, L., Gonzalez, A., Lauber, C. L., Knights, D., Ackermann, G., & Knight, R. (2013). A microbial clock provides an accurate estimate of the postmortem interval in a mouse model system. *elife*, *2*, e01104.

Michalska, A., Zadora, G., & Martyna, A. (2016). Optimized sample preparation for glass fragments by scanning electron microscopy–energy dispersive X-ray spectrometry. *Analytical Letters*, *49*(12), 1884–1895. doi:10.1080/00032719.2015.1126598

Miki, K., Oride, K., Inoue, S., Kuramochi, Y., Nayak, R. R., & Matsuoka, H. (2010). H.Harada H, Hiraoka M and Ohe K. *Biomaterials*, *31*, 934–942. doi:10.1016/j.biomaterials.2009.10.005 PMID:19853909

Minhas, R. S., Rudd, D. A., Al Hmoud, H. Z., Guinan, T. M., Kirkbride, K. P., & Voelcker, N. H. (2020). Rapid detection of anabolic and narcotic doping agents in saliva and urine by means of nanostructured silicon SALDI mass spectrometry. *ACS Applied Materials & Interfaces*, *12*(28), 31195–31204. doi:10.1021/acsami.0c07849 PMID:32551485

Mitra, S., & Basak, M. (2022). Diverse bio-sensing and therapeutic applications of plasmon enhanced nanostructures. *Materials Today*, *57*, 225–261. doi:10.1016/j.mattod.2022.05.023

Miyawaki, A. (2003). Visualization of the spatial and temporal dynamics of intracellular signaling. *Developmental Cell*, *4*(3), 295–305. doi:10.1016/S1534-5807(03)00060-1 PMID:12636912

Miyawaki, A. (2008). Green fluorescent protein glows gold. *Cell*, *135*(6), 987–990. doi:10.1016/j.cell.2008.11.025 PMID:19070562

Moawad, A. M., zaghlol, H., Abdelsalam, M. H., Abdelfattah, A., Sabry, D., & Atef, A. (2018). Differential genes expression biomarkers for menstrual and peripheral blood stains analysis. *Egyptian Journal of Forensic Sciences*, *8*(1), 41. doi:10.118641935-018-0072-8

Modena, M. M., Rühle, B., Burg, T. P., & Wuttke, S. (2019). Nanoparticle characterization: What to measure? *Advanced Materials*, *31*(32), 1901556. doi:10.1002/adma.201901556 PMID:31148285

Mohammed, A., & Abdullah, A. (2018). Scanning electron microscopy (SEM): A review. *Proceedings of the 2018 International Conference on Hydraulics and Pneumatics—HERVEX.*

Molaei, M. J. (2020). Principles, mechanisms, and application of carbon quantum dots in sensors: A review. *Analytical Methods*, *12*(10), 1266–1287. doi:10.1039/C9AY02696G

Montagner, C. C., Sodré, F. F., Acayaba, R. D., Vidal, C., Campestrini, I., Locatelli, M. A., & Jardim, W. F. (2019). Ten years-snapshot of the occurrence of emerging contaminants in drinking, surface and ground waters and wastewaters from São Paulo State, Brazil. *Journal of the Brazilian Chemical Society*, *30*, 614–632.

Montanarella, F., & Kovalenko, M. V. (2022). Three millennia of nanocrystals. *ACS Nano*, *16*(4), 5085–5102. doi:10.1021/acsnano.1c11159 PMID:35325541

Montero-Montoya, R., López-Vargas, R., & Arellano-Aguilar, O. (2018). Volatile organic compounds in air: Sources, distribution, exposure and associated illnesses in children. *Annals of Global Health*, *84*(2), 225–238. doi:10.29024/aogh.910 PMID:30873816

Moore, D. S. (2007). *Recent advances in trace explosives detection instrumentation.* Academic Press.

Moore, A., Foss, J., Juhascik, M., Botch-Jones, S., & Kero, F. (2019). Rapid screening of opioids in seized street drugs using ambient ionization high resolution time-of-flight mass spectrometry. *Forensic Chemistry (Amsterdam, Netherlands)*, *13*, 100149. doi:10.1016/j.forc.2019.100149

Moram, S. S. B., Shaik, A. K., Byram, C., Hamad, S., & Soma, V. R. (2020). Instantaneous trace detection of nitro-explosives and mixtures with nanotextured silicon decorated with Ag–Au alloy nanoparticles using the SERS technique. *Analytica Chimica Acta*, *1101*, 157–168. doi:10.1016/j.aca.2019.12.026 PMID:32029107

Morato, Y. L., Marciello, M., Chamizo, L. L., Paredes, K. O., & Filice, M. (2021). Hybrid magnetic nanoparticles for multimodal molecular imaging of cancer. In *Magnetic Nanoparticle-Based Hybrid Materials* (pp. 343–386). Elsevier. doi:10.1016/B978-0-12-823688-8.00008-9

Morris, J. A., Harrison, L. M., & Partridge, S. M. (2007). Practical and theoretical aspects of postmortem bacteriology. *Current Diagnostic Pathology*, *13*(1), 65–74. doi:10.1016/j.cdip.2006.07.005

Mosallaei, S., Hashemi, H., Hoseini, M., Dehghani, M., & Naz, A. (2023). Polycyclic Aromatic hydrocarbons (PAHs) in household dust: The association between PAHs, cancer risk and sick building syndrome. *Building and Environment*, *229*, 109966. doi:10.1016/j.buildenv.2022.109966

Moseley, P. T. (2017). Progress in the development of semiconducting metal oxide gas sensors: A review. *Measurement Science & Technology*, *28*(8), 082001. doi:10.1088/1361-6501/aa7443

Mou, Q., Ma, Y., Ding, F., Gao, X., Yan, D., Zhu, X., & Zhang, C. (2019). Two-in-One Chemogene Assembled from Drug-Integrated Antisense Oligonucleotides To Reverse Chemoresistance. *Journal of the American Chemical Society*, *141*(17), 6955–6966. doi:10.1021/jacs.8b13875 PMID:30964284

Moura, P. C., Vassilenko, V., & Ribeiro, P. A. (2023). Ion Mobility Spectrometry Towards Environmental Volatile Organic Compounds Identification and Quantification: a Comparative Overview over Infrared Spectroscopy. *Emission Control Science and Technology*, 1-22.

Mousaabadi, K. Z., Ensafi, A. A., & Rezaei, B. (2022). Simultaneous determination of some opioid drugs using Cu-hemin MOF@ MWCNTs as an electrochemical sensor. *Chemosphere*, *303*, 135149. doi:10.1016/j.chemosphere.2022.135149 PMID:35660395

Muehlethaler, C., Leona, M., & Lombardi, J. R. (2016). Review of surface enhanced Raman scattering applications in forensic science. *Analytical Chemistry*, *88*(1), 152–169. doi:10.1021/acs.analchem.5b04131 PMID:26638887

Muhammad, W., Haroon, M., Shah, M., Ullah, M. A., & Haleem, I. (2018). Potential application of nanotechnology in health care: An insight. *Nanoscale Reports*, *1*(2), 1–8. doi:10.26524/nr1821

Mulpur, P., Yadavilli, S., Mulpur, P., Kondiparthi, N., Sengupta, B., Rao, A. M., Podila, R., & Kamisetti, V. (2015). Flexible Ag–C 60 nano-biosensors based on surface plasmon coupled emission for clinical and forensic applications. *Physical Chemistry Chemical Physics*, *17*(38), 25049–25054. doi:10.1039/C5CP04268B PMID:26345678

Munir, S., Ahmed, S., Ibrahim, M., Khalid, M., & Ojha, S. C. (2020). A Spellbinding interplay between biological barcoding and nanotechnology. *Frontiers in Bioengineering and Biotechnology*, *8*, 883. doi:10.3389/fbioe.2020.00883 PMID:33014994

Munoz Barus, J. I., Febrero-Bande, M., & Cadarso-Suárez, C. (2008). Flexible regression models for estimating postmortem interval (PMI) in forensic medicine. *Statistics in Medicine*, *27*(24), 5026–5038. doi:10.1002im.3319 PMID:18618426

Munteanu, I. G., & Apetrei, C. (2023). Classification and Antioxidant Activity Evaluation of Edible Oils by Using Nanomaterial-Based Electrochemical Sensors. *International Journal of Molecular Sciences*, *24*(3), 3010. doi:10.3390/ijms24033010 PMID:36769346

Mu, Q., Hondow, N. S., Krzemiński, Ł., Brown, A. P., Jeuken, L. J., & Routledge, M. N. (2012). Mechanism of cellular uptake of genotoxic silica nanoparticles. *Particle and Fibre Toxicology*, *9*(1), 1–11. doi:10.1186/1743-8977-9-29 PMID:22823932

Murray, C. B., Kagan, C. R., & Bawendi, M. G. (2000). Synthesis and characterization of monodisperse nanocrystals and close-packed nanocrystal assemblies. *Annual Review of Materials Science*, *30*(1), 545–610. doi:10.1146/annurev.matsci.30.1.545

Murtada, K., Bowman, D., Edwards, M., & Pawliszyn, J. (2022). Thin-film microextraction combined with comprehensive two-dimensional gas chromatography time-of-flight mass spectrometry screening for presence of multiclass organic pollutants in drinking water samples. *Talanta*, *242*, 123301. doi:10.1016/j.talanta.2022.123301 PMID:35167962

Murtey, M. D., & Ramasamy, P. (2016). Sample preparations for scanning electron microscopy–life sciences. *Modern Electron Microscopy in Physical and Life Sciences, 2.*

Mwafy, E. A., & Mostafa, A. M. (2019). Multi walled carbon nanotube decorated cadmium oxide nanoparticles via pulsed laser ablation in liquid media. *Optics & Laser Technology*, *111*, 249–254. doi:10.1016/j.optlastec.2018.09.055

Nadar, S. S., Kelkar, R. K., Pise, P. V., Patil, N. P., Patil, S. P., Chaubal-Durve, N. S., Bhange, V. P., Tiwari, M. S., & Patil, P. D. (2021). The untapped potential of magnetic nanoparticles for forensic investigations: A comprehensive review. *Talanta*, *230*, 122297. doi:10.1016/j.talanta.2021.122297 PMID:33934767

Nagai, Y., Yudasaka, M., Kataura, H., & Fujigaya, T. (2019). Brighter near-IR emission of single-walled carbon nanotubes modified with a cross-linked polymer coating. *Chemical Communications (Cambridge)*, *55*(48), 6854–6857. doi:10.1039/C9CC02712B PMID:31123733

Nagar, A., & Pradeep, T. (2020). Clean water through nanotechnology: Needs, gaps, and fulfillment. *ACS Nano*, *14*(6), 6420–6435. doi:10.1021/acsnano.9b01730 PMID:32433866

Nam, J.-M., Thaxton, C. S., & Mirkin, C. A. (2003). Nanoparticle-based bio-bar codes for the ultrasensitive detection of proteins. *Science*, *301*(5641), 1884–1886. doi:10.1126cience.1088755 PMID:14512622

Narayan, R. (Ed.). (2017). *Nanobiomaterials: Nanostructured Materials for Biomedical Applications.* Elsevier.

Nasrollahzadeh, M., Sajjadi, M., Sajadi, S. M., & Issaabadi, Z. (2019). *Green nanotechnology Interface science and technology* (Vol. 28). Elsevier.

Nath, N., Kumar, A., Chakroborty, S., Soren, S., Barik, A., Pal, K., & de Souza, F. G. Jr. (2023). Carbon Nanostructure Embedded Novel Sensor Implementation for Detection of Aromatic Volatile Organic Compounds: An Organized Review. *ACS Omega*, *8*(5), 4436–4452. doi:10.1021/acsomega.2c05953 PMID:36777592

Nayak, P., Nanda, R., & Mishra, M. (2021). Role of Nanoparticles in Medical Imaging. In *Medical Imaging Methods* (pp. 107–144). CRC Press. doi:10.1201/9781003112068-5

Nazari, A. (2020). *Nanosensors for smart cities: An introduction. In Nanosensors for smart cities*. Elsevier. doi:10.1016/B978-0-12-819870-4.00001-3

Neumann, C., & Margot, P. (2009). New perspectives in the use of ink evidence in forensic science: Part I. Development of a quality assurance process for forensic ink analysis by HPTLC. *Forensic Science International, 185*(1–3), 29–37. doi:10.1016/j.forsciint.2008.11.016 PMID:19147310

Nikhar, S., Sahu, P., Rarotra, S., & Kumar, P. (2021). Biological metal organic framework (Bio-MOF) for detection of voltaic organic compounds (VOC's). *Inorganic Chemistry Communications, 130*, 108711. doi:10.1016/j.inoche.2021.108711

Nikkel, B. (2020). Fintech forensics: Criminal investigation and digital evidence in financial technologies. *Forensic Science International: Digital Investigation, 33*, 200908.

Nikolaou, P., Papoutsis, I., Dona, A., Spiliopoulou, C., & Athanaselis, S. (2014). Beware of 25C-NBOMe: An N-benzyl substituted Phenethylamine. *Journal of Forensic Toxicology and Pharmacology, 3*(4). Advance online publication. doi:10.4172/2325-9841.1000129

Nizamov, S., & Mirsky, V. M. (2018). *Wide-field surface plasmon resonance microscopy for in-situ characterization of nanoparticle suspensions In In-situ characterization techniques for nanomaterials*. Springer.

Nocerino, V., Miranda, B., Tramontano, C., Chianese, G., Dardano, P., Rea, I., & De Stefano, L. (2022). Plasmonic nanosensors: Design, fabrication, and applications in biomedicine. *Chemosensors (Basel, Switzerland), 10*(5), 150. doi:10.3390/chemosensors10050150

Nordmann, A. (2007). If and then: A critique of speculative nanoethics. *NanoEthics, 1*(1), 31–46. doi:10.100711569-007-0007-6

Nozik, A. J., Beard, M. C., Luther, J. M., Law, M., Ellingson, R. J., & Johnson, J. C. (2010). Semiconductor quantum dots and quantum dot arrays and applications of multiple exciton generation to third-generation photovoltaic solar cells. *Chemical Reviews, 110*(11), 6873–6890. doi:10.1021/cr900289f PMID:20945911

Nsibande, S. A., & Forbes, P. B. C. (2016). Fluorescence detection of pesticides using quantum dot materials–a review. *Analytica Chimica Acta, 945*, 9–22. doi:10.1016/j.aca.2016.10.002 PMID:27968720

Nugroho, D., Keawprom, C., Chanthai, S., Oh, W.-C., & Benchawattananon, R. (2022). Highly Sensitive Fingerprint Detection under UV Light on Non-Porous Surface Using Starch-Powder Based Luminol-Doped Carbon Dots (N-CDs) from Tender Coconut Water as a Green Carbon Source. *Nanomaterials (Basel, Switzerland), 12*(3), 3. Advance online publication. doi:10.3390/nano12030400 PMID:35159745

Ochekpe, N. A., Olorunfemi, P. O., & Ngwuluka, N. C. (2009). Nanotechnology and drug delivery part 1: Background and applications. *Tropical Journal of Pharmaceutical Research, 8*(3). Advance online publication. doi:10.4314/tjpr.v8i3.44546

Oliveira-Costa, J., & Mello-Patiu, C. A. d. (2004). Application of forensic entomology to estimate of the postmortem interval (PMI) in homicide investigations by the Rio de Janeiro Police Department in Brazil. *Anil Aggrawal's Internet Journal of Forensic Medicine and Toxicology, 5*(1), 40–44.

Omar, Z. M., & Ibrahim, J. (2020). An overview of Darknet, rise and challenges and its assumptions. *International Journal of Computer Science and Information Technologies, 8*, 110–116.

Ong, W.-J., Zheng, N., & Antonietti, M. (2021). *Advanced nanomaterials for energy conversion and storage: current status and future opportunities*. Academic Press.

Oosthuizen, T., & Howes, L. M. (2022). The development of forensic DNA analysis: New debates on the issue of fundamental human rights. *Forensic Science International. Genetics*, *56*, 102606. doi:10.1016/j.fsigen.2021.102606 PMID:34710822

Ortiz-Masiá, D., Cosín-Roger, J., Calatayud, S., Hernández, C., Alós, R., Hinojosa, J., Esplugues, J. V., & Barrachina, M. D. (2016). M1 Macrophages Activate Notch Signalling in Epithelial Cells: Relevance in Crohn's Disease. *Journal of Crohn's and Colitis*, *10*(5), 582–592. doi:10.1093/ecco-jcc/jjw009 PMID:26802079

Osaka, M., Mori, D., Benten, H., Ogawa, H., Ohkita, H., & Ito, S. (2017). Charge transport in intermixed regions of all-polymer solar cells studied by conductive atomic force microscopy. *ACS Applied Materials & Interfaces*, *9*(18), 15615–15622. doi:10.1021/acsami.7b00979 PMID:28437063

Owens, D. E. III, & Peppas, N. A. (2006). Opsonization, biodistribution, and pharmacokinetics of polymeric nanoparticles. *International Journal of Pharmaceutics*, *307*(1), 93–102. doi:10.1016/j.ijpharm.2005.10.010 PMID:16303268

Pachuta, S. J., & Staral, J. S. (1994). Nondestructive analysis of colorants on paper by time-of-flight secondary ion mass spectrometry. *Analytical Chemistry*, *66*(2), 276–284. doi:10.1021/ac00074a015

Paikrao Hariprasad, D. A. (2021). *Nano-Forensics a Comprehensive Review.* Crimson Publishers.

Paikrao, H. M., Tajane, D. S., Patil, A. S., & Dipale, A. D. (2022). Applications of Nanotechnology in Forensic Science. In Engineered Nanomaterials for Innovative Therapies and Biomedicine. Springer. doi:10.1007/978-3-030-82918-6_11

Paikrao, H. M., Tajane, D. S., Patil, A. S., & Dipale, A. D. (2022). *Applications of Nanotechnology in Forensic Science. In Engineered Nanomaterials for Innovative Therapies and Biomedicine*. Springer.

Pallas, G., Peijnenburg, W. J., Guinée, J. B., Heijungs, R., & Vijver, M. G. (2018). Green and clean: Reviewing the justification of claims for nanomaterials from a sustainability point of view. *Sustainability (Basel)*, *10*(3), 689. doi:10.3390u10030689

Pal, S. L., Jana, U., Manna, P. K., Mohanta, G. P., & Manavalan, R. (2011). Nanoparticle: An overview of preparation and characterization. *Journal of Applied Pharmaceutical Science*, 228–234.

Pan, D., Mi, L., Huang, Q., Hu, J., & Fan, C. (2012). Genetic analysis with nanoPCR. *Integrative Biology*, *4*(10), 1155–1163. doi:10.1039/c2ib20076g PMID:22907590

Pandey, G., Tharmavaram, M., Rawtani, D., Kumar, S., & Agrawal, Y. (2017). Multifarious applications of atomic force microscopy in forensic science investigations. *Forensic Science International*, *273*, 53–63. doi:10.1016/j.forsciint.2017.01.030 PMID:28214756

Pandya, A., & Shukla, R. (2018). New perspective of nanotechnology: Role in preventive forensic. *Egyptian Journal of Forensic Sciences*, *8*(1), 1–11. doi:10.118641935-018-0088-0

Pansare, V. J., Hejazi, S., Faenza, W. J., & Prud'homme, R. K. (2012). Review of Long-Wavelength Optical and NIR Imaging Materials: Contrast Agents, Fluorophores, and Multifunctional Nano Carriers. *Chemistry of Materials*, *24*(5), 812–827. doi:10.1021/cm2028367 PMID:22919122

Paolesse, R., Nardis, S., Monti, D., Stefanelli, M., & Di Natale, C. (2017). Porphyrinoids for chemical sensor applications. *Chemical Reviews*, *117*(4), 2517–2583. doi:10.1021/acs.chemrev.6b00361 PMID:28222604

Pardanaud, C., Dellasega, D., Passoni, M., Martin, C., Roubin, P., Addab, Y., ... Salomon, E. (2020). Post-mortem analysis of tungsten plasma facing components in tokamaks: Raman microscopy measurements on compact, porous oxide and nitride films and nanoparticles. *Nuclear Fusion*, *60*(8), 086004. doi:10.1088/1741-4326/ab9347

Park, Y. S., Kodama, S., & Sekiguchi, H. (2021). Preparation of Metal Nitride Particles Using Arc Discharge in Liquid Nitrogen. *Nanomaterials (Basel, Switzerland), 11*(9), 2214. doi:10.3390/nano11092214 PMID:34578530

Parrilla, M., Montiel, N. F., Van Durme, F., & De Wael, K. (2021). Derivatization of amphetamine to allow its electrochemical detection in illicit drug seizures. *Sensors and Actuators. B, Chemical, 337*, 129819. doi:10.1016/j.snb.2021.129819

Parrilla, M., Slosse, A., Van Echelpoel, R., Felipe Montiel, N., Langley, A. R., Van Durme, F., & De Wael, K. (2022). Rapid on-site detection of illicit drugs in smuggled samples with a portable electrochemical device. *Chemosensors (Basel, Switzerland), 10*(3), 108. doi:10.3390/chemosensors10030108

Partouche, C. F., & Bois, R. S. (2017, August). Questioned Documents. In *17th Interpol International Forensic Science Managers Symposium, Lyon 8 th-10 th October 2013* (*Vol. 526*, p. 854). Academic Press.

Parvin, N., & Mandal, T. K. (2017). Dually emissive P, N-co-doped carbon dots for fluorescent and photoacoustic tissue imaging in living mice. *Mikrochimica Acta, 184*(4), 1117–1125. doi:10.100700604-017-2108-4

Patel, S., Nanda, R., & Sahoo, S. (2015). Nanotechnology in healthcare: applications and challenges. *Med. Chem, 5*(12).

Pathak, A., Suneesh, P. V., Stanley, J., & Babu, T. G. S. (2019). Multicolor emitting N/S-doped carbon dots as a fluorescent probe for imaging pathogenic bacteria and human buccal epithelial cells. *Mikrochimica Acta, 186*(3), 10. doi:10.100700604-019-3270-7 PMID:30715615

Patolsky, F., Gill, R., Weizmann, Y., Mokari, T., Banin, U., & Willner, I. (2003). Lighting-up the dynamics of telomerization and DNA replication by CdSe– ZnS quantum dots. *Journal of the American Chemical Society, 125*(46), 13918–13919. doi:10.1021/ja035848c PMID:14611202

Patra, J. K., Das, G., Fraceto, L. F., Campos, E. V. R., Rodriguez-Torres, M. P., Acosta-Torres, L. S., Diaz-Torres, L. A., Grillo, R., Swamy, M. K., Sharma, S., Habtemariam, S., & Shin, H.-S. (2018). Nano based drug delivery systems: Recent developments and future prospects. *Journal of Nanobiotechnology, 16*(1), 71. doi:10.118612951-018-0392-8 PMID:30231877

Pautler, M., & Brenner, S. (2010). Nanomedicine: Promises and challenges for the future of public health. *International Journal of Nanomedicine*, 803–809. PMID:21042425

Pavlov, A. J., & Batova, V. N. (2014). Ensuring economic security of the innovative development of nanotechnology in the Russian Federation and foreign countries. *Life Science Journal, 11*(6), 322.

Pechal, J. L., Schmidt, C. J., Jordan, H. R., & Benbow, M. E. (2018). A large-scale survey of the postmortem human microbiome, and its potential to provide insight into the living health condition. *Scientific Reports, 8*(1), 1–15. doi:10.103841598-018-23989-w PMID:29636512

Pelliccione, C. J., Timofeeva, E. V., & Segre, C. U. (2016). Potential-resolved in situ X-ray absorption spectroscopy study of Sn and SnO2 nanomaterial anodes for lithium-ion batteries. *The Journal of Physical Chemistry C, 120*(10), 5331–5339. doi:10.1021/acs.jpcc.5b12279

Peng, G., Tisch, U., Adams, O., Hakim, M., Shehada, N., Broza, Y. Y., & Haick, H. (2009). Diagnosing lung cancer in exhaled breath using gold nanoparticles. *Nature Nanotechnology, 4*(10), 669–673. doi:10.1038/nnano.2009.235 PMID:19809459

Peng, X., Kotnala, A., Rajeeva, B. B., Wang, M., Yao, K., Bhatt, N., Penley, D., & Zheng, Y. (2021). Plasmonic Nanotweezers and Nanosensors for Point-of-Care Applications. *Advanced Optical Materials, 9*(13), 2100050. doi:10.1002/adom.202100050 PMID:34434691

Perfézou, M., Turner, A., & Merkoçi, A. (2012). Cancer detection using nanoparticle-based sensors. *Chemical Society Reviews*, *41*(7), 2606–2622. doi:10.1039/C1CS15134G PMID:21796315

Pergolizzi Jr, J. V., Dahan, A., Ann Lequang, J., & Raff, A. (2021). Overdoses due to fentanyl and its analogues (F/FAs) push naloxone to the limit. *Journal of Clinical Pharmacy and Therapeutics*, *46*(6), 1501–1504. doi:10.1111/jcpt.13462 PMID:34111307

Perhaita, I., Muresan, L. E., Tudoran, L. B., Silipas, D. T., & Borodi, G. (2020). Synthesis of silicate apatite phosphors with enhanced luminescence via optimized precipitation technique through pH control. *Journal of Sol-Gel Science and Technology*, *96*(2), 498–510. doi:10.100710971-020-05400-1

Perkins, M. J., & Langford, V. S. (2022). Multiple headspace extraction-selected ion flow tube mass spectrometry (MHE-SIFT-MS). Part 1: A protocol for method development and transfer to routine analysis. *Reviews in Separation Sciences.*

Perkins, M. J., & Langford, V. S. (2021). Application of routine analysis procedures to a direct mass spectrometry technique: Selected ion flow tube mass spectrometry (SIFT-MS). *Reviews in Separation Sciences*, *3*(1), e21003–e21003. doi:10.17145/rss.21.003

Pessoa, A. C., Sipoli, C. C., & de la Torre, L. G. (2017). Effects of diffusion and mixing pattern on microfluidic-assisted synthesis of chitosan/ATP nanoparticles. *Lab on a Chip*, *17*(13), 2281–2293. doi:10.1039/C7LC00291B PMID:28608886

Petit, T., Girard, H. A., Trouvé, A., Batonneau-Gener, I., Bergonzo, P., & Arnault, J.-C. (2013). Surface transfer doping can mediate both colloidal stability and self-assembly of nanodiamonds. *Nanoscale*, *5*(19), 8958–8962. doi:10.1039/c3nr02492j PMID:23965659

Petros, R. A., & DeSimone, J. M. (2010). Strategies in the design of nanoparticles for therapeutic applications. *Nature Reviews. Drug Discovery*, *9*(8), 615–627. doi:10.1038/nrd2591 PMID:20616808

Piccinotti, D. (2018). *Chalcogenide platforms for photonic metamaterials.* Academic Press.

Piñeiro-García, A., González-Alatorre, G., Vega-Díaz, S. M., Pérez-Pérez, M. C. I., Tristan, F., & Patiño-Herrera, R. (2020). Reduced graphene oxide coating with high performance for the solid phase micro-extraction of furfural in espresso coffee. *Journal of Food Measurement and Characterization*, *14*(1), 314–321. doi:10.100711694-019-00293-3

Pino, P., Pelaz, B., Zhang, Q., Maffre, P., Nienhaus, G. U., & Parak, W. J. (2014). Protein corona formation around nanoparticles – from the past to the future. *Materials Horizons*, *1*(3), 301–313. doi:10.1039/C3MH00106G

Pirtarighat, S., Ghannadnia, M., & Baghshahi, S. (2019). Green synthesis of silver nanoparticles using the plant extract of Salvia spinosa grown in vitro and their antibacterial activity assessment. *Journal of Nanostructure in Chemistry*, *9*(1), 1–9. doi:10.100740097-018-0291-4

Pita, R., Ehmann, F., & Papaluca, M. (2016). Nanomedicines in the EU-Regulatory Overview. *The AAPS Journal*, *18*(6), 1576–1582. doi:10.120812248-016-9967-1 PMID:27527889

Pitkethly, M. (2009). Nanotechnology and forensics. *Materials Today*, *12*(6), 6. doi:10.1016/S1369-7021(09)70167-1

Pokropivny, V., & Skorokhod, V. (2007). Classification of nanostructures by dimensionality and concept of surface forms engineering in nanomaterial science. *Materials Science and Engineering C*, *27*(5-8), 990–993. doi:10.1016/j.msec.2006.09.023

Prabakaran, E., & Pillay, K. (2021). Nanomaterials for latent fingerprint detection: A review. *Journal of Materials Research and Technology*, *12*, 1856–1885. doi:10.1016/j.jmrt.2021.03.110

Prajapati, S. K., Malaiya, A., Kesharwani, P., Soni, D., & Jain, A. (2019). Biomedical applications and toxicities of carbon nanotubes. *Drug and Chemical Toxicology, 17094*, 92. doi:10. 1080/ 01480 545 PMID:31908176

Prasad, M., Lambe, U. P., Brar, B., Shah, I., J, M., Ranjan, K., Rao, R., Kumar, S., Mahant, S., Khurana, S. K., Iqbal, H. M. N., Dhama, K., Misri, J., & Prasad, G. (2018). Nanotherapeutics: An insight into healthcare and multi-dimensional applications in medical sector of the modern world. *Biomedicine and Pharmacotherapy, 97*, 1521–1537. doi:10.1016/j.biopha.2017.11.026 PMID:29793315

Prasad, V., Lukose, S., Agarwal, P., & Prasad, L. (2020). Role of nanomaterials for forensic investigation and latent fingerprinting—A review. *Journal of Forensic Sciences, 65*(1), 26–36. doi:10.1111/1556-4029.14172 PMID:31454084

Priyonugroho, G., Zaini, J., Samoedro, E., Firmansyah, I., Nurwidya, F., Antariksa, B., ... Susanto, A. (2018). Correlation between urinary cotinine, exhaled carbon monoxide, and nicotine dependence among Indonesian individuals in the national narcotics and illicit drug rehabilitation center. *Journal of Natural Science, Biology, and Medicine, 9*(2).

Ptaszynska-Sarosiek, I., Charuta, A., Furmanek, M. A., Wardaszka, Z., & Niemcunowicz-Janica, A. (2016). Mercury Poisoning - The Cause of Death Assessed 6 years Posthumously. *Journal of Forensic Toxicology and Pharmacology, 5*(1), 1. doi:10.4172/2325-9841.1000143

Purohit, S., Pandey, G., Tharmavaram, M., Rawtani, D., & Mustansar Hussain, C. (2020). Sensors for the Detection of Illicit Drugs. *Technology in Forensic Science: Sampling, Analysis, Data and Regulations*, 221-238.

Pu, S., Gong, C., & Robertson, A. W. (2020). Liquid cell transmission electron microscopy and its applications. *Royal Society Open Science, 7*(1), 191204. doi:10.1098/rsos.191204 PMID:32218950

Qin, X. Y., Wang, J., & Yuan, Q. (2020). Synthesis and biomedical applications of lanthanides-doped persistent luminescence phosphors with NIR emissions. *Frontiers in Chemistry, 8*, 608578. doi:10.3389/fchem.2020.608578 PMID:33381494

Queralto, N., Berliner, A. N., Goldsmith, B., Martino, R., Rhodes, P., & Lim, S. H. (2014). Detecting cancer by breath volatile organic compound analysis: A review of array-based sensors. *Journal of Breath Research, 8*(2), 027112. doi:10.1088/1752-7155/8/2/027112 PMID:24862241

Quintanar-Guerrero, D., Fessi, H., Allémann, E., & Doelker, E. (1996). Influence of stabilizing agents and preparative variables on the formation of poly (D, L-lactic acid) nanoparticles by an emulsification-diffusion technique. *International Journal of Pharmaceutics, 143*(2), 133–141. doi:10.1016/S0378-5173(96)04697-2

Quinteros, E. D. (2022). *Analysis of Blue Fountain Pen Ink by UV-Vis Spectroscopy, ATR-FTIR Spectroscopy, and Thin Layer Chromatography Enhanced with RGB Spot Evaluation to Construct a Database for Comparison and Identification*. Academic Press.

Qu, J., Ge, Y., Zu, B., Li, Y., & Dou, X. (2016). Transition-Metal-Doped p-Type ZnO Nanoparticle-Based Sensory Array for Instant Discrimination of Explosive Vapors. *Small, 12*(10), 1369–1377. doi:10.1002mll.201503131 PMID:26763156

Rahman, M. M. (2020). Efficient formaldehyde sensor development based on Cu-codoped ZnO nanomaterial by an electrochemical approach. *Sensors and Actuators. B, Chemical, 305*, 127541. doi:10.1016/j.snb.2019.127541

Raina, N., Sharma, P., Slathia, P. S., Bhagat, D., & Pathak, A. K. (2020). Efficiency enhancement of renewable energy systems using nanotechnology. *Nanomaterials and Environmental Biotechnology*, 271-297.

Ramalingam, M., & Rana, D. (2015). Impact of Nanotechnology in Induced Pluripotent Stem Cells-driven Tissue Engineering and Regenerative Medicine. *Journal of Bionanoscience, 9*(1), 13–21. doi:10.1166/jbns.2015.1277

Rana, M. S., Pota, H. R., & Petersen, I. R. (2018). A survey of methods used to control piezoelectric tube scanners in high-speed AFM imaging. *Asian Journal of Control, 20*(4), 1379–1399. doi:10.1002/asjc.1728

Rao, P. K., Pandey, G., & Tharmavaram, M. (2020). Biological Evidence and Their Handling. In *Technology in Forensic Science* (pp. 35–53). https://doi.org/https://doi.org/10.1002/9783527827688.ch3

Rao, Pandey, & Tharmavaram. (2020). Biological Evidence and Their Handling. *Technology in Forensic Science.*

Ratcliff, G. C., Erie, D. A., & Superfine, R. (1998). Photothermal modulation for oscillating mode atomic force microscopy in solution. *Applied Physics Letters*, *72*(15), 1911–1913. doi:10.1063/1.121224

Raval, N., Maheshwari, R., Kalyane, D., Youngren-Ortiz, S. R., Chougule, M. B., & Tekade, R. K. (2019). *Importance of physicochemical characterization of nanoparticles in pharmaceutical product development. In Basic fundamentals of drug delivery.* Elsevier.

Rawtani, D., Tharmavaram, M., Pandey, G., & Hussain, C. M. (2019). Functionalized nanomaterial for forensic sample analysis. *Trends in Analytical Chemistry*, *120*, 115661. doi:10.1016/j.trac.2019.115661

Raza, A., & Saha, B. (2013). Silver nanoparticles doped agarose disk: Highly sensitive surface- enhanced Raman scattering substrate for in situ analysis of ink dyes. *Forensic Science International*, *233*(1–3), 21–27. doi:10.1016/j.forsciint.2013.08.004 PMID:24314497

Redasani, V. K., Tamboli, P. S., Kalal, D., & Surana, S. J. (2016). Development and validation of spectroscopic methods for the estimation of mupirocin calcium in bulk and in ointment formulation. *Journal of Forensic Toxicology and Pharmacology*, *4*(2). Advance online publication. doi:10.4172/2325-9841.1000139

Reji, R. P., Balaji, S. K. C., Sivalingam, Y., Kawazoe, Y., & Velappa Jayaraman, S. (2023). First-Principles Density Functional Theory Calculations on the Potential of Sc_2CO_2 MXene Nanosheets as a Dual-Mode Sensor for Detection of Volatile Organic Compounds in Exhaled Human Breath. *ACS Applied Nano Materials*, *6*(7), 5345–5356. doi:10.1021/acsanm.2c05474

Reshma, V. G., & Mohanan, P. V. (2019). Quantum dots: Applications and safety consequences. *Journal of Luminescence*, *205*, 287–298. doi:10.1016/j.jlumin.2018.09.015

Reul, R., Tsapis, N., Hillaireau, H., Sancey, L., Mura, S., Recher, M., Nicolas, J., Coll, J. L., & Fattal, E. (2012). Near infrared labeling of PLGA for in vivo imaging of nanoparticles. *Polymer Chemistry*, *3*(3), 694–702. doi:10.1039/c2py00520d

Reyes, L. C., López, C. E., Barrios, A. B., Soto, C. G., Ibáñez, C., & Díaz, F. J. (2018). Development and application of a new nose hairs sample collection device for GSR Particles by scanning electron microscopy with energy dispersive X-ray spectroscopy (SEM-EDS). *Forensic Science International*, *290*, 42–48. doi:10.1016/j.forsciint.2018.06.029 PMID:30015278

Rickerby, D., & Morrison, M. (2007). Nanotechnology and the environment: A European perspective. *Science and Technology of Advanced Materials*, *8*(1-2), 19–24. doi:10.1016/j.stam.2006.10.002

Riehemann, K., Schneider, S. W., Luger, T. A., Godin, B., Ferrari, M., & Fuchs, H. (2009). Nanomedicine—Challenge and perspectives. *Angewandte Chemie International Edition*, *48*(5), 872–897. doi:10.1002/anie.200802585 PMID:19142939

Righettoni, M., Schmid, A., Amann, A., & Pratsinis, S. E. (2013). Correlations between blood glucose and breath components from portable gas sensors and PTR-TOF-MS. *Journal of Breath Research*, *7*(3), 037110. doi:10.1088/1752-7155/7/3/037110 PMID:23959908

Rivers, D. B., & Dahlem, G. A. (2022). *The science of forensic entomology.* John Wiley & Sons.

Roberts, I. J., Carpenter, L. J., Shaw, M. D., & Langford, V. S. (2022). Selected Ion Flow Tube–Mass Spectrometry (SIFT-MS) study of the reactions of H3O+, NO+ and O2+ with a range of oxygenated volatile organic carbons (OVOCs). *International Journal of Mass Spectrometry*, *479*, 116892. doi:10.1016/j.ijms.2022.116892

Rocha, L. A., Campos-Junior, P. H. A., Esbenshade, J., Siqueira, R. L., Schiavon, M. A., & Ferrari, J. L. (2018). Biocompatibility and photoluminescence of Sm3+doped SiO2-Gd2O3: A promising non-toxic red phosphor to plasmatic membrane tracking. *Ceramics International, 44*(4), 3726–3733. doi:10.1016/j.ceramint.2017.11.153

Roco, M. C., Mirkin, C. A., & Hersam, M. C. (2011). *Nanotechnology research directions for societal needs in 2020: Retrospective and outlook.* Academic Press.

Roco, M. C. (2011). *The long view of nanotechnology development: the National Nanotechnology Initiative at 10 years* (Vol. 13). Springer.

Roco, M. C., & Bainbridge, W. S. (2007a). *Nanotechnology: Societal Implications: Individual Perspectives.* Citeseer. doi:10.1007/978-1-4020-5432-7

Roco, M. C., Mirkin, C. A., & Hersam, M. C. (2010). *Nanotechnology research directions for societal needs in 2020. Retrospective and Outlook. National Science Foundation.* Sponsor.

Rohatgi, R., & Dominica, a. F. (2022). Application of Nanotechnology in Forensic Science. *Journal of Forensic Research, 13*(8).

Rohde, E., Mcmanus, A. C., Vogt, C., & Heineman, W. (1997). Separation and comparison of fountain pen inks by capillary zone electrophoresis. *Journal of Forensic Sciences, 42*(6), 14252J. doi:10.1520/JFS14252J

Romero, V., Vila, V., de la Calle, I., Lavilla, I., & Bendicho, C. (2019). Turn–on fluorescent sensor for the detection of periodate anion following photochemical synthesis of nitrogen and sulphur co–doped carbon dots from vegetables. *Sensors and Actuators. B, Chemical, 280*, 290–297. doi:10.1016/j.snb.2018.10.064

Roopachandrika, C., & Reddy, D. K. A. (2023). (Forthcoming). Alerting and Tracking System of Dislocated Object. *International Journal for Innovative Engineering & Management Research.*

Roslan, M. S., Chaudhary, K. T., Doylend, N., Agam, A., Kamarulzaman, R., Haider, Z., Mazalan, E., & Ali, J. (2019). Growth of wall-controlled MWCNTs by magnetic field assisted arc discharge plasma. *Journal of Saudi Chemical Society, 23*(2), 171–181. doi:10.1016/j.jscs.2018.06.003

Rotaru, A., Dutta, S., Jentzsch, E., Gothelf, K., & Mokhir, A. (2010). Selective dsDNA-templated formation of copper nanoparticles in solution. *Angewandte Chemie International Edition in English, 49*(33), 5665–5667. doi:10.1002/anie.200907256 PMID:20629055

Rouf, Z., Dar, I. Y., Javaid, M., Dar, M. Y., & Jehangir, A. (2022). Volatile Organic Compounds Emission from Building Sector and Its Adverse Effects on Human Health. In *Ecological and Health Effects of Building Materials* (pp. 67–86). Springer. doi:10.1007/978-3-030-76073-1_5

Ruiz-Colón, K., Chavez-Arias, C., Díaz-Alcalá, J. E., & Martínez, M. (2014). *Xylazine intoxication in humans and its importance as an emerging adulterant in abused drugs: A comprehensive review of the literature.* Academic Press.

Rycroft, T., Trump, B., Poinsatte-Jones, K., & Linkov, I. (2018). *Nanotoxicology and nanomedicine: making development decisions in an evolving governance environment.* Academic Press.

Sadaf, A., Zeshan, B., Wang, Z., Zhang, R., Xu, S., Wang, C., & Cui, Y. (2012). Toxicity evaluation of hydrophilic CdTe quantum dots and CdTe@SiO2 nanoparticles in mice. *Journal of Nanoscience and Nanotechnology, 12*(11), 8287–8292. doi:10.1166/jnn.2012.6667 PMID:23421208

Saegeman, V., Cohen, M. C., Alberola, J., Ziyade, N., Farina, C., Cornaglia, G., & Fernández-Rodríguez, A. (2017). How is post-mortem microbiology appraised by pathologists? Results from a practice survey conducted by ESGFOR. *European Journal of Clinical Microbiology & Infectious Diseases*, *36*(8), 1381–1385. doi:10.100710096-017-2943-6 PMID:28236029

Safarik, M. E., Jarvis, J. P., & Nussbaum, K. E. (2002). Sexual Homicide of Elderly Females: Linking Offender Characteristics to Victim and Crime Scene Attributes. *Journal of Interpersonal Violence*, *17*(5), 500–525. doi:10.1177/0886260502017005002

Sahaym, U., & Norton, M. G. (2008). Advances in the application of nanotechnology in enabling a 'hydrogen economy'. *Journal of Materials Science*, *43*(16), 5395–5429. doi:10.100710853-008-2749-0

Sahu, G., Nayak, S. R., & Sethi, S. S. (2016). Mutilation of Body after Crime. *Journal of Forensic Toxicology and Pharmacology*, *5*(1), 1. doi:10.4172/2325-9841.1000145

Saini, V., & Shukla, R. K. (2022). *Forensic Applications of Fluorescence Microscope. In Forensic Microscopy*. CRC Press.

Saisahas, K., Soleh, A., Promsuwan, K., Saichanapan, J., Phonchai, A., Sadiq, N. S. M., ... Limbut, W. J. A. o. (2022). *Nanocoral-like Polyaniline-Modified Graphene-Based Electrochemical Paper-Based Analytical Device for a Portable Electrochemical Sensor for Xylazine Detection*. Academic Press.

Sakiyama-Elbert, S., & Hubbell, J. (2001). Functional biomaterials: Design of novel biomaterials. *Annual Review of Materials Research*, *31*(1), 183–201. doi:10.1146/annurev.matsci.31.1.183

Salem, S. S., Hammad, E. N., Mohamed, A. A., & El-Dougdoug, W. (2022). A comprehensive review of nanomaterials: Types, synthesis, characterization, and applications. *Biointerface Research in Applied Chemistry*, *13*(1), 41. doi:10.33263/BRIAC131.041

Salinas, F. M., Barros, E. F., Eisenberg, W., Smith, D. M., & Viswanathan, S. (2001). Impact of nanotechnology on telecommunications. *Sohail Anwar M. Yasin Akhtar Raja Salahuddin Qazi*, 23.

Sanjuan-Navarro, L., Moliner-Martínez, Y., & Campíns-Falcó, P. (2022). The state of art of nanocarbon black as analyte in a variety of matrices: A review. *Trends in Analytical Chemistry*, *157*, 116769. doi:10.1016/j.trac.2022.116769

Sanli, S., Ghorbani-Zamani, F., Moulahoum, H., Gumus, Z. P., Coskunol, H., Odaci Demirkol, D., & Timur, S. (2020). Application of Biofunctionalized Magnetic Nanoparticles Based-Sensing in Abused Drugs Diagnostics. *Analytical Chemistry*, *92*(1), 1033–1040. doi:10.1021/acs.analchem.9b04025 PMID:31800231

Santos-Oliveira, R. (2014). Pharmaceutical equivalence and bioequivalence of radiopharmaceuticals: Thinking the possibility of generic radiopharmaceuticals and preparing for new technology as nanotechnology drugs. *Journal of Bioequivalence & Bioavailability*, *6*(1). Advance online publication. doi:10.4172/jbb.1000175

Sarfo, D. K., Izake, E. L., O'Mullane, A. P., & Ayoko, G. A. (2019). Fabrication of nanostructured SERS substrates on conductive solid platforms for environmental application. *Critical Reviews in Environmental Science and Technology*, *49*(14), 1294–1329. doi:10.1080/10643389.2019.1576468

Sarswat, P., & Free, M. L. (2014). Quantum Dot Based Sensing Platform for Selective Detection of Volatile Organic Compounds and Biomarkers. *ECS Transactions*, *61*(19), 11–16. doi:10.1149/06119.0011ecst

Sastry, R. K., Rashmi, H. B., & Rao, N. H. (2011). Nanotechnology for enhancing food security in India. *Food Policy*, *36*(3), 391–400. doi:10.1016/j.foodpol.2010.10.012

Sato, T., Zaitsu, K., Tsuboi, K., Nomura, M., Kusano, M., Shima, N., Abe, S., Ishii, A., Tsuchihashi, H., & Suzuki, K. (2015). A preliminary study on postmortem interval estimation of suffocated rats by GC-MS/MS-based plasma metabolic profiling. *Analytical and Bioanalytical Chemistry, 407*(13), 3659–3665. doi:10.100700216-015-8584-7 PMID:25749795

Saukko, P., & Knight, B. (2016a). Sudden death in infancy. In P. Saukko & B. Knight (Eds.), *Knights's forensic pathology* (4th ed., pp. 461–473). CRC Press.

Saukko, P., & Knight, B. (2016b). The pathophysiology of death. In P. Saukko & B. Knight (Eds.), *Knights's forensic pathology* (4th ed., pp. 55–93). CRC Press.

Schaming, D., & Remita, H. (2015a). Nanotechnology: From the ancient time to nowadays. *Foundations of Chemistry, 17*(3), 187–205. doi:10.100710698-015-9235-y

Schulte, J. (2005). *Nanotechnology: Global strategies, industry trends and applications.* John Wiley & Sons. doi:10.1002/0470021071

Scorsone, E., Manai, R., Cali, K., Ricatti, M. J., Farno, S., Persaud, K., & Mucignat, C. (2021). Biosensor array based on ligand binding proteins for narcotics and explosives detection. *Sensors and Actuators. B, Chemical, 334*, 129587. doi:10.1016/j.snb.2021.129587

Seeman, N. C. (1999). DNA engineering and its application to nanotechnology. *Trends in Biotechnology, 17*(11), 437–443. doi:10.1016/S0167-7799(99)01360-8 PMID:10511701

Seiler, H. (1983). Secondary electron emission in the scanning electron microscope. *Journal of Applied Physics, 54*(11), R1–R18. doi:10.1063/1.332840

Sekimoto, K., Li, S. M., Yuan, B., Koss, A., Coggon, M., Warneke, C., & de Gouw, J. (2017). Calculation of the sensitivity of proton-transfer-reaction mass spectrometry (PTR-MS) for organic trace gases using molecular properties. *International Journal of Mass Spectrometry, 421*, 71–94. doi:10.1016/j.ijms.2017.04.006

Selli, G. I., Bonatto, A. E. T., Bonatto, F. T., Anzanello, M. J., & Bergmann, C. P. (2022). Nanosensors in Forensic Sciences. In *Environmental Applications of Nanomaterials* (pp. 239–253). Springer. doi:10.1007/978-3-030-86822-2_13

Senesac, L., & Thundat, T. (2008). *Nanosensors for trace explosive detection.* Academic Press.

Senesac, L., & Thundat, T. G. (2008). Nanosensors for trace explosive detection. *Materials Today, 11*(3), 28-36.

Seo, Y., & Jhe, W. (2007). Atomic force microscopy and spectroscopy. *Reports on Progress in Physics, 71*(1), 016101. doi:10.1088/0034-4885/71/1/016101

Serrano, E., Rus, G., & Garcia-Martinez, J. (2009). Nanotechnology for sustainable energy. *Renewable & Sustainable Energy Reviews, 13*(9), 2373–2384. doi:10.1016/j.rser.2009.06.003

Severino, M. I., Al Mohtar, A., Vieira Soares, C., Freitas, C., Sadovnik, N., Nandi, S., Mouchaham, G., Pimenta, V., Nouar, F., Daturi, M., Maurin, G., Pinto, M. L., & Serre, C. (2023). MOFs with Open Metal (III) Sites for the Environmental Capture of Polar Volatile Organic Compounds. *Angewandte Chemie International Edition, 62*(6), e202211583. doi:10.1002/anie.202211583 PMID:36468308

Shaban, S. E., Ibrahiem, N. M., El-mongy, S. A., & Elshereafy, E. E. (2013). Validation of scanning electron microscope (SEM), energy dispersive X-ray (EDX) and gamma spectrometry to verify source nuclear material for safeguards purposes. *Journal of Radioanalytical and Nuclear Chemistry, 296*(3), 1219–1224. doi:10.100710967-012-2374-x

Shahbazi, N., & Zare-Dorabei, R. (2019). A novel "off-on" fluorescence nanosensor for sensitive determination of sulfide ions based on carbon quantum dots and gold nanoparticles: Central composite design optimization. *Microchemical Journal*, *145*, 996–1002. doi:10.1016/j.microc.2018.12.022

Shahdost-Fard, F., & Roushani, M. (2017). *Designing an ultra-sensitive aptasensor based on an AgNPs/thiol-GQD nanocomposite for TNT detection at femtomolar levels using the electrochemical oxidation of Rutin as a redox probe*. Academic Press.

Shahdost-fard, F., & Roushani, M. (2017). *Impedimetric detection of trinitrotoluene by using a glassy carbon electrode modified with a gold nanoparticle@ fullerene composite and an aptamer-imprinted polydopamine*. Academic Press.

Shamsipur, M., Tabrizi, M. A., Mahkam, M., & Aboudi, J. J. E. (2015). A High Sensitive TNT Sensor Based on Electrochemically Reduced Graphene Oxide-Poly (amidoamine). *Electroanalysis*, *27*(6), 1466–1472. doi:10.1002/elan.201400634

Shang, L., Zhang, L., & Dong, S. (2009). Turn-on fluorescent cyanide sensor based on copper ion-modified CdTe quantum dots. *Analyst (London)*, *134*(1), 107–113. doi:10.1039/B812458B PMID:19082182

Shankaran, D. R., Gobi, K. V., Sakai, T., Matsumoto, K., Imato, T., Toko, K., & Miura, N. (2005). A novel surface plasmon resonance immunosensor for 2, 4, 6-trinitrotoluene (TNT) based on indirect competitive immunoreaction: A promising approach for on-site landmine detection. *IEEE Sensors Journal*, *5*(4), 616–621. doi:10.1109/JSEN.2005.848150

Sharma, S., Shekhar, S., Gautam, S., Sharma, B., Kumar, A., & Jain, P. (2020). Carbon-based nanomaterials as novel nanosensors. *Nanofabrication for smart nanosensor applications*, 323-347.

Sharma, J., Yeh, H. C., Yoo, H., Werner, J. H., & Martinez, J. S. (2011). Silver nanocluster aptamers: In situ generation of intrinsically fluorescent recognition ligands for protein detection. *Chemical Communications (Cambridge)*, *47*(8), 2294–2296. doi:10.1039/C0CC03711G

Sharma, M., Chauhan, P., Nehra, S., Sharma, R., & Kumar, D. (2023). Volatile Organic Compounds Removal by Nanofiltration from Groundwater. In *Nanofiltration Membrane for Water Purification* (pp. 119–139). Springer Nature Singapore. doi:10.1007/978-981-19-5315-6_7

Sharma, S., Jaiswal, S., Duffy, B., & Jaiswal, A. K. (2019). Nanostructured materials for food applications: Spectroscopy, microscopy and physical properties. *Bioengineering (Basel, Switzerland)*, *6*(1), 26. doi:10.3390/bioengineering6010026 PMID:30893761

Shaw, L., & Dennany, L. (2017). Applications of electrochemical sensors: Forensic drug analysis. *Current Opinion in Electrochemistry*, *3*(1), 23–28. doi:10.1016/j.coelec.2017.05.001

Sheeparamatti, B., Sheeparamatti, R., & Kadadevaramath, J. (2007). Nanotechnology: Inspiration from nature. *IETE Technical Review*, *24*(1), 5–8.

Shi, H., Xiang, W., Liu, C., Shi, H., Zhou, Y., & Gao, L. (2018). *Highly sensitive detection for cocaine using graphene oxide-aptamer based sensors in combination with tween 20*. Academic Press.

Shi, L., Hou, A., Chen, L., & Wang, Z. (2015). *Electrochemical sensor prepared from molecularly imprinted polymer for recognition of TNT*. Academic Press.

Shrestha, S., Wang, B., & Dutta, P. (2020). Nanoparticle processing: Understanding and controlling aggregation. *Advances in Colloid and Interface Science*, *279*, 102162. doi:10.1016/j.cis.2020.102162 PMID:32334131

Shreve, R. N., & Brink, J. A. Jr. (1977). *Chemical Process Industries*. McGraw-Hill Book Co.

Shukla, R., Lodha, A. S., & Pandya, A. (2016). Nanotechnology: An applied and robust approach for forensic investigation. *Forensic Research & Criminology International Journal*, *2*(1), 35–37. doi:10.15406/frcij.2016.02.00044

Simões, E. F. C., Leitão, J. M. M., & da Silva, J. C. G. E. (2016). Carbon dots prepared from citric acid and urea as fluorescent probes for hypochlorite and peroxynitrite. *Mikrochimica Acta*, *183*(5), 1769–1777. doi:10.100700604-016-1807-6

Simonenko, N. P., Musaev, A. G., Simonenko, T. L., Gorobtsov, P. Y., Volkov, I. A., Gulin, A. A., Simonenko, E. P., Sevastyanov, V. G., & Kuznetsov, N. T. (2021). Hydrothermal Synthesis of Ag Thin Films and Their SERS Application. *Nanomaterials (Basel, Switzerland)*, *12*(1), 136. doi:10.3390/nano12010136 PMID:35010086

Sim, S., & Wong, N. K. (2021). Nanotechnology and its use in imaging and drug delivery [Review]. *Biomedical Reports*, *14*(5), 42. doi:10.3892/br.2021.1418 PMID:33728048

Sinclair, R. C., Suter, J. L., & Coveney, P. V. (2019). Micromechanical exfoliation of graphene on the atomistic scale. *Physical Chemistry Chemical Physics*, *21*(10), 5716–5722. doi:10.1039/C8CP07796G PMID:30801077

Singh, K. K., Singh, A., & Rai, S. (2022). *A study on nanomaterials for water purification.* Academic Press.

Singh, R. P., Verma, R. D., Meshri, D. T., & Shreeve, J. (2006). *Energetic nitrogen-rich salts and ionic liquids.* Academic Press.

Singh, S., & Singh, M. (2003). Explosives detection systems (EDS) for aviation security. Academic Press.

Singh, S., Kumar, V., Anil, A. G., Kapoor, D., Khasnabis, S., Shekar, S., . . . Singh, J. (2021). *Adsorption and detoxification of pharmaceutical compounds from wastewater using nanomaterials: A review on mechanism, kinetics, valorization and circular economy.* Academic Press.

Singh, A., Dubey, S., & Dubey, H. K. (2019). Nanotechnology: The future engineering. *Nanotechnology*, *6*(2), 230–233.

Singh, G., Sharma, D., Pathak, D., & Dutta, S. (2022). Comparative Study on Post Mortem Analysis of Sodium and Potassium Levels of Vitreous Humour and Synovial Fluid in Determining Time Since Death. *Indian Journal of Forensic Medicine Toxicology*, *16*(3), 393–398.

Singh, J., Dutta, T., Kim, K.-H., Rawat, M., Samddar, P., & Kumar, P. (2018). 'Green'synthesis of metals and their oxide nanoparticles: Applications for environmental remediation. *Journal of Nanobiotechnology*, *16*(1), 1–24. doi:10.118612951-018-0408-4 PMID:30373622

Singh, R. K., Bansode, F. W., Sharma, S., Singh, P., Bhadauria, S., Singh, S., & Nath, C. (2015). Development of a nanotechnology based biomedicine RISUG-M as a female contraceptive in India. *Journal of Nanomedicine & Nanotechnology*, *6*(4), 1.

Singh, S., & Samal, N. (2021). Nanotechnology: A Powerful Tool in Forensic Science for Solving Criminal Cases. *Arab Journal of Forensic Sciences & Forensic Medicine*, *3*(2), 273–296. doi:10.26735/YDDJ4516

Singh, T., Shukla, S., Kumar, P., Wahla, V., Bajpai, V. K., & Rather, I. A. (2017). Application of nanotechnology in food science: Perception and overview. *Frontiers in Microbiology*, *8*, 1501. doi:10.3389/fmicb.2017.01501 PMID:28824605

Singh, V., & Tiwari, M. K. (2020). UV emitting Pb2+ doped Ca2La8(SiO4)(6)O-2 phosphors prepared by sol-gel procedure. *Optik (Stuttgart)*, *206*, 163600. doi:10.1016/j.ijleo.2019.163600

Siri, J. G., Fernando, C. A., & De Silva, S. (2020). Nanotechnology and Protection of Intellectual Property. *Emerging Trends.*, *14*(4), 307–327. PMID:32532198

Slepička, P., Slepičková Kasálková, N., Siegel, J., Kolská, Z., & Švorčík, V. (2019). Methods of gold and silver nanoparticles preparation. *Materials (Basel)*, *13*(1), 1. doi:10.3390/ma13010001 PMID:31861259

Smijs, T., Galli, F., & van Asten, A. (2016). Forensic potential of atomic force microscopy. *Forensic Chemistry (Amsterdam, Netherlands), 2*, 93–104. doi:10.1016/j.forc.2016.10.005

Smith, G. B. (2011). *Green nanotechnology.* Paper presented at the Nanostructured thin films IV. 10.1117/12.893114

Smith, C. (2019). *Neurotransmitters and the narcotic analgesics. In Chemical and Biological Aspects of Drug Dependence*. CRC Press.

Smith, D., & Španěl, P. (2022). Ternary association reactions of H3O+, NO+ and O2+• with N2, O2, CO2 and H2O; implications for selected ion flow tube mass spectrometry analyses of air and breath. *Rapid Communications in Mass Spectrometry, 36*(6), e9241. doi:10.1002/rcm.9241 PMID:34904315

Smith, J. (2022). *Forensic DNA Investigation. In Handbook of DNA Profiling*. Springer.

Smith, R. G., D'Souza, N., & Nicklin, S. (2008). A review of biosensors and biologically-inspired systems for explosives detection. *Analyst (London), 133*(5), 571–584. doi:10.1039/b717933m PMID:18427676

Soares, S., Sousa, J., Pais, A., & Vitorino, C. (2018). Nanomedicine: Principles, Properties, and Regulatory Issues. *Frontiers in Chemistry, 6*, 360. doi:10.3389/fchem.2018.00360 PMID:30177965

Sodhi, G. S., & Kaur, J. (2016). Physical developer method for detection of latent fingerprints: A review. *Egyptian Journal of Forensic Sciences, 6*(2), 44–47. doi:10.1016/j.ejfs.2015.05.001

Solaimuthu, A., Vijayan, A. N., Murali, P., & Korrapati, P. S. (2020). Nano-biosensors and their relevance in tissue engineering. *Current Opinion in Biomedical Engineering, 13*, 84–93. doi:10.1016/j.cobme.2019.12.005

So, M. K., Xu, C. J., Loening, A. M., Gambhir, S. S., & Rao, J. H. (2006). Self-illuminating quantum dot conjugates for in vivo imaging. *Nature Biotechnology, 24*(3), 339–343. doi:10.1038/nbt1188 PMID:16501578

Song, C., Xu, J., Chen, Y., Zhang, L., Lu, Y., & Qing, Z. (2019). DNA-templated fluorescent nanoclusters for metal ions detection. *Molecules (Basel, Switzerland), 24*(22), 4189. doi:10.3390/molecules24224189 PMID:31752270

Song, C., Ye, B., Xu, J., Chen, J., Shi, W., Yu, C., An, C., Zhu, J., & Zhang, W. (2022). Large-Area Nanosphere Self-Assembly Monolayers for Periodic Surface Nanostructures with Ultrasensitive and Spatially Uniform SERS Sensing. *Small, 18*(8), 2104202. doi:10.1002mll.202104202 PMID:34877766

Song, H., Su, Y., Zhang, L., & Lv, Y. (2019). Quantum dots-based chemiluminescence probes: An overview. *Luminescence, 34*(6), 530–543. doi:10.1002/bio.3633 PMID:31025479

Song, W.-K., Cui, Y.-J., Tang, A. M., Ding, W.-Q., & Tran, T. D. (2014). Experimental study on water evaporation from sand using environmental chamber. *Canadian Geotechnical Journal, 51*(2), 115–128. doi:10.1139/cgj-2013-0155

Song, Y., Chen, J., Hu, D., Liu, F., Li, P., Li, H., Chen, S., Tan, H., & Wang, L. (2015). Ratiometric fluorescent detection of biomakers for biological warfare agents with carbon dots chelated europium-based nanoscale coordination polymers. *Sensors and Actuators. B, Chemical, 221*, 586–592. doi:10.1016/j.snb.2015.07.008

Soni, S., Jain, U., Burke, D. H., & Chauhan, N. (2022). Development of Nanomaterial-Modified Impedimetric Aptasensor—A Single-Step Strategy for 3, 4-Methylenedioxymethylamphetamine Detection. *Biosensors (Basel), 12*(7), 538. doi:10.3390/bios12070538 PMID:35884341

Sonnichsen, C., Reinhard, B. M., Liphardt, J., & Alivisatos, A. P. (2005). A molecular ruler based on plasmon coupling of single gold and silver nanoparticles. *Nature Biotechnology, 23*(6), 741–745. doi:10.1038/nbt1100 PMID:15908940

Sorolla-Rosario, D., Llorca-Porcel, J., Pérez-Martínez, M., Lozano-Castelló, D., & Bueno-López, A. (2023). Microplastics' analysis in water: Easy handling of samples by a new Thermal Extraction Desorption-Gas Chromato graphy-Mass Spectrometry (TED-GC/MS) methodology. *Talanta*, *253*, 123829. doi:10.1016/j.talanta.2022.123829 PMID:36087410

Sotelo-Gonzalez, E., Fernandez-Argüelles, M. T., Costa-Fernandez, J. M., & Sanz-Medel, A. (2012). Mn-doped ZnS quantum dots for the determination of acetone by phosphorescence attenuation. *Analytica Chimica Acta*, *712*, 120–126. doi:10.1016/j.aca.2011.11.023 PMID:22177074

Sousa, F., Ferreira, D., Reis, S., & Costa, P. (2020). *Current insights on antifungal therapy: Novel nanotechnology approaches for drug delivery systems and new drugs from natural sources*. Academic Press.

Spagnolo, A. G., & Daloiso, V. (2009). *Outlining ethical issues in nanotechnologies*. Academic Press.

Spagnolo, E. V., Stassi, C., Mondello, C., Zerbo, S., Milone, L., & Argo, A. (2019). Forensic microbiology applications: A systematic review. *Legal Medicine*, *36*, 73–80. doi:10.1016/j.legalmed.2018.11.002 PMID:30419494

Spagnolo, G. S. (2006). Potentiality of 3D laser profilometry to determine the sequence of homogenous crossing lines on questioned documents. *Forensic Science International*, *164*(2–3), 102–109. doi:10.1016/j.forsciint.2005.12.004 PMID:16431054

Speller, N. C., Siraj, N., Regmi, B. P., Marzoughi, H., Neal, C., & Warner, I. M. (2015). Rational design of QCM-D virtual sensor arrays based on film thickness, viscoelasticity, and harmonics for vapor discrimination. *Analytical Chemistry*, *87*(10), 5156–5166. doi:10.1021/ac5046824 PMID:25913394

Squire, K. J., Sivashanmugan, K., Zhang, B., Kraai, J., Rorrer, G., & Wang, A. X. (2020). Multiscale photonic crystal enhanced core–shell plasmonic nanomaterial for rapid vapor-phase detection of explosives. *ACS Applied Nano Materials*, *3*(2), 1656–1665. doi:10.1021/acsanm.9b02399

Srividya, B. (2016). Nanotechnology in forensics and its application in forensic investigation. *Res. Rev. J. Pharm. Nanotechnol*, *4*(2), 1–7.

Srividya, B. (2016). Nanotechnology in forensics and its application in forensic investigation. *Res. Rev. J. Pharm. Nanotechnology*, *4*(2), 1–7.

Steinegger, A., Wolfbeis, O. S., & Borisov, S. M. (2020). Optical sensing and imaging of pH values: Spectroscopies, materials, and applications. *Chemical Reviews*, *120*(22), 12357–12489. doi:10.1021/acs.chemrev.0c00451 PMID:33147405

Steinhauer, S. (2021). Gas sensors based on copper oxide nanomaterials: A review. *Chemosensors (Basel, Switzerland)*, *9*(3), 51. doi:10.3390/chemosensors9030051

Stern, S. T., & McNeil, S. E. (2008). Nanotechnology safety concerns revisited. *Toxicological Sciences*, *101*(1), 4–21. doi:10.1093/toxsci/kfm169 PMID:17602205

Stetefeld, J., McKenna, S. A., & Patel, T. R. (2016). Dynamic light scattering: A practical guide and applications in biomedical sciences. *Biophysical Reviews*, *8*(4), 409–427. doi:10.100712551-016-0218-6 PMID:28510011

Stewart, A. J., Hendry, J., & Dennany, L. (2015). Whole blood electrochemiluminescent detection of dopamine. *Analytical Chemistry*, *87*(23), 11847–11853. doi:10.1021/acs.analchem.5b03345 PMID:26530000

Stöber, W., Fink, A., & Bohn, E. (1968). Controlled growth of monodisperse silica spheres in the micron size range. *Journal of Colloid and Interface Science*, *26*(1), 62–69. doi:10.1016/0021-9797(68)90272-5

Sufian, M. M., Khattak, J. Z. K., Yousaf, S., & Rana, M. S. (2017). Safety issues associated with the use of nanoparticles in human body. *Photodiagnosis and Photodynamic Therapy*, *19*, 67–72. doi:10.1016/j.pdpdt.2017.05.012 PMID:28552731

Sunagawa, K., & Sugitani, M. (2017). Post-mortem detection of bacteremia using pairs of blood culture samples. *Legal Medicine*, *24*, 92–97. doi:10.1016/j.legalmed.2016.12.009 PMID:28081798

Sun, Q., Luo, Y., Zhang, Q., Yang, X., & Xu, C. (2016). How much can a forensic laboratory do to discriminate questioned ink entries? *Journal of Forensic Sciences*, *61*(4), 1116–1121. doi:10.1111/1556-4029.13067 PMID:27364295

Sun, X., Shao, K., & Wang, T. (2016). Detection of volatile organic compounds (VOCs) from exhaled breath as noninvasive methods for cancer diagnosis. *Analytical and Bioanalytical Chemistry*, *408*(11), 2759–2780. doi:10.100700216-015-9200-6 PMID:26677028

Sun, Z., Wu, S., Ma, J., Shi, H., Wang, L., Sheng, A., Yin, T., Sun, L., & Li, G. (2019). Colorimetric Sensor Array for Human Semen Identification Designed by Coupling Zirconium Metal–Organic Frameworks with DNA-Modified Gold Nanoparticles. *ACS Applied Materials & Interfaces*, *11*(40), 36316–36323. doi:10.1021/acsami.9b10729 PMID:31522499

Suravi Pandit, Dewan, & Ahmed. (2016). Nanotechnology based biosensors and its application. *The Pharma Innovation*, *6*(5), 18–25.

Sutter, P., & Sutter, E. (2014). Thickness determination of few-layer hexagonal boron nitride films by scanning electron microscopy and Auger electron spectroscopy. *APL Materials*, *2*(9), 092502. doi:10.1063/1.4889815

Swierczewska, M., Lee, S., & Chen, X. Y. (2011). Inorganic Nanoparticles for Multimodal Molecular Imaging. *Molecular Imaging*, *10*(1), 3–16. doi:10.2310/7290.2011.00001 PMID:21303611

Syedmoradi, L., Daneshpour, M., Alvandipour, M., Gomez, F. A., Hajghaseem, H., & Omidfar, K. (2017). Point of care testing: The impact of nanotechnology. *Biosensors & Bioelectronics*, *87*, 373–387. doi:10.1016/j.bios.2016.08.084 PMID:27589400

Takahashi, F., Nitta, S., Shimizu, R., & Jin, J. (2018). Electrochemiluminescence and voltammetry of tris (2, 2′-bipyridine) ruthenium (II) with amphetamine-type stimulants as coreactants: An application to the discrimination of methamphetamine. *Forensic Toxicology*, *36*(1), 185–191. doi:10.100711419-017-0388-3

Tamers, S. L., Streit, J., Pana-Cryan, R., Ray, T., Syron, L., Flynn, M. A., ... Guerin, R. (2020). Envisioning the future of work to safeguard the safety, health, and well-being of the workforce: A perspective from the CDC's National Institute for Occupational Safety and Health. *American Journal of Industrial Medicine*, *63*(12), 1065–1084. doi:10.1002/ajim.23183 PMID:32926431

Tan, N. H. (2014). *Rock art research in Southeast Asia: a synthesis.* Paper Presented at the Arts. 10.3390/arts3010073

Tang, C., He, Z., Liu, H., Xu, Y., Huang, H., Yang, G., ... Deng, Y. (2020). Application of magnetic nanoparticles in nucleic acid detection. *Journal of Nanobiotechnology*, *18*(1), 1–19. doi:10.118612951-020-00613-6 PMID:32316985

Tang, L. B., Ji, R. B., Cao, X. K., Lin, J. Y., Jiang, H. X., Li, X. M., Teng, K. S., Luk, C. M., Zeng, S. J., Hao, J. H., & Lau, S. P. (2012). Deep ultraviolet photoluminescence of water-soluble self-passivated graphene quantum dots. *ACS Nano*, *6*(6), 5102–5110. doi:10.1021/nn300760g PMID:22559247

Tang, M., Ren, G., Zhu, B., Yu, L., Liu, X., Chai, F., Wu, H., & Wang, C. (2019). Correction: Facile synthesis of orange emissive carbon dots and their application for mercury ion detection and fast fingerprint development. *Analytical Methods*, *11*(18), 2502–2502. doi:10.1039/C9AY90065A

Tang, R. K., Liu, Y., Liu, Y. Z., Zhu, S. M., Huang, W., Zhao, P., & Li, J. B. (2013). Evaluation of post-mortem heart blood culture in a Chinese population. *Forensic Science International*, *231*(1-3), 229–233. doi:10.1016/j.forsciint.2013.05.020 PMID:23890642

Tantra, R., & Knight, A. (2011). Cellular uptake and intracellular fate of engineered nanoparticles: A review on the application of imaging techniques. *Nanotoxicology*, *5*(3), 381–392. doi:10.3109/17435390.2010.512987 PMID:20846020

Tao, H. Q., Yang, K., Ma, Z., Wan, J. M., Zhang, Y. J., Kang, Z. H., & Liu, Z. (2012). In vivo NIR fluorescence imaging, biodistribution, and toxicology of photoluminescent carbon dots produced from carbon nanotubes and graphite. *Small*, *8*(2), 281–290. doi:10.1002mll.201101706 PMID:22095931

Tao, Z. M., Hong, G. S., Shinji, C., Chen, C. X., Diao, S., Antaris, A. L., Zhang, B., Zou, Y. P., & Dai, H. J. (2013). Biological Imaging Using Nanoparticles of Small Organic Molecules with Fluorescence Emission at Wavelengths Longer than 1000 nm. *Angewandte Chemie International Edition*, *52*(49), 13002–13006. doi:10.1002/anie.201307346 PMID:24174264

Taubin, D., Wilson, J. C., & Wilens, T. E. (2022). ADHD and Substance Use Disorders in Young People: Considerations for Evaluation, Diagnosis, and Pharmacotherapy. *Child and Adolescent Psychiatric Clinics of North America*, *31*(3), 515–530. doi:10.1016/j.chc.2022.01.005 PMID:35697399

Taz, H., Ruther, R., Malasi, A., Yadavali, S., Carr, C., Nanda, J., & Kalyanaraman, R. (2015). In situ localized surface plasmon resonance (LSPR) spectroscopy to investigate kinetics of chemical bath deposition of CdS thin films. *The Journal of Physical Chemistry C*, *119*(9), 5033–5039. doi:10.1021/jp512738b

Teymourian, H., Parrilla, M., Sempionatto, J. R., Montiel, N. F., Barfidokht, A., Van Echelpoel, R., & Wang, J. (2020). Wearable electrochemical sensors for the monitoring and screening of drugs. *ACS Sensors*, *5*(9), 2679–2700. doi:10.1021/acssensors.0c01318 PMID:32822166

Theis, T. N. (2001). Information Technology Based on a Mature Nanotechnology: Some Societal Implications. *Societal Implications of Nanoscience and Nanotechnology*, 60.

Thorn, J. L., Gilchrist, K. B., Sobonya, R. E., Gaur, N. K., Lipke, P. N., & Klotz, S. A. (2010). Postmortem candidaemia: Marker of disseminated disease. *Journal of Clinical Pathology*, *63*(4), 337–340. doi:10.1136/jcp.2009.070607 PMID:19939858

Timbrell, J., & Barile, F. A. (2023). *Introduction to toxicology*. CRC Press. doi:10.1201/9781003188575

Tinkle, S., McNeil, S. E., Mühlebach, S., Bawa, R., Borchard, G., Barenholz, Y. C., Tamarkin, L., & Desai, N. (2014). Nanomedicines: Addressing the scientific and regulatory gap. *Annals of the New York Academy of Sciences*, *1313*(1), 35–56. doi:10.1111/nyas.12403 PMID:24673240

Tiruwa, R. (2016). A review on nanoparticles-preparation and evaluation parameters. *Indian Journal of Pharmaceutical and Biological Research*, *4*(2), 27.

Tiwari, S. B., & Amiji, M. M. (2006). A review of nanocarrier-based CNS delivery systems. *Current Drug Delivery*, *3*(2), 219–232. doi:10.2174/156720106776359230 PMID:16611008

Tizro, P., Choi, C., & Khanlou, N. (2019). Sample preparation for transmission electron microscopy. *Biobanking: Methods and Protocols*, 417-424.

Todaro, B., & Santi, M. (2022). *Characterization and Functionalization Approaches for the Study of Polymeric Nanoparticles: The State of the Art in Italian Research.* Paper Presented at the Micro. 10.3390/micro3010002

To, K. C., Ben-Jaber, S., & Parkin, I. P. (2020). Recent developments in the field of explosive trace detection. *ACS Nano*, *14*(9), 10804–10833. doi:10.1021/acsnano.0c01579 PMID:32790331

Tonndorf, R., Aibibu, D., & Cherif, C. (2021). *Isotropic and anisotropic scaffolds for tissue engineering: Collagen, conventional, and textile fabrication technologies and properties.* Academic Press.

Torres-Sangiao, E., Leal Rodriguez, C., & García-Riestra, C. (2021). Application and perspectives of MALDI–TOF mass spectrometry in clinical microbiology laboratories. *Microorganisms*, *9*(7), 1539. doi:10.3390/microorganisms9071539 PMID:34361974

Tosi, G., Bondioli, L., Ruozi, B., Badiali, L., Severini, G. M., Biffi, S., De Vita, A., Bortot, B., Dolcetta, D., Forni, F., & Vandelli, M. A. (2011). NIR-labeled nanoparticles engineered for brain targeting: In vivo optical imaging application and fluorescent microscopy evidences. *Journal of Neural Transmission (Vienna, Austria)*, *118*(1), 145–153. doi:10.100700702-010-0497-1

Trefz, P., Rösner, L., Hein, D., Schubert, J. K., & Miekisch, W. (2013). Evaluation of needle trap micro-extraction and automatic alveolar sampling for point-of-care breath analysis. *Analytical and Bioanalytical Chemistry*, *405*(10), 3105–3115. doi:10.100700216-013-6781-9 PMID:23388692

Trotta, F., Zanetti, M., & Cavalli, R. (2012). Cyclodextrin-based nanosponges as drug carriers. *Beilstein Journal of Organic Chemistry*, *8*, 2091–2099. doi:10.3762/bjoc.8.235 PMID:23243470

Tsai, L.-C., Su, C.-W., Lee, J. C.-I., Lu, Y.-S., Chen, H.-C., Lin, Y.-C., Linacre, A., & Hsieh, H.-M. (2018). The detection and identification of saliva in forensic samples by RT-LAMP. *Forensic Science, Medicine, and Pathology*, *14*(4), 469–477. doi:10.100712024-018-0008-5 PMID:30058014

Tsai, W. T. (2023). A Survey on Toxic Volatile Organic Compounds (VOCs): Toxicological Profiles, Health Exposure Risks, and Regulatory Strategies for Mitigating Emissions from Stationary Sources in Taiwan. *Atmosphere (Basel)*, *14*(2), 242. doi:10.3390/atmos14020242

Tsutsumi, A., Yamamoto, Y., Okamura, K., Oki, Y., Hamada, K., & Ishizu, H. (1984). LDH (lactate dehydrogenase) isoenzyme in human semen and seminal stains. *Hanzaigaku Zassi*, *50*, 41–48.

Tully, G., Cohen, N., Compton, D., Davies, G., Isbell, R., & Watson, T. (2020). Quality standards for digital forensics: Learning from experience in England & Wales. *Forensic Science International: Digital Investigation*, *32*, 200905.

Tummala, S., Huang, W. A., Wu, B. H., Chang, K. C., & Ho, Y. P. (2020). Fluorescent mesoporous nanoparticles for beta-lactamase screening assays. *ChemistryOpen*, *9*(10), 1074–1081. doi:10.1002/open.202000221 PMID:33117628

Tunaboylu, B., Toprak, B., Binark, A. K., Öztürk, O., & Zaim, S. (2021). *Manufacturing process development for thin film filaments as a new product.* Paper presented at the Digital Conversion on the Way to Industry 4.0: Selected Papers from ISPR2020. 10.1007/978-3-030-62784-3_44

van Dongen, R. M., Alderliefste, G. J., Onderwater, G. L., Ferrari, M. D., & Terwindt, G. M. (2021). Migraine prevalence in visual snow with prior illicit drug use (hallucinogen persisting perception disorder) versus without. *European Journal of Neurology*, *28*(8), 2631–2638. doi:10.1111/ene.14914 PMID:33979006

van Landeghem, F. K., Maier-Hauff, K., Jordan, A., Hoffmann, K.-T., Gneveckow, U., Scholz, R., Thiesen, B., Brück, W., & von Deimling, A. (2009). Post-mortem studies in glioblastoma patients treated with thermotherapy using magnetic nanoparticles. *Biomaterials*, *30*(1), 52–57. doi:10.1016/j.biomaterials.2008.09.044 PMID:18848723

Vandan Prasada, S. L. (2016). Emerging Forensic Applications of Nanotechnology. *Int J Eng Allied Sci,* 1-8.

Vargas, A., Pegaz, B., Debefve, E., Konan-Kouakou, Y., Lange, N., Ballini, J.-P., van den Bergh, H., Gurny, R., & Delie, F. (2004). Improved photodynamic activity of porphyrin loaded into nanoparticles: An in vivo evaluation using chick embryos. *International Journal of Pharmaceutics*, *286*(1-2), 131–145. doi:10.1016/j.ijpharm.2004.07.029 PMID:15501010

Velichko, E., Makarov, S., Nepomnyashchaya, E., & Dong, G. (2020). Molecular Aggregation in immune system activation studied by dynamic light scattering. *Biology (Basel)*, *9*(6), 123. doi:10.3390/biology9060123 PMID:32545635

Verhagen, A., & Kelarakis, A. (2020). Carbon dots for forensic applications: A critical review. *Nanomaterials (Basel, Switzerland), 10*(8), 1535. doi:10.3390/nano10081535 PMID:32764380

Verma, P., & Maheshwari, S. K. (2018). Preparation of Sliver and Selenium Nanoparticles and Its Characterization by Dynamic Light Scattering and Scanning Electron Microscopy. *Journal of Microscopy and Ultrastructure, 6*(4), 182–187. doi:10.4103/jmau.jmau_3_18 PMID:30464890

Vieira, S., Vial, S., Reis, R. L., & Oliveira, J. M. (2017). Nanoparticles for bone tissue engineering. *Biotechnology Progress, 33*(3), 590–611. doi:10.1002/btpr.2469 PMID:28371447

Vikrant, K., Bhardwaj, N., Bhardwaj, S. K., Kim, K.-H., & Deep, A. (2019). Nanomaterials as efficient platforms for sensing DNA. *Biomaterials, 214*, 119215. doi:10.1016/j.biomaterials.2019.05.026 PMID:31146176

Vinoda, B. M., Vinuth, M., Bodke, Y., & Manjanna, J. (2015). Photocatalytic degradation of toxic methyl red dye using silica nanoparticles synthesized from rice husk ash. *Journal of Environmental & Analytical Toxicology, 5*(336), 2161–0525.

Virkler, K., & Lednev, I. K. (2009). Analysis of body fluids for forensic purposes: From laboratory testing to non-destructive rapid confirmatory identification at a crime scene. *Forensic Science International, 188*(1), 1–17. doi:10.1016/j.forsciint.2009.02.013 PMID:19328638

Viswanath, D. S., Ghosh, T. K., & Boddu, V. M. (2018). *Emerging energetic materials: Synthesis, physicochemical, and detonation properties*. Springer. doi:10.1007/978-94-024-1201-7

Walia, S., & Acharya, A. (2014). Fluorescent cadmium sulfide nanoparticles for selective and sensitive detection of toxic pesticides in aqueous medium. *Journal of Nanoparticle Research, 16*(12), 1–10. doi:10.100711051-014-2778-3

Walters, G. D. (2018). Resistance to peer influence and crime desistance in emerging adulthood: A moderated mediation analysis. *Law and Human Behavior, 42*(6), 520–530. doi:10.1037/lhb0000293 PMID:29927278

Wang, J. B., Ren, Z., Hou, Y., Yan, X. L., Liu, P. Z., Zhang, H., Zhang, H. X., & Guo, J. J. (2020). A review of graphene synthesis at low temperatures by CVD methods. *New Carbon Materials, 35*(3), 193–208. doi:10.1016/S1872-5805(20)60484-X

Wang, J., & Yan, B. (2019). Improving covalent organic frameworks fluorescence by triethylamine pinpoint surgery as selective biomarker sensor for diabetes mellitus diagnosis. *Analytical Chemistry, 91*(20), 13183–13190. doi:10.1021/acs.analchem.9b03534 PMID:31529947

Wang, L. Y., Kan, X. W., Zhang, M. C., Zhu, C. Q., & Wang, L. (2002). Fluorescence for the determination of protein with functionalized nano-ZnS. *Analyst (London), 127*(11), 1531–1534. doi:10.1039/b206587h PMID:12475047

Wang, N., Ernle, L., Bekö, G., Wargocki, P., & Williams, J. (2022). Emission rates of volatile organic compounds from humans. *Environmental Science & Technology, 56*(8), 4838–4848. doi:10.1021/acs.est.1c08764 PMID:35389619

Wang, Q., Ishikawa, T., Michiue, T., Zhu, B.-L., Guan, D.-W., & Maeda, H. (2012). Stability of endogenous reference genes in postmortem human brains for normalization of quantitative real-time PCR data: Comprehensive evaluation using geNorm, NormFinder, and BestKeeper. *International Journal of Legal Medicine, 126*(6), 943–952. doi:10.100700414-012-0774-7 PMID:23010907

Wang, Q., Lin, H., Xu, J., Huang, P., & Wang, Z. (2018). Current Research and Prospects on Postmortem Interval Estimation. *Fa Yi Xue Za Zhi, 34*(5), 459–467. PMID:30468046

Wang, Q., Zhang, Y., Lin, H., Zha, S., Fang, R., Wei, X., Fan, S., & Wang, Z. (2017). Estimation of the late postmortem interval using FTIR spectroscopy and chemometrics in human skeletal remains. *Forensic Science International, 281*, 113–120. doi:10.1016/j.forsciint.2017.10.033 PMID:29127892

Wang, T., Zheng, H., You, C., & Ju, J. (2023). A Texture-Hidden Anti-Counterfeiting QR Code and Authentication Method. *Sensors (Basel)*, *23*(2), 795. doi:10.339023020795 PMID:36679589

Wang, W., Kang, S., Zhou, W., & Vikesland, P. J. (2023). Environmental routes of virus transmission and the application of nanomaterial-based sensors for virus detection. *Environmental Science. Nano*, *10*(2), 393–423. doi:10.1039/D2EN00600F

Wang, X., Liao, T., Wang, H., Hao, H., Yang, Q., Zhou, H., Ma, Y., Zhi, M., Wang, J., & Fan, R. (2022). Novel Organic-Inorganic Hybrid Polystyrene Nanoparticles with Trichromatic Luminescence for the Detection of Latent Fingerprints. *International Journal of Analytical Chemistry*, *2022*, 2022. doi:10.1155/2022/2230360 PMID:35295922

Wang, Y., & Gu, H. C. (2015). Core-shell-type magnetic mesoporous silica nanocomposites for bioimaging and therapeutic agent delivery. *Advanced Materials*, *27*(3), 576–585. doi:10.1002/adma.201401124 PMID:25238634

Wang, Y., & Liu, B. (2007). Silica nanoparticle assisted DNA assays for optical signal amplification of conjugated polymer based fluorescent sensors. *Chemical Communications*, (34), 3553–3555. doi:10.1039/b705936a PMID:18080542

Wang, Y., Wan, D., Xie, S., Xia, X., Huang, C. Z., & Xia, Y. (2013). Synthesis of silver octahedra with controlled sizes and optical properties via seed-mediated growth. *ACS Nano*, *7*(5), 4586–4594. doi:10.1021/nn401363e PMID:23631674

Wang, Z., Zhang, Y., Zhang, B., & Lu, X. (2018). Mn2+ doped ZnS QDs modified fluorescence sensor based on molecularly imprinted polymer/sol-gel chemistry for detection of Serotonin. *Talanta*, *190*, 1–8. doi:10.1016/j.talanta.2018.07.065 PMID:30172484

Wani, S. U. D., Ali, M., Masoodi, M. H., Khan, N. A., Zargar, M. I., Hassan, R., ... Osmani, R. M. A. (2022). A review on nanoparticles categorization, characterization and applications in drug delivery systems. *Vibrational Spectroscopy*. doi:10.1016/j.vibspec.2022.103407

Warheit, D. B. (2018). Hazard and risk assessment strategies for nanoparticle exposures: How far have we come in the past 10 years? *F1000 Research*, *7*, 7. doi:10.12688/f1000research.12691.1 PMID:29636906

Warheit, D. B., Hoke, R. A., Finlay, C., Donner, E. M., Reed, K. L., & Sayes, C. M. (2007). Development of a base set of toxicity tests using ultrafine TiO2 particles as a component of nanoparticle risk management. *Toxicology Letters*, *171*(3), 99–110. doi:10.1016/j.toxlet.2007.04.008 PMID:17566673

Warnement, M. R., Tomlinson, I. D., & Rosenthal, S. J. (2007). Fluorescent Imaging Applications of Quantum Dot Probes. *Current Nanoscience*, *3*(4), 273–284. doi:10.2174/157341307782418595

Wasson, M. C., Otake, K.-i., Gong, X., Strathman, A. R., Islamoglu, T., Gianneschi, N. C., & Farha, O. K. (2020). Modulation of crystal growth and structure within cerium-based metal–organic frameworks. *CrystEngComm*, *22*(47), 8182–8188. doi:10.1039/D0CE01223H

Wells, K., & Bradley, D. (2012). *A review of X-ray explosives detection techniques for checked baggage*. Academic Press.

Wenger, W. N., Bates, F. S., & Aydil, E. S. (2017). Functionalization of cadmium selenide quantum dots with poly (ethylene glycol): Ligand exchange, surface coverage, and dispersion stability. *Langmuir*, *33*(33), 8239–8245. doi:10.1021/acs.langmuir.7b01924 PMID:28768415

Weng, S., Dong, R., Zhu, Z., Zhang, D., Zhao, J., Huang, L., & Liang, D. (2018). Dynamic surface-enhanced Raman spectroscopy and Chemometric methods for fast detection and intelligent identification of methamphetamine and 3, 4-Methylenedioxy methamphetamine in human urine. *Spectrochimica Acta. Part A: Molecular and Biomolecular Spectroscopy*, *189*, 1–7. doi:10.1016/j.saa.2017.08.004 PMID:28783586

Wester, N., Mynttinen, E., Etula, J., Lilius, T., Kalso, E., Kauppinen, E. I., . . . Koskinen, J. (2019). *Simultaneous detection of morphine and codeine in the presence of ascorbic acid and uric acid and in human plasma at nafion single-walled carbon nanotube thin-film electrode*. Academic Press.

Wester, N., Mynttinen, E., Etula, J., Lilius, T., Kalso, E., Kauppinen, E. I., Laurila, T., & Koskinen, J. (2019). Simultaneous detection of morphine and codeine in the presence of ascorbic acid and uric acid and in human plasma at nafion single-walled carbon nanotube thin-film electrode. *ACS Omega*, *4*(18), 17726–17734. doi:10.1021/acsomega.9b02147 PMID:31681878

Whetstone, Z., & Kearfott, K. (2014). *A review of conventional explosives detection using active neutron interrogation*. Academic Press.

Widjonarko, N. E. (2016). Introduction to advanced X-ray diffraction techniques for polymeric thin films. *Coatings*, *6*(4), 54. doi:10.3390/coatings6040054

Wiener, E., Brechbiel, M., Brothers, H., Magin, R., Gansow, O., Tomalia, D., & Lauterbur, P. (1994). Dendrimer-based metal chelates: A new class of magnetic resonance imaging contrast agents. *Magnetic Resonance in Medicine*, *31*(1), 1–8. doi:10.1002/mrm.1910310102 PMID:8121264

Wilczewska, A. Z., Niemirowicz, K., Markiewicz, K. H., & Car, H. (2012). Nanoparticles as drug delivery systems. *Pharmacological Reports*, *64*(5), 1020–1037. doi:10.1016/S1734-1140(12)70901-5 PMID:23238461

Wilde, M. J., Cordell, R. L., Salman, D., Zhao, B., Ibrahim, W., Bryant, L., & Monks, P. S. (2019). Breath analysis by two-dimensional gas chromatography with dual flame ionisation and mass spectrometric detection–method optimisation and integration within a large-scale clinical study. *Journal of Chromatography. A*, *1594*, 160–172. doi:10.1016/j.chroma.2019.02.001 PMID:30755317

Williams, R., & Alivisatos, P. (2013). *Nanotechnology research directions: iwgn Workshop Report: vision for nanotechnology in the next decade*. Springer Science & Business Media.

Willig, K. I., Rizzoli, S. O., Westphal, V., Jahn, R., & Hell, S. W. (2006). STED microscopy reveals that synaptotagmin remains clustered after synaptic vesicle exocytosis. *Nature*, *440*(7086), 935–939. doi:10.1038/nature04592 PMID:16612384

Wongkaew, N., Simsek, M., Griesche, C., & Baeumner, A. J. (2019). Functional nanomaterials and nanostructures enhancing electrochemical biosensors and lab-on-a-chip performances: Recent progress, applications, and future perspective. *Chemical Reviews*, *119*(1), 120–194. doi:10.1021/acs.chemrev.8b00172 PMID:30247026

Wonner, K., Evers, M. V., & Tschulik, K. (2018). Simultaneous opto-and spectro-electrochemistry: Reactions of individual nanoparticles uncovered by dark-field microscopy. *Journal of the American Chemical Society*, *140*(40), 12658–12661. doi:10.1021/jacs.8b02367 PMID:29995398

Wu, D., Wu, Q., Lu, Y., Wang, C., Yv, S., Wang, L., ... Gao, S. (2022). A novel approach for forensic identification of automotive paints using optical coherence tomography and multivariate statistical methods. *Journal of Forensic Sciences*, *67*(6), 2253–2266. doi:10.1111/1556-4029.15114 PMID:35913098

Wu, J., Zhang, L., Huang, F., Ji, X., Dai, H., & Wu, W. (2020). Surface enhanced Raman scattering substrate for the detection of explosives: Construction strategy and dimensional effect. *Journal of Hazardous Materials*, *387*, 121714. doi:10.1016/j.jhazmat.2019.121714 PMID:31818672

Wu, P., Xu, C., Hou, X., Xu, J.-J., & Chen, H.-Y. (2015). Dual-emitting quantum dot nanohybrid for imaging of latent fingerprints: Simultaneous identification of individuals and traffic light-type visualization of TNT. *Chemical Science (Cambridge)*, *6*(8), 4445–4450. doi:10.1039/C5SC01497B PMID:30155001

Wu, X., Li, W., Li, C., Yin, J., Wu, T., Zhang, M., & Li, Z. (2022). Discrimination and characterization of the volatile organic compounds of Acori tatarinowii rhizoma based on headspace-gas chromatography-ion mobility spectrometry and headspace solid phase microextraction-gas chromatography-mass spectrometry. *Arabian Journal of Chemistry*, *15*(11), 104231. doi:10.1016/j.arabjc.2022.104231

Wu, Z., Wang, H., Wang, X., Zheng, H., Chen, Z., & Meng, C. (2020). Development of electronic nose for qualitative and quantitative monitoring of volatile flammable liquids. *Sensors (Basel)*, *20*(7), 1817. doi:10.339020071817 PMID:32218148

Xiao, D., Yuan, D., He, H., & Gao, M. (2013). Microwave assisted one-step green synthesis of fluorescent carbon nanoparticles from ionic liquids and their application as novel fluorescence probe for quercetin determination. *Journal of Luminescence*, *140*, 120–125. doi:10.1016/j.jlumin.2013.02.032

Xiao, Z., Kong, L. B., Ruan, S., Li, X., Yu, S., Li, X., Jiang, Y., Yao, Z., Ye, S., Wang, C., Zhang, T., Zhou, K., & Li, S. (2018). Recent development in nanocarbon materials for gas sensor applications. *Sensors and Actuators. B, Chemical*, *274*, 235–267. doi:10.1016/j.snb.2018.07.040

Xia, Z., Zhai, X., Liu, B., & Mo, Y. (2015). Determination of Electrical Conductivity of Cadaver Skeletal Muscle: A promising method for the estimation of late postmortem interval. *Journal of Forensic Science and Medicine*, *1*(1), 16. doi:10.4103/2349-5014.155554

Xing, L. I. U., & Yang, L. U. O. (2014). Surface modifications technology of quantum dots based biosensors and their medical applications. *Chinese Journal of Analytical Chemistry*, *42*(7), 1061–1069. doi:10.1016/S1872-2040(14)60753-2

Xu, K., Huang, J., Ye, Z., Ying, Y., & Li, Y. (2009). Recent Development of Nano-Materials Used in DNA Biosensors. In Sensors (Vol. 9, Issue 7, pp. 5534–5557). doi:10.339090705534

Xue, C., Zhu, H., Tan, D., Ren, H., Gu, X., Zhao, Y., Zhang, P., Sun, Z., Yang, Y., Gu, J., Gu, Y., & Gu, J. (2018). Electrospun silk fibroin-based neural scaffold for bridging a long sciatic nerve gap in dogs. *Journal of Tissue Engineering and Regenerative Medicine*, *12*(2), e1143–e1153. doi:10.1002/term.2449 PMID:28485084

Xue, W., Tan, X., Oo, M. K. K., Kulkarni, G., Ilgen, M. A., & Fan, X. (2020). Rapid and sensitive detection of drugs of abuse in sweat by multiplexed capillary based immuno-biosensors. *Analyst (London)*, *145*(4), 1346–1354. doi:10.1039/C9AN02498K PMID:31967116

Xu, H., Zhang, K., Liu, Q., Liu, Y., & Xie, M. (2017). Visual and fluorescent detection of mercury ions by using a dually emissive ratiometric nanohybrid containing carbon dots and CdTe quantum dots. *Mikrochimica Acta*, *184*(4), 1199–1206. doi:10.100700604-017-2099-1

Xu, X., Liu, X., Nie, Z., Pan, Y., Guo, M., & Yao, S. (2011). Label-free fluorescent detection of protein kinase activity based on the aggregation behavior of unmodified quantum dots. *Analytical Chemistry*, *83*(1), 52–59. doi:10.1021/ac102786c PMID:21128608

Xu, X., Ray, R., Gu, Y., Ploehn, H. J., Gearheart, L., Raker, K., & Scrivens, W. A. (2004). Electrophoretic analysis and purification of fluorescent single-walled carbon nanotube fragments. *Journal of the American Chemical Society*, *126*(40), 12736–12737. doi:10.1021/ja040082h PMID:15469243

Xu, Z. Q., Broza, Y. Y., Ionsecu, R., Tisch, U., Ding, L., Liu, H., & Haick, H. (2013). A nanomaterial-based breath test for distinguishing gastric cancer from benign gastric conditions. *British Journal of Cancer*, *108*(4), 941–950. doi:10.1038/bjc.2013.44 PMID:23462808

Yacobi, B. G., & Holt, D. B. (2013). *Cathodoluminescence microscopy of inorganic solids*. Springer Science & Business Media.

Yadav, S. K., Khan, Z. A., & Mishra, B. (2013). Impact of nanotechnology on socio-economic aspects: An overview. *Reviews in Nanoscience and Nanotechnology*, *2*(2), 127–142. doi:10.1166/rnn.2013.1026

Yadav, T. P., Yadav, R. M., & Singh, D. P. (2012). Mechanical milling: A top down approach for the synthesis of nanomaterials and nanocomposites. *Nanoscience and Nanotechnology*, *2*(3), 22–48. doi:10.5923/j.nn.20120203.01

Yáñez-Sedeño, P., Agüí, L., Campuzano, S., & Pingarrón, J. M. (2019). What Electrochemical Biosensors Can Do for Forensic Science? Unique Features and Applications. In Biosensors (Vol. 9, Issue 4). doi:10.3390/bios9040127

Yang, W., Gong, Y., & Li, W. (2021). *A review: Electrode and packaging materials for neurophysiology recording implants*. Academic Press.

Yang, L., Li, Y., Sun, Y., Wang, W., & Shao, Z. (2022). Perovskite oxides in catalytic combustion of volatile organic compounds: Recent advances and future prospects. *Energy & Environmental Materials*, *5*(3), 751–776. doi:10.1002/eem2.12256

Yang, N., Chen, X., Ren, T., Zhang, P., & Yang, D. (2015). Carbon nanotube based biosensors. *Sensors and Actuators. B, Chemical*, *207*, 690–715. doi:10.1016/j.snb.2014.10.040

Yang, X., Gao, L., Zhao, S., Pan, G., Fan, G., Xia, Z., & Jin, X. (2023). Volatile Organic Compounds in the North China Plain: Characteristics, Sources, and Effects on Ozone Formation. *Atmosphere (Basel)*, *14*(2), 318. doi:10.3390/atmos14020318

Yang, Y. K., Chang, Y. Y., Guo, Y. Y., Yu, L. G., Zhang, G. H., Zhai, D. D., Wang, X. M., & Sun, X. T. (2019). Fluorometric microplate-based dimethoate assay using CdSe/ZnS quantum dots coated with a molecularly imprinted polymer. *Mikrochimica Acta*, *186*(8), 10. doi:10.100700604-019-3649-5 PMID:31367777

Yang, Y., Cui, J., Zheng, M., Hu, C., Tan, S., Xiao, Y., Yang, Q., & Liu, Y. (2012). One-step synthesis of amino-functionalized fluorescent carbon nanoparticles by hydrothermal carbonization of chitosan. *Chemical Communications (Cambridge)*, *48*(3), 380–382. doi:10.1039/C1CC15678K PMID:22080285

Yan, X., Cui, X., & Li, L. (2010). Synthesis of large, stable colloidal grapheme quantum dots with tunable size. *Journal of the American Chemical Society*, *132*(17), 5944–5945. doi:10.1021/ja1009376 PMID:20377260

Yıldırım, M. Ş., Akçan, R., Kaliszan, M., & Saglam, N. (2021). Microbial Nanotechnology in Reference to Postmortem Diagnosis. *Nanotechnology Applications in Health and Environmental Sciences*, 333-345.

Yıldırım, M., & Sinan, S., Akçan, R., Aysun Balseven, O., & Tümer, A. (2015). Use of microbiological methods in postmortem interval estimation. *The Bulletin of Legal Medicine*, *20*(1).

Yonar, F. C., Karatas, O., Erkan, I., & Yukseloglu, E. H. (2022). Nanoscience and forensic genetics. *Understanding Crime Through Forensic Sciences*, 54.

Yoon, T., Jun, J., Kim, D. Y., Pourasad, S., Shin, T. J., Yu, S. U., . . . Kim, K. (2018). *An ultra-sensitive, flexible and transparent gas detection film based on well-ordered flat polypyrrole on single-layered graphene*. Academic Press.

Yuan, W., Tang, Y., Yang, X., & Wan, Z. (2012). Porous metal materials for polymer electrolyte membrane fuel cells–A review. *Applied Energy*, *94*, 309–329. doi:10.1016/j.apenergy.2012.01.073

Yue, K. X., Li, H. L., Pi, Z. F., Song, F. R., Wang, Y. P., & Liu, Z. Y. (2022). Multiresidue screening of pesticides in Panax Ginseng CA Meyer by ultra-high-performance liquid chromatography with quadrupole time-of-flight mass spectrometry. *Journal of Separation Science*, *45*(10), 1702–1710. doi:10.1002/jssc.202100933 PMID:35263500

Yu, J. H., Choi, S. M., Richards, C. I., Antoku, Y., & Dickson, R. M. (2008). Live Cell Surface Labeling with Fluorescent Ag Nanocluster Conjugates †. *Photochemistry and Photobiology*, *84*(6), 1435–1439. doi:10.1111/j.1751-1097.2008.00434.x PMID:18764887

Yu, L., Chen, H. X., Yue, J., Chen, X. F., Sun, M. T., Hou, J., Alamry, K. A., Marwani, H. M., Wang, X. K., & Wang, S. H. (2020). Europium metal-organic framework for selective and sensitive detection of doxycycline based on fluorescence enhancement. *Talanta*, *207*, 8. doi:10.1016/j.talanta.2019.120297 PMID:31594583

Yun, K., Nam, H.-S., & Kim, S. (2020). Categorization of atomic mixing patterns in bimetallic nanoparticles by the energy competition. *Physical Chemistry Chemical Physics*, *22*(15), 7787–7793. doi:10.1039/C9CP06967D PMID:32239065

Yu, Y., Jiang, Q., Wan, H., Li, R., Sun, Y., Zhang, Z., Mao, Z., Cao, Y., & Chen, F. (2022). In Situ Identification of Unknown Crystals in Acute Kidney Injury Using Raman Spectroscopy. *Nanomaterials (Basel, Switzerland)*, *12*(14), 2395. doi:10.3390/nano12142395 PMID:35889619

Yu, Z., Liu, C., Niu, H., Wu, M., Gao, W., Zhou, Z., & Li, X. (2020). Real time analysis of trace volatile organic compounds in ambient air: A comparison between membrane inlet single photon ionization mass spectrometry and proton transfer reaction mass spectrometry. *Analytical Methods*, *12*(35), 4343–4350. doi:10.1039/D0AY01102A PMID:32844845

Zala, A., & Patel, H. (2022). Dendrimer Enhanced Fingerprint and Explosive Detection: A Critical Review. *Engineered Science, 20*.

Zamora-Perez, P., Tsoutsi, D., & Xu, R., & Rivera Gil, P. (2018). Hyperspectral-enhanced dark field microscopy for single and collective nanoparticle characterization in biological environments. *Materials (Basel)*, *11*(2), 243. doi:10.3390/ma11020243 PMID:29415420

Zarić, B. L., Petrović, S., Bjekić, M., Rajić, I., Popović, A. R., & Đorđević, D. S. (2014). Analysis of human exhaled breath in a population of young volunteers. *Archives of Biological Sciences*, *66*(4), 1529–1538. doi:10.2298/ABS1404529Z

Zhang, K., Rossi, C., Ardila Rodriguez, G. A., Tenailleau, C., & Alphonse, P. (2007). *Development of a nano-Al/ Cu O based energetic material on silicon substrate*. Academic Press.

Zhang, X., Guo, Q., & Cui, D. (2009). Recent Advances in Nanotechnology Applied to Biosensors. In Sensors (Vol. 9, Issue 2, pp. 1033–1053). doi:10.339090201033

Zhang, C. Y., & Hu, J. (2010). Single quantum dot-based nanosensor for multiple DNA detection. *Analytical Chemistry*, *82*(5), 1921–1927. doi:10.1021/ac9026675 PMID:20121246

Zhang, H., Zhang, L., Liang, R. P., Huang, J., & Qiu, J. D. (2013). Simultaneous determination of Concanavalin A and Peanut agglutinin by dual-color quantum dots. *Analytical Chemistry*, *85*(22), 10969–10976. doi:10.1021/ac402496e PMID:24128387

Zhang, L. Z., Lin, Y., Zhang, Y. J., Chen, R., Zhu, Z. S., Wu, W., & Jiang, X. Q. (2012). Fluorescent Micelles Based on Star Amphiphilic Copolymer with a Porphyrin Core for Bioimaging and Drug Delivery. *Macromolecular Bioscience*, *12*(1), 83–92. doi:10.1002/mabi.201100197 PMID:22052617

Zhang, L., Yang, H., Tang, Y., Xiang, W., Wang, C., Xu, T., & Zhang, J. (2022). High-performance CdSe/CdS@ ZnO quantum dots enabled by ZnO sol as surface ligands: A novel strategy for improved optical properties and stability. *Chemical Engineering Journal*, *428*, 131159. doi:10.1016/j.cej.2021.131159

Zhang, Q., Bao, X., Liang, Q., Sun, Q., Xu, W., Zou, X., & Chu, Y. (2022). Evaluation of a New DC-Ion Funnel Drift Tube for Use in Proton Transfer Reaction Mass Spectrometry. *Analytical Chemistry*, *94*(20), 7174–7180. doi:10.1021/acs.analchem.1c05086 PMID:35536750

Zhang, Y., Chan, K. F., Wang, B., Chiu, P. W. Y., & Zhang, L. (2018). Spore-derived color-tunable multi-doped carbon nanodots as sensitive nanosensors and intracellular imaging agents. *Sensors and Actuators. B, Chemical*, *271*, 128–136. doi:10.1016/j.snb.2018.05.112

Zhang, Y., Liu, L., & Ren, L. (2020). Liquid chromatography-tandem mass spectrometry (LC-MS/MS) determination of cantharidin in biological specimens and application to postmortem interval estimation in cantharidin poisoning. *Scientific Reports*, *10*(1), 10438. doi:10.103841598-020-67278-x PMID:32591575

Zhang, Y., Tang, H., Chen, W., & Zhang, J. (2022). Nanomaterials Used in Fluorescence Polarization Based Biosensors. *International Journal of Molecular Sciences*, *23*(15), 8625. doi:10.3390/ijms23158625 PMID:35955779

Zhang, Z., & Peng, D. (2022). Recent Advances in Enhancement Techniques for Blood Fingerprints. *Critical Reviews in Analytical Chemistry*, 1–20. PMID:36045560

Zhao, J., Zhang, X., Yonzon, C. R., Haes, A. J., & Van Duyne, R. P. (2006). *Localized surface plasmon resonance biosensors*. Academic Press.

Zhao, H., Tu, N., Zhang, W. B., Zhang, M., & Wang, J. (2021). Novel synthesis of Silicon/Carbon nanotubes microspheres as anode additives through chemical vapor deposition in fluidized bed reactors. *Scripta Materialia*, *192*, 49–54. doi:10.1016/j.scriptamat.2020.09.048

Zhao, Q. L., Zhang, Z. L., Huang, B. H., Peng, J., Zhang, M., & Pang, D. W. (2008). Facile preparation of low cytotoxicity fluorescent carbon nanocrystals by electrooxidation of graphite. *Chemical Communications (Cambridge)*, *41*(41), 5116–5118. doi:10.1039/b812420e PMID:18956040

Zhao, X.-C., Wang, L., Sun, J., Jiang, B.-W., Zhang, E.-L., & Ye, J. (2016). Isolating Sperm from Cell Mixtures Using Magnetic Beads Coupled with an Anti-PH-20 Antibody for Forensic DNA Analysis. *PLoS One*, *11*(7), e0159401. doi:10.1371/journal.pone.0159401 PMID:27442128

Zhao, Y., Xie, Y., Zhou, Q., Wang, P., Chang, Y., & Lin, C. (2020). Automatic Sensing Setup for Methamphetamine Based on the Reactional Wettability Variation Strategy. *ACS Omega*, *6*(3), 2045–2051. doi:10.1021/acsomega.0c04995 PMID:33521443

Zhao, Z., Dai, J., Wang, T., Niu, G., He, F., & Duan, Y. (2020). Development of microwave plasma proton transfer reaction mass spectrometry (MWP-PTR-MS) for on-line monitoring of volatile organic compounds: Design, characterization and performance evaluation. *Talanta*, *208*, 120468. doi:10.1016/j.talanta.2019.120468 PMID:31816683

Zheng, H., Smith, R. K., Jun, Y.-w., Kisielowski, C., Dahmen, U., & Alivisatos, A. P. (2009). Observation of single colloidal platinum nanocrystal growth trajectories. *Science, 324*(5932), 1309-1312.

Zheng, X., Zhang, P., Fu, Z., Meng, S., & Dai, L. (2021). *Applications of nanomaterials in tissue engineering*. doi:10.1039/D1RA01849C

Zheng, G., Patolsky, F., Cui, Y., Wang, W. U., & Lieber, C. M. (2005). Multiplexed electrical detection of cancer markers with nanowire sensor arrays. *Nature Biotechnology*, *23*(10), 1294–1301. doi:10.1038/nbt1138 PMID:16170313

Zhou, J., Liu, Y., Tang, J., & Tang, W. (2017). Surface ligands engineering of semiconductor quantum dots for chemosensory and biological applications. *Materials Today*, *20*(7), 360–376. doi:10.1016/j.mattod.2017.02.006

Zhou, L., Cai, M., Tong, T., & Wang, H. (2017). Progress in the correlative atomic force microscopy and optical microscopy. *Sensors (Basel)*, *17*(4), 938. doi:10.339017040938 PMID:28441775

Zhou, X., Hu, Y., Cao, Y., Liu, Y., & Qian, T. (2022). A novel signal-on fluorometric sensor based on metal ion-mediated carbon dots for formaldehyde determination and lysosome-targeted bioimaging. *New Journal of Chemistry*, *46*(36), 17540–17547. doi:10.1039/D2NJ03599E

Zibbell, J. E., Clarke, S. D., Kral, A. H., Richardson, N. J., Cauchon, D., & Aldridge, A. (2022). Association between law enforcement seizures of illicit drugs and drug overdose deaths involving cocaine and methamphetamine, Ohio, 2014–2019. *Drug and Alcohol Dependence*, *232*, 109341. doi:10.1016/j.drugalcdep.2022.109341 PMID:35134733

Zięba-Palus, J. (2017). Forensic examinations of micro-traces. *Forensic Science and Criminology*, *2*(2), 1–3. doi:10.15761/FSC.1000113

Zięba-Palus, J., & Trzcińska, B. M. (2011). Establishing of chemical composition of printing ink. *Journal of Forensic Sciences*, *56*(3), 819–821. doi:10.1111/j.1556-4029.2011.01734.x PMID:21361954

Zięba, S., Wiergowski, M., Cieślik, B. M., Anand, J. S., & Krzyżanowska, M. (2023). Uncertainty of Postmortem Time Estimation Based on Potassium Ion Determination in Vitreous Humor Using Potentiometric Ion-Selective Electrode and Microwave-Induced Plasma with Optical Emission Spectrometry Methods. *Separations*, *10*(3), 201. doi:10.3390eparations10030201

Zilberman, Y., Ionescu, R., Feng, X. L., Mullen, K., & Haick, H. (2011). Nanoarray of polycyclic aromatic hydrocarbons and carbon nanotubes for accurate and predictive detection in real-world environmental humidity. *ACS Nano*, *5*(8), 6743–6753. doi:10.1021/nn202314k PMID:21774511

Ziyade, N., Sirin, G., Elgormus, N., & Das, T. (2015). Detection of human bocavirus DNA by multiplex PCR analysis: Postmortem case report. *Balkan Medical Journal*, *32*(2), 226–229. doi:10.5152/balkanmedj.2015.15254 PMID:26167351

Zlotnick, J., & Smith, F. (1999). *Chromatographic and electrophoretic approaches in ink analysis*. Academic Press.

Zogka, A. G., Romanias, M. N., & Thevenet, F. (2022). Formaldehyde and glyoxal measurement deploying a selected ion flow tube mass spectrometer (SIFT-MS). *Atmospheric Measurement Techniques*, *15*(7), 2001–2019. doi:10.5194/amt-15-2001-2022

Zong, J., Zhu, Y., Yang, X., Shen, J., & Li, C. (2011). Synthesis of photoluminescent carbogenic dots using mesoporous silica spheres as nanoreactors. *Chemical Communications (Cambridge)*, *47*(2), 764–766. doi:10.1039/C0CC03092A PMID:21069221

Zor, F., Selek, F. N., Orlando, G., & Williams, D. F. (2019). Biocompatibility in regenerative nanomedicine. *Nanomedicine (London)*, *14*(20), 2763–2775. doi:10.2217/nnm-2019-0140 PMID:31612774

Zuo, P. L., Lu, X. H., Sun, Z. G., Guo, Y. H., & He, H. (2016). A review on syntheses, properties, characterization and bioanalytical applications of fluorescent carbon dots. *Mikrochimica Acta*, *183*(2), 519–542. doi:10.100700604-015-1705-3

About the Contributors

Allah Rakha is professor and head of forensic team at University of Health Sciences, Lahore. He completed his doctorate from Xi'an Jiaotong University, Xian with shared supervision from University of Central Lancashire, UK in 2008. After doctorate, he joined Younsei University, school of medicine as postdoc fellow till 2010. He started applied forensics as Forensic DNA expert by joining National Forensic Science Agency, National Police Bureau, Islamabad, Pakistan. In 2011, he joined University of Health Sciences, Lahore and started academic/research programs in Forensic medicine and forensic genetics. In 2016, he was awarded second postdoc from Chinese Academy of Sciences. His research areas of interest include analysis of biomarkers for postmortem interval, novel markers for human identification, phylogenetics and Forensic DNA phenotyping.

Anam Munawar is an Assistant Professor at University of Health Sciences, Lahore. She completed her Ph. D in 2020, from Pakistan Institute of Engineering and Applied Sciences (PIEAS), Islamabad. She participated in two National Hackathon and got distinctive position by presenting her ideas to solve agriculture problems and hepatitis C detection. She got a project based on the "Texonomy Challenge" organized by British Council Project no: (RLCC 714700770). She has published 16 research articles with an impact factor more than 70 and three book chapters. Her area of interest is Biosensors and biomimetic sensors technology; system miniaturization and applications (bio)chemical molecular recognition layers for separation and sensing applications, artificial receptor materials via molecular imprinting technique, development of sensitive and selective coatings for mass-sensitive, microelectrode, and electrochemical devices; Substantiating developed materials and surfaces by using various characterizing techniques. Establishing mechanistic studies for their potential applications in real environment.

Virat Khanna is an Associate Professor and Head in the Department of Mechanical Engineering at Maharaja Agrasen University, HP, India and has 12 years of teaching experience. He is Ph.D. in Mechanical Engineering from Maharaja Agrasen University, HP, India. He Completed his M Tech in Mechanical Engineering in 2013 and B Tech in Mechanical Engineering in 2008 from I. K. Gujral Punjab Technical University, Jalandhar, India. His areas of research are Aluminium Metal matrix nanocomposite (Graphene, CNT, Processing technique, optimization and mathematical modelling techniques). He has published and presented more than 20 papers in Journals and conferences of national and international repute. He is also the editor of several books and conference proceedings with various international publishers like Elsevier, CRC Press, IGI Global, Bentham Books, etc.

Suneev Anil Bansal has more than 15 years of diverse experience in the industry, research laboratories and academia. He has served in key positions in key projects like automobile development at Hero Motor Corp., fighter aircraft development at DRDO, and nanomaterials & manufacturing development at various universities. His research interest is in the fields of micro/nano-manufacturing, materials, 2D materials, graphene, polymer/metal composite materials, sensing, etc. He has published several research papers, holds 1 patent and has served as a subject expert for a number of journals.

* * *

Waseem Abbas has a PhD from Shanghai Jiao Tong University, Shanghai, China Working as assistant Professor of Physics at institution of Physics, Bahauddin Zakriy University, Multan, Pakistan.

Ahtisham Shuja Abbasi is currently enrolled in M.Phil Forensic Sciences from University of Health Sciences Lahore. He has done bachelors in Biotechnology from University of Health Sciences Lahore. His area of interest is in Forensic Sciences, Molecular Biology and application of Nano materials in molecular Biology and Forensic Sciences to explore their hidden features.

Amna Anwar is currently enrolled in M.Phil Forensic Sciences. She has synthesized different nanomaterials and exploring their application the area of forensic sciences. Her area of interest in the hydrothermal synthesis of nanomaterials and sensors development.

Maryam Arshad was born in 1993 in Islamabad to a well known educated family with research background. Completed her early education from Mexico, while her father pursuing Doctorate and Post-doctorate. After completion of her Intermediate from Islamabad Model College for girls F-7/2 in Pre-Medical, she admitted to BS Biosciences from COMSATS University Islamabad. After her Graduation she pursued her post- graduation from COMSATS University Islamabad. Her research topics in Graduation and Post graduation includes Cancer Genetics, therapeutics and Nano Materials. Along with COMSATS her collaboration with NILOP, Islamabad and Nanophotomedicine Laboratory, Pakistan Institute of Engineering and Applied Sciences (PIEAS), bring diversity and multi-latereal exposure in research. After Post-Graduation, her Doctoral studies is from COMSATS University too in Collaboration from PIEAS, Islamabad and NIH Islamabad. Her Doctoral research is about polymeric and extra vesicular based nano carriers for drug delivery system in Cancer Therapy.

Rahat Bukhari is currently pursuing her M.Phil degree in Forensic Sciences from the University of Health Sciences, Lahore, Pakistan.

Tahir Farooq has been serving at the university level for 11 years. Currently, he is serving as Associate Professor in the Department of Applied Chemistry, Government College University, Faisalabad, Pakistan.

Mubeen Khalid is a young researcher with experience of working in forensic laboratory. A post graduate from University of Health Sciences, Lahore, Pakistan in Forensic Sciences believes that the perfect balance of classical and modern techniques yield the best results. Mubeen recently completed the project on Forensic Anthropology and Archaeology and the laboratory experience includes PMI, DNA Analysis, Latent Fingerprints and Crime Scene Investigation.

Marriam Zargham Currently pursuing her M.Phil. degree in Forensic Science form University of Health Sciences, Lahore Pakistan. Previously she has done BS Hons, in Biotechnology from University of Lahore Pakistan. She had served as Molecular Biologist in Lab One pvt. limited Lahore Pakistan from 2017 - 2020. Her research area include, Biotechnology, Nanotechnology, Molecular Biology, Medical Diagnosis and Forensic Sciences.

Muhammad Farhan Khan is currently pursuing his PhD Degree in Forensic Sciences from the University of Health Sciences, Lahore, Pakistan. Previously he has competed M.Phil in Forensic Sciences from the University of Veterinary and Animal Sciences- UVAS Lahore. Currently, he is serving as lecturer in the Department of Forensic Medicine/Medical Jurisprudence at University of Health Sciences, Lahore, Pakistan. His research area includes Medical Sciences, Behavioural Sciences, Medical Diagnoses, Forensic Sciences, Data Mining and Image Processing.

Saba Naz is a PhD scholar. She is doing her PhD under the kind supervision of Dr Shumaila Kiran. She has more than 10 publications. She has written many international book chapters under the kind guidance of Dr Shumaila Kiran.

Shahid Nazir (PhD) is a young enthusiastic researcher, academician, and detailed oriented Forensic Scientist with vast experience working in crime Laboratory, teaching, and Training. Working knowledge of more than 10 years in Crime Scene investigation, DeathScene investigation, Latent Fingerprint, Forensic DNA, Genetics, Medical Laboratory Sciences, Biostatics, mtDNA, Y-STR, and other Forensic markers used for identification.

Malika Rani has been working since last seven years as Assistant professor has now been appointed as Associate professor of Physics and Head of Department of Physics, The Women University Multan. She has achieved 03 projects including Higher Education Commission of Pakistan (HEC) Grant under START UP Research Grant Program at total cost of 0.3723 Million, Higher Education Commission of Pakistan (HEC) under National Research Program for Universities (NRPU) at total cost of Rs. 4.841,654 Million and "A study of optical properties of some strongly correlated electron systems" funded by the Women University Multan at total cost of Rs 0.195 Million. She has 27 publications in W category journal and only 03 publications in X category journal. Her research area includes Energy storage devices including supercapacitors including 2D materials such as Graphene Oxide and MXene, MOFs, Binary and Ternary nanocomposites. She has worked on device fabrication including RE-RAM structures. She has been Invited speaker in 14 International Conferences . She has organized 05 International conferences funded be HEC/PHECs and established Environment and Energy Development Laboratory. She is also Editor in Chief of "Journal of Nanoscope". She has registered 05 patents and have expertise in various nanomaterials characterizations techniques including XRD, SEM, FTIR, UV-Vis, Raman, Photoluminescence, I-V characteristics, Zeta potential measurements and electrochemical investigations such as CV, GCD and EIS Spectroscopy.

Danish Rao is a student at Xi'an Jiaotong University, China.

Rahat Rehman has obtained her Ph.D in Biological Sciences from School of Biological Sciences, University of the Punjab in 2013. She has been serving as Forensic Scientist and Supervisor of DNA and

Serology section in Punjab forensic science agency, Lahore. Being at PFSA, she got extensive training in Forensic serology and Forensic DNA Typing. She herself has analyzed more than 5000 Forensic DNA and serology cases during her stay at PFSA. She got training of Forensic Advanced DNA training from University of Canberra, Australia. There she went through mixture interpretations and report writing. She has also received Quality management training from UKAS, London where she learned ISO 17025 standard requirements for "General requirements for the competence of testing and calibration laboratories". She is well proficient in Forensic Serology testing and results interpretation, Forensic DNA profiling, Random Match Probability, Mixture interpretations, Parentage Index, Relationship (Kinship) testing, Forensic Report writing and interpretations.

Rubia Shafique has recently completed PhD under supervision of Associate Professor/HOD Physics department, The Women University Multan with 11 publications in both internationally and nationally HEC recognized Journals. The major contributing field is to study the electrochemical behavior of Nano-structural electrodes for energy storage applications especially for Supercapacitors.

Iqra Zareef is a Master's degree student at the University of Health Sciences. She received a bachelor's degree in Biochemistry from University of Central Punjab, Lahore. Her current enrollment is in Department of Forensic Sciences. She is interested in interpreting crime scene details, collecting, analyzing and preserving evidence as well as in nanotechnology and medicine.

Index

P

Q

R

S

T

V

www.ingramcontent.com/pod-product-compliance
Lightning Source LLC
LaVergne TN
LVHW062119300926
849161LV00010B/2491